This book comes with access to more content online.

Quiz yourself, track your progress, and improve your grade!

Register your book or ebook at **www.dummies.com/go/getaccess.**

Select your product, and then follow the prompts to validate your purchase.

You'll receive an email with your PIN and instructions.

Statistics

ALL-IN-ONE

Statistics

ALL-IN-ONE

by Deborah J. Rumsey, PhD

Statistics All-in-One For Dummies®

Published by: **John Wiley & Sons, Inc.,** 111 River Street, Hoboken, NJ 07030-5774, www.wiley.com

Published simultaneously in Canada

For general information on our other products and services, please contact our Customer Care Department within the U.S. at 877-762-2974, outside the U.S. at 317-572-3993, or fax 317-572-4002. For technical support, please visit https://hub.wiley.com/community/support/dummies.

Wiley publishes in a variety of print and electronic formats and by print-on-demand. Some material included with standard print versions of this book may not be included in e-books or in print-on-demand. If this book refers to media such as a CD or DVD that is not included in the version you purchased, you may download this material at http://booksupport.wiley.com. For more information about Wiley products, visit www.wiley.com.

Library of Congress Control Number: 2022945239

ISBN 978-1-119-90256-0 (pbk); ISBN 978-1-119-90257-7 (ebk); ISBN 978-1-119-90258-4 (ebk)

SKY10036123_092722

Contents at a Glance

Table of Contents

Introduction

You get hit with an incredible amount of statistical information on a daily basis. You know what I'm talking about: charts, graphs, tables, and headlines that talk about the results of the latest poll, survey, experiment, or other scientific study. The purpose of this book is to develop and sharpen your skills in sorting through, analyzing, and evaluating all that info, and to do so in a clear, fun, and pain-free way with tons of opportunities to practice. You also gain the ability to decipher and make important decisions about statistical results (for example, the results of the latest medical studies), while being ever aware of the ways that people can mislead you with statistics. And you see how to do it right when it's your turn to design the study, collect the data, crunch the numbers, and/or draw the conclusions.

This book is also designed to help those of you who are looking to get a solid foundation in introductory statistics or those taking a statistics class and wanting some backup. You'll gain a working knowledge of the big ideas of statistics and gather a boatload of tools and tricks of the trade that'll help you get ahead of the curve, especially for taking exams.

This book is chock-full of real examples from real sources that are relevant to your everyday life — from the latest medical breakthroughs, crime studies, and population trends to the latest U.S. government reports. I even address a survey on the worst cars of the millennium! By reading this book, you'll understand how to collect, display, and analyze data correctly and effectively, and you'll be ready to critically examine and make informed decisions about the latest polls, surveys, experiments, and reports that bombard you every day. You will even find out how to use crickets to gauge temperature!

You will also get to climb inside the minds of statisticians to see what's worth taking seriously and what isn't to be taken so seriously. After all, with the right skills and knowledge, you don't have to be a professional statistician to understand introductory statistics. You can be a data guru in your own right.

About This Book

This book departs from traditional statistics texts, references, supplemental books, and study guides in the following ways:

- It includes practical and intuitive explanations of statistical concepts, ideas, techniques, formulas, and calculations found in an introductory statistics course.
- It shows you clear and concise step-by-step procedures that explain how you can intuitively work through statistics problems.
- It features interesting real-world examples relating to your everyday life and workplace.

» It contains plenty of excellent practice problems crafted in a straightforward manner to lead you down the path of success.

» It offers not only answers, but also clear, complete explanations of the answers. Explanations help you know exactly how to approach a problem, what information you need to solve it, and common problems you need to avoid.

» It includes tips, strategies, and warnings based on my vast experience with students of all backgrounds and learning styles.

» It gives you upfront and honest answers to your questions like, "What does this really mean?" and "When and how will I ever use this?"

As you work your way through the lessons and problems in this book, you should be aware of four conventions that I've used.

» **Dual use of the word *statistics:*** In some situations, I refer to statistics as a subject of study or as a field of research, so the word is a singular noun. For example, "Statistics is really quite an interesting subject." In other situations, I refer to statistics as the plural of *statistic,* in a numerical sense. For example, "The most commonly used statistics are the mean and the standard deviation."

» **Use of the word *data:*** You're probably unaware of the debate raging among statisticians about whether the word *data* should be singular ("data is ") or plural ("data are "). It got so bad that one group of statisticians had to develop two versions of a statistics T-shirt: "Messy Data Happens" and "Messy Data Happen." I go with the plural version of the word *data* in this book.

» **Use of the term *standard deviation:*** When I use the term *standard deviation,* I mean *s,* the sample standard deviation. (When I refer to the population standard deviation, I let you know.)

» **Use of *italics:*** I use *italics* to let you know a new statistical term is appearing on the scene. Look for a definition accompanying its first appearance.

Foolish Assumptions

I don't assume that you've had any previous experience with statistics, other than the fact that you're a member of the general public who gets bombarded every day with statistics in the form of numbers, percents, charts, graphs, "statistically significant" results, "scientific" studies, polls, surveys, experiments, and so on.

What I do assume is that you can do some of the basic mathematical operations and understand some of the basic notation used in algebra, such as the variables x and y, summation signs (Σ), taking the square root, squaring a number, and so on. If you need to brush up on your algebra skills, check out *U Can Algebra I For Dummies* by Mary Jane Sterling (Wiley).

I don't want to mislead you: You do encounter formulas in this book, because statistics does involve a bit of number crunching. But don't let that worry you. I take you slowly and carefully through each step of any calculations you need to do, explaining things both with notation and without. I also provide practice questions for you to work so you can become familiar and comfortable with the calculations and make them your own.

Icons Used in This Book

You'll see the following five icons throughout the book:

EXAMPLE

Each example is a stat question based on the discussion and explanation, followed by a solution. Work through these examples, and then refer to them to help you solve the practice problems that follow them as well as the quiz questions at the end of the chapter.

REMEMBER

This icon points out important information that you need to focus on. Make sure you understand this information fully before moving on. You can skim through these icons when reading a chapter to make sure you remember the highlights.

TIP

Tips are hints that can help speed you along when answering a question. See whether you find them useful when working on practice problems.

WARNING

This icon flags common mistakes that students make if they're not careful. Take note and proceed with caution!

YOUR TURN

When you see this icon, it's time to put on your thinking cap and work out a few practice problems on your own. The answers and detailed solutions are available so you can feel confident about your progress.

Beyond the Book

In addition to the material in the print or e-book you're reading right now, this book also comes with a handy online Cheat Sheet. Use it when you need a quick refresher on a formula or the next step in conducting a hypothesis test. To get this Cheat Sheet, simply go to `www.dummies.com` and type **Statistics All in One For Dummies Cheat Sheet** in the Search box.

You'll also have access to online quizzes related to each chapter, beginning with Unit 2, Chapter 4. These quizzes provide a whole new set of problems for practice and confidence-building. To access the quizzes, follow these simple steps:

1. **Register your book or ebook at Dummies.com to get your PIN. Go to `www.dummies.com/go/getaccess`.**
2. **Select your product from the drop-down list on that page.**
3. **Follow the prompts to validate your product, and then check your email for a confirmation message that includes your PIN and instructions for logging in.**

If you do not receive this email within two hours, please check your spam folder before contacting us through our Technical Support website at http://support.wiley.com or by phone at 877-762-2974.

Now you're ready to go! You can come back to the practice material as often as you want — simply log on with the username and password you created during your initial login. No need to enter the access code a second time.

Your registration is good for one year from the day you activate your PIN.

Where to Go from Here

This book is written in such a way that you can start anywhere and still be able to understand what's going on. So you can take a peek at the table of contents or the index, look up the information that interests you, and flip to the page listed. However, if you have a specific topic in mind and are eager to dive into it, here are some directions:

- To work on interpreting graphs, charts, means or medians, and the like, head to Unit 2.
- To find info on the normal, *Z*-, *t*-, or binomial distributions or the Central Limit Theorem, see Unit 3.
- To focus on confidence intervals and hypothesis tests of all shapes and sizes, flip to Unit 4.
- To delve into surveys, experiments, regression, and two-way tables, see Unit 5.

Or if you aren't sure where you want to start, start with Chapter 1 for the big picture and then plow your way through the rest of the book. Ready, set, go!

1 Getting Started with Statistics

In This Unit . . .

IN THIS CHAPTER

» Raising questions about statistics you see in everyday life

» Encountering statistics in the workplace

Chapter 1
The Statistics of Everyday Life

Today's society is completely taken over by numbers. Numbers are everywhere you look, from billboards showing the on-time statistics for a particular airline, to sports shows discussing the Las Vegas odds for upcoming football games. The evening news is filled with stories focusing on crime rates, the expected life span of junk-food junkies, and the president's approval rating. On a normal day, you can run into 5, 10, or even 20 different statistics (with many more on election night). Just by reading a Sunday newspaper all the way through, you come across literally hundreds of statistics in reports, advertisements, and articles covering everything from soup (how much does an average person consume per year?) to nuts (almonds are known to have positive health effects — what about other types of nuts?).

In this chapter we discuss the statistics that often appear in your life and work, and talk about how statistics are presented to the general public. After reading this chapter, you'll realize just how often the media hits you with numbers and how important it is to be able to unravel the meaning of those numbers. Like it or not, statistics are a big part of your life. So, if you can't beat 'em, join 'em. And if you don't want to join 'em, at least try to understand 'em.

Statistics and the Media: More Questions than Answers?

Open a newspaper and start looking for examples of articles and stories involving numbers. It doesn't take long before those numbers begin to pile up. Readers are inundated with results of studies, announcements of breakthroughs, statistical reports, forecasts, projections, charts, graphs, and summaries. The extent to which statistics occur in the media is mind-boggling. You may not even be aware of how many times you're hit with numbers nowadays.

This section looks at just a few examples from one Sunday paper's worth of news that I read the other day. When you see how frequently statistics are reported in the news without providing all the information you need, you may find yourself getting nervous, wondering what you can and can't believe anymore. Relax! That's what this book is for — to help you sort out the good information from the bad (the chapters in Unit 2 give you a great start on that).

Probing popcorn problems

The first article I came across that dealt with numbers was "Popcorn plant faces health probe," with the subheading: "Sick workers say flavoring chemicals caused lung problems." The article describes how the Centers for Disease Control (CDC) expressed concern about a possible link between exposure to chemicals in microwave popcorn flavorings and some cases of fixed obstructive lung disease. Eight people from one popcorn factory alone contracted this lung disease, and four of them were awaiting lung transplants.

According to the article, similar cases were reported at other popcorn factories. Now, you may be wondering, what about the folks who eat microwave popcorn? According to the article, the CDC found "no reason to believe that people who eat microwave popcorn have anything to fear." (Stay tuned.) The next step is to evaluate employees in more depth, including conducting surveys to determine health and possible exposures to the flavoring chemicals, checks of lung capacity, and detailed air samples. The question here is: How many cases of this lung disease constitute a real pattern, compared to mere chance or a statistical anomaly? (You find out more about this in Chapter 15.)

Venturing into viruses

A second article discussed a recent cyber attack: A wormlike virus made its way through the Internet, slowing down web browsing and email delivery around the world. How many computers were affected? The experts quoted in the article said that 39,000 computers were infected, and they in turn affected hundreds of thousands of other systems.

Questions: How did the experts get that number? Did they check each computer out there to see whether it was affected? The fact that the article was written less than 24 hours after the attack suggests the number is a guess. Then why say 39,000 and not 40,000 — to make it seem less like a guess? To find out more on how to guesstimate with confidence (and how to evaluate someone else's numbers), see Chapter 14.

Comprehending crashes

Next in the paper was an alert about the soaring number of motorcycle fatalities. Experts said that the *fatality rate* — the number of fatalities per 100,000 registered vehicles — for motorcyclists has been steadily increasing, as reported by the National Highway Traffic Safety Administration (NHTSA). In the article, many possible causes for the increased motorcycle death rate were discussed, including age, gender, size of engine, whether the driver had a license, alcohol use, and state helmet laws (or lack thereof). The report was very comprehensive, showing various tables and graphs with the following titles:

- Motorcyclists killed and injured, and fatality and injury rates by year, per number of registered vehicles, and per millions of vehicle miles traveled
- Motorcycle rider fatalities by state, helmet use, and blood alcohol content
- Occupant fatality rates by vehicle type (motorcycles, passenger cars, light trucks), per 10,000 registered vehicles and per 100 million vehicle miles traveled
- Motorcyclist fatalities by age group
- Motorcyclist fatalities by engine size (displacement)
- Previous driving records of drivers involved in fatal traffic crashes by type of vehicle (including previous crashes, DUI convictions, speeding convictions, and license suspensions and revocations)

This article was very informative and provided a wealth of detailed information regarding motorcycle fatalities and injuries in the U.S. However, the onslaught of so many tables, graphs, rates, numbers, and conclusions can be overwhelming and confusing and lead you to miss the big picture. With a little practice, and help from Unit 2, you'll be better able to sort out graphs, tables, and charts and all the statistics that go along with them. For example, some important statistical issues come up when you see rates versus counts (such as death rates versus number of deaths). As I address in Chapter 2, counts can give you misleading information if they're used when rates would be more appropriate.

Mulling malpractice

Further along in the newspaper was a report about a recent medical malpractice insurance study: Malpractice cases affect people in terms of the fees doctors charge and the ability to get the healthcare they need. The article indicates that one in five Georgia doctors have stopped doing risky procedures (such as delivering babies) because of the ever-increasing malpractice insurance rates in the state. This is described as a "national epidemic" and a "health crisis" around the country. Some brief details of the study are included, and the article states that of the 2,200 Georgia doctors surveyed, 2,800 of them — which they say represent about 18 percent of those sampled — were expected to stop providing high-risk procedures.

Wait a minute! That can't be right. Out of 2,200 doctors, 2,800 don't perform the procedures, and that is supposed to represent 18 percent? That's impossible! You can't have a bigger number on the top of a fraction, and still have the fraction be under 100 percent, right? This is one of many examples of errors in media reporting of statistics. So what's the real percentage? There's no way to tell from the article. Chapter 4 nails down the particulars of calculating these kinds of statistics so you can know what to look for and immediately tell when something's not right.

Belaboring the loss of land

In the same Sunday paper was an article about the extent of land development and speculation across the United States. Knowing how many homes are likely to be built in your neck of the woods is an important issue to get a handle on. Statistics are given regarding the number of acres of farmland being lost to development each year. To further illustrate how much land is being lost, the area is also listed in terms of football fields. In this particular example, experts said that the mid-Ohio area is losing 150,000 acres per year, which is 234 square miles, or 115,385 football fields (including end zones). How do people come up with these numbers, and how accurate are they? And does it help to visualize land loss in terms of the corresponding number of football fields? I discuss the accuracy of data collected in more detail in Chapter 17.

Scrutinizing schools

The next topic in the paper was school proficiency — specifically, whether extra school sessions help students perform better. The article stated that 81.3 percent of students in this particular district who attended extra sessions passed the writing proficiency test, whereas only 71.7 percent of those who didn't participate in the extra school sessions passed it. But is this enough of a difference to account for the $386,000 price tag per year? And what's happening in these sessions to cause an improvement? Are students in these sessions spending more time just preparing for those exams rather than learning more about writing in general? And here's the big question: Were the participants in the extra sessions student volunteers who may be more motivated than the average student to try to improve their test scores? The article didn't say.

Studies like this appear all the time, and the only way to know what to believe is to understand what questions to ask and to be able to critique the quality of the study. That's all part of statistics! The good news is, with a few clarifying questions, you can quickly critique statistical studies and their results. Chapter 18 helps you do just that.

Scanning sports

The sports section is probably the most numerically jam-packed section of the newspaper. Beginning with game scores, the win/loss percentages for each team, and the relative standing for each team, the specialized statistics reported in the sports world are so deep that they require wading boots to get through. For example, basketball statistics are broken down by team, by quarter, and by player. For each player, you get minutes played, field goals, free throws, rebounds, assists, personal fouls, turnovers, blocks, steals, and total points.

Who needs to know this stuff, besides the players' mothers? Apparently, many fans do. Statistics are something that sports fans can never get enough of and players often can't stand to hear about. Stats are the substance of water-cooler debates and the fuel for armchair quarterbacks around the world.

STUDYING SURVEYS OF ALL SHAPES AND SIZES

Surveys and polls are among the most visible mechanisms used by today's media to grab your attention. It seems that everyone, including market managers, insurance companies, TV stations, community groups, and even students in high school classes, wants to do a survey. Here are just a few examples of survey results that are part of today's news:

With the aging of the American workforce, companies are planning for their future leadership. (How do they know that the American workforce is aging, and if it is, by how much is it aging?) A recent survey shows that nearly 67 percent of human resource managers polled said that planning for succession had become more important in the past five years than it had been in the past. The survey also says that 88 percent of the 210 respondents said they usually or often fill senior positions with internal candidates. But how many managers did not respond, and is 210 respondents really enough people to warrant a story on the front page of the business section? Believe it or not, when you start looking for them, you'll find numerous examples in the news of surveys based on far fewer participants than 210. (To be fair, however, 210 can actually be a good number of subjects in some situations. The issues of what sample size is large enough and what percentage of respondents is big enough are addressed in full detail in Chapter 17.)

Some surveys are based on current interests and trends. For example, a Harris-Interactive survey found that nearly half (47 percent) of U.S. teens say their social lives would end or be worsened without their cellphones, and 57 percent go as far as to say that their cellphones are the key to their social life. The study also found that 42 percent of teens say that they can text while blindfolded (how do you really test this?). Keep in perspective, though, that the study did not tell you what percentage of teens actually have cellphones or what demographic characteristics those teens have compared to teens who do *not* have cellphones. And remember that data collected on topics like this aren't always accurate, because the individuals who are surveyed may tend to give biased answers (who wouldn't want to say they can text blindfolded?). For more information on how to interpret and evaluate the results of surveys, see Chapter 17.

Fantasy sports have also made a huge impact on the sports money-making machine. Fantasy sports are games where participants act as owners to build their own teams from existing players in a professional league. The fantasy team owners then compete against each other. What is the competition based on? Statistical performance of the players and teams involved, as measured by rules set up by a "league commissioner" and an established point system. According to the Fantasy Sports Trade Association, the number of people age 12 and up who are involved in fantasy sports is more than 30 million, and the amount of money spent is $3 to 4 billion per year. (And even here, you can ask how the numbers were calculated — the questions never end, do they?)

Banking on business news

The business section of the newspaper provides statistics about the stock market. In one week, the market went down 455 points; is that decrease a lot or a little? You need to calculate a percentage to really get a handle on that.

The business section of my paper contained reports on the highest yields nationwide on every kind of certificate of deposit (CD) imaginable. (By the way, how do they know those yields are the highest?) I also found reports about rates on 30-year fixed loans, 15-year fixed loans, 1-year adjustable rate loans, new car loans, used car loans, home equity loans, and loans from your grandmother (well, actually no, but if grandma read these statistics, she might increase her cushy rates).

Finally, I saw numerous ads for those beloved credit cards — ads listing the interest rates, the annual fees, and the number of days in the billing cycle. How do you compare all the information about investments, loans, and credit cards in order to make a good decision? What statistics are most important? The real question is: Are the numbers reported in the paper giving the whole story, or do you need to do more detective work to get at the truth? Chapters 17 and 18 help you start tearing apart these numbers and making decisions about them.

Touring the travel news

You can't even escape the barrage of numbers by heading to the travel section. For example, there I found that the most frequently asked question coming in to the Transportation Security Administration's response center (which receives about 2,000 telephone calls, 2,500 email messages, and 200 letters per week on average — would you want to be the one counting all of those?) is, "Can I carry this on a plane?" *This* can refer to anything from an animal to a wedding dress to a giant tin of popcorn. (I wouldn't recommend the tin of popcorn. You have to put it in the overhead compartment horizontally, and because things shift during a flight, the cover will likely open; and when you go to claim your tin at the end of the flight, you and your seatmates will be showered. Yes, I saw it happen once.)

The number of reported responses in this case leads to an interesting statistical question: How many operators are needed at various times of the day to field those calls, emails, and letters coming in? Estimating the number of anticipated calls is your first step, and being wrong can cost you money (if you overestimate it) or a lot of bad PR (if you underestimate it). These kinds of statistical challenges are tackled in Chapter 14.

Surveying sexual stats

In today's age of info-overkill, it's very easy to find out what the latest buzz is, including the latest research on people's sex lives. An article in my paper reported that married people have 6.9 more sexual encounters per year than people who have never been married. That's nice to know, I guess, but how did someone come up with this number? The article I'm looking at doesn't say (maybe some statistics are better left unsaid?).

If someone conducts a survey by calling people on the phone asking for a few minutes of their time to discuss their sex lives, who will be the most likely to want to talk about it? And what are they going to say in response to the question, "How many times a week do you have sex?" Are they going to report the honest truth, tell you to mind your own business, or exaggerate a little? Self-reported surveys can be a real source of bias and can lead to misleading statistics. But how would you recommend people go about finding out more about this very personal subject? Sometimes, research is more difficult than it seems. (Chapter 17 discusses biases that come up when collecting certain types of survey data.)

Breaking down weather reports

Weather reports provide another mass of statistics, with forecasts of the next day's high and low temperatures (how do they decide it'll be 16 degrees and not 15 degrees?) along with reports of the day's UV factor, pollen count, pollution standard index, and water quality and quantity. (How do they get these numbers — by taking samples? How many samples do they take, and where do they take them?) You can find out what the weather is right now anywhere in the world. You can get a forecast looking ahead three days, a week, a month, or even a year! Meteorologists collect and record tons and tons of data on the weather each day. Not only do these numbers help you decide whether to take your umbrella to work, but they also help weather researchers to better predict longer-term forecasts and even global climate changes over time.

Even with all the information and technologies available to weather researchers, how accurate are weather reports these days? Given the number of times you get rained on when you were told it was going to be sunny, it seems they still have work to do on those forecasts. What the abundance of data really shows, though, is that the number of variables affecting weather is almost overwhelming, not just to you, but for meteorologists, too.

REMEMBER

Statistical computer models play an important role in making predictions about major weather-related events, such as hurricanes, earthquakes, and volcano eruptions. Scientists still have some work to do before they can predict tornados before they begin to form, or tell you exactly where and when a hurricane is going to hit land, but that's certainly their goal, and they continue to get better at it. For more on modeling and statistics, see Chapter 19.

Using Statistics at Work

Now let's put down the Sunday newspaper and move on to the daily grind of the workplace. If you're working for an accounting firm, of course numbers are part of your daily life. But what about people like nurses, portrait studio photographers, store managers, newspaper reporters, office staff, or construction workers? Do numbers play a role in those jobs? You bet. This section gives you a few examples of how statistics creep into *every* workplace.

REMEMBER

You don't have to go far to see how statistics weaves its way in and out of your life and work. The secret is being able to determine what it all means and what you can believe, and to be able to make sound decisions based on the real story behind numbers so you can handle and become used to the statistics of everyday life.

Delivering babies — and information

Sue works as a nurse during the night shift in the labor and delivery unit at a university hospital. She takes care of several patients in a given evening, and she does her best to accommodate everyone. Her nursing manager has told her that each time she comes on shift she should identify herself to the patient, write her name on the whiteboard in the patient's room, and ask whether the patient has any questions. Why? Because a few days after each mother leaves with her baby, the hospital gives her a phone call asking about the quality of care, what was missed, what it could do to improve its service and quality of care, and what the staff could do to ensure that the hospital is chosen over other hospitals in town. For example, surveys show that patients who know the names of their nurses feel more comfortable, ask more questions, and have a more positive experience in the hospital than those who don't know the names of their nurses. Sue's salary raises depend on her ability to follow through with the needs of new mothers. No doubt the hospital has also done a lot of research to determine the factors involved in quality of patient care well beyond nurse-patient interactions. (See Chapter 18 for in-depth info concerning medical studies.)

Posing for pictures

Carol works as a photographer for a department store portrait studio; one of her strengths is working with babies. Based on the number of photos purchased by customers over the years, this store has found that people buy more posed pictures than natural-looking ones. As a result, store managers encourage their photographers to take posed shots.

A mother comes in with her baby and has a special request: "Could you please not pose my baby too deliberately? I just like his pictures to look natural." If Carol says, "Can't do that, sorry. My raises are based on my ability to pose a child well," you can bet that the mother is going to fill out that survey on quality service after this session — and not just to get $2.00 off her next sitting (if she ever comes back). Instead, Carol should show her boss the information in Chapter 17 about collecting data on customer satisfaction.

Poking through pizza data

Terry is a store manager at a local pizzeria that sells pizza by the slice. He is in charge of determining how many workers to have on staff at a given time, how many pizzas to make ahead of time to accommodate the demand, and how much cheese to order and grate, all with minimal waste of wages and ingredients. Friday night at midnight, the place is dead. Terry has five workers left and has five large pans of pizza he could throw in the oven, making about 40 slices of pizza each. Should he send two of his workers home? Should he put more pizza in the oven or hold off?

The store owner has been tracking the demand for weeks now, so Terry knows that every Friday night things slow down between 10 p.m. and 12 a.m., but then the bar crowd starts pouring in around midnight and doesn't let up until the doors close at 2:30 a.m. So Terry keeps the workers on, puts in the pizzas in 30-minute intervals from midnight on, and is rewarded with a profitable night, with satisfied customers and a happy boss. For more information on how to make good estimates using statistics, see Chapter 14.

Statistics in the office

D.J. is an administrative assistant for a computer company. How can statistics creep into her office workplace? Easy. Every office is filled with people who want to know answers to questions, and they want someone to "Crunch the numbers," to "Tell me what this means," to "Find out if anyone has any hard data on this," or to simply say, "Does this number make any sense?" They need to know everything from customer satisfaction figures to changes in inventory during the year; from the percentage of time employees spend on email to the cost of supplies for the last three years. Every workplace is filled with statistics, and D.J.'s marketability and value as an employee could go up if she's the one the head honchos turn to for help. Every office needs a resident statistician — why not let it be you?

IN THIS CHAPTER

» Examining the extent of statistics abuse

» Feeling the impact of statistics gone wrong

Chapter 2
Taking Control: So Many Numbers, So Little Time

The sheer amount of statistics in daily life can leave you feeling overwhelmed and confused. This chapter gives you a tool to help you deal with statistics: skepticism! Not radical skepticism like "I can't believe anything anymore," but healthy skepticism like "Hmm, I wonder where that number came from?" and "I need to find out more information before I believe these results." To develop healthy skepticism, you need to understand how the chain of statistical information works.

Statistics end up on your TV and in your newspaper as a result of a process. First, the researchers who study an issue generate results; this group is composed of pollsters, doctors, marketing researchers, government researchers, and other scientists. They are considered the *original sources* of the statistical information.

After they get their results, these researchers naturally want to tell people about them, so they typically either put out a press release or publish a journal article. Enter the journalists or reporters, who are considered the *media sources* of the information. Journalists hunt for interesting press releases and sort through journals, basically searching for the next headline. When reporters complete their stories, statistics are immediately sent out to the public through all forms of media. Now the information is ready to be taken in by the third group — the *consumers* of the information (you). You and other consumers of information are faced with the task of listening to and reading the information, sorting through it, and making decisions about it.

At any stage in the process of doing research, communicating results, or consuming information, errors can take place, either unintentionally or by design. The tools and strategies you find in this chapter give you the skills to be a good detective.

Detecting Errors, Exaggerations, and Just Plain Lies

Statistics can go wrong for many different reasons. First, a simple, honest error can occur. This can happen to anyone, right? Other times, the error is something other than a simple, honest mistake. In the heat of the moment, because someone feels strongly about a cause and because the numbers don't quite bear out the point that the researcher wants to make, statistics get tweaked, or, more commonly, exaggerated, either in their values or how they're represented and discussed.

Another type of error is an *error of omission* — information that is missing that would have made a big difference in terms of getting a handle on the real story behind the numbers. That omission makes the issue of correctness difficult to address, because you're lacking information to go on.

You may even encounter situations in which the numbers have been completely fabricated and can't be repeated by anyone because they never happened. This section gives you tips to help you spot errors, exaggerations, and lies, along with some examples of each type of error that you, as an information consumer, may encounter.

Checking the math

The first thing you want to do when you come upon a statistic or the result of a statistical study is to ask, "Is this number correct?" Don't assume it is! You'd probably be surprised at the number of simple arithmetic errors that occur when statistics are collected, summarized, reported, or interpreted.

TIP

To spot arithmetic errors or omissions in statistics:

- **Check to be sure everything adds up.** In other words, do the percents in the pie chart add up to 100 (or close enough due to rounding)? Do the number of people in each category add up to the total number surveyed?
- **Double-check even the most basic calculations.**
- **Always look for a total so you can put the results into proper perspective.** Ignore results based on tiny sample sizes.
- **Examine whether the projections are reasonable.** For example, if three deaths due to a certain condition are said to happen per minute, that adds up to over 1.5 million such deaths in a year. Depending on what condition is being reported, this number may be unreasonable.

Uncovering misleading statistics

By far, the most common abuses of statistics are subtle, yet effective, exaggerations of the truth. Even when the math checks out, the underlying statistics themselves can be misleading if they exaggerate the facts. Misleading statistics are harder to pinpoint than simple math errors, but they can have a huge impact on society, and, unfortunately, they occur all the time.

Breaking down statistical debates

Crime statistics are a great example of how statistics are used to show two sides of a story, only one of which is really correct. Crime is often discussed in political debates, with one candidate (usually the incumbent) arguing that crime has gone down during her tenure, and the challenger often arguing that crime has gone up (giving the challenger something to criticize the incumbent for). How can two candidates make such different conclusions based on the same data set? Turns out, depending on the way you measure crime, getting either result can be possible.

Table 2-1 shows the population of the United States for 2000 to 2008, along with the number of reported crimes and the crime *rates* (crimes per 100,000 people), calculated by taking the number of crimes divided by the population size and multiplying by 100,000.

Table 2-1 Number of Crimes, Estimated Population Size, and Crime Rates in the U.S.

Year	Number of Crimes	Population Size	Crime Rate per 100,000 People
2000	11,608,072	281,421,906	4,124.8
2001	11,876,669	285,317,559	4,162.6
2002	11,878,954	287,973,924	4,125.0
2003	11,826,538	290,690,788	4,068.4
2004	11,679,474	293,656,842	3,977.3
2005	11,565,499	296,507,061	3,900.6
2006	11,401,511	299,398,484	3,808.1
2007	11,251,828	301,621,157	3,730.5
2008	11,149,927	304,059,784	3,667.0

Source: U.S. Crime Victimization Survey.

Now, compare the number of crimes and the crime rates for 2001 and 2002 in Table 2-1. In column 2, you see that the *number of crimes* increased by 2,285 from 2001 to 2002 (11,878,954 – 11,876,669). This represents an increase of 0.019 percent (dividing the difference, 2,285, by the number of crimes in 2001, 11,876,669). Note the population size (column 3) also increased from 2001 to 2002, by 2,656,365 people (287,973,924 – 285,317,559), or 0.931 percent (dividing this difference by the population size in 2001). However, in column 4, you see the crime *rate* decreased from 2001 to 2002 from 4,162.6 (per 100,000 people) in 2001 to 4,125.0 (per 100,000) in 2002. How did the crime rate decrease? Although the number of crimes and the number of people both went up, the number of crimes increased at a slower rate than the increase in population size (0.019 percent compared to 0.931 percent).

So how should the crime trend be reported? Did crime actually go up or down from 2001 to 2002? Based on the crime rate — which is a more accurate gauge — you can conclude that crime decreased during that year. But be watchful of the politician who wants to show that the incumbent didn't do their job; they will be tempted to look at the number of crimes and claim that crime went up, creating an artificial controversy and resulting in confusion (not to mention skepticism) on behalf of the voters. (Aren't election years fun?)

To create an even playing field when measuring how often an event occurs, you convert each number to a percent by dividing by the total to get what statisticians call a *rate*. Rates are usually better than count data because rates allow you to make fair comparisons when the totals are different.

Untwisting tornado statistics

Which state has the most tornados? It depends on how you look at it. If you just count the number of tornados in a given year (which is how I've seen the media report it most often), the top state is Texas. But think about it. Texas is the second-biggest state (after Alaska). Yes, Texas is in that part of the U.S. called "Tornado Alley," and yes, it gets a lot of tornados, but it also has a huge surface area for those tornados to land and run.

A more fair comparison, and how meteorologists look at it, is to look at the number of tornados per 10,000 square miles. Using this statistic (depending on your source), Florida comes out on top, followed by Oklahoma, Indiana, Iowa, Kansas, Delaware, Louisiana, Mississippi, and Nebraska, and finally Texas weighs in at number 10 (although I'm sure this is one statistic they are happy to rank low on — as opposed to their AP rankings in NCAA football).

Other tornado statistics that are measured and reported include the state with the highest percentage of killer tornadoes as a percentage of all tornados (Tennessee), and the total length of tornado paths per 10,000 square miles (Mississippi). Note that each of these statistics is reported appropriately as a *rate* (amount per unit).

Before believing statistics indicating "the highest XXX" or "the lowest XXX," take a look at how the variable is measured to see whether it's fair and whether there are other statistics that should be examined to get the whole picture. Also make sure the units are appropriate for making fair comparisons.

Zeroing in on what the scale tells you

Charts and graphs are useful for making a quick and clear point about your data. Unfortunately, in many cases the charts and graphs accompanying everyday statistics aren't done correctly or fairly. One of the most important elements to watch for is the way that the chart or graph is scaled. The *scale* of a graph is the quantity used to represent each tick mark on the axis of the graph. Do the tick marks increase by 1s, 10s, 20s, 100s, 1,000s, or what? The scale can make a big difference in terms of the way the graph or chart looks.

For example, the Kansas Lottery routinely shows its recent results from the Pick 3 Lottery. One of the statistics reported is the number of times each number (0 through 9) is drawn among the three winning numbers. Table 2-2 shows a chart of the number of times each number was drawn during 1,613 total Pick 3 games (4,839 single numbers drawn). It also reports the percentage of times that each number was drawn. Depending on how you choose to look at these results, you can again make the statistics appear to tell very different stories.

Table 2-2 Numbers Drawn in the Pick 3 Lottery

Number Drawn	No. of Times Drawn out of 4,839	Percentage of Times Drawn (No. of Times Drawn ÷ 4,839)
0	485	10.0%
1	468	9.7%
2	513	10.6%
3	491	10.1%
4	484	10.0%
5	480	9.9%
6	487	10.1%
7	482	10.0%
8	475	9.8%
9	474	9.8%

The way lotteries typically display results like those in Table 2-2 is shown in Figure 2-1a. Notice that in this chart, it seems that the number 1 doesn't get drawn nearly as often (only 468 times) as number 2 does (513 times). The difference in the height of these two bars appears to be very large, exaggerating the difference in the number of times these two numbers were drawn. However, to put this in perspective, the actual difference here is $513 - 468 = 45$ out of a total of 4,839 numbers drawn. In terms of percentages, the difference between the number of times the number 1 and the number 2 are drawn is $45 \div 4,839 = 0.009$, or only nine-tenths of 1 percent.

What makes this chart exaggerate the differences? Two issues come to mind. First, notice that the vertical axis, which shows the number of times (or frequency) that each number is drawn, goes up by increments of 5. So a difference of 5 out of a total of 4,839 numbers drawn appears significant. Stretching the scale so that differences appear larger than they really are is a common trick used to exaggerate results. Second, the chart starts counting at 465, not at 0. Only the top part of each bar is shown, which also exaggerates the results. In comparison, Figure 2-1b graphs the *percentage* of times each number was drawn. Normally the shape of a graph wouldn't change when going from counts to percentages; however, this chart uses a more realistic scale than the one in Figure 2-1a (going by 2 percent increments) and starts at 0, both of which make the differences appear as they really are — not much different at all. Boring, huh?

Maybe the lottery folks thought so, too. In fact, maybe they use Figure 2-1a rather than Figure 2-1b because they want you to think that some "magic" is involved in the numbers — and you can't blame them; that's their business.

REMEMBER

Looking at the scale of a graph or chart can really help you keep the reported results in proper perspective. Stretching the scale out or starting the *y*-axis at the highest possible number makes differences appear larger than they really are; squeezing down the scale or starting the *y*-axis at a much lower value than needed makes differences appear smaller.

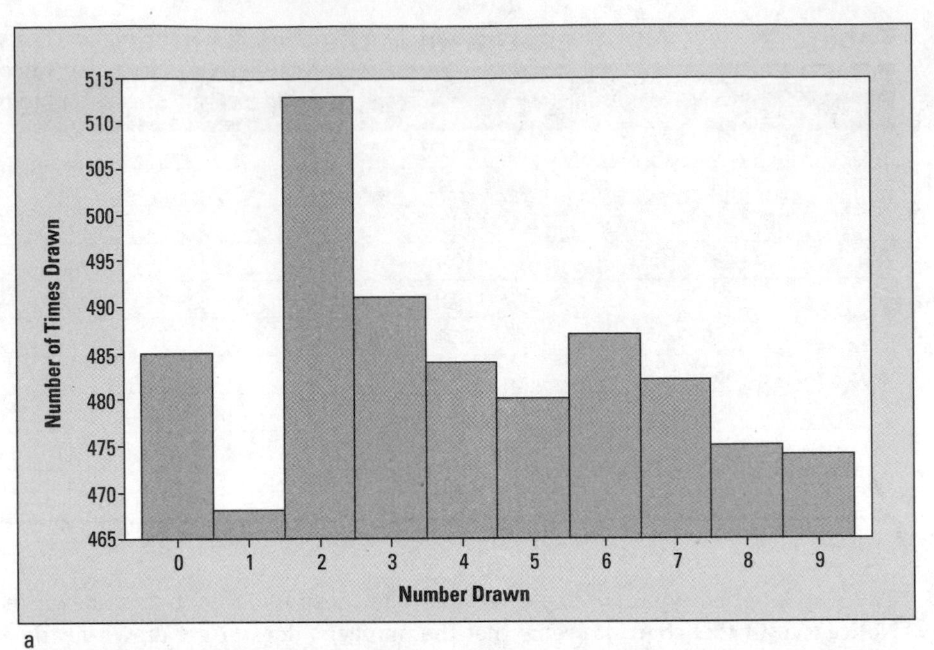

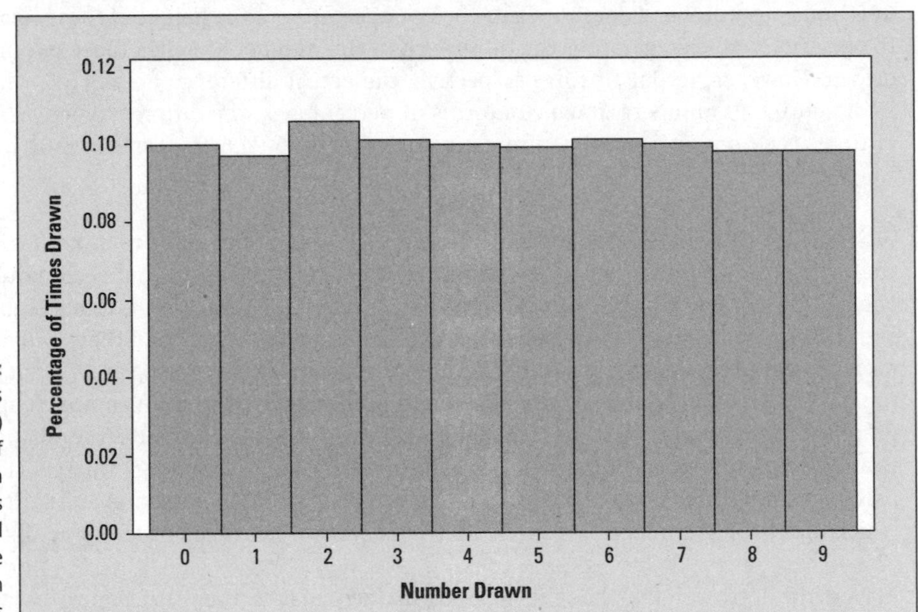

FIGURE 2-1: Bar charts showing a) number of times each number was drawn; and b) percentage of times each number was drawn.

Checking your sources

When examining the results of any study, check the source of the information. The best results are often published in reputable journals that are well known by the experts in the field. For example, in the world of medical science, the *Journal of the American Medical Association* (JAMA), the *New England Journal of Medicine*, *The Lancet*, and the *British Medical Journal* are all reputable journals that doctors use to publish results and read about new findings.

TIP

Consider the source and who financially supported the research. Many companies finance research and use it for advertising their products. Although that in itself isn't necessarily a bad thing, in some cases a conflict of interest on the part of researchers can lead to biased results. And if the results are very important to you, ask whether more than one study was conducted, and if so, ask to examine all the studies that were conducted, not just those whose results were published in journals or appeared in advertisements.

Counting on sample size

Sample size isn't everything, but it does count for a great deal in surveys and studies. If the study is designed and conducted correctly, and if the participants are selected randomly (that is, with no bias), then sample size is an important factor in determining the accuracy and repeatability of the results. (See Chapters 17 and 18 for more information on designing and carrying out studies including random samples.)

Many surveys are based on large numbers of participants, but that isn't always true for other types of research, such as carefully controlled experiments. Because of the high cost of some types of research in terms of time and money, some studies are based on a small number of participants or products. Researchers have to find the appropriate balance when determining sample size.

REMEMBER

The most unreliable results are those based on *anecdotes,* stories that talk about a single incident in an attempt to sway opinion. Have you ever told someone not to buy a product because you had a bad experience with it? Remember that an anecdote (or story) is really a nonrandom sample whose size is only one.

Considering cause and effect

Headlines often simplify or skew the "real" information, especially when the stories involve statistics and the studies that generated the statistics.

A study conducted a few years back evaluated videotaped sessions of 1,265 patient appointments with 59 primary-care physicians and 6 surgeons in Colorado and Oregon. This study found that physicians who had not been sued for malpractice spent an average of 18 minutes with each patient, compared to 16 minutes for physicians who *had* been sued for malpractice. The study was reported by the media with the headline, "Bedside manner fends off malpractice suits." However, this study seemed to say that if you are a doctor who gets sued, all you have to do is spend more time with your patients, and you're off the hook. (Now when did bedside manner get characterized as time spent?)

Beyond that, are we supposed to believe that a doctor who has been sued needs only add a couple more minutes of time with each patient to avoid being sued in the future? Maybe what the doctor does during that time counts much more than how much time the doctor actually spends with each patient. You tackle the issues of cause-and-effect relationships between variables in Chapter 19.

Finding what you want to find

You may wonder how two political candidates can discuss the same topic and get two opposing conclusions, both based on "scientific surveys." Even small differences in a survey can create big differences in results.

One common source of skewed survey results comes from question wording. Here are three different questions that are trying to get at the same issue — public opinion regarding the line-item veto option available to the president:

- Should the line-item veto be available to the president to eliminate waste (yes/no/no opinion)?
- Does the line-item veto give the president too much individual power (yes/no/no opinion)?
- What is your opinion on the presidential line-item veto? Choose 1–5, with 1 = strongly opposed and 5 = strongly support.

The first two questions are misleading and will lead to biased results in opposite directions. The third version will draw results that are more accurate in terms of what people really think. However, not all surveys are written with the purpose of finding the truth; many are written to support a certain viewpoint.

REMEMBER

Research shows that even small changes in wording affect survey outcomes, leading to results that conflict when different surveys are compared. If you can tell from the wording of the question how they want you to respond to it, you know you're looking at a leading question; and leading questions lead to biased results.

Looking for lies in all the right places

Every once in a while, you hear about someone who faked his data, or "fudged the numbers." Probably the most commonly committed lie involving statistics and data is when people throw out data that don't fit their hypothesis, don't fit the pattern, or appear to be outliers. In cases when someone has clearly made an error (for example, someone's age is recorded as 200), removing that erroneous data point or trying to correct the error makes sense. Eliminating data for any other reason is ethically wrong; yet it happens.

Regarding missing data from experiments, a commonly used phrase is "Among those who completed the study." What about those who didn't complete the study, especially a medical one? Did they get tired of the side effects of the experimental drug and quit? If so, the loss of this person will create results that are biased toward positive outcomes.

REMEMBER

Before believing the results of a study, check out how many people were chosen to participate, how many finished the study, and what happened to all the participants, not just the ones who experienced a positive result.

Surveys are not immune to problems from missing data, either. For example, it's known by statisticians that the opinions of people who respond to a survey can be very different from the opinions of those who don't. In general, the lower the percentage of people who respond to a survey (the response rate), the less credible the results will be. For more about surveys and missing data, see Chapter 17.

Feeling the Impact of Misleading Statistics

You make decisions every day based on statistics and statistical studies that you've heard about or seen, many times without even realizing it. Misleading statistics affect your life in small or large ways, depending on the type of statistics that cross your path and what you choose to do with the information you're given. Here are some little everyday scenarios where statistics slip in:

- "Gee, I hope Rex doesn't chew up my rugs again while I'm at work. I heard somewhere that dogs on Prozac deal better with separation anxiety. How did they figure that out? And what would I tell my friends?"
- "I thought everyone was supposed to drink eight glasses of water a day, but now I hear that too much water could be bad for me; what should I believe?"
- "A study says that people spend two hours a day at work checking and sending personal emails. How is that possible? No wonder my boss is paranoid."

You may run into other situations involving statistics that can have a larger impact on your life, and you need to be able to sort it all out. Here are some examples:

- A group lobbying for a new skateboard park tells you 80 percent of the people surveyed agree that taxes should be raised to pay for it, so you should too. Will you feel the pressure to say yes?
- The radio news at the top of the hour says cellphones cause brain tumors. Your spouse uses their cellphone all the time. Should you panic and throw away all cellphones in your house?
- You see an advertisement that tells you a certain drug will cure your particular illness. Do you run to your doctor and demand a prescription?

REMEMBER

Although not all statistics are misleading and not everyone is out to get you, you do need to be vigilant. By sorting out the good information from the suspicious and bad information, you can steer clear of statistics that go wrong. The tools and strategies in this chapter are designed to help you to stop and say, "Wait a minute!" so you can analyze and critically think about the issues and make good decisions.

IN THIS CHAPTER

» Seeing statistics as a process, not just as numbers

» Gaining success with statistics in your everyday life, your career, and in the classroom

» Becoming familiar with some basic statistical jargon

Chapter 3
Tools of the Trade

In today's world, the buzzword is data, as in, "Do you have any data to support your claim?" "What data do you have on this?" "The data supported the original hypothesis that," "Statistical data show that," and "The data bear this out." But the field of statistics is not just about data.

REMEMBER

Statistics is the entire process involved in gathering evidence to answer questions about the world, in cases where that evidence happens to be data.

In this chapter, I give you an overview of the role statistics plays in today's data-packed society and what you can do to not only survive but also thrive. You see firsthand how statistics works as a process and where the numbers play their part. You're also introduced to the most commonly used forms of statistical jargon, and you find out how these definitions and concepts all fit together as part of that process. You get a much broader view of statistics as a partner in the scientific method — designing effective studies, collecting good data, organizing and analyzing information, interpreting results, and making appropriate conclusions. (And you thought statistics was just number-crunching!)

Thriving in a Statistical World

It's hard to get a handle on the flood of statistics that affect your daily life in large and small ways. It begins the moment you wake up in the morning and check the news and listen to the meteorologist give you their predictions for the weather, based on their statistical analyses of past data and present weather conditions. You pore over nutritional information on the side of your cereal box while you eat breakfast. At work, you pull numbers from charts and tables, enter data into spreadsheets, run diagnostics, take measurements, perform calculations, estimate expenses, make decisions using statistical baselines, and order inventory based on past sales data.

At lunch, you go to the number-one restaurant based on a survey of 500 people. You eat food that is priced based on marketing data. You go to your doctor's appointment, where they take your blood pressure, temperature, and weight, and do a blood test; after all the information is collected, you get a report showing your numbers and how you compare to the statistical norms.

You head home in your car that's been serviced by a computer running statistical diagnostics. When you get home, you turn on the news and hear the latest crime statistics, see how the stock market performed, and discover how many people visited the zoo last week.

At night, you brush your teeth with toothpaste that's been statistically proven to fight cavities, you read a few pages of your *New York Times* bestseller (based on statistical sales estimates), and you go to sleep — only to start all over again the next morning. But how can you be sure that all those statistics you encounter and depend on each day are correct? In Chapter 1, I discuss a few examples of how statistics is involved in your life and workplace, what its impact is, and how you can raise your awareness of it.

WARNING

Some statistics are vague, inappropriate, or just plain wrong. You need to become more aware of the statistics you encounter each day and train your mind to stop and say, "Wait a minute!" Then you sift through the information, ask questions, and raise red flags when something's not quite right. In Chapter 2, you see ways in which you can be misled by bad statistics, and you develop skills to think critically and identify problems before automatically believing results.

Like any other field, statistics has its own set of jargon, and I outline and explain some of the most commonly used statistical terms in this chapter. Knowing the language increases your ability to understand and communicate statistics at a higher level without being intimidated. It raises your credibility when you use precise terms to describe what's wrong with a statistical result (and why). And your presentations involving statistical tables, graphs, charts, and analyses will be informational and effective. (Heck, if nothing else, you need the jargon because I use it throughout this book. Don't worry, though; I always review it.)

In the following sections, you see how statistics is involved in each phase of the scientific method.

Statistics: More than Just Numbers

Statisticians don't just "do statistics." Although the rest of the world views them as number crunchers, they think of themselves as the keepers of the scientific method. Of course, statisticians work with experts in other fields to satisfy their need for data, because people cannot live by statistics alone, but crunching someone's data is only a small part of a statistician's job. (In fact, if that's all they did all day, they'd quit their day jobs and moonlight as casino consultants.) In reality, statistics is involved in every aspect of the *scientific method* — formulating good questions, setting up studies, collecting good data, analyzing the data properly, and making appropriate conclusions. But aside from analyzing the data properly, what do any of these aspects have to do with statistics? In this chapter you find out.

All research starts with a question, such as:

- Is it possible to drink too much water?
- What's the cost of living in San Francisco?
- Who will win the next presidential election?
- Do herbs really help maintain good health?
- Will my favorite TV show get renewed for next year?

None of these questions asks anything directly about numbers. Yet each question requires the use of data and statistical processes to come up with the answer.

Suppose a researcher wants to determine who will win the next U.S. presidential election. To answer with confidence, the researcher has to follow several steps:

1. **Determine the population to be studied.**

   In this case, the researcher intends to study registered voters who plan to vote in the next election.

2. **Collect the data.**

   This step is a challenge, because you can't go out and ask every person in the United States whether they plan to vote, and if so, for whom they plan to vote. Beyond that, suppose someone says, "Yes, I plan to vote." Will that person *really* vote come Election Day? And will that same person tell you whom they actually plan to vote for? And what if that person changes their mind later on and votes for a different candidate?

3. **Organize, summarize, and analyze the data.**

   After the researcher has gone out and collected the data they need, getting it organized, summarized, and analyzed helps them answer their questions. This step is what most people recognize as the business of statistics.

4. **Take all the data summaries, charts, graphs, and analyses, and draw conclusions from them to try to answer the researcher's original question.**

   Of course, the researcher will not be able to have 100 percent confidence that their answer is correct, because not every person in the United States was asked. But they can get an answer that they are *nearly* 100 percent sure is correct. In fact, with a sample of about 2,500 people who are selected in a fair and *unbiased* way (that is, every possible sample of size 2,500 had an equal chance of being selected), the researcher can get accurate results within plus or minus 2.5 percent (if all the steps in the research process are done correctly).

WARNING

In making conclusions, the researcher has to be aware that every study has limits and that — because the chance for error always exists — the results could be wrong. A numerical value can be reported that tells others how confident the researcher is about the results and how accurate these results are expected to be. (See Chapter 13 for more information about margin of error.)

REMEMBER

After the research is done and the question has been answered, the results typically lead to even more questions and even more research. For example, if young voters appear to favor one candidate but older voters favor the opponent, the next questions may be: "Who goes to the polls more often on Election Day — young voters or older voters — and what factors determine whether they will vote?"

The field of statistics is really the business of using the scientific method to answer research questions about the world. Statistical methods are involved in every step of a good study, from designing the research to collecting the data, organizing and summarizing the information, doing an analysis, drawing conclusions, discussing limitations, and, finally, designing the next study in order to answer new questions that arise. Statistics is more than just numbers — it's a process.

Designing Appropriate Studies

Everyone's asking questions, from drug companies to biologists, from marketing analysts to the U.S. government. And ultimately, all those people will use statistics to help them answer their questions. In particular, many medical and psychological studies are done because someone wants to know the answer to a question. For example,

- Will this vaccine be effective in preventing the flu?
- What do Americans think about the state of the economy?
- Does an increase in the use of social networking websites cause depression in teenagers?

The first step after a research question has been formed is to design an effective study to collect data that will help answer that question. This step amounts to figuring out what process you'll use to get the data you need. In this section, I give an overview of the two major types of studies — surveys and experiments — and explore why it's so important to evaluate how a study was designed before you believe the results.

Surveys (Polls)

A *survey* (more commonly known as a *poll*) is a questionnaire; it's most often used to gather people's opinions along with some relevant demographic information. Because so many policymakers, marketers, and others want to "get at the pulse of the American public" and find out what the average American is thinking and feeling, many people now feel that they cannot escape the barrage of requests to take part in surveys and polls. In fact, you've probably received many requests to participate in surveys, and you may even have become numb to them, simply throwing away surveys received in the mail or saying "no" when asked to participate in a telephone survey.

If done properly, a good survey can really be informative. People use surveys to find out what TV programs Americans (and others) like, how consumers feel about Internet shopping, and whether the United States should allow someone under 35 to become president. Surveys are used by companies to assess the level of satisfaction their customers feel, to find out what products their customers want, and to determine who is buying their products. TV stations use surveys to get instant reactions to news stories and events, and movie producers use them to determine how to end their movies.

However, if statisticians had to choose one word to assess the general state of surveys in the media today, they'd say it's *quantity* rather than *quality.* In other words, you'll find no shortage of bad surveys. But in this book you find no shortage of good tips and information for analyzing, critiquing, and understanding survey results, and for designing your own surveys to do the job right. (To take off with surveys, head to Chapter 17.)

Experiments

An *experiment* is a study that imposes a treatment (or control) on the subjects (participants), controls their environment (for example, restricting their diets, giving them certain dosage levels of a drug or placebo, or asking them to stay awake for a prescribed period of time), and records the responses. The purpose of most experiments is to pinpoint a cause-and-effect relationship between two factors (such as alcohol consumption and impaired vision, or dosage level of a drug and intensity of side effects). Here are some typical questions that experiments try to answer:

- Does taking zinc help reduce the duration of a cold? Some studies show that it does.
- Does the shape and position of your pillow affect how well you sleep at night? The Emory Spine Center in Atlanta says yes.
- Does shoe heel height affect foot comfort? A study done at UCLA says up to 1-inch heels are better than flat soles.

In this section, I discuss some additional definitions of words that you may hear when someone is talking about experiments — Chapter 18 is entirely dedicated to the subject. For now, just concentrate on basic experiment lingo.

Treatment group versus control group

Most experiments try to determine whether some type of experimental treatment (or important factor) has a significant effect on an outcome. For example, does zinc help to reduce the length of a cold? Subjects who are chosen to participate in the experiment are typically divided randomly into two groups: a treatment group and a control group. (More than one treatment group is possible.)

- The *treatment group* consists of participants who receive the experimental treatment whose effect is being studied (in this case, zinc tablets).
- The *control group* consists of participants who do not receive the experimental treatment being studied. Instead, they get a placebo (a fake treatment; for example, a sugar pill); a standard, nonexperimental treatment (such as vitamin C, in the zinc study); or no treatment at all, depending on the situation.

In the end, the responses of those in the treatment group are compared with the responses from the control group to look for differences that are statistically significant (unlikely to have occurred just by chance assuming the subjects were chosen in a fair and unbiased way).

The fact that they were randomly assigned makes the samples similar, and any present biases wash out, so to speak. If the groups were not randomly assigned, the end results may not be telling us what we think they are.

Placebo

A *placebo* is a fake treatment, such as a sugar pill. Placebos are given to the control group to account for a psychological phenomenon called the *placebo effect,* in which patients receiving a fake treatment still report having a response, as if it were the real treatment. For example, after taking a sugar pill, a patient experiencing the placebo effect might say, "Yes, I feel better already," or "Wow, I *am* starting to feel a bit dizzy." By measuring the placebo effect in the control group, you can tease out what portion of the reports from the treatment group is real and what portion is likely due to the placebo effect. (Experimenters assume that the placebo effect affects both the treatment and control groups in the same way.)

Blind and double-blind

A *blind experiment* is one in which the subjects who are participating in the study are not aware of whether they're in the treatment group or the control group. In the zinc example, the vitamin C tablets and the zinc tablets would be made to look exactly alike, and patients would not be told which type of pill they were taking. A blind experiment attempts to control for bias on the part of the participants.

A *double-blind experiment* controls for potential bias on the part of both the patients *and* the researchers. Neither the patients nor the researchers collecting the data know which subjects received the treatment and which didn't. So who does know what's going on as far as who gets what treatment? Typically a third party (someone not otherwise involved in the experiment) puts together the pieces independently. A double-blind study is best, because even though researchers may claim to be unbiased, they often have a special interest in the results — otherwise, they wouldn't be doing the study!

Collecting Quality Data

After a study has been designed, be it a survey or an experiment, the individuals who will participate have to be selected, and a process must be in place to collect the data. This phase of the process is critical to producing credible data in the end, and this section hits the highlights.

Sample, random, or otherwise

When you sample some soup, what do you do? You stir the pot, reach in with a spoon, take out a little bit of the soup, and taste it. Then you draw a conclusion about the whole pot of soup, without actually having tasted all of it. If your sample is taken in a fair way (for example, you don't just grab all the good stuff), you will get a good idea of how the soup tastes without having to eat it all. Taking a sample works the same way in statistics. Researchers want to find out something about a population, but they don't have the time or money needed to study every single individual in the population. So they select a subset of individuals from the population, study those individuals, and use that information to draw conclusions about the whole population. This subset of the population is called a *sample.*

Although the idea of a selecting a sample seems straightforward, it's anything but. The way a sample is selected from the population can mean the difference between results that are correct and fair and results that are garbage. Here's an example: Suppose you want a sample of teenagers' opinions on whether they're spending too much time on the Internet. If you send out a survey using text messaging, your results have a higher chance of representing the teen population than if you call them; research shows that most teenagers have access to texting and are more likely to answer a text than a phone call.

Some of the biggest culprits of statistical misrepresentation caused by bad sampling are surveys done on the Internet. You can find thousands of surveys on the Internet that are done by having people log on to a particular website and give their opinions. But even if 50,000 people in the U.S. complete a survey on the Internet, it doesn't represent the population of all Americans. It represents only those folks who have Internet access, who logged on to that particular website, and who were interested enough to participate in the survey (which typically means that they have strong opinions about the topic in question). The result of all these problems is *bias* — systematic favoritism of certain individuals or certain outcomes of the study.

How do you select a sample in a way that avoids bias? The key word is *random.* A *random sample* is a sample selected by equal opportunity; that is, every possible sample the same size as yours has an equal chance to be selected from the population. What *random* really means is that no group in the population is favored in or excluded from the selection process.

Nonrandom (in other words, *biased*) *samples* are selected in such a way that some type of favoritism and/or automatic exclusion of a part of the population is involved. A classic example of a nonrandom sample comes from polls for which the media asks you to phone in your opinion on a certain issue ("call-in" polls). People who choose to participate in call-in polls do not represent the population at large because they had to be watching that program, and they had to feel strongly enough to call in. They technically don't represent a sample at all, in the statistical sense of the word, because no one selected them beforehand — they selected themselves to participate, creating a *volunteer* or *self-selected* sample. The results will be skewed toward people with strong opinions.

To take an authentic random sample, you need a randomizing mechanism to select the individuals. For example, the Gallup Organization starts with a computerized list of all telephone exchanges in America, along with estimates of the number of residential households that have those exchanges. The computer uses a procedure called *random digit dialing* (RDD) to randomly create phone numbers from those exchanges, and then selects samples of telephone numbers from those created numbers. So what really happens is that the computer creates a list of *all possible* household phone numbers in America and then selects a subset of numbers from that list for Gallup to call.

Another example of random sampling involves the use of random number generators. In this process, the items in the sample are chosen using a computer-generated list of random numbers, where each sample of items has the same chance of being selected. Researchers also may use this type of randomization to assign patients to a treatment group versus a control group in an experiment (see Chapter 20). This process is equivalent to drawing names out of a hat or drawing numbers in a lottery.

TIP

No matter how large a sample is, if it's based on nonrandom methods, the results will not represent the population that the researcher wants to draw conclusions about. Don't be taken in by large samples — first, check to see how they were selected. Look for the term *random sample.* If you see that term, dig further into the fine print to see how the sample was actually selected, and use the preceding definition to verify that the sample was, in fact, selected randomly. A small random sample is better than a large non-random one. Then second, look at how the population was defined (what is the target population?) and how the sample was selected. Does the sample represent the target population?

Bias

Bias is a word you hear all the time, and you probably know that it means something bad. But what really constitutes bias? *Bias* is systematic favoritism that is present in the data collection process, resulting in lopsided, misleading results. Bias can occur in any of a number of ways.

- **In the way the sample is selected:** For example, if you want to estimate how much holiday shopping people in the United States plan to do this year, and you take your clipboard and head out to a shopping mall on the day after Thanksgiving to ask customers about their shopping plans, you have bias in your sampling process. Your sample tends to favor those die-hard shoppers at that particular mall who were braving the massive crowds on that day known to retailers and shoppers as "Black Friday."
- **In the way data are collected:** Poll questions are a major source of bias. Because researchers are often looking for a particular result, the questions they ask can often reflect and lead to that expected result. For example, the issue of a tax levy to help support local schools is something every voter faces at one time or another. A poll question asking, "Don't you think it would be a great investment in our future to support the local schools?" has a bit of bias. On the other hand, so does "Aren't you tired of paying money out of your pocket to educate other people's children?" Question wording can have a huge impact on results.

Other issues that result in bias with polls are timing, length, level of question difficulty, and the manner in which the individuals in the sample were contacted (phone, mail, house-to-house, and so on). See Chapter 17 for more information on designing and evaluating polls and surveys.

TIP

When examining polling results that are important to you or that you're particularly interested in, find out what questions were asked and exactly how the questions were worded before drawing your conclusions about the results.

Grabbing Some Basic Statistical Jargon

Every trade has a basic set of tools, and statistics is no different. If you think about the statistical process as a series of stages that you go through to get from question to answer, you may guess that at each stage you'll find a group of tools and a set of terms (or jargon) to go along with it. Now, if the hair is beginning to stand up on the back of your neck, don't worry. No one is asking you to become a statistics expert and plunge into the heavy-duty stuff, or to turn into a statistics nerd who uses this jargon all the time. Hey, you don't even have to carry a calculator and pocket protector in your shirt pocket (because statisticians really don't do that; it's just an urban myth).

But as the world becomes more numbers-conscious, statistical terms are thrown around more in the media and in the workplace, so knowing what the language really means can give you a leg up. Also, if you're reading this book because you want to find out more about how to calculate some statistics, understanding basic jargon is your first step. So, in this section, you get a taste of statistical jargon; I send you to the appropriate chapters elsewhere in the book to get details.

Data

Data are the actual pieces of information that you collect through your study. For example, say that you ask five of your friends how many pets they own, and they give you the following data: 0, 2, 1, 4, 18. (The fifth friend counted each of their aquarium fish as a separate pet.) Not all data are numbers; you also record the gender of each of your friends, giving you the following data: male, male, female, male, female.

Most data fall into one of two groups: numerical or categorical. (I present the main ideas about these variables here; see Chapters 4 and 5 for more details.)

- **Numerical data:** These data have meaning as a measurement, such as a person's height, weight, IQ, or blood pressure; or they're a count, such as the number of stock shares a person owns, how many teeth a dog has, or how many pages you can read of your favorite book before you fall asleep. (Statisticians also call numerical data *quantitative data.*)

  Numerical data can be further divided into two types: discrete and continuous.

  - *Discrete data* represent items that can be counted; they take on possible values that can be listed out. The list of possible values may be fixed (also called *finite*); or it may go from 0, 1, 2, on to infinity (making it *countably infinite*). For example, the number of heads in 100 coin flips takes on values from 0 through 100 (finite case), but the number of flips needed to get 100 heads takes on values from 100 (the fastest scenario) on up to infinity. Its possible values are listed as 100, 101, 102, 103, (representing the countably infinite case).
  - *Continuous data* represent measurements; their possible values cannot be counted and can only be described using intervals on the real number line. For example, the exact amount of gas purchased at the pump for cars with 20-gallon tanks represents nearly continuous data from 0.00 gallons to 20.00 gallons, represented by the interval [0, 20>, inclusive. (Okay, you *can* count all these values, but why would you want to? In cases like these, statisticians bend the definition of continuous a wee bit.) The lifetime of a

C battery can technically be anywhere from zero to infinity, with all possible values in between. Granted, you don't expect a battery to last more than a few hundred hours, but no one can put a cap on how long it can go (remember the Energizer bunny?).

- **Categorical data:** Categorical data represent characteristics such as a person's marital status, hometown, or the types of movies they like. Categorical data can take on numerical values (such as "1" indicating married, "2" for single, "3" for divorced, and "4" for domestic partner), but those numbers don't have any meaning. You couldn't add them together, for example. (Other names for categorical data are *qualitative data*, or *Yes/No data.*)

REMEMBER

Ordinal data mixes numerical and categorical data. The data fall into categories, but the numbers placed on the categories have meaning. For example, rating a restaurant on a scale from 0 to 4 stars gives you ordinal data. Ordinal data are often treated as categorical, where the groups are ordered when graphs and charts are made. I don't address them separately in this book.

Data set

A *data set* is the collection of all the data taken from your sample. For example, if you measure the weights of five packages, and those weights are 12, 15, 22, 68, and 3 pounds, then those five numbers (12, 15, 22, 68, 3) constitute your data set. If you only record the general size of the package (for example, small, medium, or large), then your data set may look like this: medium, medium, medium, large, small.

Variable

A *variable* is any characteristic or numerical value that varies from individual to individual. A variable can represent a count (for example, the number of pets you own); or a measurement (the time it takes you to wake up in the morning). Or the variable can be categorical, where each individual is placed into a group (or category) based on certain criteria (for example, political affiliation, race, or marital status). Actual pieces of information recorded on individuals regarding a variable are the data.

Population

For virtually any question you may want to investigate about the world, you have to center your attention on a particular group of individuals (such as a group of people, cities, animals, rock specimens, exam scores, and so on). For example:

- What do Americans think about the president's foreign policy?
- What percentage of planted crops in Wisconsin did deer destroy last year?
- What's the prognosis for breast cancer patients taking a new experimental drug?
- What percentage of all cereal boxes get filled according to specification?

In each of these examples, a question is posed. And in each case, you can identify a specific group of individuals being studied: the American people, all planted crops in Wisconsin, all breast cancer patients, and all cereal boxes that are being filled, respectively. The group of individuals you want to study in order to answer your research question is called a *population.* Populations, however, can be hard to define. In a good study, researchers define the population very clearly, whereas in a bad study, the population is poorly defined.

The question of whether babies sleep better with music is a good example of how difficult defining the population can be. Exactly how would you define a baby? Under three months old? Under a year? And do you want to study babies only in the United States, or all babies worldwide? The results may be different for older and younger babies, for American versus European versus African babies, and so on.

REMEMBER

Many times, researchers want to study and make conclusions about a broad population, but in the end — to save time, money, or just because they don't know any better — they study only a narrowly defined population. That shortcut can lead to big trouble when conclusions are drawn. For example, suppose a college professor, Dr. Lewis, wants to study how TV ads persuade consumers to buy products. Her study is based on a group of her own students who participated to get five points extra credit. This test group may be convenient, but her results can't be generalized to any population beyond her own students, because no other population is represented in her study.

Statistic

A *statistic* is a number that summarizes the data collected from a sample. People use many different statistics to summarize data. For example, data can be summarized as a percentage (60 percent of U.S. households sampled own more than two cars), an average (the average price of a home in this sample is), a median (the median salary for the 1,000 computer scientists in this sample was), or a percentile (your baby's weight is at the 90th percentile this month, based on data collected from over 10,000 babies).

The type of statistic calculated depends on the type of data. For example, percentages are used to summarize categorical data, and means are used to summarize numerical data. The price of a home is a numerical variable, so you can calculate its mean or standard deviation. However, the color of a home is a categorical variable; finding the standard deviation or median of color makes no sense. In this case, the important statistics are the percentages of homes of each color.

REMEMBER

Not all statistics are correct or fair, of course. Just because someone gives you a statistic, nothing guarantees that the statistic is scientific or legitimate. You may have heard the saying, "Figures don't lie, but liars figure."

Parameter

Statistics are based on sample data, not on population data. If you collect data from the entire population, that process is called a *census.* If you then summarize all of the census information from one variable into a single number, that number is a *parameter,* not a statistic. Most of the time, researchers are trying to estimate the parameters using statistics. The U.S. Census Bureau wants to report the total number of people in the U.S., so it conducts a census. However, due to logistical problems in doing such an arduous task (such as being able to contact homeless folks), the census numbers can only be called *estimates* in the end, and they're adjusted upward to account for people the census missed.

Mean (Average)

The mean, also referred to by statisticians as the *average*, is the most common statistic used to measure the center, or middle, of a numerical data set. The *mean* is the sum of all the numbers divided by the total number of numbers. The mean of the entire population is called the *population mean*, and the mean of a sample is called the *sample mean*. (See Chapter 5 for more on the mean.)

WARNING

The mean may not be a fair representation of the data, because the average is easily influenced by *outliers* (very small or large values in the data set that are not typical).

Median

The median is another way to measure the center of a numerical data set. A statistical median is much like the median of an interstate highway. On many highways, the median is the middle, and an equal number of lanes lay on either side of it. In a numerical data set, the *median* is the point at which there are an equal number of data points whose values lie above and below the median value. Thus, the median is truly the middle of the data set. See Chapter 5 for more on the median.

REMEMBER

The next time you hear an average reported, look to see whether the median is also reported. If not, ask for it! The average and the median are two different representations of the middle of a data set and can often give two very different stories about the data, especially when the data set contains *outliers* (very large or small numbers that are not typical).

Standard deviation

Have you heard anyone report that a certain result was found to be "two standard deviations above the mean"? More and more, people want to report how significant their results are, and the number of standard deviations above or below average is one way to do it. But exactly what is a standard deviation?

The *standard deviation* is a measurement statisticians use for the amount of variability (or spread) among the numbers in a data set. As the term implies, a standard deviation is a standard (or typical) amount of deviation (or distance) from the average (or mean, as statisticians like to call it). So the standard deviation, in very rough terms, is the average distance from the mean.

The formula for standard deviation (denoted by s) is as follows, where n equals the number of values in the data set, each x represents a number in the data set, and $\bar{x}$ is the average of all the data:

$$s = \sqrt{\sum_{i=1}^{n} \frac{(x_i - \bar{x})^2}{n-1}}$$

For detailed instructions on calculating the standard deviation, see Chapter 5.

The standard deviation is also used to describe where most of the data should fall, in a relative sense, compared to the average. For example, if your data have the form of a bell-shaped curve (also known as a *normal distribution*), about 95 percent of the data lie within two standard deviations of the mean. (This result is called the *Empirical Rule*, or the *68–95–99.7% rule*. See Chapter 5 for more on this.)

WARNING

The standard deviation is an important statistic, but it is often absent when statistical results are reported. Without it, you're getting only part of the story about the data. Statisticians like to tell the story about a man named Dustin who had one foot in a bucket of ice water and the other foot in a bucket of boiling water. He said on average he felt just great! But think about the variability in the two temperatures for each of his feet. Closer to home, the average house price, for example, tells you nothing about the range of house prices you may encounter when house-hunting. The average salary may not fully represent what's really going on in your company, if the salaries are extremely spread out.

REMEMBER

Don't be satisfied with finding out only the average; be sure to ask for the standard deviation as well. Without a standard deviation, you have no way of knowing how spread out the values may be. (If you're talking starting salaries, for example, this could be very important!)

Percentile

You've probably heard references to percentiles before. If you've taken any kind of standardized test, you know that when your score was reported, it was presented to you with a measure of where you stood compared to the other people who took the test. This comparison measure was most likely reported to you in terms of a percentile. The *percentile* reported for a given score is the percentage of values in the data set that fall below that certain score. For example, if your score was reported to be at the 90th percentile, that means that 90 percent of the other people who took the test with you scored lower than you did (and 10 percent scored higher than you did). The median is right in the middle of a data set, so it represents the 50th percentile. For more specifics on percentiles, see Chapter 5.

REMEMBER

Percentiles are used in a variety of ways for comparison purposes and to determine *relative standing* (that is, how an individual data value compares to the rest of the group). Babies' weights are often reported in terms of percentiles, for example. Percentiles are also used by companies to see where they stand compared to other companies in terms of sales, profits, customer satisfaction, and so on.

Standard score

The standard score is a slick way to put results in perspective without having to provide a lot of details — something that the media loves. The *standard score* represents the number of standard deviations above or below the mean (without caring what that standard deviation or mean actually is).

For example, suppose Pat took the statewide 10th-grade test recently and scored 400. What does that mean? Not much, because you can't put 400 into perspective. But knowing that Pat's standard score on the test is +2 tells you everything. It tells you that Pat's score is two standard deviations above the mean. (Bravo, Pat!) Now suppose Emily's standard score is –2. In this case, this is not good (for Emily), because it means her score is two standard deviations *below* the mean.

The process of taking a number and converting it to a standard score is called *standardizing.* For the details on calculating and interpreting standard scores when you have a normal (bell-shaped) distribution, see Chapter 10.

Distribution and normal distribution

The *distribution* of a data set (or a population) is a listing or function showing all the possible values (or intervals) of the data and how often they occur. When a distribution of categorical data is organized, you see the number or percentage of individuals in each group. When a distribution of numerical data is organized, they're often ordered from smallest to largest, broken into reasonably sized groups (if appropriate), and then put into graphs and charts to examine the shape, center, and amount of variability in the data.

The world of statistics includes dozens of different distributions for categorical and numerical data; the most common ones have their own names. One of the most well-known distributions is called the *normal distribution,* also known as the *bell curve.* The normal distribution is based on numerical data that is continuous; its possible values lie on the entire real number line. Its overall shape, when the data are organized in graph form, is a symmetric bell shape. In other words, most of the data are centered around the mean (giving you the middle part of the bell), and as you move farther out on either side of the mean, you find fewer and fewer values (representing the downward-sloping sides on either side of the bell).

The mean (and hence the median) is directly in the center of the normal distribution due to symmetry, and the standard deviation is measured by the distance from the mean to the *inflection point* (where the curvature of the bell changes from concave up to concave down). Figure 3-1 shows a graph of a normal distribution with mean 0 and standard deviation 1 (this distribution has a special name: the *standard normal distribution* or *Z-distribution*). The shape of the curve resembles the outline of a bell.

Because every distinct population of data has a different mean and standard deviation, an infinite number of different normal distributions exists, each with its own mean and its own standard deviation to characterize it. See Chapter 10 for plenty more on the normal and standard normal distributions.

Central Limit Theorem

The normal distribution is also used to help measure the accuracy of many statistics, including the mean, using an important tool in statistics called the *Central Limit Theorem.* This theorem gives you the ability to measure how much your sample mean will vary, without having to take any other sample means to compare it with (thankfully!). By taking this variability into account, you can now use your data to answer questions about the population, such as, "What's the mean household income for the whole U.S.?" or "This report said 75 percent of all gift cards go unused; is that really true?" (These two particular analyses made possible by the Central Limit Theorem are called *confidence intervals* and *hypothesis tests,* respectively, and are described in Chapters 14 and 15, respectively.)

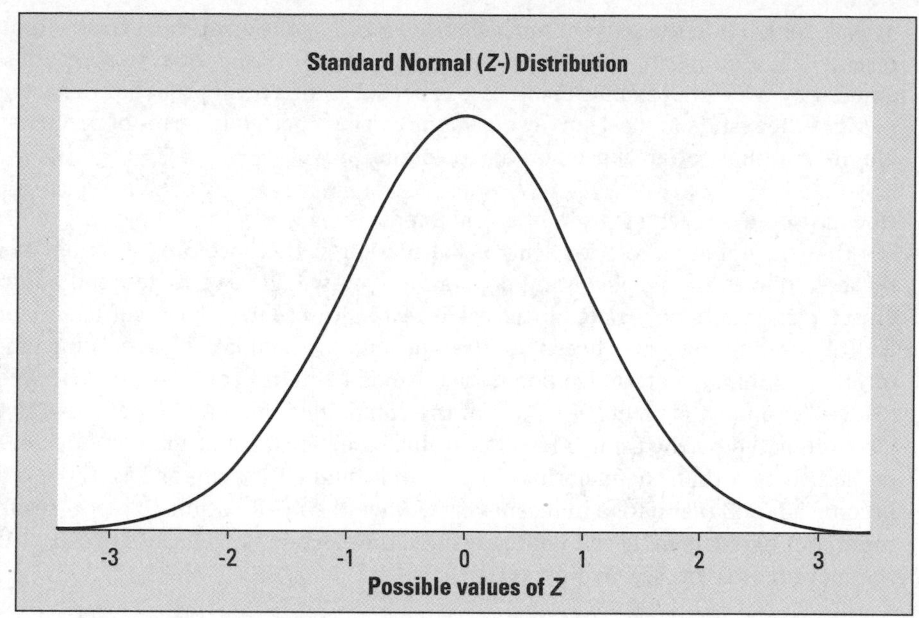

FIGURE 3-1: A standard normal (*Z*-) distribution has a bell-shaped curve with mean 0 and standard deviation 1.

The Central Limit Theorem (CLT for short) basically says that for non-normal data, your sample mean has an approximate normal distribution, no matter what the distribution of the original data looks like (as long as your sample size was large enough). And it doesn't just apply to the sample mean; the CLT is also true for other sample statistics, such as the sample proportion (see Chapters 14 and 15). Because statisticians know so much about the normal distribution (see the preceding section), these analyses are much easier. See Chapter 12 for more on the Central Limit Theorem, known by statisticians as the "crown jewel in the field of all statistics." (Should you even bother to tell them to get a life?)

z-values

If a data set has a normal distribution, and you standardize all the data to obtain standard scores, those standard scores are called *z*-values. All *z*-values have what is known as a standard normal distribution (or *Z*-distribution). The *standard normal distribution* is a special normal distribution with a mean equal to 0 and a standard deviation equal to 1.

The standard normal distribution is useful for examining the data and determining statistics like percentiles, or the percentage of the data falling between two values. So if researchers determine that the data have a normal distribution, they usually first standardize the data (by converting each data point into a *z*-value) and then use the standard normal distribution to explore and discuss the data in more detail. See Chapter 10 for more details on *z*-values.

Margin of error

You've probably heard or seen results like this: "This survey had a margin of error of plus or minus 3 percentage points." What does this mean? Most surveys (except a census) are based on information collected from a sample of individuals, not the entire population. A certain amount of error is bound to occur — not in the sense of calculation error (although there may be some of that, too), but in the sense of *sampling error,* which is the error that occurs simply because the researchers aren't asking everyone. The *margin of error* is supposed to measure the maximum amount by which the sample results are expected to differ from those of the actual population. Because the results of most survey questions can be reported in terms of percentages, the margin of error most often appears as a percentage, as well.

How do you interpret a margin of error? Suppose you know that 51 percent of people sampled say that they plan to vote for Senator Calculation in the upcoming election. Now, projecting these results to the whole voting population, you would have to add and subtract the margin of error and give a range of possible results in order to have sufficient confidence that you're bridging the gap between your sample and the population. Assuming a margin of error of plus or minus 3 percentage points, you would be pretty confident that between 48 percent (51% – 3%) and 54 percent (51% + 3%) of the population will vote for Senator Calculation in the election, based on the sample results. In this case, Senator Calculation may get slightly more or slightly less than the majority of votes and could either win or lose the election. This has become a familiar situation in recent years, when the media wants to report results on election night, but based on early exit polling results, the election is "too close to call." For more about the margin of error, see Chapter 13.

REMEMBER

The margin of error measures accuracy; it does not measure the amount of bias that may be present (you find a discussion of bias earlier in this chapter). Results that look numerically scientific and precise don't mean anything if they are collected in a biased way.

Confidence interval

One of the biggest uses of statistics is to estimate a population parameter using a sample statistic. In other words, you use a number that summarizes a sample to help you guesstimate the corresponding number that summarizes the whole population (the definitions of *parameter* and *statistic* appear earlier in this chapter). You're looking for a population parameter in each of the following questions:

- What's the average household income in America? (Population = all households in America; parameter = average household income.)
- What percentage of all Americans watched the Academy Awards this year? (Population = all Americans; parameter = percentage who watched the Academy Awards this year.)
- What's the average life expectancy of a baby born today? (Population = all babies born today; parameter = average life expectancy.)
- How effective is this new drug on adults with Alzheimer's? (Population = all adults who have Alzheimer's; parameter = percentage of these people who see improvement when taking this drug.)

It's not possible to find these parameters exactly; they each require an estimate based on a sample. You start by taking a random sample from a population (say a sample of 1,000 households in America) and then finding the corresponding statistic from that sample (the sample's mean household income). Because you know that sample results vary from sample to sample, you need to add a "plus or minus something" to your sample results if you want to draw conclusions about the whole population (all households in America). This "plus or minus" that you add to your sample statistic in order to estimate a parameter is the margin of error.

When you take a sample statistic (such as the sample mean or sample percentage) and you add/subtract a margin of error, you come up with what statisticians call a *confidence interval.* A confidence interval represents a range of likely values for the population parameter, based on your sample statistic. For example, suppose the average time it takes you to drive to work each day is 35 minutes, with a margin of error of plus or minus 5 minutes. You estimate that the average time driving to work would be anywhere from 30 to 40 minutes. This estimate is a confidence interval.

REMEMBER

Some confidence intervals are wider than others (and wide isn't good, because it means less accuracy). Several factors influence the width of a confidence interval, such as sample size, the amount of variability in the population being studied, and how confident you want to be in your results. (Most researchers are happy with a 95 percent level of confidence in their results.) For more on factors that influence confidence intervals, as well as instructions for calculating and interpreting confidence intervals, see Chapter 14.

Hypothesis testing

Hypothesis test is a term you probably haven't run across in your everyday dealings with numbers and statistics. But I guarantee that hypothesis tests have been a big part of your life and your workplace, simply because of the major role they play in industry, medicine, agriculture, government, and a host of other areas. Any time you hear someone talking about their study showing a "statistically significant result," you're encountering a hypothesis test. (A statistically significant result is one that is unlikely to have occurred by chance, based on using a well-selected sample. See Chapter 15 for the full scoop.)

Basically, a *hypothesis test* is a statistical procedure in which data are collected from a sample and measured against a claim about a population parameter. For example, if a pizza delivery chain claims to deliver all pizzas within 30 minutes of placing the order, on average, you could test whether this claim is true by collecting a random sample of delivery times over a certain period and looking at the average delivery time for that sample. To make your decision, you must also take into account the amount by which your sample results can change from sample to sample (which is related to the margin of error).

Because your decision is based on a sample (even a well-selected one) and not the entire population, a hypothesis test can sometimes lead you to the wrong conclusion. However, statistics are all you have, and if done properly, they can give you a good chance of being correct. For more on the basics of hypothesis testing, see Chapter 15.

A variety of hypothesis tests are done in scientific research, including *t*-tests (comparing two population means), paired *t*-tests (looking at before/after data), and tests of claims made about proportions or means for one or more populations. For specifics on these hypothesis tests, see Chapter 16.

p-values

Hypothesis tests are used to test the validity of a claim that is made about a population. This claim that's on trial, in essence, is called the *null hypothesis*. The *alternative hypothesis* is the one you would conclude if the null hypothesis is rejected. The evidence in the trial is your data and the statistics that go along with it. All hypothesis tests ultimately use a *p*-value to weigh the strength of the evidence (what the data are telling you about the population). The *p*-value is a number between 0 and 1 and is interpreted in the following way:

- A small *p*-value (such as less than 0.05) indicates strong evidence against the null hypothesis, so you reject it.
- A large *p*-value (greater than 0.05) indicates weak evidence against the null hypothesis, so you fail to reject it.
- *p*-values very close to the cutoff (0.05) are considered to be marginal (could go either way). Always report the *p*-value so your readers can draw their own conclusions.

For example, suppose a pizza place claims its delivery times are 30 minutes or less on average but you think it's more than that. You conduct a hypothesis test because you believe the null hypothesis, H_0, that the mean delivery time is 30 minutes max, is incorrect. Your alternative hypothesis (H_a) is that the mean time is greater than 30 minutes. You randomly sample some delivery times and run the data through the hypothesis test, and your *p*-value turns out to be 0.001, which is much less than 0.05. You conclude that the pizza place is wrong; their delivery times are in fact more than 30 minutes on average, and you want to know what they're gonna do about it! (Of course, you could be wrong by having sampled an unusually high number of late pizzas just by chance; but whose side am I on?) For more on *p*-values, head to Chapter 15.

Statistical significance

Whenever data are collected to perform a hypothesis test, the researcher is typically looking for something out of the ordinary. (Unfortunately, research that simply confirms something that was already well known doesn't make headlines.) Statisticians measure the amount by which a result is out of the ordinary using hypothesis tests (see Chapter 15). They use *probability*, the chance of how likely or unlikely some event is to occur, to put a number on how ordinary their result is. They define a *statistically significant* result as a result with a very small probability of happening just by chance, and provide a number called a *p*-value to reflect that probability (see the previous section on *p*-values).

For example, if a drug is found to be more effective at treating breast cancer than the current treatment, researchers say that the new drug shows a statistically significant improvement in the survival rate of patients with breast cancer. That means that based on their data, the difference in the overall results from patients using the new drug compared to those using the old treatment is so big that it would be hard to say it was just a coincidence. However, proceed with caution: You can't say that these results necessarily apply to each individual or to each individual in the same way. For full details on statistical significance, see Chapter 15.

REMEMBER

When you hear that a study's results are statistically significant, don't automatically assume that the study's results are important. *Statistically significant* means the results were unusual, but unusual doesn't always mean important. For example, would you be excited to learn that cats move their tails more often when lying in the sun than when lying in the shade, and that those results are statistically significant? This result may not even be important to the cat, much less anyone else!

Sometimes statisticians make the wrong conclusion about the null hypothesis because a sample doesn't represent the population (just by chance). For example, a positive effect that was experienced by a sample of people who took the new treatment may have just been a fluke; or in the example in the preceding section, the pizza company really may have been delivering those pizzas on time and you just got an unlucky sample of slow ones. However, the beauty of research is that as soon as someone gives a press release saying that they found something significant, the rush is on to try to replicate the results, and if the results can't be replicated, this probably means that the original results were wrong for some reason (including being wrong just by chance). Unfortunately, a press release announcing a "major breakthrough" tends to get a lot of play in the media, but follow-up studies refuting those results often don't show up on the front page.

REMEMBER

One statistically significant result shouldn't lead to quick decisions on anyone's part. In science, what most often counts is not a single remarkable study, but a body of evidence that is built up over time, along with a variety of well-designed follow-up studies. Take any major breakthroughs you hear about with a grain of salt and wait until the follow-up work has been done before using the information from a single study to make important decisions in your life. The results may not be replicable, and even if they are, you can't know whether they necessarily apply to each individual.

Correlation, regression, and two-way tables

One of the most common goals of research is to find links between variables. For example,

- Which lifestyle behaviors increase or decrease the risk of cancer?
- What side effects are associated with this new drug?
- Can I lower my cholesterol by taking this new herbal supplement?
- Does spending a large amount of time on the Internet cause a person to gain weight?

Finding links between variables is what helps the medical world design better drugs and treatments, provides marketers with information on who is more likely to buy their products, and gives politicians information on which to build arguments for and against certain policies.

In the mega-business of looking for relationships between variables, you find an incredible number of statistical results — but can you tell what's correct and what's not? Many important decisions are made based on these studies, and it's important to know what standards need to be met in order to deem the results credible, especially when a cause-and-effect relationship is being reported.

Chapter 19 breaks down all the details and nuances of plotting data from two numerical variables (such as dosage level and blood pressure), finding and interpreting *correlation* (the strength and direction of the linear relationship between *x* and *y*), finding the equation of a line that best fits the data (and when doing so is appropriate), and how to use these results to make predictions for one variable based on another (called *regression*). You also gain tools for investigating when a line fits the data well and when it doesn't, and what conclusions you can make (and shouldn't make) in the situations where a line does fit.

I cover methods used to look for and describe links between two categorical variables (such as the number of doses taken per day and the presence or absence of nausea) in detail in Chapter 20. I also provide info on collecting and organizing data into *two-way tables* (where the possible values of one variable make up the rows and the possible values for the other variable make up the columns), interpreting the results, analyzing the data from two-way tables to look for relationships, and checking for independence. And, as I do throughout this book, I give you strategies for critically examining results of these kinds of analyses for credibility.

Drawing Credible Conclusions

To perform statistical analyses, researchers use statistical software that depends on formulas. But formulas don't know whether they are being used properly, and they don't warn you when your results are incorrect. At the end of the day, computers can't tell you what the results mean; you have to figure it out. Throughout this book you see what kinds of conclusions you can and can't make after the analysis has been done. The following sections provide an introduction to drawing appropriate conclusions.

Reeling in overstated results

Some of the most common mistakes made in conclusions are overstating the results or generalizing the results to a larger group than was actually represented by the study. For example, Professor Lewis wants to know which Super Bowl commercials viewers liked best. He gathers 100 students from his class on Super Bowl Sunday and asks them to rate each commercial as it is shown. A top-five list is formed, and he concludes that all Super Bowl viewers liked those five commercials the best. But he really only knows which ones *his students* liked best — he didn't study any other groups, so he can't draw conclusions about all viewers.

Questioning claims of cause and effect

One situation in which conclusions cross the line is when researchers find that two variables are related (through an analysis such as regression; see the earlier section, "Correlation, regression, and two-way tables," for more info) and then automatically leap to the conclusion that those two variables have a cause-and-effect relationship.

For example, suppose a researcher conducted a health survey and found that people who took vitamin C every day reported having fewer colds than people who didn't take vitamin C every day. Upon finding these results, the researcher wrote a paper and gave a press release saying vitamin C prevents colds, using this data as evidence.

Now, while it may be true that vitamin C does prevent colds, this researcher's study can't claim that. This study was observational, which means they didn't assign who would take vitamin C. For example, people who take vitamin C every day may be more health-conscious overall, washing their hands more often, exercising more, and eating better foods; all these behaviors may be helpful in reducing colds.

REMEMBER

A controlled experiment can give you a cause-and-effect conclusion based on relationships you find. Observational studies can find cause-and-effect connections too, but it takes a larger body of evidence and more complex methods. (I discuss experiments in more detail earlier in this chapter.)

Becoming a Sleuth, Not a Skeptic

Statistics is about much more than numbers. To really "get" statistics, you need to understand how to make appropriate conclusions from studying data and be savvy enough to not believe everything you hear or read until you find out how the information came about, what was done with it, and how the conclusions were drawn. That's something I discuss throughout the book.

TIP

Much of our advice is based on understanding the big picture as well as the details of tackling statistical problems and coming out a winner on the other side.

REMEMBER

Becoming skeptical or cynical about statistics is very easy, especially after finding out what's going on behind the scenes; don't let that happen to you. You can find a lot of good information out there that can affect your life in a positive way. Find a good channel for your skepticism by setting two personal goals:

1. To become a well-informed consumer of the statistical information you see every day.
2. To establish job security by being the statistics "go-to" person who knows when and how to help others and when to find a statistician.

Through reading and using the information in this book, you'll be confident in knowing you can make good decisions about statistical results. You'll conduct your own statistical studies in a credible way. And you'll be ready to tackle your next office project, critically evaluate that annoying political ad, or ace your next exam!

2 Number-Crunching Basics

In This Unit . . .

IN THIS CHAPTER

» Summarizing categorical data effectively

» Interpreting commonly used statistics

» Realizing what statistics do and don't say

Chapter 4 Crunching Categorical Data

Every data set has a story, and if statistics are used properly, they do a good job of uncovering and reporting that story. Statistics that are improperly used can tell a different story, or only part of it, so knowing how to make good decisions about the information you're given is very important.

A *descriptive statistic* (or *statistic* for short) is a number that summarizes or describes some characteristic about a set of data. In this chapter and the next one, you see some of the most common descriptive statistics and how they are used, and you find out how to calculate them, interpret them, and put them together to get a good picture of a data set. You also find out what these statistics say and what they don't say about the data.

Summing Up Data with Descriptive Statistics

Descriptive statistics take a data set and boil it down to a set of basic information. Summarized data are often used to provide people with information that is easy to understand and that helps answer their questions. Picture your boss coming to you and asking, "What's our client base like these days, and who's buying our products?" How would you like to answer that question? With a long, detailed, and complicated stream of numbers that are sure to glaze their eyes over? Probably not. You want clean, clear, and concise statistics that sum up the client base for them, so they can see how brilliant you are and then send you off to collect even more data to see how they can include more people in the client base. (That's what you get for being efficient.)

Summarizing data has other purposes, as well. After all the data have been collected from a survey or some other kind of study, the next step is for the researcher to try to make sense out of the data. Typically, the first step researchers take is to run some basic statistics on the data to get a rough idea about what's happening in it. Later in the process, researchers can do more analyses to formulate or test claims made about the population the data came from, estimate certain characteristics about the population (like the mean), look for links between variables they measured, and so on.

Another big part of research is reporting the results, not only to their peers, but also to the media and the general public. Although a researcher's peers may be anxiously waiting to hear about all the complex analyses that were done on a data set, the general public is neither ready for nor interested in that. What does the public want? Basic information. Statistics that make a point clearly and concisely are typically used to relay information to the media and to the public.

WARNING

If you really need to learn more from data, a quick statistical overview isn't enough. In the statistical world, less is not more, and sometimes the real story behind the data can get lost in the shuffle. To be an informed consumer of statistics, you need to think about which statistics are being reported, what these statistics really mean, and what information is missing. This chapter focuses on these issues.

Crunching Categorical Data: Tables and Percents

Categorical data (also known as *qualitative data*) capture qualities or characteristics about an individual, such as a person's eye color, gender, political party, or opinion on some issue (using categories such as Agree, Disagree, or No Opinion). Categorical data tend to fall into groups or categories pretty naturally. "Political Party," for example, typically has four groups in the United States: Democrat, Republican, Independent, and Other. Categorical data often come from survey data, but they can also be collected in experiments. For example, in an experimental test of a new medical treatment, researchers may use three categories to assess the outcome of the experiment: Did the patient get better, worse, or stay the same while undergoing the treatment?

Counting on the frequency

One way to summarize categorical data is to simply count, or tally up, the number of individuals that fall into each category. The number of individuals in any given category is called the *frequency* for that category. If you list all the possible categories along with the frequency for each of them, you create a *frequency table*. The total of all the frequencies should equal the size of the sample (because you place each individual in one category).

EXAMPLE

Q. Suppose you take a sample of ten people and ask them all whether or not they own the latest smartphone. Each person falls into one of two categories: yes or no. The data is shown in the following table:

Person #	Latest Smartphone	Person #	Latest Smartphone
1	Y	6	Y
2	N	7	Y
3	Y	8	Y
4	N	9	N
5	Y	10	Y

a. Summarize this data in a frequency table.

b. What's an advantage of summarizing categorical data?

A. Data summaries boil down the data quickly and clearly.

a. The frequency table for this data is shown here:

Own the Latest Smartphone?	Frequency
Y	7
N	3
Total	10

b. A data summary allows you to see patterns in the data, which aren't clear if you look only at the original data.

YOUR TURN

1 You survey 20 shoppers to see what type of soft drink they like best, Brand A or Brand B. The results are: A, A, B, B, B, B, B, B, A, A, A, B, A, A, A, A, B, B, A, A. Which brand do the shoppers prefer? Make a frequency table and explain your answer.

2 A local city government asks voters to vote on a tax levy for the local school district. A total of 18,726 citizens vote on the issue. The Yes count comes in at 10,479, and the rest of the voters say No.

a. Show the results in a frequency table.

b. Why is it important to include the total number at the bottom of a frequency table?

A zoo asks 1,000 people whether they've been to the zoo in the last year. The pollsters count the results and find that 592 say yes, 198 say no, and 210 don't respond.

a. Show the results in a frequency table.

b. Explain why you need to include the people who don't respond.

Relating with percentages

Another way to summarize categorical data is to show the percentage of individuals that fall into each category, thereby creating a relative frequency. The *relative frequency* of a given category is the frequency (number of individuals in that category) divided by the total sample size, multiplied by 100 to get the percentage; hence, the calculated value is relative to the total surveyed. For example, if you survey 50 people and 10 are in favor of a certain issue, the relative frequency of the "in-favor" category is 10 ÷ 50 = 0.20 times 100, which gives you 20 percent. If you list all the possible categories along with their relative frequencies, you create a *relative frequency table.* The total of all the relative frequencies should equal 100 percent (subject to possible round-off error).

EXAMPLE

Q. Using the cellphone data from the following table, make a relative frequency table and interpret the results.

Person #	Latest Smartphone	Person #	Latest Smartphone
1	Y	6	Y
2	N	7	Y
3	Y	8	Y
4	N	9	N
5	Y	10	Y

A. Following is a relative frequency table for the cellphone data. Seventy percent of the people sampled reported owning the latest smartphones, and 30 percent admitted to being technologically just a little behind the times.

Own the Latest Smartphone?	Relative Frequency
Y	70%
N	30%

You get the 70 percent by taking $7 \div 10 \times 100$, and you calculate the 30 percent by taking $3 \div 10 \times 100$.

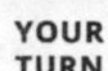

YOUR TURN

4 You survey 20 shoppers to see what type of soft drink they like best, Brand A or Brand B. The results are: A, A, B, B, B, B, B, B, A, A, A, B, A, A, A, A, B, B, A, A. Which brand do the shoppers prefer?

a. Use a relative frequency table to determine the preferred brand.

b. In general, if you had to choose, which is easier to interpret: frequencies or relative frequencies? Explain your answer.

5 A local city government asked voters in the last election to vote on a tax levy for the local school district. A record 18,726 voted on the issue. The Yes count came in at 10,479, and the rest of the voters checked the No box. Show the results in a relative frequency table.

6 A zoo surveys 1,000 people to find out whether they've been to the zoo in the last year. The pollsters count the results and find that 592 say yes, 198 say no, and 210 don't respond. Make a relative frequency table and use it to find the *response rate* (percentage of people who respond to the survey).

7 Suppose that instead of showing the number in each group, you show just the percentage (that is, the relative frequency).

a. What's one advantage a relative frequency table has over a frequency table?

b. Name one disadvantage that comes with creating a relative frequency table compared to using a frequency table.

Two-way tables: Summarizing multiple measures

You can break down categorical data further by creating two-way tables. *Two-way tables* (also called *crosstabs*) are tables with rows and columns. They summarize the information from two categorical variables at once, such as gender and political party, so you can see (or easily calculate) the percentage of individuals in each combination of categories and use them to make comparisons between groups.

For example, if you had data about the gender and political party of your respondents, you would be able to look at the percentage of Republican females, Republican males, Democratic females, Democratic males, and so on. In this example, the total number of possible combinations in your table would be $2 * 4 = 8$, or the total number of gender categories times the total number of party affiliation categories. (See Chapter 20 for the full scoop, and then some, on two-way tables.)

The U.S. government calculates and summarizes loads of categorical data using crosstabs. Typical age and gender data, reported by the U.S. Census Bureau for a survey conducted in 2020, are shown in Table 4-1. (Normally age would be considered a numerical variable, but the way the U.S. government reports it, age is broken down into categories, making it a categorical variable.)

Table 4-1 U.S. Population, Broken Down by Age and Gender

Age Group	Total (in Mil.)	%	# Males (in Mil.)	%	# Females (in Mil.)	%
< 5	19.30	5.86	9.86	6.07	9.44	5.65
5–9	20.24	6.14	10.35	6.38	9.89	5.91
10–14	20.75	6.30	10.59	6.53	10.16	6.08
15–19	20.96	6.36	10.69	6.59	10.27	6.14
20–24	21.59	6.55	11.03	6.80	10.56	6.32
25–29	23.24	7.05	11.88	7.32	11.36	6.79
30–34	22.84	6.93	11.57	7.13	11.27	6.74
35–39	21.83	6.63	10.94	6.74	10.89	6.51
40–44	20.31	6.16	10.11	6.23	10.20	6.10
45–49	19.97	6.06	9.87	6.08	10.10	6.04
50–54	20.39	6.19	10.05	6.19	10.34	6.18
55–59	21.60	6.56	10.51	6.48	11.09	6.63
60–64	20.80	6.32	9.98	6.15	10.82	6.47
65–69	17.87	5.42	8.39	5.17	9.48	5.67
70–74	14.67	4.45	6.79	4.19	7.88	4.71
75–79	9.98	3.03	4.47	2.76	5.51	3.30
80–84	6.47	1.97	2.75	1.70	3.72	2.22
85–	6.65	2.02	2.41	1.49	4.24	2.54
Total	329.46	100	162.24	100	167.22	100

You can examine many different facets of the U.S. population by looking at and working with different numbers from Table 4-1. For example, looking at gender, you notice that women slightly outnumber men — the population in 2020 was 50.76 percent female (divide the total number of females by the total population size and multiply by 100 percent) and 49.24 percent male (divide the total number of males by the total population size and multiply by 100 percent). You can also look at age: The percentage of the entire population that is under 5 years old was 5.86 percent (divide the total number under age 5 by the total population size and multiply by 100 percent). The largest group belongs to the 25- to 29-year-olds, who made up 7.05 percent of the population.

Next, you can explore a possible relationship between gender and age by comparing various parts of the table. You can compare, for example, the percentage of females to males in the 80-and-over age group. Because these data are reported in 5-year increments, you have to do a little math in order to get your answer, though. The percentage of the population that's female and aged 80 and above (looking at column 7 of Table 4-1) is $2.22\% + 2.54\% = 4.76\%$. The percentage of males aged 80 and over (looking at column 5 of Table 4-1) is $1.70\% + 1.49\% = 3.19\%$. This shows that the 80-and-over age group for the females is about 49 percent larger than for the males (because $[4.76 - 3.19] \div 3.19 = 0.49$).

These data confirm the widely accepted notion that women tend to live longer than men. However, the gap between men and women is narrowing over time. According to the U.S. Census Bureau, back in 2001 the percentage of women who were 80 years old and over was 4.36, compared to 2.31 for men. The females in this age group outnumbered the males by a whopping 89 percent back in 2001 (noting that $[4.36 - 2.31] \div 2.31 = 0.89$).

After you have the crosstabs that show the breakdown of two categorical variables, you can conduct hypothesis tests to determine whether a significant relationship or link between the two variables exists, taking into account the fact that data vary from sample to sample. Chapter 15 gives you all the details on hypothesis tests.

Interpreting counts and percents with caution

Not all summaries of categorical data are fair and accurate. Knowing what to look for can help you keep your eyes open for misleading and incomplete information.

Instructors often ask you to "interpret the results." Your instructor wants you to use the statistics available to talk about how they relate to the given situation. In other words, what do the results mean to the person who collects the data?

With relative frequency tables, don't forget to check whether all categories sum to 1, or 100 percent (subject to round-off error), and remember to look for some indicator as to total sample size. See the following for an example of critiquing a data summary.

EXAMPLE

Q. You watch a commercial where the manufacturer of a new cold medicine ("Nocold") compares it to the leading brand. The results are shown in the following table.

How Nocold Compares	Percentage
Much better	47%
At least as good	18%

a. What kind of table is this?

b. Interpret the results. (Did the new cold medicine beat out the leading brand?)

c. What important details are missing from this table?

A. Much like the cold medicines we always take, the table about "Nocold" does "Nogood."

a. This table is an incomplete relative frequency table. The remaining category is "not as good" for the Nocold brand, and the advertiser doesn't show it. But you can do the math and see that 100 percent − (47 percent + 18 percent) = 35 percent of the people say that the leading brand is better.

b. If you put the two groups in the table together, 65 percent of the patients say that Nocold is at least as good as the leading brand, and almost half of the patients say Nocold is much better.

c. What's missing? The remaining percentage (to keep all possible results in perspective). But more importantly, the total sample size is missing. You don't know whether the surveyors sampled 10 people, 100 people, or 1,000 people. This means the precision of the results is unknown. (Precision means how consistent the results will be from sample to sample; it's related to sample size, as you see in Chapter 13.)

YOUR TURN

8 Suppose you ask 1,000 people to identify from a list of five vacation spots which ones they've already visited. The frequencies you receive are as follows: Disney World, 216; New Orleans, 312; Las Vegas, 418; New York City, 359; and Washington, D.C., 188.

a. Explain why creating a traditional relative frequency table doesn't make sense here.

b. How can you summarize this data with percents in a way that makes sense?

---

9 If you have only a frequency table, can you find the corresponding relative frequency table? Conversely, if you have only a relative frequency table, can you find the corresponding frequency table? Explain your answer.

Practice Questions Answers and Explanations

1. **Eleven shoppers prefer Brand A, and nine shoppers prefer Brand B.** The survey results are shown in the following frequency table. Brand A got more votes, but the results are pretty close.

Brand Preferred	Frequency
A	11
B	9
Total	20

2. Frequencies are fine for summarizing data as long as you keep the total number in perspective.

   a. The results are shown in the following frequency table. Because the total is 18,726, and the Yes count is 10,479, the No count is the difference between the two, which is $18,726 - 10,479 = 8,247$.

Vote	Frequency
Y	10,479
N	8,247
Total	18,726

   b. **The total is important because it helps keep the frequencies in perspective when you compare them to each other.**

3. This problem shows the importance of reporting not only the results of participants who responded, but also what percentage of the total actually responded.

   a. The results are shown in the following frequency table:

Gone to the Zoo in the Last Year?	Frequency
Y	592
N	198
Nonrespondents	210
Total	1,000

   b. **If you don't show the nonrespondents, the total doesn't add up to 1,000 (the number surveyed).** An alternative way to show the data is to base it on only the respondents, but the results would be biased. You can't definitively say that the nonrespondents would respond the same way as the respondents.

4 Relative frequencies do just what they say: They help you relate the results to each other (by finding percentages).

a. Eleven shoppers out of the twenty prefer Brand A, and nine shoppers out of the twenty prefer Brand B. The survey results are shown in the following relative frequency table. Brand A got more votes, but the results are pretty close, with 55 percent of the shoppers preferring Brand A and 45 percent preferring Brand B.

Brand Preferred	Relative Frequency
A	55%
B	45%

b. **You often have an easier time interpreting percents**, because when you need to interpret counts, you have to put them in perspective in terms of "out of how many?"

5 The results are shown in the following relative frequency table. The Yes percentage is $10{,}479 \div 18{,}726 * 100 = 55.96$ percent. Because the total is 100 percent, the No percentage is 100 percent $- 55.96$ percent $= 44.04$ percent.

Vote	Relative Frequency
Y	55.96%
N	44.04%

6 You can see the relative frequency table following this answer. Knowing the response rate is critical for interpreting the results of a survey. (The higher the response rate, the better.) The response rate is 59.2 percent $+ 19.8$ percent $= 79.0$ percent, that is, the total percentage of people who responded in any way (yes or no) to the survey. Note that 21 percent is the non-response rate.

Gone to the Zoo in the Last Year?	Relative Frequency
Y	$592 \div 1{,}000 = 0.592 = 59.2\%$
N	19.8%
Nonrespondents	21.0%

7 Showing the percentages rather than counts means making a relative frequency table rather than a frequency table.

a. **One advantage of a relative frequency table is that everything sums to 100 percent, making it easier to interpret the results, especially if you have a large number of categories.**

b. **One disadvantage of a relative frequency table is that if you see only the percents, you don't know how many people participated in the study; therefore, you don't know how precise the results are.** Remember the commercial about 'four out of five dentists surveyed'? Maybe the company only asked five dentists! You can get around this problem by putting the total sample size somewhere at the top or bottom of your relative frequency table.

REMEMBER

When making a relative frequency table, include the total sample size somewhere on the table.

Be careful how you interpret tables where an individual can be in more than one category at the same time.

a. **The frequencies don't sum to 1,000, because people have the option to choose multiple locations or none at all, so each person doesn't end up in exactly one group.** If you take the grand total of all the frequencies (1,493) and divide each frequency by 1,493 to get a relative frequency, the relative frequencies sum to 1 (or 100 percent). But what does that mean? It makes it hard to interpret these percents because they don't account for the total number of people.

b. **One way you can summarize this data is by showing the percentage of people who have been at each location separately (compared to the percentage who haven't been there before).** These percents add up to 1 for each location. The following table shows the results summarized with this method. ***Note:*** The table isn't a relative frequency table; however, it uses relative frequencies.

Location	% Who Have Been There	% Who Haven't Been There
Disney World	$216 \div 1{,}000 = 21.6\%$	$100\% - 21.6\% = 78.4\%$
New Orleans	$312 \div 1{,}000 = 31.2\%$	$100\% - 31.2\% = 68.8\%$
Las Vegas	$418 \div 1{,}000 = 41.8\%$	$100\% - 41.8\% = 58.2\%$
New York City	$359 \div 1{,}000 = 35.9\%$	$100\% - 35.9\% = 64.1\%$
Washington, D.C.	$188 \div 1{,}000 = 18.8\%$	$100\% - 18.8\% = 81.2\%$

REMEMBER

Not all tables involving percents should sum to 1. Don't force tables to sum to 1 when they shouldn't; do make sure you understand whether each individual can fall under more than one category. In those cases, a typical relative frequency table isn't appropriate.

9 **You can always sum all the frequencies to get a total and then find each relative frequency by taking the frequency divided by the total.** However, if you have only the percents, you can't go back and find the original counts unless you know the total number of individuals.

Suppose you know that 80 percent of the people in a survey like ice cream. How many actual people in the survey like ice cream? If the total number of respondents is 100, then $100 * 0.80 = 80$ people like ice cream. If the total is 50, then you're looking at $50 * 0.80 = 40$ positive answers. If the total is 5, then you're dealing only with 4 respondents ($5 * 0.80 = 4$). This illustrates why relative frequency tables need to have the total sample size somewhere.

REMEMBER

Watch for total sample sizes when you're given a relative frequency table. Don't be misled by percentages alone, thinking they're always based on large sample sizes. Many are not.

If you're ready to test your skills a bit more, take the following chapter quiz that incorporates all the chapter topics.

Whaddya Know? Chapter 4 Quiz

Quiz time! Complete each problem to test your knowledge on the various topics covered in this chapter. You can then find the solutions and explanations in the next section.

A researcher orders 10 of the exact same pizzas for delivery from 10 different pizzerias on a given night. They note whether or not each pizza came on time and record the results in the following table.

Pizza #	On Time?	Pizza #	On Time?
1	N	6	N
2	N	7	Y
3	Y	8	Y
4	N	9	N
5	Y	10	Y

a. Organize the results in a relative frequency table.

b. What does the researcher conclude about these 10 pizzas? Interpret the results.

Bob surveys 10 of his real-estate clients at random to see whether they are thinking of selling in the next year. The results are shown in the following table.

Person #	Selling	Person #	Selling
1	N	6	N
2	N	7	N
3	Y	8	Y
4	N	9	N
5	Y	10	Y

a. Make a frequency table that summarizes the results.

b. Interpret the results.

a. What do you get when you sum the frequencies in a frequency table?

b. What do you get when you sum the relative frequencies in a relative frequency table?

Is the following table a frequency table or a relative frequency table? Why?

Voted ($n = 1,000$ people surveyed)	Results
Y	32%
N	68%

4. Ten winter days were randomly chosen in the month of January in Columbus, Ohio. Each day, it was noted whether the temperature dipped below 32 or not. On 30% of the days, it did drop below 32, and on 70% of the days, it didn't.

   a. Make a frequency table of the data.

   b. Show an example of what the data might have looked like for those 10 days.

5. Professor Charleston's final grades for his class turned out to have 10 As, 8 Bs, 5 Cs, 2 Ds, and 0 Fs.

   a. Make a relative frequency table of the final grades.

   b. Interpret the results.

6. Explain how a pie chart is similar to a relative frequency table.

7. Susan gave a survey to 200 randomly chosen students in her statistics class. She asked the students whether they planned to go away from home for spring break. One hundred students said yes, 60 students said no, and 40 students did not respond.

   a. What was the response rate of Susan's survey?

   b. Show the results in a relative frequency table.

   c. Interpret the results.

8. Hasan the quality control manager takes a random sample of 20 packages of frozen shrimp off the shelf. He determines whether each package has expired (yes, no), and checks to see if it has the proper weight (yes, no). He places the results in a table. What type of table will Hasan's results be summarized in?

9. A survey asks for political party (Republican, Democrat, Independent, Other) and age group (18–30, 31–50, over 50) of 150 randomly chosen registered voters. How many categories (cells) will be in the resulting two-way table?

Answers to Chapter 4 Quiz

1. **5 / 10 = 50% Yes; 5 / 10 = 50% No.** Relative frequency table:

$n = 10$ Pizzas	On Time? Yes	On Time? No
Percent (Relative Frequency)	50%	50%

b. Fifty percent of the 10 pizzas were delivered on time and 50% were not.

2.

a.

$n = 10$ Clients	Thinking of Selling? Yes	Thinking of Selling? No
Count (Frequency)	4	6

b. Four of the 10 clients are thinking of selling and 6 are not.

3.

a. When you sum the frequencies in a frequency table, you get the total sample size, *n*.

b. When you sum the relative frequencies in a relative frequency table, you get 1.

4. The table is a relative frequency table because the results are shown in percentage form, and sum to $100\% = 1$.

5.

a.

$n = 10$ Days	Dropped below 32? Yes	Dropped below 32? No
Count (Frequency)	$30\% * 10 = 0.30 * 10 = 3$	$70\% * 10 = 0.70 * 10 = 7$

b. yes, yes, no, no, no, no, no, no, yes, no. Note that any combination of 3 yeses and 7 nos works here.

6.

a. The total number of grades is 25; divide each frequency by 25 to get the relative frequency. The results should sum to 1.

Grade	A	B	C	D	F
Relative Frequency	$10 / 25 = 0.40$	$8 / 25 = 0.32$	$5 / 25 = 0.20$	$2 / 25 = 0.08$	$0 / 25 = 0.00$

b. The 25-member class had 40% As, 32% Bs, 20% Cs, 8% Ds, and 0% Fs.

7. A pie chart is similar to a relative frequency table because each slice of the pie is a group and each group has a percentage attached to it that indicates the percentage of individuals in that group. The percents sum to 1. A pie chart is a graphical representation of a relative frequency table.

8.

a. Response rate $= (100+60)/200 = 160/200 = 0.80$ or 80%.

b.

$n = 200$	Yes	No	No Response
Percent (Relative Frequency)	$100/200 = 0.50 = 50\%$	$60/200 = 0.30 = 30\%$	$40/200 = 0.20 = 20\%$

c. The survey showed 50% of the students plan to go away from home for spring break; 30% do not plan to; and 20% did not respond.

9. A two-way table with rows and columns representing expired (yes, no), and proper weight (yes, no). One possible two-way table is the following:

	Proper Weight? Yes	Proper Weight? No
Expired? Yes		
Expired? No		

10. The number of cells will be 4 (political parties) * 3 (age groups) = 12 category combinations = 12 cells.

IN THIS CHAPTER

» Summarizing numerical data effectively

» Looking at means, medians, and standard deviation

» Working with percentages

Chapter 5
Means, Medians, and More

In Chapter 4, you see a few statistics used to tell the story of data that falls into specific categories. The topic of descriptive statistics continues in this chapter, in which you focus on ways to summarize numerical data.

Measuring the Center with Mean and Median

With *numerical data,* measurable characteristics such as height, weight, IQ, age, or income are represented by numbers that make sense within the context of the problem (for example, in units of feet, dollars, or people). Because the data have numerical meaning, you can summarize them in more ways than are possible with categorical data. The most common way to summarize a numerical data set is to describe where the center is. One way of thinking about what the center of a data set means is to ask, "What's a typical value?" or, "Where is the middle of the data?" The center of a data set can actually be measured in different ways, and the method chosen can greatly influence the conclusions people make about the data. This section hits on measures of center.

Averaging out to the mean

NBA players make a lot of money, right? You often hear about players like Kobe Bryant or LeBron James, who make tens of millions of dollars a year. But is that what the typical NBA player makes? Not really (although I don't exactly feel sorry for the others, given that they still make more money than most of us will ever make). Tens of millions of dollars is the kind of money you can command when you are a superstar among superstars, which is what these elite players are.

So how much money does the typical NBA player make? One way to answer this is to look at the average (the most commonly used statistic of all time).

The *average*, also called the *mean* of a data set, is denoted $\bar{x}$. The formula for finding the mean is

$$\bar{x} = \frac{\sum x}{n}$$

where each individual value in the data set is denoted by an x and $\sum x$ represents the sum of the n numbers in the data set.

Here's how you calculate the mean of a data set:

1. **Add up all the numbers in the data set.**
2. **Divide by the number of numbers in the data set, *n*.**

The mean I discuss here applies to a sample of data and is technically called the *sample mean.* The mean of an entire population of data is denoted with the Greek letter μ and is called the *population mean.* It's found by summing up all the values in the population and dividing by the population size, denoted N (to distinguish it from a sample size, n). Typically the population mean is unknown, and you use a sample mean to estimate it (plus or minus a margin of error; see all the details in Chapter 14).

For example, player salary data for the 13 players on the 2021–2022 NBA Los Angeles Lakers are shown in Table 5-1.

The mean of all the salaries on this team is $163,766,023 / 23 = $7,120,261.87. That's a pretty nice average salary, isn't it? But notice that Russell Westbrook, LeBron James, and Anthony Davis really stand out at the top of this list, and they should because they are such excellent players. If you remove these three players from the equation, the average salary of all the Lakers players besides Russell, LeBron, and Anthony becomes $43,012,973 / 20 = $2,150,648.65 — a difference of around $5 million.

This new mean is still a hefty amount, but it's significantly lower than the mean salary of all the players including the three superstars. (Fans would tell you that this reflects their importance to the team, and others would say no one is worth that much money; this issue is but the tip of the iceberg of the never-ending debates that sports fans love to have about statistics.)

Table 5-1 Salaries for L.A. Lakers NBA Players (2021–2022)

Player	Salary ($)
Russell Westbrook	**$44,211,146**
LeBron James	**$41,180,544**
Anthony Davis	**$35,361,360**
Talen Horton-Tucker	**$9,500,000**
Luol Deng	**$5,000,000**
Kendrick Nunn	**$5,000,000**
Carmelo Anthony	**$2,641,691**
Trevor Ariza	**$2,641,691**
Avery Bradley	**$2,641,691**
Wayne Ellington	**$2,641,691**
Dwight Howard	**$2,641,691**
DeAndre Jordan	**$2,641,691**
Kent Bazemore	**$2,401,537**
Malik Monk	**$1,789,256**
Stanley Johnson	**$1,248,915**
Austin Reaves	**$925,258**
Mason Jones	**$295,125**
Sekou Doumbouya	**$294,461**
Jay Huff	**$225,997**
Darren Collison	**$151,821**
Isiah Thomas	**$151,821**
Chaundee Brown	**$93,058**
Jemerrio Jones	**$85,578**
TOTAL	$163,766,023

Bottom line: The mean doesn't always tell the whole story. In some cases it may be a bit misleading, and this is one of those cases. That's because every year, a few top-notch players (like Russell, LeBron, and Anthony) make much more money than anybody else, and their salaries pull up the overall average salary.

WARNING

Numbers in a data set that are extremely high or extremely low compared to the rest of the data are called *outliers*. Because of the way the average is calculated, high outliers tend to drive the average upward (as the three superstars' salaries did in the preceding example). Low outliers tend to drive the average downward.

EXAMPLE

Q. Find the mean of the following data set: 1, 6, 5, 7, 3, 2.5, 2, –2, 1, 0.

A. **2.55.** The mean is $1+6+5+7+3+2.5+2+-2+1+0=25.5$, divided by 10 (because you have 10 numbers), or 2.55.

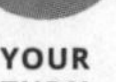

YOUR TURN

1 Does the mean have to be one of the numbers in the data set? Explain your answer.

Suppose you have an *outlier* in a data set (a number that stands out away from the rest). How does an outlier affect the mean of that data set?

Suppose you find the mean for a certain data set.

(a) Depending on what the data actually are, the mean should always lie between the largest and smallest values of the data set. Explain why.

(b) When can the mean be the largest value in the data set?

Splitting your data down the median

Remember in school when you took an exam, and you and most of the rest of the class did badly, but a couple of nerds got 100? Remember how the teacher didn't curve the scores to reflect the poor performance of most of the class? Your teacher was probably using the average, and the average in that case didn't really represent what statisticians might consider the best measure of center for the students' scores.

What can you report, other than the average, to show what the salary of a "typical" NBA player would be or what the test score of a "typical" student in your class was? Another statistic used to measure the center of a data set is called the median. The median is still an unsung hero of statistics in the sense that it isn't used nearly as often as it should be, although people are beginning to report it more nowadays.

The *median* of a data set is the value that lies exactly in the middle when the data have been ordered. It's denoted in different ways; some people use M and some use $\tilde{x}$. Here are the steps for finding the median of a data set:

1. **Order the numbers from smallest to largest.**
2. **If the data set contains an odd number of numbers, choose the one that is exactly in the middle. You've found the median.**
3. **If the data set contains an even number of numbers, take the two numbers that appear in the middle and average them to find the median.**

The salaries for the Los Angeles Lakers during the 2021–2022 season (refer to Table 5-1) are ordered from lowest (at the bottom) to highest (at the top). Because the list contains the names and salaries of 23 players, the middle salary is the 12th one from the bottom: DeAndre Jordan, who earned $2.64 million that season from the Lakers. DeAndre is at the median.

WARNING

This median salary ($2.64 million) is well below the average of $7.12 million for the 2021–2022 Lakers team. Notice that only 4 players of the 23 earned more than the average Lakers salary of $7.12 million. Because the average includes outliers (like the salary of Russell Westbrook), the median salary is more representative of the center for the team salaries. The median isn't affected by the salaries of those players who are way out there on the high end the way the average is.

Note: By the way, the lowest Lakers' salary for the 2021–2022 season was $85,578 — a decent amount of money by most people's standards, but peanuts compared to what you imagine when you think of an NBA player's salary!

TIP

The U.S. government most often uses the median to represent the center with respect to income data, again, because the median is not affected by outliers. For example, the U.S. Census Bureau reported that in 2021, the median household income was $67,500 while the mean was found to be $78,500. That's quite a difference!

EXAMPLE

Q. Find the median of the following data set: 1, 6, 5, 7, 3, 2.5, 2, –2, 1, 0.

A. **2.25.** To find the median, order the numbers: –2, 0, 1, 1, 2, 2.5, 3, 5, 6, 7. Now move to the middle and find the two middle values: 2 and 2.5. Take the average: $(2+2.5)\div 2 = 4.5\div 2 = 2.25$.

YOUR TURN

4 Does the median have to be one of the numbers in the data set? Explain your answer.

5 Why do you have to order the data to calculate the median but not the mean?

6 Suppose you have an *outlier* in a data set (a number that stands out away from the rest). How does an outlier affect the median of that data set?

7 Give an example of two different data sets containing *three* numbers each that both have the same median and mean. Explain why the median isn't enough to tell the whole story about a data set.

8 Suppose the mean and median salary at a company is $50,000, and all employees get a $1,000 raise.

(a) What happens to the mean?

(b) What happens to the median?

9 Suppose the mean and median salary at a company is $50,000, and all employees get a 10 percent raise.

(a) What happens to the mean?

(b) What happens to the median?

Comparing means and medians: Histograms

Sometimes the mean versus median debate can get quite interesting. Suppose you're part of an NBA team trying to negotiate salaries. If you represent the owners, you want to show how much everyone is making and how much money you're spending, so you want to take into account those superstar players and report the average. But if you're on the side of the players, you would want to report the median, because that's more representative of what the players in the middle are making. Fifty percent of the players make a salary above the median, and 50 percent make a salary below the median. To sort it all out, it's best to find and compare both the mean and the median. A graph showing the shape of the data is a great place to start.

REMEMBER

One of the graphs you can make to illustrate the shape of numerical data (how many values are close to or far from the mean, where the center is, how many outliers there might be) is a histogram. A *histogram* is a graph that organizes and displays numerical data in picture form, showing groups of data and the number or percentage of the data that fall into each group. It gives you a nice snapshot of the data set. (See Chapter 7 for more information on histograms and other types of data displays.)

Data sets can have many different shapes; here is a sampling of three shapes that are commonly discussed in introductory statistics courses:

- If most of the data are on the left side of the histogram but a few larger values are on the right, the data are said to be skewed to the right.

  Histogram A in Figure 5-1 shows an example of data that are skewed to the right. The few larger values bring the mean upwards but don't really affect the median. So when data are skewed right, the mean is larger than the median. An example of such data is NBA salaries.

- If most of the data are on the right, with a few smaller values showing up on the left side of the histogram, the data are skewed to the left.

  Histogram B in Figure 5-1 shows an example of data that are skewed to the left. The few smaller values bring the mean down, and again the median is minimally affected (if at all). An example of skewed-left data is the amount of time students use to take an exam; some students leave early, more of them stay later, and many stay until the bitter end (some would stay forever if they could!). When data are skewed left, the mean is smaller than the median.

- If the data are symmetric, they have about the same shape on either side of the middle. In other words, if you fold the histogram in half, it looks about the same on both sides.

  Histogram C in Figure 5-1 shows an example of symmetric data in a histogram. With symmetric data, the mean and median are close together.

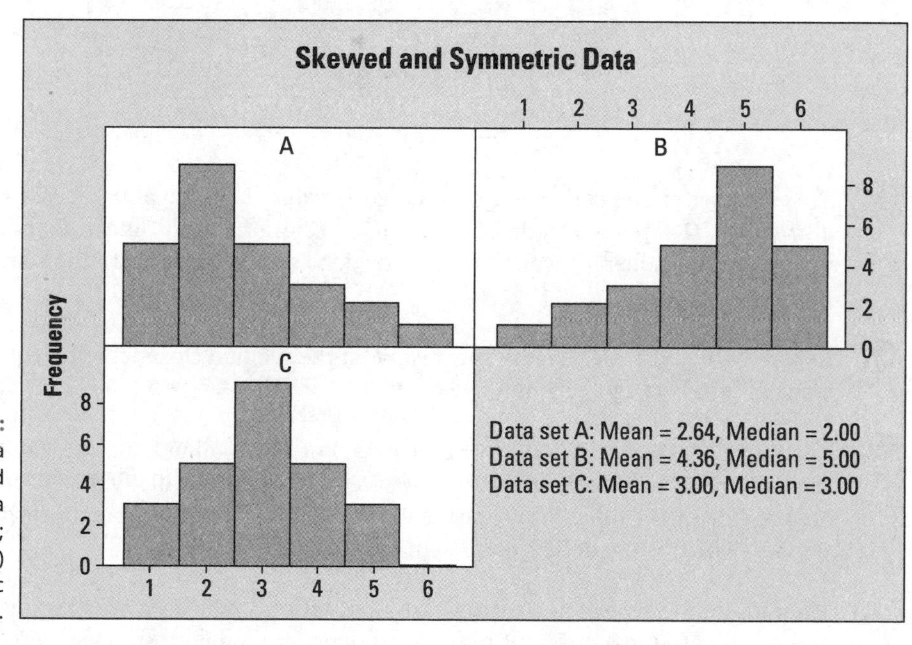

FIGURE 5-1: A) Data skewed right; B) data skewed left; and C) symmetric data.

TIP

By looking at Histogram A in Figure 5-1 (whose shape is skewed right), you can see that the "tail" of the graph (where the bars are getting shorter) is to the right, while the "tail" is to the left in Histogram B (whose shape is skewed left). By looking at the direction of the tail of a skewed distribution, you determine the direction of the skewness. Always add the direction when describing a skewed distribution.

Histogram C is symmetric (it has about the same shape on each side). However, not all symmetric data has a bell shape like Histogram C. As long as the shape is approximately the same on both sides, you say that the shape is symmetric.

REMEMBER

The average (or mean) of a data set is affected by outliers, but the median is not. In statistical lingo, if a statistic is not affected by a certain characteristic of the data (such as outliers, or skewness), then you say that statistic is *resistant* to that characteristic. In this case, the median is resistant to outliers; the mean is not. If someone reports the average value, also ask for the median so that you can compare the two statistics and get a better feel for what's actually going on in the data and what's truly typical.

EXAMPLE

Q. Describe the shape of the data shown in the following histogram.

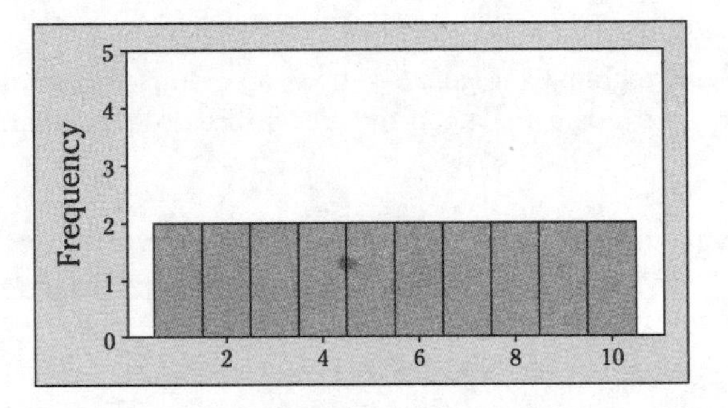

A. This data set is evenly distributed among the values 1–10, meaning it has a *uniform distribution.* (Keep in mind that a uniform distribution doesn't mean that all the data values are the same; it means that the frequencies for the data are the same across the groups.)

Q. Explain why a data set having the mean and the median close to being equal will have a shape that is roughly symmetric.

A. The mean is affected by outliers in the data, but the median is not. If the mean and median are close to each other, the data aren't skewed and likely don't contain outliers on one side or the other. That means the data look about the same on each side of the middle, which is the definition of symmetric data.

REMEMBER

The fact that the mean and median are close tells you the data are roughly symmetric and can be used in a different type of test question. Suppose someone asks you whether the data are symmetric, and you don't have a histogram, but you do have the mean and median. Compare the two values of the mean and median, and if they are close, the data are symmetric. If they aren't, the data are not symmetric.

10 Is the data set shown in the following figure symmetric or skewed? How many modes does this data set have?

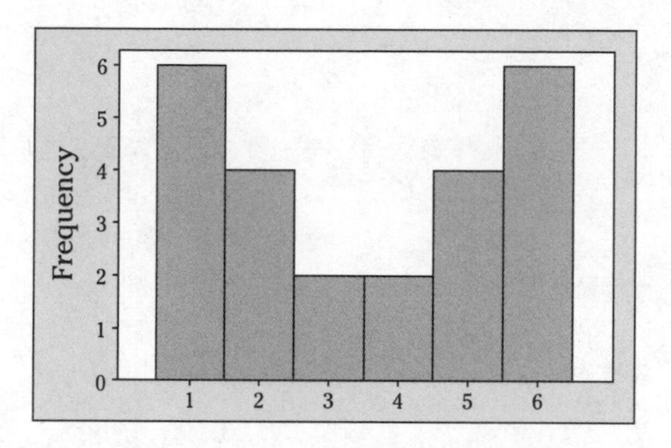

11 Describe the shape of the data set shown in the following figure.

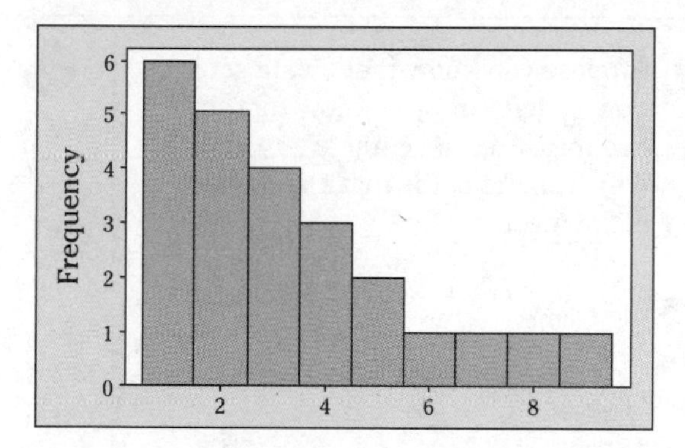

12 Describe the shape of the data set shown in the following figure.

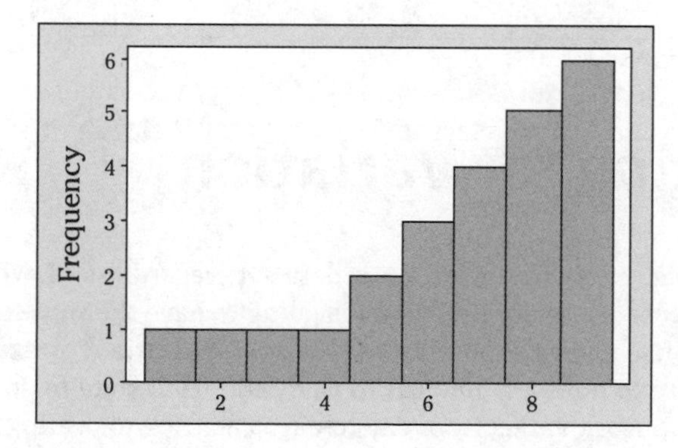

13 Give an example of a symmetric histogram with two modes.

14 Suppose the average salary at a certain company is $100,000, and the median salary is $40,000.

(a) What do these figures tell you about the shape of the histogram of salaries at this company?

(b) Which measure of center is more appropriate here?

(c) Suppose the company goes through a salary negotiation. How can people on each side use these summary statistics to their advantage?

15 Suppose you know that a data set is skewed left, and you know that the two measures of center are 19 and 38. Which figure is the mean and which is the median?

16 Can the mean of a data set be higher than most of the values in the set? If so, how? Can the median of a set be higher than most of the values? If so, how?

Accounting for Variation

Variation always exists in a data set, regardless of which characteristics you're measuring, because not every individual is going to have the same exact value for every variable. Variation is what makes the field of statistics what it is. For example, the price of homes varies from house to house, from year to year, and from state to state. The amount of time it takes you to get to work varies from day to day. The trick to dealing with variation is to be able to measure that variation in a way that best captures it.

Reporting the standard deviation

By far the most common measure of variation for numerical data is the standard deviation. The *standard deviation* measures how concentrated the data are around the mean; the more concentrated, the smaller the standard deviation. It's not reported nearly as often as it should be, but when it is, you often see it in parentheses: $(s = 2.68)$.

Calculating standard deviation

The formula for the sample standard deviation of a data set (s) is

$$s = \sqrt{\frac{\Sigma(x - \bar{x})^2}{n-1}}$$

To calculate s, do the following steps:

1. **Find the average of the data set, $\bar{x}$.**
2. **Take each number in the data set (x) and subtract the mean from it to get $(x - \bar{x})$.**
3. **Square each of the differences, $(x - \bar{x})^2$.**
4. **Add up all of the results from Step 3 to get the sum of squares: $\Sigma(x - \bar{x})^2$.**
5. **Divide the sum of squares (found in Step 4) by the number of numbers in the data set minus 1; that is, $(n - 1)$. Now you have:**

   $$\frac{\Sigma(x - \bar{x})^2}{n-1}$$

6. **Take the square root to get**

   $$s = \sqrt{\frac{\Sigma(x - \bar{x})^2}{n-1}}$$

   which is the sample standard deviation, s. Whew!

WARNING

At the end of Step 5, you have found a statistic called the *sample variance*, denoted by s^2. The variance is another way to measure variation in a data set; its downside is that it's in square units. If your data are in dollars, for example, the variance would be in square dollars — which makes no sense. That's why I proceed to Step 6. Standard deviation has the same units as the original data.

Look at the following example: Suppose you have four quiz scores: 1, 3, 5, and 7. The mean is $16 \div 4 = 4$ points. Subtracting the mean from each number, you get $(1-4) = -3$, $(3-4) = -1$, $(5-4) = +1$, and $(7-4) = +3$. Squaring each of these results, you get 9, 1, 1, and 9. When you add them up, the sum is 20. In this example, $n = 4$, and therefore $n - 1 = 3$, so you divide 20 by 3 to get 6.67. The units here are "points squared," which obviously makes no sense. Finally, you take the square root of 6.67, to get 2.58. The standard deviation for these four quiz scores is 2.58 points.

Because calculating the standard deviation involves many steps, in most cases you have a computer calculate it for you. However, knowing how to calculate the standard deviation helps you better interpret this statistic and can help you figure out when the statistic may be wrong.

Statisticians divide by $n-1$ instead of by n in the formula for s so the results have nicer properties that operate on a theoretical plane that's beyond the scope of this book. (Not the *Twilight Zone* but close; trust me, that's more than you want to know about *that!*)

The *standard deviation of an entire population of data* is denoted with the Greek letter σ. When I use the term *standard deviation,* I mean s, the sample standard deviation. (When I refer to the population standard deviation, I let you know.)

Interpreting standard deviation

Standard deviation can be difficult to interpret as a single number on its own. Basically, a small standard deviation means that the values in the data set are close to the mean of the data set, on average, and a large standard deviation means that the values in the data set are farther away from the mean, on average.

A small standard deviation can be a goal in certain situations where the results are restricted, for example, in product manufacturing and quality control. A particular type of car part that has to be 2 centimeters in diameter to fit properly had better not have a very big standard deviation during the manufacturing process. A big standard deviation in this case would mean that lots of parts end up in the trash because they don't fit right; either that or the cars will have problems down the road.

But in situations where you just observe and record data, a large standard deviation isn't necessarily a bad thing; it just reflects a large amount of variation in the group that is being studied. For example, if you look at salaries for everyone in a certain company, including everyone from the student intern to the CEO, the standard deviation may be very large. On the other hand, if you narrow the group down by looking only at the student interns, the standard deviation is smaller, because the individuals within this group have salaries that are less variable. The second data set isn't better, it's just less variable.

Similar to the mean, outliers affect the standard deviation (after all, the formula for standard deviation includes the mean). In the NBA salaries example, the salaries of the L.A. Lakers in the 2021–2022 season (shown in Table 5-1) range from the highest, $23,034,375 (Russell Westbrook), down to $85,578 (Jemerrio Jones). Lots of variation, to be sure! The standard deviation of the salaries for this team turns out to be $13,363,152.03; it's almost twice as large as the average salary. However, as you may guess, if you remove Russell, LeBron, and Anthony's salaries from the data set, the standard deviation decreases because the remaining salaries are more concentrated around the mean. The standard deviation then becomes $2,308,873.81.

Watch for the units when determining whether a standard deviation is large. For example, a standard deviation of 2 in units of years is equivalent to a standard deviation of 24 in units of months. Also look at the value of the mean when putting standard deviation into perspective. If the average number of Internet newsgroups that a user posts to is 5.2 and the standard deviation is 3.4, that's a lot of variation, relatively speaking. But if you're talking about the age of the newsgroup users where the mean is 25.6 years, that same standard deviation of 3.4 is comparatively smaller.

Understanding properties of standard deviation

Here are some properties that can help you when interpreting a standard deviation:

- The standard deviation can never be a negative number, due to the way it's calculated and the fact that it measures a distance (distances are never negative numbers).
- The smallest possible value for the standard deviation is zero, and that happens only in contrived situations where every single number in the data set is exactly the same (no deviation).
- The standard deviation is affected by outliers (extremely low or extremely high numbers in the data set). That's because the standard deviation is based on the distance from the *mean.* And remember, the mean is also affected by outliers.
- The standard deviation has the same units as the original data.

Lobbying for standard deviation

The standard deviation is a commonly used statistic, but it doesn't often get the attention it deserves. Although the mean and median are out there in full view in the everyday media, you rarely see them accompanied by any measure of how diverse that data set was, so you are getting only part of the story. In fact, you could be missing the most interesting part of the story.

Without standard deviation, you can't get a handle on whether the data are close to the average (as are the diameters of car parts that come off of a conveyor belt when everything is operating correctly) or whether the data are spread out over a wide range (as are house prices and income levels in the U.S.).

For example, if someone told you that the average starting salary for someone working at Company Statistix is $70,000, you may think, "Wow! That's great." But if the standard deviation for starting salaries at Company Statistix is $20,000, that's a lot of variation in terms of how much money you can make, so the average starting salary of $70,000 isn't as informative in the end, is it?

On the other hand, if the standard deviation is only $5,000, you have a much better idea of what to expect for a starting salary at that company. Which is more appealing? That's a decision each person has to make; however, it'll be a much more informed decision once you realize standard deviation matters.

Without the standard deviation, you can't compare two data sets effectively. Suppose two sets of data have the same average; does that mean that the data sets must be exactly the same? Not at all. For example, the data sets (199, 200, 201) and (0, 200, 400) both have the same average (200), yet they have very different standard deviations. The first data set has a *very* small standard deviation ($s = 1$) compared to the second data set ($s = 200$).

References to the standard deviation may become more commonplace in the media as more and more people (like you, for example) discover what the standard deviation can tell them about a set of results and start asking for it. In your career, you're more likely to see the standard deviation reported and used in the future as its value becomes more apparent to people.

Being out of range

The range is another statistic that some folks use to measure diversity in a data set. The *range* is the largest value in the data set minus the smallest value in the data set. It's easy to find; all you do is put the numbers in order (from smallest to largest) and do a quick subtraction. Maybe that's why the range is used so often; it certainly isn't because of its interpretative value.

TIP

The range of a data set is almost meaningless. It depends on only two numbers in the data set, both of which may reflect extreme values (outliers). My advice is to ignore the range and find the standard deviation, which is a more informative measure of the variation in the data set because it involves all the values. Or you can also calculate another statistic called the *interquartile range,* which is similar to the range with an important difference: it eliminates outlier and skewness issues by only looking at the middle 50 percent of the data and finding the range for those values. The section, "Exploring interquartile range," at the end of this chapter gives you more details.

EXAMPLE

Q. Find and interpret the standard deviation of the following data set: 1, 2, 3, 4, 5.

A. **1.58.** First, the mean of this data set is 3 (see the previous section, "Averaging out to the mean," for mean info). After you calculate the mean, find the deviations from the mean and square them: $1-3=-2$, and –2 squared equals 4; $2-3=-1$, and –1 squared equals 1; $3-3=0$, and 0 squared equals 0; $4-3=1$, and 1 squared equals 1; and finally, $5-3=2$, and 2 squared equals 4. Sum these values up to get $4+1+0+1+4=10$. Divide 10 by 5 – 1 (because $n=5$) to get $10 \div 4 = 2.5$. The final step is to take the square root of 2.5, which gives you $s=1.58$. This answer means the data are, on average, about 1.58 steps from the mean (3).

Q. Find and interpret the range of the following data set: 1, 2, 3, 4, 5.

A. **4.** The largest value in the data set is 5, and the smallest value is 1. Thus, the range is $5-1=4$. This means that all the values of the data set span a distance of 4, but it doesn't say how the values in the middle are spread out.

YOUR TURN

17 What's the smallest standard deviation you can figure, and when would that happen?

18 Choose four numbers from 1 to 5, with repetitions allowed, to create the largest standard deviation possible.

19 Suppose the mean salary at a company is \$50,000, and all employees get a \$1,000 raise. What happens to the standard deviation?

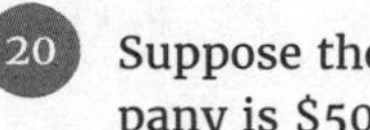

Suppose the mean salary at a company is $50,000, and all employees get a 10 percent raise. What happens to the standard deviation?

Is the standard deviation affected by skewed data? If so, how?

Examining the Empirical Rule (68-95-99.7)

Putting a measure of center (such as the mean or median) together with a measure of variation (such as standard deviation or interquartile range) is a good way to describe the values in a population. In the case where the data are in the shape of a bell curve (that is, they have a normal distribution; see Chapter 10), the population mean and standard deviation are the combination of choice, and a special rule links them together to get some pretty detailed information about the population as a whole.

The *Empirical Rule* says that if a population has a normal distribution with population mean μ and standard deviation σ, then

- About 68 percent of the values lie within 1 standard deviation of the mean (or between the mean minus 1 times the standard deviation, and the mean plus 1 times the standard deviation). In statistical notation, this is represented as $\mu \pm 1\sigma$.
- About 95 percent of the values lie within 2 standard deviations of the mean (or between the mean minus 2 times the standard deviation, and the mean plus 2 times the standard deviation). The statistical notation for this is $\mu \pm 2\sigma$.
- About 99.7 percent of the values lie within 3 standard deviations of the mean (or between the mean minus 3 times the standard deviation and the mean plus 3 times the standard deviation). Statisticians use the following notation to represent this: $\mu \pm 3\sigma$.

TIP

The Empirical Rule is also known as the *68-95-99.7 Rule*, corresponding with those three properties. It's used to describe a population rather than a sample, but you can also use it to help you decide whether a sample of data came from a normal distribution. If a sample is large enough and you can see that its histogram looks close to a bell-shape, you can check to see whether the data follow the 68-95-99.7 percent specifications. If they do, it's reasonable to conclude the data came from a normal distribution. This is huge because the normal distribution has lots of perks, as you can see in Chapter 10.

Figure 5-2 illustrates all three components of the Empirical Rule.

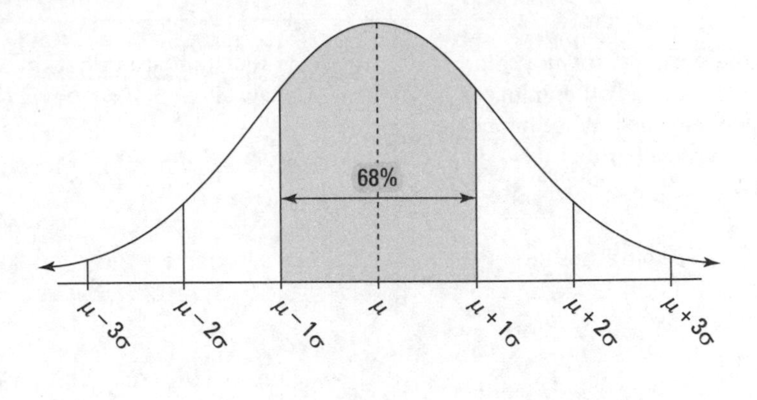

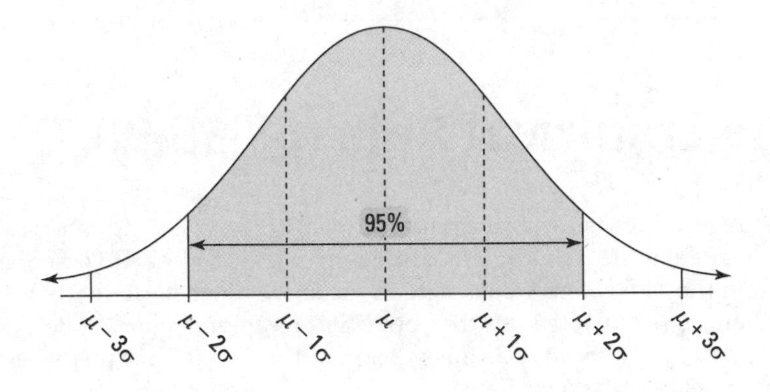

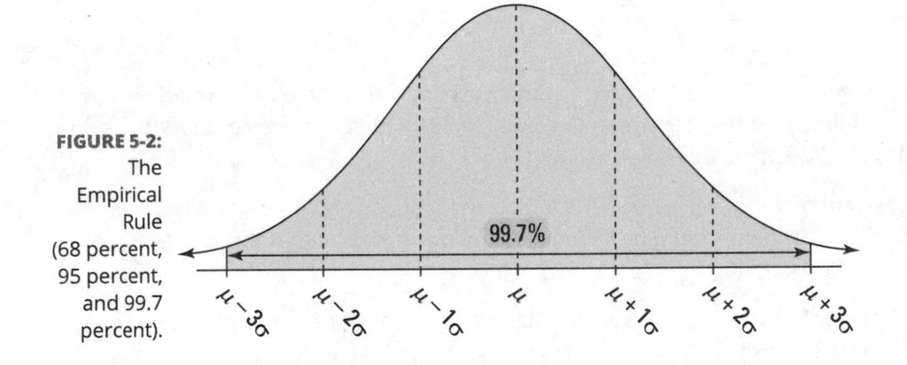

FIGURE 5-2: The Empirical Rule (68 percent, 95 percent, and 99.7 percent).

The reason that so many (about 68 percent) of the values lie within 1 standard deviation of the mean in the Empirical Rule is because when the data are bell-shaped, the majority of the values are mounded up in the middle, close to the mean (as Figure 5-2 shows).

Adding another standard deviation on either side of the mean increases the percentage from 68 to 95, which is a big jump and gives a good idea of where "most" of the data are located. Most researchers stay with the 95 percent range (rather than 99.7 percent) for reporting their results, because increasing the range to 3 standard deviations on either side of the mean (rather than just 2) doesn't seem worthwhile, just to pick up that last 4.7 percent of the values.

REMEMBER

The Empirical Rule tells you about what percentage of values are within a certain range of the mean, and I need to stress the word *about*. These results are approximations only, and they only apply if the data follow a normal distribution. However, the Empirical Rule is important in statistics because the concept of "going out about two standard deviations to get about 95 percent of the values" is one that you see mentioned often with confidence intervals and hypothesis tests (see Chapters 14 and 15).

Here's an example of using the Empirical Rule to better describe a population whose values have a normal distribution: In a study of how people make friends in cyberspace using newsgroups, the age of the users of an Internet newsgroup was reported to have a mean of 31.65 years, with a standard deviation of 8.61 years. Suppose the data were graphed using a histogram and were found to have a bell-shaped curve similar to what's shown in Figure 5-2.

According to the Empirical Rule, about 68 percent of the newsgroup users had ages within 1 standard deviation (8.61 years) of the mean (31.65 years). So about 68 percent of the users were between ages $31.65 - 8.61$ years and $31.65 + 8.61$ years, or between 23.04 and 40.26 years. About 95 percent of the newsgroup users were between the ages of $31.65 - 2(8.61)$, and $31.65 + 2(8.61)$, or between 14.43 and 48.87 years. Finally, about 99.7 percent of the newsgroup users' ages were between $31.65 - 3(8.61)$ and $31.65 + 3(8.61)$, or between 5.82 and 57.48 years.

This application of the rule gives you a much better idea about what's happening in this data set than just looking at the mean, doesn't it? As you can see, the mean and standard deviation, when used together, add value to your results; plugging these values into the Empirical Rule allows you to report ranges for "most" of the data yourself.

REMEMBER

The condition for being able to use the Empirical Rule is that the data have a normal distribution. If that's not the case (or if you don't know what the shape actually is), you can't use it. To describe your data in these cases, you can use percentiles, which represent certain cutoff points in the data (see the later section, "Gathering a five-number summary").

EXAMPLE

Q. Does the Empirical Rule apply to the data set in the following figure? Explain your answer.

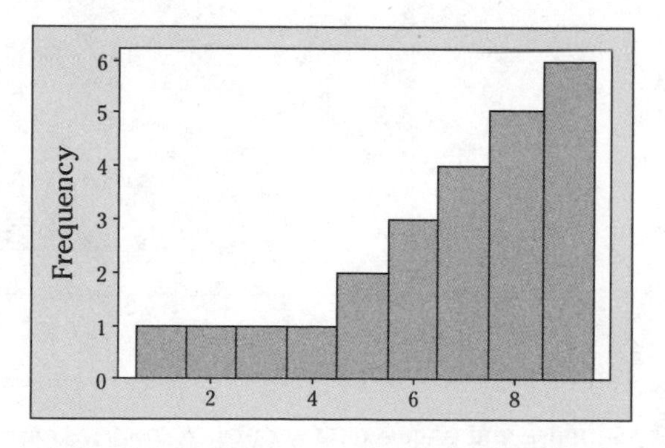

A. **Yes,** the Empirical Rule applies, because the data set is mound-shaped.

Q. Suppose a driver's test has a mean score of 7 (out of 10 points) and a standard deviation of 0.5.

(a) Explain why you can reasonably assume that the data set of the test scores is mound-shaped.

(b) For the drivers taking this particular test, where should 68 percent of them score?

(c) Where should 95 percent of them score?

(d) Where should 99.7 percent of them score?

Q. Sometimes, you have to make reasonable assumptions before you proceed to solve a problem.

(a) Test scores from a properly written and calibrated test are likely to be bell-shaped, with equal numbers of people scoring below the mean as scoring above the mean.

(b) The Empirical Rule says that about 68 percent of the test takers should score within one standard deviation of the mean — in other words, $7 \pm (1 \times 0.5) = 7 \pm 1(0.5)$. The lower limit is $7 - 0.5$, which is 6.5, and the upper limit is $7 + 0.5$, which is 7.5.

(c) The Empirical Rule says that about 95 percent of test takers should score within two standard deviations of the mean — in other words, $7 \pm 2(0.5)$. The lower limit is $7 - 1$, which is 6, and the upper limit is $7 + 1$, which is 8.

(d) The Empirical Rule says that about 99.7 percent of test takers should score within three standard deviations of the mean — in other words, $7 \pm 3(0.5)$. The lower limit is $7 - 1.5$, which is 5.5, and the upper limit is $7 + 1.5$, which is 8.5.

YOUR TURN

22 Does the Empirical Rule apply to the data set shown in the following figure? Explain your answer.

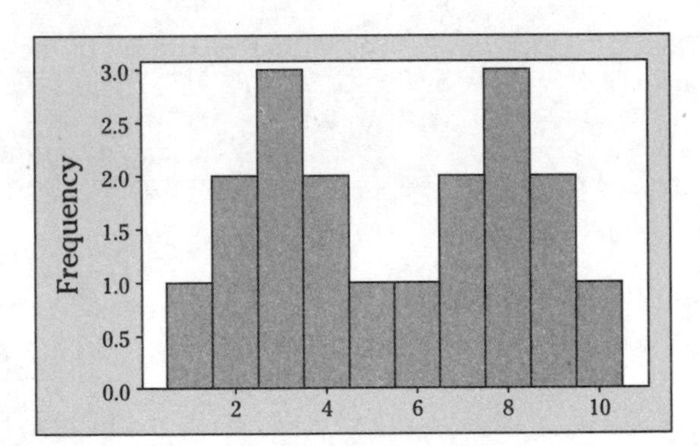

23 Suppose you have a data set of 1, 2, 2, 3, 3, 3, 4, 4, 5, and you assume this sample represents a population.

(a) Explain why you can apply the Empirical Rule to this data set.

(b) Where would "most of the values" in the population fall, based on this data set?

 Suppose a mound-shaped data set has a mean of 10 and a standard deviation of 2.

a. About what percentage of the data should lie between 8 and 12?

b. About what percentage of the data should lie above 10?

c. About what percentage of the data should lie above 12?

 Suppose a mound-shaped data set has a mean of 10 and a standard deviation of 2.

a. About what percentage of the data should lie between 6 and 12?

b. About what percentage of the data should lie between 4 and 6?

c. About what percentage of the data should lie below 4?

 Explain how you can use the Empirical Rule to find out whether a data set is mound-shaped, using only the values of the data themselves (no histogram available).

Measuring Relative Standing with Percentiles

Sometimes the precise values of the mean, median, and standard deviation just don't matter, and all you are interested in is where you stand compared to the rest of the herd. In this situation, you need a statistic that reports *relative standing,* and that statistic is called a percentile. The k-*th percentile* is a number in the data set that splits the data into two pieces: The lower piece contains *k* percent of the data, and the upper piece contains the rest of the data (which amounts to [100 – *k*] percent, because the total amount of data is 100 percent). ***Note:*** *k* is any number between 1 and 100.

TIP

The median is the 50th percentile: the point in the data where 50 percent of the data fall below that point, and 50 percent fall above it.

In this section, you find out how to calculate, interpret, and put together percentiles to help you uncover the story behind a data set.

Calculating percentiles

To calculate the *k*-th percentile (where *k* is any number between 1 and 100), do the following steps:

1. **Order all the numbers in the data set from smallest to largest.**
2. **Multiply *k* percent times the total number of numbers, *n*.**

3. **If the result from Step 2 is not a whole number, round it up to the nearest whole number and go to Step 4a. If the result from Step 2 is a whole number, go to Step 4b.**

4a. **Count the numbers in your data set from left to right (from the smallest to the largest number) until you reach the value indicated by Step 3.** The corresponding value in your data set is the k-th percentile.

4b. **Count the numbers in your data set from left to right until you reach the one indicated by Step 3.** The k-th percentile is the average of that corresponding value in your data set and the value that directly follows it.

EXAMPLE

Q. Suppose you have 25 test scores, and in order from lowest to highest they look like this: 43, 54, 56, 61, 62, 66, 68, 69, 69, 70, 71, 72, 77, 78, 79, 85, 87, 88, 89, 93, 95, 96, 98, 99, 99. What is the 90th percentile of the data set?

A. **98.** To find the 90th percentile for these (ordered) scores, start by multiplying 90 percent times the total number of scores, which gives you $90\% \times 25 = 0.90 \times 25 = 22.5$. Rounding up to the nearest whole number, you get 23.

Counting from left to right (from the smallest to the largest number in the data set), you go until you find the 23rd number in the data set. That number is 98, and it's the 90th percentile for this data set.

WARNING

There is no single definitive formula for calculating percentiles. The formula here is designed to make finding the percentile easier and more intuitive, especially if you're doing the work by hand; however, other formulas are used when you're working with technology. The results you get using various methods may differ, but not by much.

YOUR TURN

27 Suppose you have measured the height in inches of 11 randomly selected adults. In order from lowest to highest, they look like this: 59, 61, 65, 66, 67, 67, 69, 70, 72, 73, 75.

(a) What height marks the 15th percentile of the data set?

(b) What is the 80th percentile of the data set?

28 Suppose you have surveyed 8 randomly selected adults for their typical commute times to work. The responses in minutes look like this: 8, 13, 25, 16, 20, 15, 5, 35.

(a) What commute time marks the 60th percentile of the data set?

(b) What is the 33rd percentile of the data set?

Interpreting percentiles

Percentiles report the relative standing of a particular value within a data set. If that's what you're most interested in, the actual mean and standard deviation of the data set are not important, and neither is the actual data value. What's important is where you stand — not in relation to the mean, but in relation to everyone else: That's what a percentile gives you.

For example, in the case of exam scores, who cares what the mean is, as long as you scored better than most of the class? Who knows, it may have been an impossible exam and 40 points out of 100 was a great score. In this case, your score itself is meaningless, but your percentile tells you everything.

Suppose your exam score is better than 90 percent of the rest of the class. That means your exam score is at the 90th percentile (so $k = 90$), which hopefully gets you an A. Conversely, if your score is at the 10th percentile (which would never happen to you, because you're such an excellent student), then $k = 10$; that means only 10 percent of the other scores are below yours, and 90 percent of them are above yours; in this case, an A is not in your future.

A nice property of percentiles is they have a universal interpretation: Being at the 95th percentile means the same thing, whether you are looking at exam scores or weights of packages sent through the postal service; the 95th percentile always means that 95 percent of the other values lie below yours, and 5 percent lie above it. This also allows you to fairly compare two data sets that have different means and standard deviations (like ACT scores in reading versus math). It evens the playing field and gives you a way to compare apples to oranges, so to speak.

REMEMBER

A percentile is *not* a percent; a percentile is a number (or the average of two numbers) in the data set that marks a certain percentage of the way through the data. Suppose your score on the GRE was reported to be the 80th percentile. This doesn't mean you scored 80 percent of the questions correctly. It means that 80 percent of the students' scores were lower than yours and 20 percent of the students' scores were higher than yours.

WARNING

A high percentile doesn't always constitute a good thing. For example, if your city is at the 90th percentile in terms of crime rate compared to cities of the same size, that means 90 percent of cities similar to yours have a crime rate that is lower than yours, which is not good for you. Another example is golf scores; a low score in golf is a good thing, so let's just say that being at the 80th percentile with your score wouldn't qualify you for the PGA tour.

Comparing household incomes

The U.S. government often reports percentiles among its data summaries. For example, the U.S. Census Bureau reported the median (the 50th percentile) household income for 2014 to be $53,700, and in 2021 it was reported to be $67,463. The Bureau also reports various percentiles for household income for each year, including the 10th, 20th, 50th, 80th, 90th, and 95th. Table 5-2 shows the values of each of these percentiles for both 2014 and 2021.

Looking at the percentiles for 2014 in Table 5-2, you can see that the bottom half of the incomes are closer together than the top half of the incomes. The difference between the 20th percentile and the 50th percentile is about $33,400, whereas the spread between the 50th percentile and the 80th percentile is more like $46,300. The difference between the 10th and 50th percentiles is only about $41,400, whereas the difference between the 50th and the 90th percentiles is a whopping $103,800.

Table 5-2 U.S. Household Income (2014 versus 2021)

Percentile	2014 Household Income	2021 Household Income
10th	$12,300	$15,600
20th	$20,291	$27,012
50th	$53,700	$67,463
80th	$100,000	$141,100
90th	$157,500	$201,052
95th	$206,600	$273,850

The percentiles for 2021 are all higher than the percentiles for 2014 (which is a good thing!). They are also more spread out. For 2021, the difference between the 20th and 50th percentiles is around $40,000, and from the 50th to the 80th it's approximately $73,600; both of these differences are larger than for 2014. Similarly, the 10th percentile is farther from the 50th (about $52,000 difference) in 2021 compared to 2014, and the 50th is farther from the 90th (by about $133,600) in 2021, compared to 2014. These results tell you that incomes are increasing in general at all levels between 2014 and 2021, but the gap is widening between those levels. For example, the 10th percentile for income in 2014 was $12,300 (as shown in Table 5-2), compared to $15,600 in 2021; this represents about a 27 percent increase (subtract the two and divide by 12,300). Now compare the 95th percentiles for 2014 versus 2021; the increase is almost 32.5 percent. Now, technically, you may want to adjust the values for inflation, but you get the basic idea.

WARNING

Percentage changes affect the variability in a data set. For example, when salary raises are given on a percentage basis, the diversity in the salaries also increases; it's the "rich get richer" idea. The guy making $30,000 gets a 10 percent raise and his salary goes up to $33,000 (an increase of $3,000); but the guy making $300,000 gets a 10 percent raise and now makes $330,000 (a difference of $30,000). So when you first get hired for a new job, negotiate the highest possible salary you can because your raises that follow will also net a higher amount.

Examining ACT Scores

Each year, millions of U.S. high school students take a nationally administered ACT exam as part of the process of applying for colleges. The test is designed to assess college readiness in the areas of English, Math, Reading, and Science. Each test has a possible score of 36 points.

ACT does not release the average or standard deviation of the test scores for a given exam. (That would be a real hassle if they did, because these statistics can change from exam to exam, and people would complain that this exam was harder than that exam when the actual scores are not relevant.) To avoid these issues, and for other reasons, ACT reports test results using percentiles.

Percentiles are usually reported in the form of a predetermined list. For example, the U.S. Census Bureau reports the 10th, 20th, 50th, 80th, 90th, and 95th percentiles for household income (as shown previously in Table 5-2). However, ACT uses percentiles in a different way. Rather than reporting the exam scores corresponding to a premade list of percentiles, they list each possible exam score and report its corresponding percentile, whatever that turns out to be. That way, to find out where you stand, you just look up your score to find out your percentile.

Table 5-3 shows the percentiles for the scores on recent Mathematics and Reading ACT exams. To interpret an exam score, find the row corresponding to the score and the column for the exam area (for example, Reading). Intersect the row and column, and you find out which percentile your score represents; in other words, you see what percentage of your fellow exam-taking comrades scored lower than you.

Table 5-3 Percentiles for All Possible ACT Exam Scores in Math and Reading

ACT Score	Mathematics Percentile	Reading Percentile
34–36	99	99
33	98	97
32	97	95
31	96	93
30	95	91
29	93	88
28	91	85
27	88	81
26	84	78
25	79	74
24	74	70
23	68	65
22	62	59
21	57	54
20	52	47
19	47	41
18	40	34
17	33	30
16	24	24
15	14	19
14	06	14
13	02	09
12	01	06
11	01	03
1–10	01	01

For example, suppose you scored 30 on the Math exam; in Table 5-3 you look at the row for 30 in the column for Math; you see your score is at the 95th percentile. In other words, 95 percent of the students scored lower than you, and only 5 percent scored higher than you.

Now suppose you also scored a 30 on the Reading exam. Just because a score of 30 represents the 95th percentile for Math doesn't necessarily mean a score of 30 is at the 95th percentile for Reading as well. (It's probably reasonable to expect that fewer people score 30 or higher on the Math exam than on the Reading exam.)

To test this theory, look at column 3 of Table 5-3 in the row for a score of 30. You see that a score of 30 on the Reading exam puts you at the 91st percentile — not quite as great as your position on the Math exam, but certainly not a bad score.

Gathering a five-number summary

Beyond reporting a single measure of center and/or a single measure of spread, you can create a group of statistics and put them together to get a more detailed description of a data set. The Empirical Rule (as you see in the section, "Examining the Empirical Rule [68-95-99.7]," earlier in this chapter) uses the mean and standard deviation in tandem to describe a bell-shaped data set. In the case where your data are not bell-shaped, you use a different set of statistics (based on percentiles) to describe the big picture of your data. This method involves cutting the data into four pieces (with an equal amount of data in each piece) and reporting the resulting five cutoff points that separate these pieces. These cutoff points are represented by a set of five statistics that describe how the data are laid out.

The *five-number summary* is a set of five descriptive statistics that divide the data set into four equal sections. The five numbers in a five-number summary are

1. The *minimum* (smallest) number in the data set
2. The *25th percentile* (also known as *the first quartile*, or Q_1)
3. The *median* (50th percentile)
4. The *75th percentile* (also known as *the third quartile*, or Q_3)
5. The *maximum* (largest) number in the data set

EXAMPLE

Q. Find the five-number summary of the following 25 (ordered) exam scores: 43, 54, 56, 61, 62, 66, 68, 69, 69, 70, 71, 72, 77, 78, 79, 85, 87, 88, 89, 93, 95, 96, 98, 99, 99.

A. **43, 68, 77, 89, and 99.** The minimum is 43, the maximum is 99, and the median is the number directly in the middle, 77. To find Q_1 and Q_3, you use the steps shown in the section, "Calculating percentiles," with $n = 25$. Step 1 is done because the data are ordered. For Step 2, because Q_1 is the 25th percentile, you multiply $0.25 \times 25 = 6.25$. This is not a whole number, so Step 3a says to round it up to 7 and proceed to Step 3b.

Following Step 3b, you count from left to right in the data set until you reach the 7th number, 68; this is Q_1. For Q_3 (the 75th percentile), you multiply $0.75 \times 25 = 18.75$, which you round up to 19. The 19th number on the list is 89, so that's Q_3. Putting it all together, the five-number summary for these 25 test scores is 43, 68, 77, 89, and 99. To best interpret a five-number summary, you can use a boxplot; see Chapter 7 for details.

YOUR TURN

29 Suppose you have measured the height in inches of 11 randomly selected adults. In order from lowest to highest, they look like this: 59, 61, 65, 66, 67, 67, 69, 70, 72, 73, 75. Find the five-number summary for this data set.

30 Suppose you have surveyed eight randomly selected adults for their typical commute times to work. The responses in minutes look like this: 8, 13, 25, 16, 20, 15, 5, 35. Find the five-number summary for this collection of commute times.

Exploring interquartile range

The purpose of the five-number summary is to give descriptive statistics for center, variation, and relative standing all in one shot. The measure of center in the five-number summary is the median, and the first quartile, median, and third quartiles are measures of relative standing.

To obtain a measure of variation based on the five-number summary, you can find what's called the *inter-quartile range* (or *IQR*). The IQR equals $Q_3 - Q_1$ (that is, the 75th percentile minus the 25th percentile) and reflects the distance taken up by the innermost 50 percent of the data. If the IQR is small, you know a lot of data are close to the median. If the IQR is large, you know the data are more spread out from the median. The IQR for the test scores data set is $89 - 68 = 21$, which is fairly large, seeing as how test scores only go from 0 to 100.

TIP

The interquartile range is a much better measure of variation than the regular range (maximum value minus minimum value; see the section, "Being out of range," earlier in this chapter). That's because the interquartile range doesn't take outliers into account; it cuts them out of the data set by only focusing on the distance within the middle 50 percent of the data (that is, between the 25th and 75th percentiles).

REMEMBER

Descriptive statistics that are well chosen and used correctly can tell you a great deal about a data set, such as where the center is located, how diverse the data are, and where a good portion of the data lies. However, descriptive statistics can't tell you everything about the data, and in some cases they can be misleading. Be on the lookout for situations where a different statistic would be more appropriate (for example, the median describes center more fairly than the mean when the data is skewed), and keep your eyes peeled for situations where critical statistics are missing (for example, when a mean is reported without a corresponding standard deviation).

Practice Questions Answers and Explanations

1. **The mean (or average) doesn't have to be one of the numbers in the data set, but it can be.** For example, in the data set 1, 2, the mean is 1.5, which isn't in the data set; however, in the data set 1, 2, 3, the mean is 2.

2. **Outliers attract the mean toward them and away from the rest of the data.** For example, the mean of the data set 1, 2, 3 is 2. Suppose the largest value changes so that you have the data set 1, 2, 297. The mean is now $1+2+297$ divided by 3, which is $300 \div 3 = 100$. In this case, the mean of 100 doesn't really represent a "typical" value in the data set.

3. This problem gives you one way to check your answer to see whether it makes sense.

   a. **Because it averages out all the data in the set, the mean has to be somewhere between the largest and smallest values in the data set.**

   b. **The mean could equal the maximum value in a data set if all the values in the data set are the same;** otherwise, any other value that isn't at the maximum pulls the mean down.

4. **The median will be one of the numbers in the data set if the set has an odd number of values in it,** because the set has one distinct middle value in that case. If the set has an even number of values, you find the median by averaging the two middle values, and the answer may or may not be one of the values in the data set. For example, if the data set is 1, 2, 3, 4, the median is 2.5, which isn't included in the data set; however, if the data set is 1, 3, 3, 4, the median is $(3+3) \div 2 = 3$, which is included.

5. **If you don't order the data to find the median, you get a different answer.** For example, look at the data set 1, 5, 2. The median is 2, but if you don't order the data, it would be 5. And, if you reorder the same data set to be 2, 1, 5, you get a different answer for the median: 1. So, you should always order the data from smallest to largest to always get the same answer for the median.

   For the mean, you add up all the values in the data set and divide by the size of the data set. Using the commutative property for addition (and you thought you'd never use algebra later in life!), you know that $a+b=b+a$. Even if you reorder the data, you still get the same sum. So you don't have to order the data to always get the same answer for the mean of a given data set.

6. **Outliers affect the mean, but they don't affect the median.** For example, the mean and the median of the data set 1, 2, 3 is 2. Suppose the largest value changes so that you have the data set 1, 2, 297. The mean is now 100. However, the median of the data set 1, 2, 297 is still 2.

   REMEMBER

   Outliers affect the mean, but they don't affect the median. The mean gets pulled in the direction of the outlier and may not truly represent a "typical" value in the data set.

7. Many answers are possible. The key is to put the same number in the middle. **One possible answer is data set 1: 100, 200, 300; data set 2: 199, 200, 201.** The mean and median of both data sets is 200. These two data sets have the same center with totally different ranges (or spreads). If you want to tell the story about a data set, the center isn't enough because it can't distinguish between two data sets with different spreads.

8. This problem really points out what happens to the measures of center when you add any constant to all the values in the data set.

a. **The mean also increases by $1,000 to $51,000,** because you literally pick up all the salaries, move them up $1,000 on the number line, and put them back down, which moves the mean by the same amount.

b. **The median also increases by the same amount, to $51,000,** for the same reason.

TIP

Adding or subtracting a constant to or from all the values in a data set changes the mean and median by that same constant. Be careful — that constant could be negative as well as positive.

9. This scenario highlights what happens when you multiply all the data by a constant. Here the constant is 1.1, because you take the old salary, call it *X*, and add 10 percent of the *X* to it: $X + 0.10(X)$. But $X + 0.10(X) = 1.10X$; so, in other words, 1.10 times the original salary gets you the new salary.

a. **The mean also increases by 10 percent to become $\$50,000 \times 1.1 = \$55,000$,** because you multiply each value in the data set by 1.1.

b. **The median also increases by 10 percent to become $55,000,** for the same reason.

10. This data set is known as "U-shaped" because it looks like the letter U. **The data is symmetric,** because if you draw a vertical line down the middle, the picture is the same on each side. **The data set has two modes, or peaks, at the values 1 and 6** — the values that appear most often in the data set.

REMEMBER

Sometimes a data set has no peaks, as in a uniform distribution, which is flat (refer to the figure in the first example problem in this chapter for an illustration). You may think that every data value is its own peak, but that doesn't mean very much; so, in that case, you should say that no modes exist.

11. **This data set is skewed.** The skew is in the direction that the tail goes (the part that trails off). The set is skewed to the right because the tail trails off to the right. The skew means that a few of the values in the data set are higher than the rest of the values. Data that skew right are also called *positively skewed.*

12. **This data set is skewed to the left because the tail trails off to the left.** The skew means that a few of the values in the data set are lower than the rest of the values. Data that skew left are also called *negatively skewed.*

REMEMBER

When data are skewed, you can easily confuse the direction of the skew. Look at where the tail trails off to find the direction of the skew (not where the mound in the data set is).

13. **The histogram shown in the following figure is symmetric because it looks the same on each side when you draw a line down the middle.** It also has two peaks, or modes. When a histogram has two modes, the data have a *bimodal distribution.* ***Note:*** There are many possible correct answers to this problem.

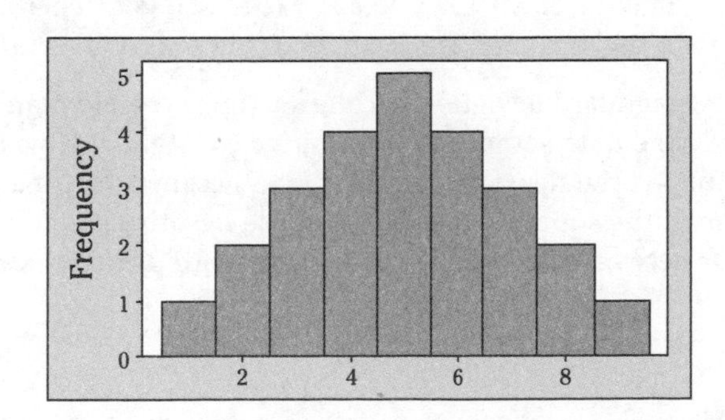

14 Just knowing the mean and median can tell you a lot about a data set.

a. **This data set is skewed to the right.** A few people in the company have very large salaries compared to the others, driving the mean up and away from most of the data, yet the median remains unaffected by the skew, staying in the "middle" of the data.

b. **The median is the most appropriate measure of center to use when the data set is skewed because outliers don't affect it.**

c. **The employees would want to report the median salary because it's lower and better represents most employees in the company. The employers may try to use the mean salary, because it's higher and indicates what they actually have to pay overall for their employees.**

TIP

If the data are skewed right, the mean is higher than the median. Conversely, if the data are skewed left, the mean is lower than the median. If the data are symmetric, the mean and median are the same (or very close).

15 Because the data set is skewed left, you have a few values that are lower than the rest, which drives down the mean. **So, the mean must be 19, and the median must be 38.**

REMEMBER

This is another point that instructors hammer home: the relationship between the mean and median for skewed data. Remember that *skewed right* means a few large values, and *skewed left* means a few small values. This helps you think about how the mean and median compare.

16 Where the mean shows up among the data depends on the shape of the data. **The mean can be higher than most of the values; in this case, the data are skewed to the right. The median can't be higher than most of the values; it's higher than 50 percent of the values and lower than 50 percent of the values, putting it right in the middle.**

17 The standard deviation can't be negative because of the squaring that goes on in its calculation. **However, it can be 0, although it happens only when the data set has no deviation in it — in other words, when all the data are exactly the same value.** For example, 1, 1, 1 or 2, 2, 2, 2, 2 are two data sets with a standard deviation of 0.

18 **If you choose 1, 1, 5, 5, you get the largest standard deviation possible,** because these numbers are as far as possible from the mean (which is 3).

19 **Adding a constant to the data doesn't change the standard deviation,** because you just relocate the data in a different spot on the number line; you don't change how far apart the values are from the mean.

20 **Multiplying by a constant changes the standard deviation.** If you multiply an entire data set by 1.1, the spread increases. Suppose two employees have salaries of $30,000 and $50,000 — right now the figures are $20,000 apart. With a 10 percent raise, they become $33,000 and $55,000, making them $22,000 apart (the rich get richer and the poor get less rich). If you recalculate the standard deviation, you find that it goes up here by a factor of 1.1 as well.

TIP

The new standard deviation becomes c times the old standard deviation, when you multiply the data set by a non-negative constant c. If you multiply the data by a negative constant, $-c$, the new standard deviation becomes $|c|$ times the old standard deviation (again, because of the squaring that goes on, the negative sign disappears). Also note that if c is a number between zero and 1, the new standard deviation gets smaller than the old one.

21 The standard deviation is based on the average distance from the mean. Outliers (which often appear in skewed data) influence the mean, and, therefore, influence the standard deviation. **Outliers drive up the standard deviation, making the average distance from the mean seem larger than it is for most of the data.** How to get around this? You can report something called the *interquartile range,* the difference between the 75th and 25th percentiles (which I discuss later in this chapter).

TIP

How can you tell whether a point in the data set is an official outlier? Statisticians have a few general rules, but nothing really hard and fast. You can look at the histogram and calculate your statistics with and without the outliers to see how the numbers change. The obvious outliers are the ones that change the statistics the most.

No. Because the data are skewed, the data set isn't mound-shaped. It has one mode, but it isn't symmetric.

WARNING

The Empirical Rule works only if the data is mound-shaped. It doesn't apply to skewed data. Check the shape of the data first before you attempt to apply the Empirical Rule. Don't get caught applying something that doesn't fit the data. (If you learned Chebyshev's Theorem, use that for data that isn't mound- or bell-shaped.)

23 The histogram for this data is shown in the following figure.

WARNING

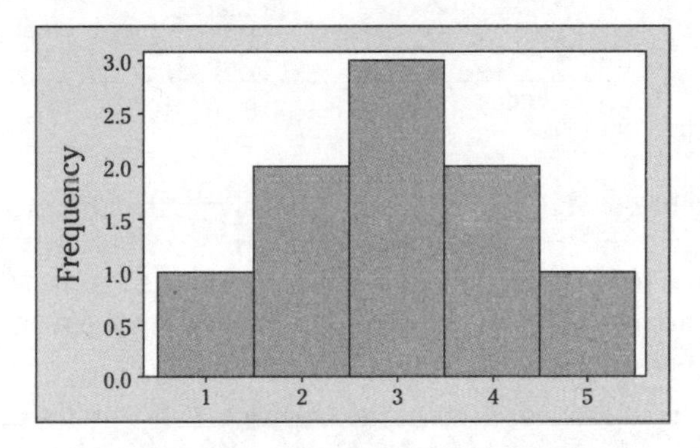

a. **This data set is mound-shaped, so the Empirical Rule applies.**

b. **Because this sample is representative of the population, you can say that most of the values in the population should lie within two standard deviations of the mean of the sample.** The mean is 3, and the standard deviation is 1.225. This indicates that 95 percent of the values in the population should lie between $3-(2\times1.225)$ and $3+(2\times1.225)$; in other words, they should lie between 0.55 and 5.45.

"Most of the data" can mean 68 percent, 95 percent, or 99.7 percent, but most statisticians would say that 95 percent is the magic number when you want to discuss where "most" of the data lies. Be sure you define what you mean by "most" if you're asked to describe where most of the data is.

24. This problem combines the Empirical Rule with the idea that the total of all the relative frequencies of a data set have to equal 1.

    a. Because 8 and 12 are both one standard deviation from the mean, the percentage of data lying between them according to the Empirical Rule is **about 68 percent.**

    b. Because mound-shaped data are symmetric, half the data should lie above the mean and half the data should lie below the mean. **So the answer is 50 percent.**

    c. Because about 68 percent of the data lie between 8 and 12, $68 \div 2 = 34$ percent of the data lie between 8 and 10, and the other 34 percent lie between 10 and 12. Because half of the data lie above 10, and 34 percent of the data lie between 10 and 12, 50 percent – 34 percent = **16 percent of the data should lie above 12.** Take what you know and subtract the part you don't need in order to get the part that you want. See the following figure for an illustration of these ideas.

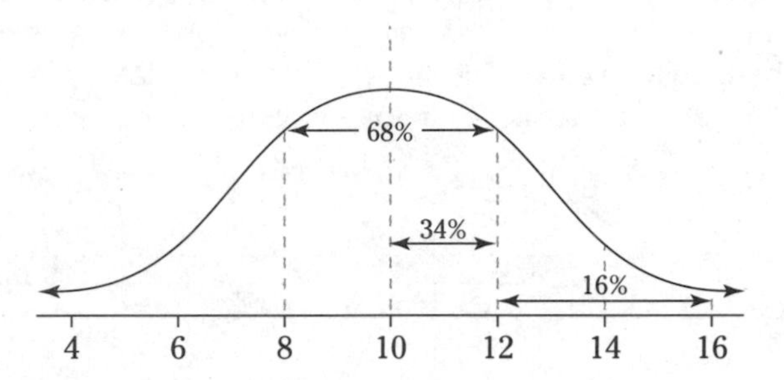

TIP

A picture may be very helpful for problems involving the Empirical Rule. It helps you to stay organized in your thinking, and it helps your professor see what you're doing, in case you make a mistake in your calculations somewhere along the line. Mark off the mean and the standard deviations to the left and right (go out three of them on each side), and then shade in the area for which you want to find the percentage.

25. This problem uses different parts of the Empirical Rule in tandem.

    a. The number 6 is two standard deviations below the number 10 because $10 - 2(2) = 6$. You know by the Empirical Rule that about 95 percent of the data set lies between 6 and 14, so half of that (or 47.5 percent) lies between 6 and 10 by symmetry. That gives you one piece of what you need. Now you need the part from 10 to 12. Twelve is one standard deviation away from the mean, and you know that about 68 percent of the values lie between 8 and 12, so 34 percent of the data lie between 10 and 12. This gives you the other piece that you need. Add the two pieces together to determine that 47.5 percent + 34 percent = **81.5 percent of the data should lie between 6 and 12.**

REMEMBER

When using the Empirical Rule, add the two pieces together when you want the percentage of data that fall between the two numbers when one is below the mean and the other is above the mean.

    b. This time you find the two pieces and subtract them, because you want the area between them. Because 6 is two standard deviations below 10, $10 - 2(2) = 6$, so you know that about $95 \div 2 = 47.5$ percent of the data lie between 6 and 10. Because 4 is three standard deviations below 10, $10 - 3(2) = 4$, so about 99.7 percent of the data lie between 4 and 16. That means half of the data, or 49.85 percent, should lie between 4 and 10. You want the area between 4 and 6, so take the area between 4 and 10 (49.85 percent) and subtract the area you don't want, which is the area between 6 and 10 (47.5 percent). You determine that 49.85 percent – 47.5 percent = **2.35 percent of the data should lie between 4 and 6.**

REMEMBER

Percentages are never negative, so make sure you take the bigger area first and subtract the smaller area that you don't want to find the percentage of the data that fall between two numbers. This subtracting is done only when both numbers are on the same side of the mean and the Empirical Rule is being used.

c. To get the area below 4, take the area from 4 to 10, which is about 99.7 percent ÷ 2 percent = 49.85 percent, and subtract that from 50 percent (0.5) because half of the data lies below 10 by symmetry. So the answer is about 50 percent – 49.85 percent = **0.15 percent of the data should lie below 4.**

(26) **You can calculate the mean plus or minus one standard deviation to get the upper and lower limits.** According to the Empirical Rule, if the data are mound-shaped, about 68 percent of the data should lie between these two values. Count how many data points are actually in this interval and divide by the sample size. If this percentage isn't close to 68, the data isn't mound-shaped. If it is, move on to the next standard deviation. Take the mean plus or minus two standard deviations and find those limits. Find the percentage of data lying between those two numbers and compare it to 95 percent. If the figure is close, go on to the third step (99.7 percent). All three criteria have to be met in order to say the data are mound-shaped.

(27) To calculate the percentiles, simply follow the steps. Fortunately, I have one less thing to do because the data is already ordered.

a. To find the 15th percentile for these (ordered) heights, start by multiplying 15 percent times the total number of heights, which gives you $15\% \times 11 = 0.15 \times 11 = 1.65$. Rounding up to the nearest whole number, you get 2. Counting from left to right, go until you find the second number in the data set. **That number is 61 inches,** and it's the 15th percentile for this data set.

b. To find the 80th percentile for these heights, start by multiplying 80 percent times the total number of heights, which gives you $80\% \times 11 = 0.80 \times 11 = 8.8$. Rounding up to the nearest whole number, you're looking for the ninth number in the data set. Counting from left to right, you find that's **a height of 72 inches**.

(28) These aren't in order! Remember to line them up from smallest to largest before doing anything else.

a. In order from smallest to largest, the values are 5, 8, 13, 15, 16, 20, 25, 35. Now multiply 60 percent times the total number of values, which gives you $60\% \times 8 = 0.60 \times 8 = 4.8$. Rounding up to the nearest whole number, you get 5, and the number in the fifth position is 16. **So, the 60th percentile for this data set is 16 minutes.**

b. Once the values are in order, multiply 33 percent times the total number of values. This gives you $33\% \times 8 = 0.33 \times 8 = 2.64$. Rounding up to the nearest whole number, you get 3, and the number in the third position is 13. **So, the 33rd percentile for this data set is 13 minutes.**

29 The minimum is 59, and the maximum is 75. Because this data set contains an odd number of values, the median is the value smack dab in the middle: 67. To find Q_1 and Q_3, you use the steps to find the 25th and 75th percentiles with $n = 11$. Step 1 is already complete because the data are ordered. Because Q_1 is the 25th percentile, you multiply $0.25 \times 11 = 2.75$. This is not a whole number, so round it up to 3 and count from left to right in the data set until you reach the 3rd number, 65; this is Q_1. For Q_3 (the 75th percentile) you multiply $0.75 \times 11 = 8.25$, which you round up to 9. The 9th number on the list is 72, so that's Q_3. **Putting it all together, the five-number summary for these 11 heights is 59, 65, 67, 72, and 75.**

30 Remember to line up the commute times from smallest to largest before doing anything else.

In order from smallest to largest, the values are 5, 8, 13, 15, 16, 20, 25, 35. The minimum is 5, and the maximum is 35. Because this data set contains an even number of values, the median is the average of the middle two: $(15 + 16) \div 2 = 31 \div 2 = 15.5$.

To find Q_1 and Q_3, you use the steps to find the 25th and 75th percentiles with $n = 8$. For Q_1 (the 25th percentile), multiply $0.25 \times 8 = 2$. Count from left to right in the data set until you reach the 2nd number, which is 8. For Q_3 (the 75th percentile), you multiply $0.75 \times 8 = 6$. Count from left to right until you reach the 6th number, which is 20. **Thus, the five-number summary for these eight commute times is 5, 8, 15.5, 20, and 35.**

If you're ready to test your skills a bit more, take the following chapter quiz that incorporates all the chapter topics.

Whaddya Know? Chapter 5 Quiz

Quiz time! Complete each problem to test your knowledge on the various topics covered in this chapter. You can then find the solutions and explanations in the next section.

1. Suppose you start with a set of 10 salaries that are almost the same, then add an unusually small salary to the group. Will the standard deviation increase, decrease, or stay the same?

2. Suppose you start with a set of 10 salaries that are almost the same, then add an unusually small salary to the group. Will the median increase, decrease, or stay about the same?

3. Suppose you take a data set of 10 numbers and cut them all in half. What will happen to the mean?

4. Suppose you take a data set of 10 numbers and cut them all in half. What will happen to the median?

5. Give a situation where the standard deviation of a five-number data set is 0, and the mean is 6.

6. If a histogram is skewed left, the mean is ____________ than the median.

7. If the standard deviation is large, the mean must also be large. True or False?

8. A five-number summary includes the mean. True or False?

9. The interquartile range is the 75th percentile minus the 25th percentile. True or False?

10. Suppose a class of students took an exam and most students were done up to halfway through the test, while a few students stayed until the near the end, and a couple of them stayed until the very end. Are the exam-taking times skewed left, skewed right, or symmetric?

11. Is it possible for Q_3 and the median to be equal to each other: Yes or No?

12. Suppose exam scores are mound-shaped with a mean of 80 and a standard deviation of 5. About what percentage of the test scores lie between 75 and 80?

13. Suppose exam scores are mound-shaped with a mean of 80 and a standard deviation of 5. About what percentage of the test scores lie between 85 and 90?

14. Which set of 4 numbers has the largest standard deviation: Data set 1 (1, 2, 3, 4); Data set 2 (1, 1, 4, 4); or Data set 3 (1, 1, 1, 4)?

15. What is wrong with this sentence? Bob's test score is at the 80th percentile; that means he scored 80 on the test.

Answers to Chapter 5 Quiz

1. **The standard deviation will increase,** because the unusually small salary will make the other numbers further from the mean and will also add more distance to the mean on its own.

2. **The median will stay about the same.** The one usually small number that was added to the group will go away in the new calculation of the median, and the rest of the numbers are about the same. The median won't be affected.

3. **The mean will also be cut in half,** because all of the values are cut in half.

4. **The median will also be cut in half,** because all of the values are cut in half.

5. **The only answer is 6, 6, 6, 6, 6.** All the numbers are the same, so the standard deviation is 0, and the mean is 6 because all the numbers are 6.

6. **If a histogram is skewed left, the mean is less than the median.** This is because the few extra small numbers compared to the others in the skewed-left data set pull the mean toward them, bringing it below the median.

7. **False.** If the standard deviation is large, it only means the concentration of data around the mean must be large; it has nothing to do with what the mean actually is.

8. **False.** The measure of center that is included in the five-number summary is the median. The mean is not part of the five-number summary.

9. **True.** The interquartile range is $Q_3 - Q_1$. The value of Q_3 is the same as the 75th percentile, and the value of Q_1 is the same as the 25th percentile.

10. **Skewed right.** A few students stayed longer than the rest of the students, making for a few long test times, compared to everyone else. Those students make the histogram of exam-taking times skewed right.

11. **Yes.** If the median and Q_3 are equal, that means all the numbers between the 50th percentile in the data set and the 75th percentile in the data set are the same.

12. **About 34 percent.** You know by the Empirical Rule that about 68 percent of the scores lie within one standard deviation $(5 * 1)$ of the mean (80), so about 68 percent of the values lie between 75 and 85. You want the percentage that lie between 75 and 80, which is half of about 68, or about 34 percent.

13. **About 13.5 percent.** You know by the Empirical Rule that about 95 percent of the scores lie within two standard deviations $(5 * 2)$ of the mean (80), so about 95 percent of the values lie between 70 and 90. Half of this (47.50 percent) represents the area between 80 and 90. You don't want the area between 80 and 85, which is half of 68, because 85 is one standard deviation above the mean, and 68 percent of the values lie between one standard deviation below the mean (75) and one standard deviation above the mean (85). So, subtract $47.50\% - 34\% = 13.5\%$.

14. **Data set 2 (1, 1, 4, 4) has the largest standard deviation of the data sets presented.** It has the largest concentration of data away from its mean (2.5), pushing the numbers as far from the center as they can possibly get. In data set 1 (1, 2, 3, 4), the numbers are farthest from each other, but not as far from the mean (2.5) as data set 2. And data set 3 (1, 1, 1, 4) has a mean of 1.75, bringing the 1's closer to it than data set 2.

15. **The conclusion of the sentence is wrong.** Bob's test score is at the 80th percentile. That means 80 percent of the scores lie below Bob's score. Bob's score could be anything at this point; you only know that 80 percent of the other test scores are lower than Bob's.

IN THIS CHAPTER

» Making data displays for categorical data

» Interpreting and critiquing charts and graphs

Chapter 6
Getting the Picture: Graphing Categorical Data

Data displays, especially charts and graphs, seem to be everywhere, showing everything from election results, broken down by every conceivable characteristic, to how the stock market has fared over the past few years (months, weeks, days, minutes). We're living in an instant gratification, fast-information society; everyone wants to know the bottom line and be spared the details.

The abundance of graphs and charts is not necessarily a bad thing, but you have to be careful; some of them are incorrect or even misleading (sometimes intentionally and sometimes by accident), and you have to know what to look for.

This chapter is about graphs involving *categorical data* (data that places individuals into groups or categories, such as gender, opinion, or whether a patient takes medication every day.) Here you find out how to read and make sense of these data displays and get some tips for evaluating them and spotting problems. (***Note:*** Data displays for *numerical data,* such as weight, exam score, or the *number* of pills taken by a patient each day, are covered in Chapter 7.)

The most common types of data displays for categorical data are pie charts and bar graphs. In this chapter, I present examples of each type of data display and share some thoughts on interpretation and tips for critically evaluating each type.

Take Another Little Piece of My Pie Chart

A pie chart takes categorical data and breaks them down by group, showing the percentage of individuals that fall into each group. Because a pie chart takes on the shape of a circle, the "slices" that represent each group can easily be compared and contrasted.

TIP

Because each individual in the study falls into one and only one category, the sum of all the slices of the pie should be 100 percent or close to it (subject to a bit of rounding off). However, just in case, keep your eyes open for pie charts whose percentages just don't add up.

Before you make a pie chart, you can first summarize the data in table format. A frequency table shows how many individuals fall into each category (the sum of which is the *total sample size*). A relative frequency table shows what percentage of individuals fall into each category by taking the frequencies and dividing by the total sample size. The relative frequencies in the table and then in the pie chart should sum to 1, or 100 percent (subject to possible round-off error).

Tallying personal expenses

When you spend your money, what do you spend it on? What are your top three expenses? According to the U.S. Bureau of Labor Statistics Consumer Expenditure Survey, the top six sources of consumer expenditures in the U.S. were housing (33.9 percent), transportation (17.0 percent), food (12.8 percent), personal insurance and pensions (11.1 percent), healthcare (5.9 percent), and entertainment (5.6 percent). These six categories make up over 85 percent of average consumer expenses. (Although the exact percentages change from year to year, the list of the top six items remains the same.)

Figure 6-1 summarizes the U.S. expenditures in a pie chart. Notice that the "other" category appears a bit large in this chart (13.7 percent). However, with so many other possible expenditures out there (including this book), each one would only get a tiny slice of the pie for itself, and the resulting pie chart would be a mess. In this case, it is too difficult to break "other" down further. (But in many other cases, you can.)

TIP

Ideally, a pie chart shouldn't have too many slices because a large number of slices distracts the reader from the main point(s) the pie chart is trying to relay. However, lumping the remaining categories into one slice that's one of the largest in the whole pie chart leaves readers wondering what's included in that particular slice. With charts and graphs, doing it right is a delicate balance.

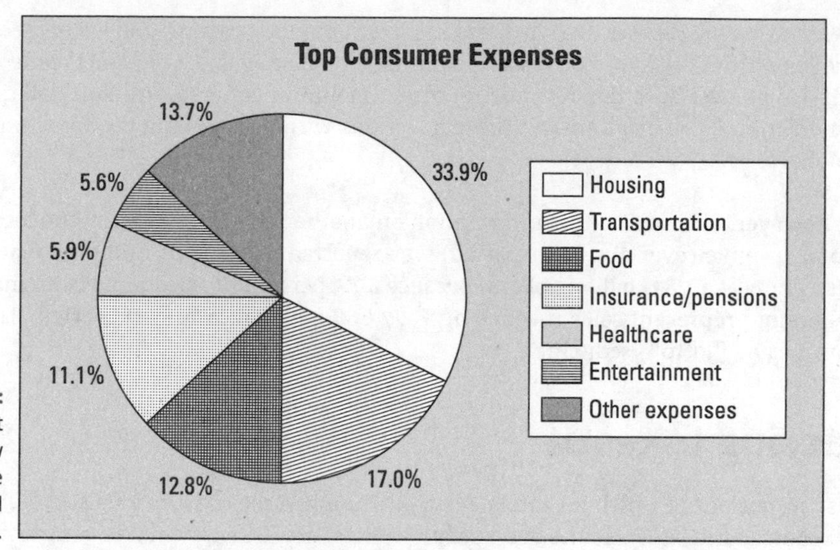

FIGURE 6-1: Pie chart showing how people in the U.S. spend their money.

Bringing in a lotto revenue

State lotteries bring in a great deal of revenue, and they also return a large portion of the money received, with some of the revenues going to prizes and some being allocated to state programs such as education. Where does lottery revenue come from? Figure 6-2 is a pie chart showing the types of games and their percentage of revenue as recently reported by Ohio's state lottery. (Note that the slices don't sum to 100 percent exactly due to a slight rounding error.)

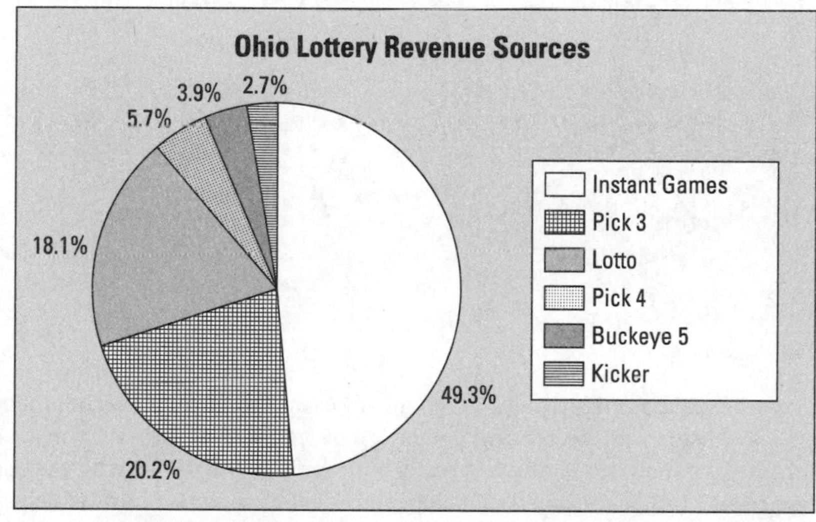

FIGURE 6-2: Pie chart breaking down a state's lottery revenue.

You can see by the pie chart in Figure 6-2 that 49.3 percent of the lottery sales revenue comes from the instant (scratch-off) games. The rest comes from various lottery-type games in which players choose a set of numbers and win if a certain number of their numbers match those chosen by the lottery.

Notice that this pie chart doesn't tell you *how much* money came in, only *what percentage* of the money came from each type of game. About half the money (49.3 percent) came from instant scratch-off games; does this revenue represent 1 million dollars, 2 million dollars, 10 million dollars, or more? You can't answer these questions without knowing the total amount of revenue dollars.

I was, however, able to find this information on another chart provided by the lottery website: The total revenue (over a 10-year period) was reported as "1,983.1 million dollars" — which you also know as 1.9831 billion dollars. Because 49.3 percent of sales came from instant games, they therefore represent sales revenue of $977,668,300 over a 10-year period. That's a lot of (or dare I say a "lotto") scratching.

Ordering takeout

It's also important to watch for totals when examining a pie chart from a survey. A newspaper I read reported the latest results of a "people poll." They asked, "What is your favorite night to order takeout for dinner?" The results are shown in the pie chart in Figure 6-3.

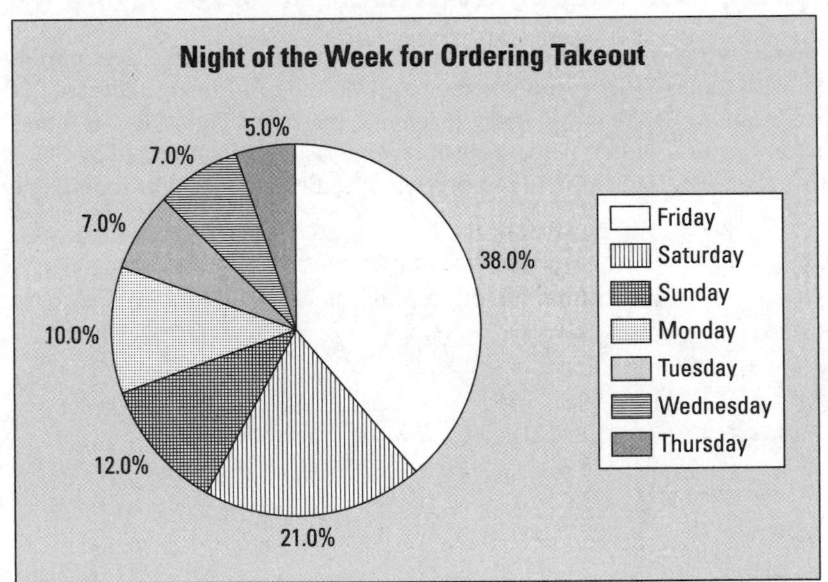

FIGURE 6-3: Pie chart for takeout food survey results.

You can clearly see that Friday night is the most popular night for ordering takeout (and that result makes sense), with decreasing demand moving from Saturday through Monday. The actual percentages shown in Figure 6-3 really only apply to the people who were surveyed; how close these results mimic the population depends on many factors, one of which is sample size. But unfortunately, sample size is not included as part of this graph. (For example, it would be nice to see "n = XXX" below the title, where n represents sample size.)

Even though you have a good sample selected (i.e. it's representative of the target population and is random), without knowing the sample size, you can't tell how accurate the information is. Which results would you find to be more accurate: those based on 25 people, 250 people, or 2,500 people? When you see "10 percent," you don't know if it's 10 out of 100, 100 out of 1,000, or even 1 out of 10. To statisticians, 1 ÷ 10 is not the same as 100 ÷ 1,000, even though they both represent 10 percent. (Don't tell that to mathematicians — they'll think you're nuts!)

REMEMBER

Pie charts often don't include the total sample size. Always check for the sample size, especially if the results are very important to you; don't assume it's large! If you don't see the sample size, go to the source of the data and ask for it.

Projecting age trends

The U.S. Census Bureau provides an almost unlimited amount of data, statistics, and graphics about the U.S. population, including the past, the present, and projections for the future. It often makes comparisons between years in order to look for changes and trends.

One recent Census Bureau population report looked at what it calls the "older U.S. population" (by the government's definition, this means people 65 years old or over). Age was broken into the following groups: 65–69 years, 70–74 years, 75–79 years, 80–84 years, and 85 and over. The Bureau calculated and reported the percentage in each age group for the year 2020 and made projections for the percentage in each age group for the year 2050.

I made side-by-side pie charts for the years 2020 versus 2050 (projections) to make comparisons; you can see the results in Figure 6-4. The percentage of the older population in each age group for 2020 is shown in one pie chart, and alongside it is a pie chart of the projected percentage for each age group for 2050 (based on the current age of the entire U.S. population, birth and death rates, and other variables).

If you compare the sizes of the slices from one graph to the other in Figure 6-4, you see that the slices for corresponding age groups are larger for the 2050 projections (compared to 2020) among the older age groups, and the slices are smaller for the 2050 projections (compared to 2020) among the younger age groups. For example, the 65–69 age group decreases from 30 percent in 2020 to a projected 25 percent in 2050; while the 85-and-over age group increases from 14 percent in 2020 to 19 percent projected for 2050.

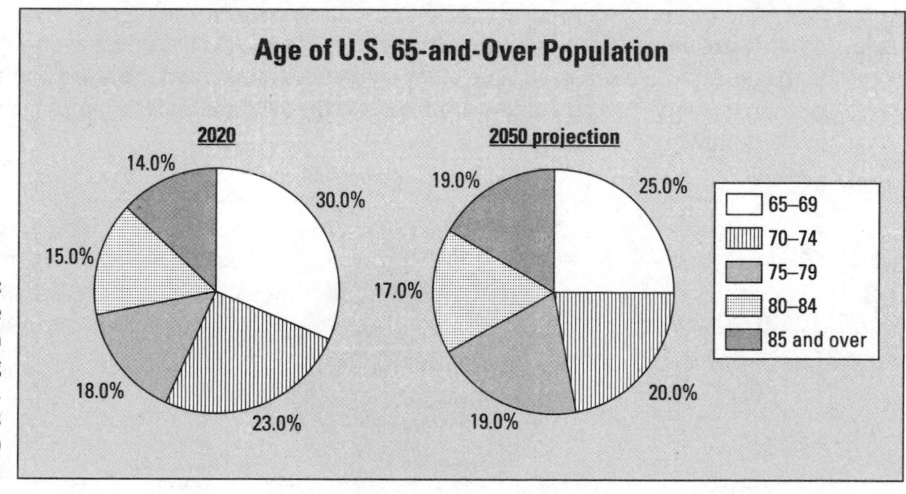

FIGURE 6-4: Side-by-side pie charts on the aging population, 2020 versus 2050 projections.

EVALUATING A PIE CHART

The following tips help you taste-test a pie chart for statistical correctness:

- Check to be sure the percentages add up to 100 percent or very close to it (any round-off error should be very small).
- Beware of slices of the pie called "other" that are larger than many of the other slices.
- Look for a reported total number of units (people, dollar amounts, and so on) so that you can determine (in essence) how "big" the pie was before being divided into the slices that you're looking at.
- Avoid three-dimensional pie charts; they don't show the slices in their proper proportions. The slices in front look larger than they should.
- Make sure the size of the slice of the pie is actually correct – yes we've seen pie charts where it says the slice is 25% of the pie but it shows half of the pie – that can't be right.

The results from Figure 6-4 indicate a shift in the ages of the population toward the older categories. From there, the medical and social research communities can examine the ramifications of this trend in terms of healthcare, assisted living, social security, and so on.

The operative words here are *if the trend continues.* As you know, many variables affect population size, and you need to take those into account when interpreting these projections into the future. The U.S. government always points out caveats like this in its reports; it is very diligent about that.

The pie charts in Figure 6-4 work well for comparing groups because they are side by side on the same graph, they use the same coding for the age groups in each chart, and their slices are in the same order for both charts as you move clockwise around them. They aren't all scrambled up on each chart so you have to hunt for a certain age group on each chart separately.

Q. A hardware store wants to know what percentage of its customers are women. The manager takes a random sample of 76 customers who enter the store and records their gender. Twenty-two customers are females; the rest are males. I summarize the results in the following pie chart.

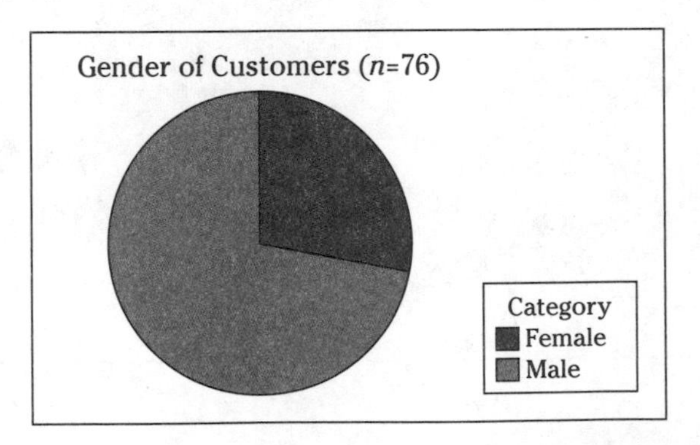

(a) Describe the results.

(b) How can this pie chart be improved?

A. Apparently, the DIY craze is popular with women, too.

(a) The results of the pie chart show that the percentage of female customers appears to be around 1/3 (or around 33 percent).

(b) You can improve the chart by showing the exact percentages in each slice. (The actual percentages are females: 28.9 percent; males: 71.1 percent.)

YOUR TURN

Suppose 375 individuals are asked what type of vehicle they own: SUV, truck, or car. The results are shown in the following frequency table.

Category	Frequency
SUV	150
Truck	125
Car	100
Total	375

(a) Make a relative frequency table of these results.

(b) Make a pie chart of these results.

(c) Interpret the results.

Suppose Lewis, a restaurant owner, keeps track of data on when his customers patronize his restaurant: breakfast, lunch, dinner, or other times. For a month, he takes time to check off which category each customer falls into. He records data on 1,000 customers for the month. The pie chart in the following figure shows his results.

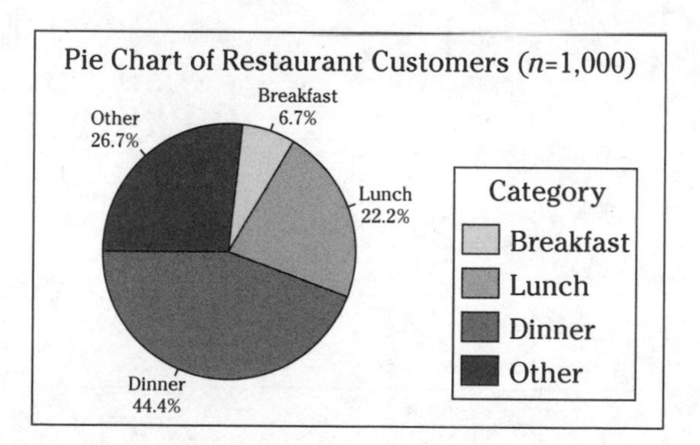

(a) What does this information tell him?

(b) Can you spot a problem with the "other" category? How can this study be improved in the future?

---

3 Suppose Susan, an office manager, wants to try to figure out a better way to multitask. She notices that answering emails is one of her most time-consuming duties, so she decides to categorize her emails into five groups: 1) highest priority, 2) medium priority, 3) very low priority, 4) personal, and 5) spam that she can delete immediately. You can see her results over a two-week period in the following frequency table.

Category	Frequency
Highest	60
Medium	120
Lowest	20
Personal	50
Spam	150
Total	400

(a) Make a relative frequency table of this data.

(b) Make a pie chart of this data.

(c) Interpret the results for the office manager.

4 Suppose a survey is conducted to see what types of pets people own. The survey of 100 adults finds that 40 of the people own a dog, 60 own a cat, 20 own fish, and 10 own some sort of rodent (hamster, gerbil, mouse, and so on). Can this data be organized in a pie chart? Explain your answer.

5 Suppose as part of a driver's education program, students have to observe drivers in the real world and see how consistently they come to a complete stop at intersections. The students sit at an intersection for four hours and record whether each driver comes to a complete stop, rolls through the stop sign slowly, or runs the stop sign altogether. You can see the data from the study in the following pie chart.

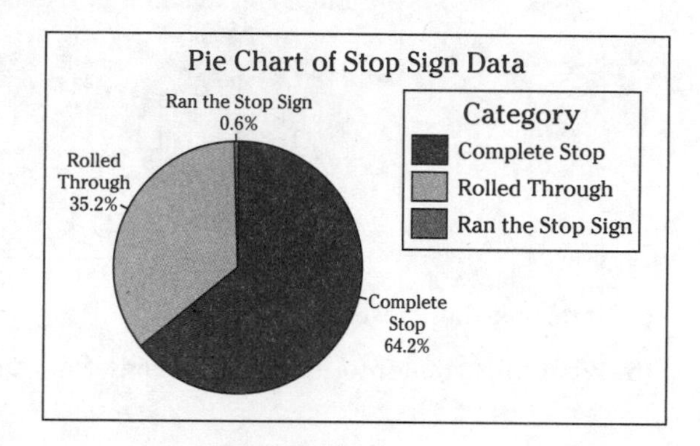

(a) Interpret the results.

(b) Do you see any issues with this pie chart?

(c) Is it a big deal if you don't know the sample size?

(d) Can you make generalizations about all drivers from this data?

6 A survey is conducted to determine whether 20 office employees of a certain company would prefer to work at home, if given the chance. Of the ten women surveyed, seven say they would prefer to work at home and three say no. Of the ten men surveyed, eight say no and two say yes. Compare the results by using two pie charts. Does gender seem to be associated with one's preference to work at home? Explain your answer.

Give an example of categorical data that you can't summarize correctly by using a pie chart.

8 An employer/employee study website conducts a survey that asks employers and employees if they think surfing non-work-related websites compromises employee productivity. I summarize the results with the following pie charts.

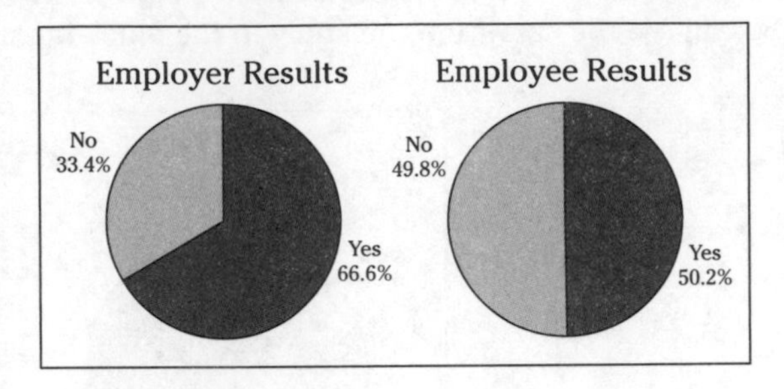

(a) Interpret these results.

(b) What important information is missing from these pie charts?

Raising the Bar on Bar Graphs

A *bar graph* (or *bar chart*) is perhaps the most common data display used by the media. Like a pie chart, a bar graph breaks categorical data down by group. Unlike a pie chart, it represents these amounts by using bars of different lengths; whereas a pie chart most often reports the amount in each group as percentages, a bar graph typically uses either the number of individuals in each group (also called the *frequency*) or the percentage in each group (called the *relative frequency*).

Tracking transportation expenses

How much of their income do people in the United States spend on transportation to get to and from work? It depends on how much money they make. The Bureau of Transportation Statistics (did you know such a department existed?) recently conducted a study on transportation in the U.S., and many of its findings are presented as bar graphs like the one shown in Figure 6-5.

This particular bar graph shows how much money is spent on transportation by people in different household-income groups. It appears that as household income increases, the total expenditures on transportation also increase. This makes sense, because the more money people have, the more they have available to spend.

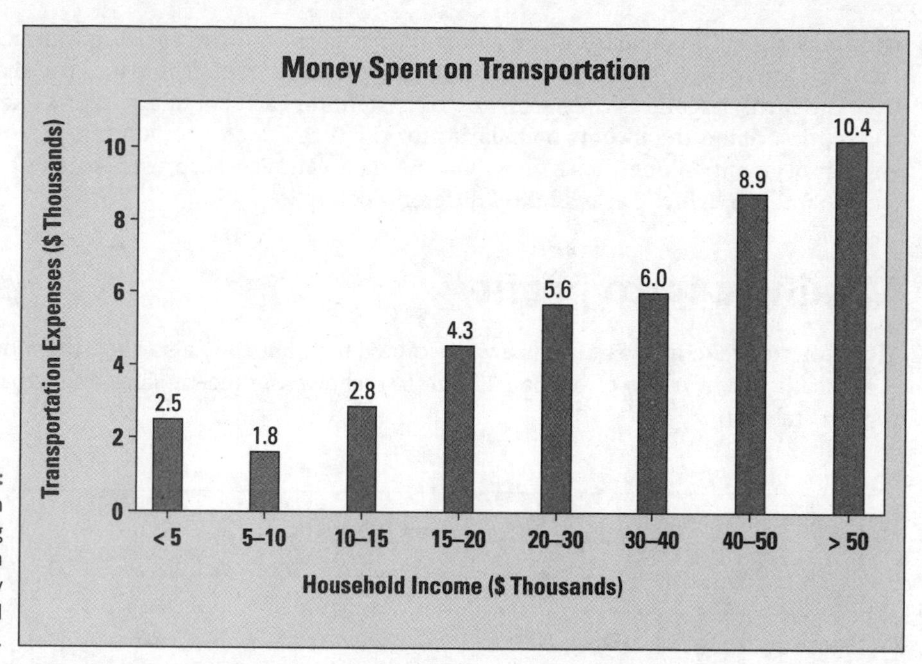

FIGURE 6-5: Bar graph showing transportation expenses by household income group.

But would the bar graph change if you looked at transportation expenditures not in terms of total dollar amounts, but as the percentage of household income? The households in the first group make less than $5,000 a year and have to spend $2,500 of that income on transportation. (***Note:*** The label reads "2.5," but because the units are in thousands of dollars, the 2.5 translates into $2,500.)

This $2,500 represents 50 percent of the annual income of those who make $5,000 per year; the percentage of the total income is even higher for those who make less than $5,000 per year. The households earning $30,000–$40,000 per year pay $6,000 per year on transportation, which is between 15 percent and 20 percent of their household income. So, although the people making more money spend more dollars on transportation, they don't spend more as a percentage of their total income. Depending on how you look at expenditures, the bar graph can tell two somewhat different stories.

Another point to check out is the groupings on the graph. The categories for household income as shown aren't equivalent. For example, each of the first four bars represents household incomes in intervals of $5,000, but the next three groups increase by $10,000 each, and the last group contains every household making more than $50,000 per year. Bar graphs using different-sized intervals to represent numerical values (such as Figure 6-5) make true comparisons between groups more difficult. (However, I'm sure the government has its reasons for reporting the numbers this way; for example, this may be the way income is broken down for tax-related purposes.)

One last thing: Notice that the numerical groupings in Figure 6-5 overlap on the boundaries. For example, $30,000 appears in both the fifth and sixth bars of the graph. So, if you have a household income of $30,000, which bar do you fall into? (You can't tell from Figure 6-5, but I'm sure the instructions are buried in a huge report in the basement of some building in Washington, D.C.) This kind of overlap appears quite frequently in graphs, but you need to know how the borderline values are being treated. For example, the rule may be "Any data lying exactly on a boundary value automatically goes into the bar to its immediate right." (Looking at Figure 6-5, that puts a household with a $30,000 income into the sixth bar rather than the fifth.) As long as they are being consistent for each boundary, that's okay. The alternative, describing the income boundaries for the fifth bar as "20,000 to $29,999.99," is not an improvement. Along those lines, income data can also be presented using a histogram (see Chapter 7), which has a slightly different look to it.

Making a lotto profit

That lotteries rake in the bucks is a well-known fact; but they also shell them out. How does it all shake out in terms of profits? Figure 6-6 shows the recent sales and expenditures of a certain state lottery.

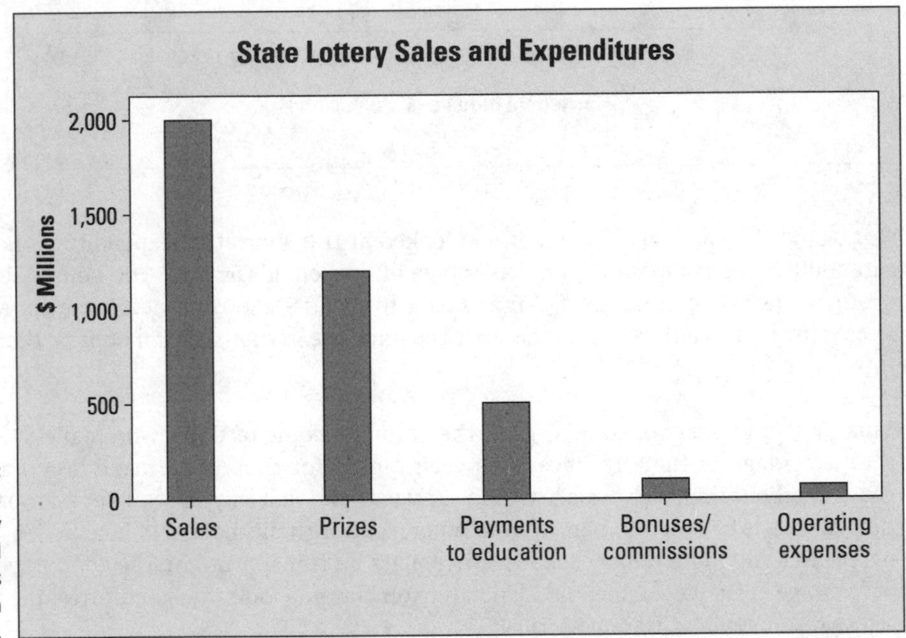

FIGURE 6-6: Bar graph of lottery sales and expenditures for a certain state.

In my opinion, this bar graph needs some additional info from behind the scenes to make it more understandable. The bars in Figure 6-6 don't represent similar types of entities. The first bar represents sales (a form of revenue), and the other bars represent expenditures. The graph would be much clearer if the first bar weren't included; for example, the total sales could be listed as a footnote.

Tipping the scales on a bar graph

Another way a graph can be misleading is through its choice of scale on the frequency/relative frequency axis (that is, the axis where the amounts in each group are reported) and/or its starting value.

By using a "stretched-out" scale (for example, having each half-inch of a bar represent 10 units versus 50 units), you can stretch the truth, make differences look more dramatic, or exaggerate values. Truth-stretching can also occur if the frequency axis starts out at a number that's very close to where the differences in the heights of the bars start; you are, in essence, chopping off the bottom of the bars (the less exciting part) and just showing their tops — emphasizing (in a misleading way) where the action is. Not every frequency axis has to start at zero, but watch for situations that elevate the differences.

A good example of a graph with a stretched-out scale is shown in Chapter 2, regarding the results of numbers drawn in the "Pick 3" lottery. (You choose three one-digit numbers and if they all match what's drawn, you win.) In Chapter 2, the percentage of times each number (from 0–9) was drawn is shown in Table 2-2, and the results are displayed in a bar graph in Figure 2-1a. The scale on the graph is stretched and starts at 465, making the differences in the results look larger than they really are; for example, it looks like the number 1 was drawn much less often, whereas the number 2 was drawn much more often, when in reality there is no statistical difference between the percentage of times each number was drawn. (I checked.)

Why was the graph in Figure 2-1a made this way? It might lead people to think they've got an inside edge if they choose the number 2 because it's "on a hot streak," or they might be led to choose the number 1 because it's "due to come up." Both of these theories are wrong, by the way, because the numbers are chosen at random; what happened in the past doesn't matter. In Figure 2-1b you see a graph that's been made correctly. (For more examples of where your intuition can go wrong with probability and what the scoop really is, see *Probability For Dummies* by Deborah Rumsey, also published by Wiley.)

Alternatively, by using a "squeezed-down" scale (for example, having each half-inch of a bar represent 50 units versus 10 units), you can downplay differences, making results look less dramatic than they actually are. For example, maybe a politician doesn't want to draw attention to a big increase in crime from the beginning to the end of their term, so they may have the number of crimes of each type shown where each half-inch of a bar represents 500 crimes versus 100 crimes. This squeezes the numbers together and makes differences less noticeable. Their opponent in the next election would go the other way and use a stretched-out scale to emphasize a crime increase in dramatic fashion, and voilà! (Now you know the answer to the question, "How can two people talk about the same data and get two different conclusions?" Welcome to the world of politics.)

TIP

With a pie chart, however, the scale can't be changed to over-emphasize (or downplay) the results. No matter how you slice up a pie chart, you're always slicing up a circle, and the proportion of the total pie belonging to any given slice won't change, even if you make the pie bigger or smaller.

Pondering pet peeves

A recent survey of 100 people with office jobs asked them to report their biggest pet peeves in the workplace. (Before going on, you may want to jot down a couple of yours, just for fun.) A bar graph of the results of the survey is shown in Figure 6-7. Poor time management looks to be the number-one issue for these workers (I hope they didn't do this survey on company time).

EVALUATING A BAR GRAPH

To raise the statistical bar on bar graphs, check out these tips:

- Bars that divide values of a numerical variable (such as income) should be equal in width (if possible) for fair comparison.
- Be aware of the scale of the bar graph and determine whether it's an appropriate representation of the information.
- Some bar graphs don't sum to 1 because they are showing the results of more than one variable; make sure it's clear what's being summarized.
- Check whether the results are shown as the percentage within each group (relative frequencies) or the number in each group (frequencies).
- If you see relative frequencies, check for the total sample size — it matters. If you see frequencies, divide each one by the total sample size to get percentages, which are easier to compare.

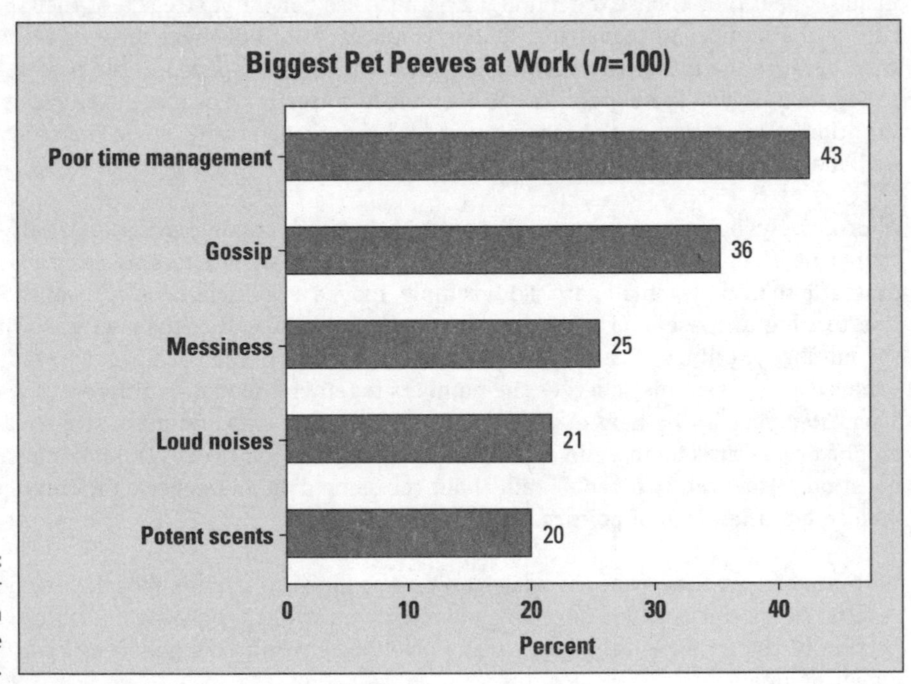

FIGURE 6-7: Bar graph for survey data with multiple responses.

TIP

If you take a look at the percentages shown for each pet peeve listed, you see they don't sum to 1. That tells you that each person surveyed was allowed to choose more than one pet peeve (like that would be hard to do); perhaps they were asked to name their top three pet peeves, for example. For this data set and others like it that allow for multiple responses, a pie chart wouldn't be possible (unless you made one for every single pet peeve on the list).

Note that Figure 6-7 is a *horizontal bar graph* (its bars go side to side) as opposed to a *vertical bar graph* (in which bars go up and down, as shown earlier in Figure 6-6). Either orientation is fine; use whichever one you prefer when you make a bar graph. Do, however, make sure that you label the axes appropriately and include proper units (such as gender, opinion, or day of the week) where appropriate.

EXAMPLE

Q. Following are a pie chart and bar graph (respectively) of scores from a quiz of ten questions, where the data shows the number of questions answered correctly. Name one advantage each has over the other.

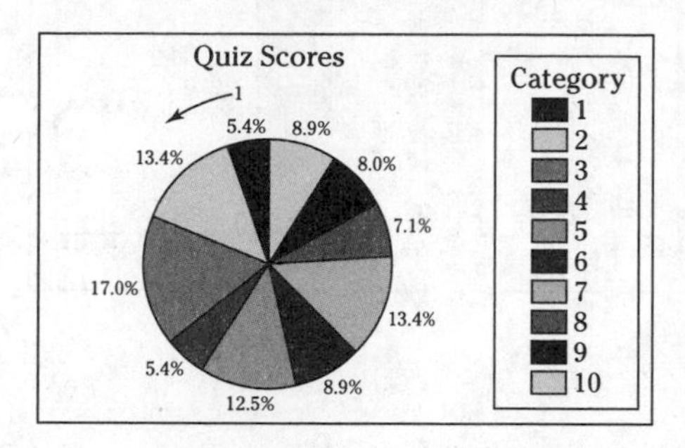

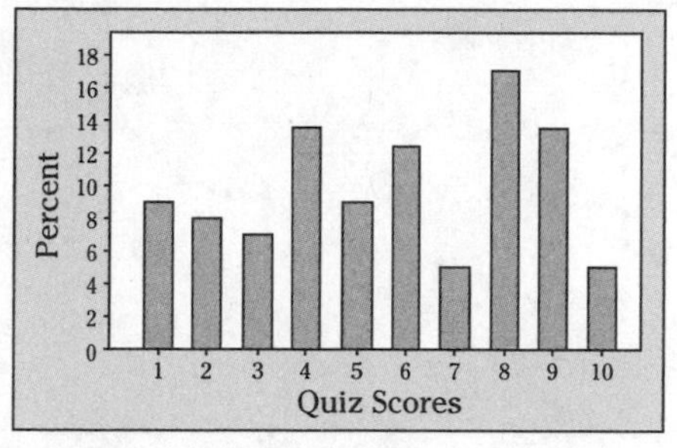

A. The pie chart shows everything as part of a whole, so you can make relative comparisons, and you know it all sums to 100 percent. The bar graph, however, makes it easier to compare the groups to each other. (And if the pie chart doesn't show the percentages, you have a much harder time estimating the percent in each group.)

YOUR TURN

9 The following figure shows a frequency bar graph of 500 people who make up three categories (1: support a smoking ban; 2: oppose a smoking ban; and 3: no opinion).

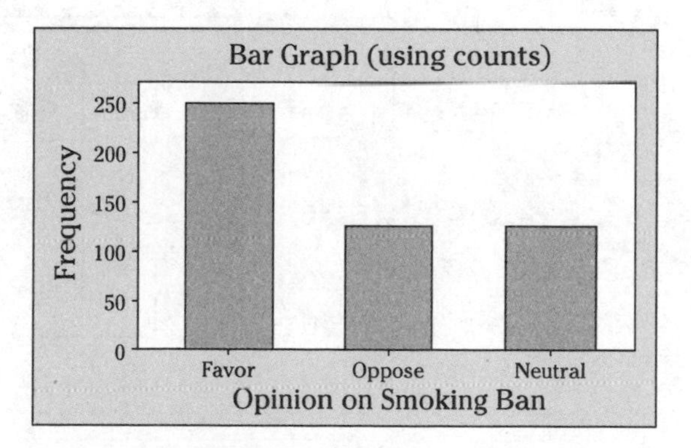

(a) Make a relative frequency table of this data.

(b) Use the relative frequency table to make a bar graph of this data.

(c) Interpret the results. (How do people in the sample feel about the smoking ban?)

10 Suppose a health club asks 30 customers to rate the services as very good (1), good (2), fair (3), or poor (4). You can see the results in the following bar graph. What percentage of the customers rated the services as good?

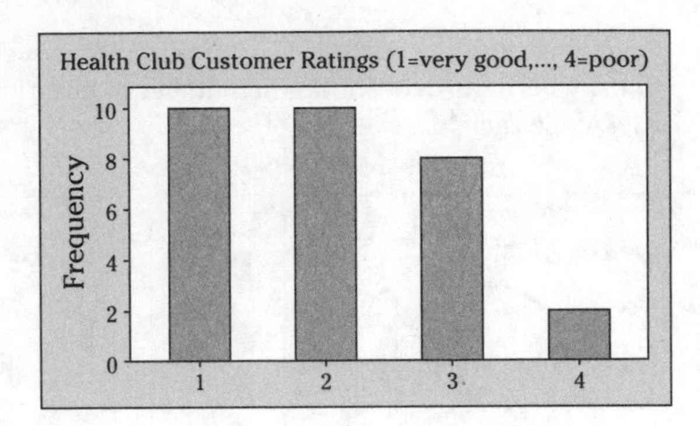

---

11 A polling organization wants to find out what voters think of Issue X. It chooses a random sample of voters and asks them for their opinions of Issue X: yes, no, or no opinion. I organize the results in the following bar graph.

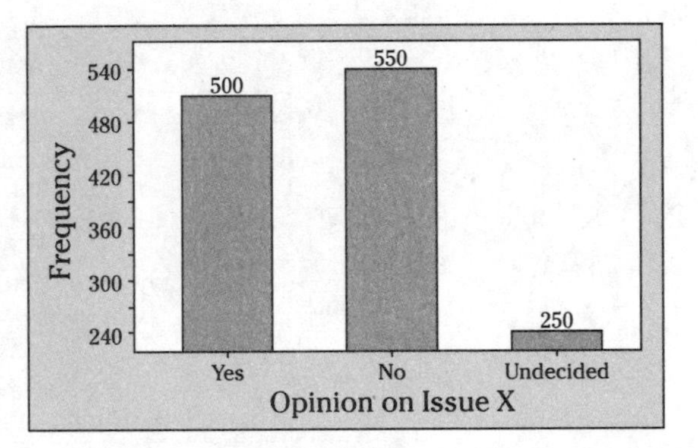

(a) Make a frequency table of these results (including the total number).

(b) Evaluate the bar graph as to whether or not it fairly represents the results.

---

12 Suppose that a random sample of 270 graduating seniors are asked what their immediate priorities are, including whether or not buying a house is a priority. The results are shown in the following bar graph.

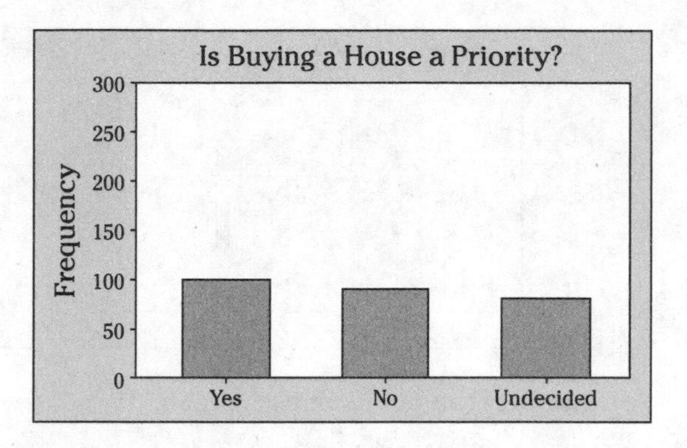

(a) The bar graph is misleading; explain why.

(b) Make a new bar graph that more fairly presents the results. Note that 100 said Yes, 90 said No, and 80 said Undecided.

---

13 A car dealership specializing in minivan sales conducts a survey to find out more about who their customers are. One of the variables the company measures is gender; the results of this part of the survey are shown in the following bar graph.

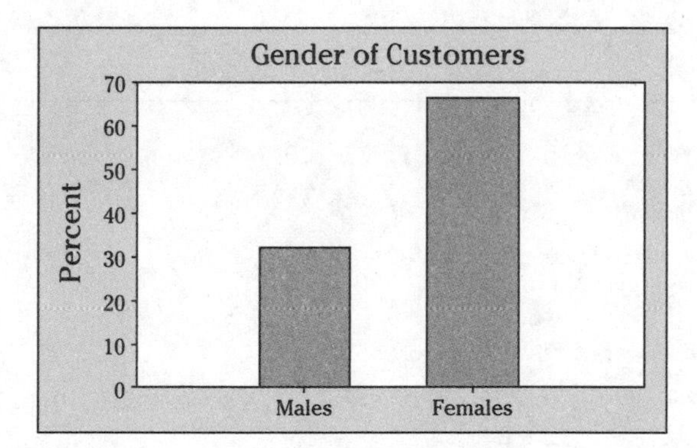

(a) Interpret these results.

(b) Explain whether or not you think the bar graph is a fair and accurate representation of this data.

14 A survey is conducted to determine whether 20 office employees of a certain company would prefer to work at home, if given the chance. The overall results are shown in the first bar graph, and the results broken down by gender are presented in the second.

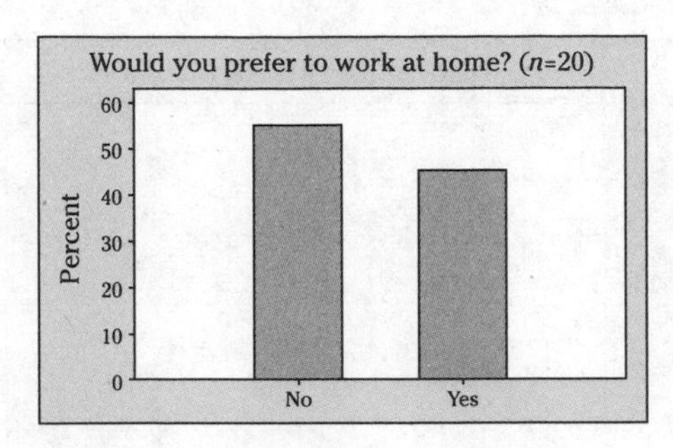

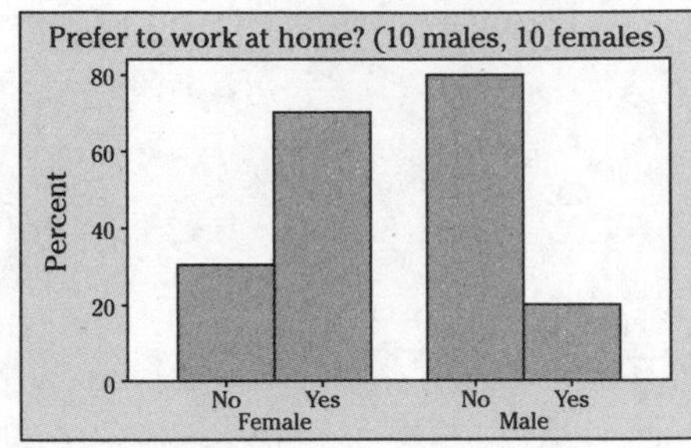

(a) Interpret the results of each graph.

(b) Discuss the added value in including gender in the second bar graph. (The second bar graph in this problem is called a *side-by-side bar graph* and is often used to show results broken down by two or more variables.)

(c) Compare the side-by-side bar graph with the two pie charts that you make for Problem 1. Which of the two methods is best for comparing two groups, in your opinion?

Practice Questions Answers and Explanations

1. Organizing the data in a relative frequency table before you make a pie chart is often helpful.

   a. See the following relative frequency table.

Category	Relative Frequency
SUV	$150 \div 375 = 0.400$ or 40.0%
Truck	$125 \div 375 = 0.333$ or 33.3%
Car	$100 \div 375 = 0.267$ or 26.7%
Total	$375 \div 375 = 1.00$ or 100%

   b. The following pie chart shows the results using Minitab. If you make your pie chart by hand, it should look similar, but it may not be exactly the same because it can be hard to gauge how big the slices should be.

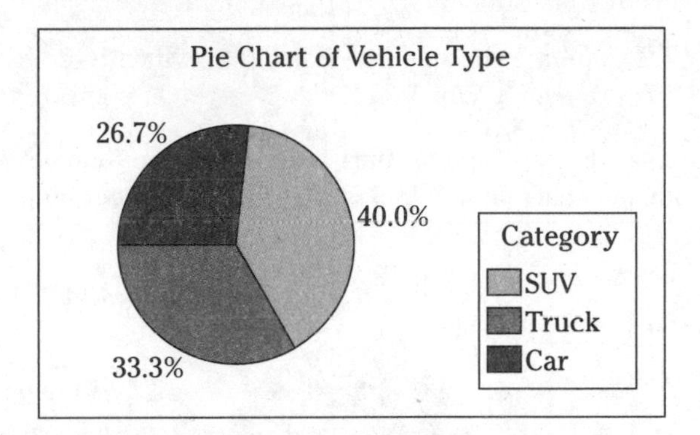

TIP

Although making pie charts by hand works fine (and on exams, you'll likely have to do so), using a computer is the easiest and most accurate way to make a pie chart. The problem is getting the size of the slices just right. You can take the percentage for the category and multiply by 360 degrees to figure out how big of an angle to make, if you remember how to do that sort of thing from trigonometry. Or you can divide the pie into quarters (25 percent each) and divide each quarter into eighths (12.5 percent each), using dotted lines, to make your best estimate from there. Remembering that half the pie chart represents 50 percent can also be helpful.

   c. From these results, you can see that 40 percent of the individuals have SUVs. Trucks and cars split up the remainder of the group, with 26.7 percent owning cars and the rest owning trucks.

2. Watch out for pie charts that have large, ambiguous "other" categories.

   a. **The pie chart shows that of the three meals, dinner brings in the most business** (44 percent for dinner compared to only 22 percent for lunch and around 7 percent for breakfast).

   b. The pie chart does have a problem. **The "other" group is very large, comprising almost 27 percent of the business, but he has no idea when those people come in, so he has little information to work from.** He may have received better results if he had broken the "other" category into more categories, such as between breakfast and lunch, between lunch and dinner, after dinner, bakery purchases on the go, and so on.

WARNING

Beware of slices of the pie labeled "other" or "miscellaneous" that become larger than many of the other slices. This discrepancy is a clue that the creator should have added more categories.

3. Pie charts do a nice job of summarizing data accurately and quickly.

a. It didn't take long for the office manager to realize that spam floods her inbox and crowds out the more important emails. The relative frequency table explains why.

Category	Relative Frequency
Highest	$60 \div 400 = 15\%$
Medium	$120 \div 400 = 30\%$
Lowest	$20 \div 400 = 5\%$
Personal	$50 \div 400 = 12.5\%$
Spam	$150 \div 400 = 37.5\%$
Total	$400 \div 400 = 100\%$

b. The corresponding pie chart is shown in the following figure using statistical software. Your pie chart should look similar if you draw it by hand.

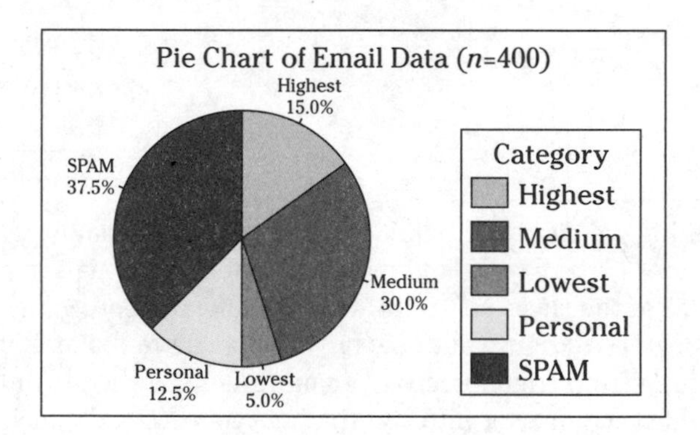

TIP

If you plan to draw a pie chart by hand, try starting with the largest slice and working your way down. Your results should look similar to charts drawn by any computer software package.

c. **The results tell the office manager that she gets a great deal of spam.** She also gets quite a bit of personal email, which she can save for breaks and lunch to maximize her work time. She also sees that 15 percent + 30 percent = 45 percent of her emails are high to medium priority, which can cause some stress.

4. **No.** The data appears to report $40 + 60 + 20 + 10 = 130$ pet owners, but the total number of people surveyed was only 100. Why? Because some folks are counted more than once if they own more than one different type of pet. The total doesn't add up to n (the sample size), and the percents don't add up to 100 if you divide each frequency by n (which is what you should do). Therefore, a pie chart doesn't work for this survey. (A bar chart is a good alternative.)

WARNING

All the percentages in a pie chart must add up to 100 percent or close to it (subject to round-off error).

5. Interpreting pie charts can seem so easy that you may be tempted to go too far at times.

   a. **The pie chart shows that 64.2 percent of the drivers who approached the intersection came to a complete stop, 35.2 percent rolled through the stop sign, and 0.6 percent (or 0.006) actually ran the stop sign.**

   b. **You have no indication of how many cars the students examined** (*n*, the sample size, isn't known). They may have seen a small number of vehicles.

   c. **Yes.** Not knowing the sample size upon which a pie chart is based can lead to imprecise or even misleading results.

   d. **No.** The data came from only one intersection on a single day for a four-hour period, so you can't make generalizations about all drivers from this very limited data set.

REMEMBER

Look for the total sample size, which is related to the precision of your results.

6. The pie charts are shown in the following figure. **Yes, gender does seem to be related to the preference to work at home (for this company).** More females at this company prefer to work at home (70 percent) than males (20 percent).

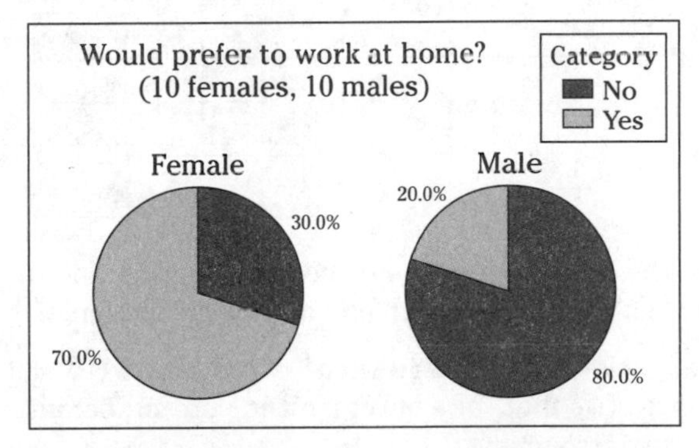

7. Any example where the percentages don't sum to 100 — where an individual or object can be in more than one group at the same time — can't be summarized accurately in a pie chart. **For example, suppose a group of adults are asked what kinds of activities they like to do on a Friday night: 41 percent say watch television, 50 percent say go to a movie, and 60 percent say go out to eat.** The surveyor doesn't ask what the people like to do best, so they can choose more than one at a time. A pie chart doesn't make sense here.

8. A missing sample size is a common error with graphs and charts.

   a. **More employers feel employee surfing compromises productivity by a 2-to-1 margin (67 percent yes to 34 percent no). But employees are equally split on the issue (about 50 percent yes to 50 percent no).** That's not surprising, is it?

   b. **The sample sizes are missing.** You don't know whether 10 people or 10,000 people responded, which affects the precision of the results. The date of the survey is also missing. (Turns out the Internet company surveyed 451 employees and 670 employers.)

A pie chart should stand alone; all the necessary information should be included and labeled within the chart.

REMEMBER

9. Relative frequencies (percents) allow you to make easy comparisons between groups.

a. See the following relative frequency table.

Category	Relative Frequency
Support smoking ban	$250 \div 500 = 50\%$
Oppose smoking ban	$125 \div 500 = 25\%$
No opinion	$125 \div 500 = 25\%$
Total	$500 \div 500 = 100\%$

b. The following figure shows the bar graph.

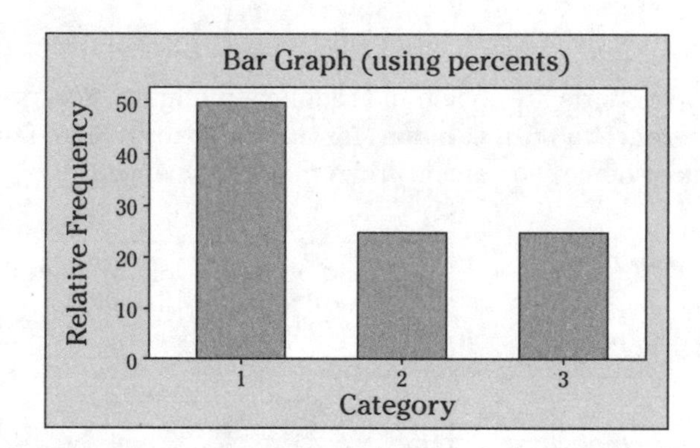

c. Half the individuals support the smoking ban, and the rest of the people are evenly split between opposing the ban and having no opinion.

10. **The bar graph shows that 10 out of 30 customers ($10 \div 30 = 33.3$ percent) rated the services as good.** Notice that the answer isn't 10 percent, because frequencies appeared on the *y*-axis, not relative frequencies.

When interpreting the results of a bar graph, be sure that you know what the graph is meant to report: counts (frequencies) or percents (relative frequencies).

REMEMBER

11. Starting point, scale, and sample size are three major ways a bar graph can mislead you.

a. See the following frequency table of the results of the opinion poll on Issue X. The frequencies come from the bar graph and represent the height of each bar. A total of 1,300 people were surveyed.

Opinion	Frequency
Yes	500
No	550
No opinion	250
Total	1,300

b. **The bar graph appears to be misleading.** Notice that the bar for the "no opinion" group is much less than 1/2 of the length of the bar for the "yes" group, because the frequency axis starts at 240 and not at 0. The following figure shows a better bar graph. The scale on this second bar graph is a little different (it uses increments of 100 rather than 30, which appear on the original bar graph). The new scale makes the graph easier to read, because it results in "nice" numbers on the tick marks like 300 and 400. The scale changes quite a bit, but the biggest problem with the original bar graph isn't the scale; it's the starting point. The total sample size is now included on the *x*-axis, making the results easier to interpret.

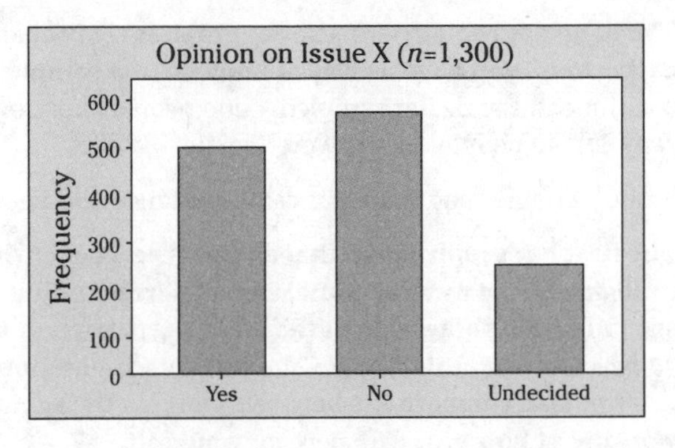

TIP

Watch the starting point on the counts/percents axis. If it doesn't start at 0, the differences in the bar lengths may appear larger than they really are.

12. Scale and ending-point errors can squeeze the space in between bars and create extra, unnecessary space in the graph. Each makes the graph misleading.

a. **The bar graph is misleading because it makes the differences in the bars look smaller than they truly are.** You can attribute this deception to two things. First, the scale on the *y*-axis is 50, which is bigger than it should be, squeezing the bars closer together in length and making differences appear relatively small. Also, the *y*-axis goes all the way to 300, when it could easily stop at half that distance. This creates a large portion of unused space in the graph, again making the lengths of the bars look smaller than they should be. The extra space also makes it appear that the sample size is smaller than it actually is (as if not very many people were in each group).

b. A fairer and more informative bar graph is shown in the following figure.

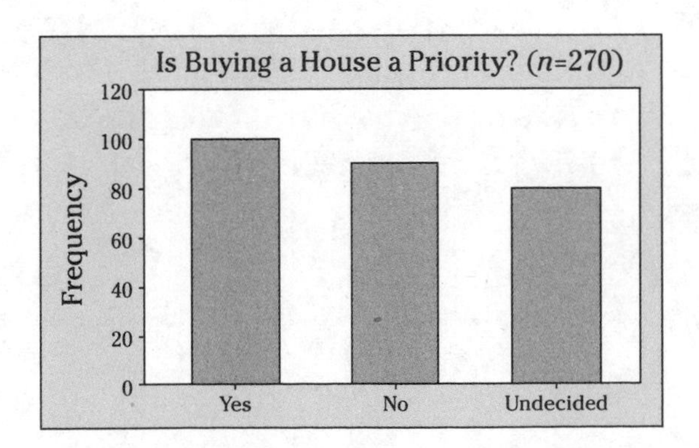

TIP

Watch the scale (size of the increments or tick marks) on the counts/percents axis. Look for scales that show differences as less or more dramatic than they should be.

(13) Although percents allow you to make relative comparisons, you still need to have the sample sizes to determine how precise the results are.

a. **The bar graph shows that about twice as many of the minivan customers were females compared to males.** The percentage of females seems to be around 67 percent (about 2/3), compared to around 33 percent males (about 1/3).

b. **The biggest problem with the bar graph is that because you see only the percentages in each group, you have no way of knowing the sample size.** Sixty-seven percent females could mean the dealer sampled 3,000 people and 2,000 were female, or it could mean they sampled 30 people and 20 were female.

(14) A second variable may be of critical importance and shouldn't be left out.

a. **The first bar graph shows that about 55 percent of the office employees of this company would prefer to work at home, and 45 percent enjoy an at-work environment. The second graph shows that of the females, 70 percent would prefer to work at home, and 30 percent wouldn't.** (Notice the percents sum to 100 for each group separately, allowing you to make comparisons between groups.) **Of the males, only 20 percent would prefer working at home, and 80 percent wouldn't.**

b. **The second bar graph is more interesting, because it shows that results differ depending on gender.**

c. **The side-by-side bars are my personal choice over the two pie charts, because the former are shown using the same scale, making it easier to visually see the differences.** Pie charts usually have different slices for all the groups, making it hard to compare them without an obvious difference.

If you're ready to test your skills a bit more, take the following chapter quiz that incorporates all the chapter topics.

Whaddya Know? Chapter 6 Quiz

Quiz time! Complete each problem to test your knowledge on the various topics covered in this chapter. You can then find the solutions and explanations in the next section.

1. A pie chart is a graphical version of what type of table: a frequency table or a relative frequency table?

2. What are three ways in which a bar graph can mislead you?

3. Why is it so important to include the sample size (*n*) somewhere with your pie chart?

4. Can you make a pie chart out of the following data table?

Favorite Car Colors (choose all that apply)	Percentage of People Choosing That Color
Red	50%
White	20%
Black	40%
Yellow	10%
Orange	20%
Blue	70%
Brown	30%
Green	25%
Purple	15%
Other	5%

5. Name two important items to add to your pie chart, other than the names of the slices and the percentage for each slice.

6. What is the impact of changing the scale on a bar graph?

7. The starting point on a bar graph can affect how the bar graph looks. How?

8. Is it wrong to have an "other" category in a pie chart? Explain your answer.

9. Three-dimensional pie charts are a great way to show categorical data. True or False?

10. You want to avoid having too many slices on your pie chart. Why is that?

Answers to Chapter 6 Quiz

1. A pie chart is a graphical version of a **relative frequency table** because both show percents and not counts.

2. **Starting point, scale, and sample size are three ways in which a bar graph can mislead you.**

3. **Because pie charts only show the percentage in each group; it doesn't tell you exactly how many are in each group.** For example, the result 3/10 is based on much fewer data than the result 30/100, even though they both come out to 0.30.

4. **No.** This is because the percents add up to way more than 1. You could make a separate pie chart for each color, showing the percentage that did vote for it versus the percentage that didn't vote for it. But you cannot make one single pie chart out of a data set whose percents sum to more than 1.

5. **Each pie chart needs a title and the sample size,** so the information can stand alone and the reader does not have to wonder what the pie chart is showing, or how much data went into it.

6. **The scale of a bar graph affects how spread out the bars look, or how close together they are.** If you have a scale in large increments, the bars will appear closer together in length or height, and if you have a scale in small increments, the bars will appear further apart in length or height.

7. Watch the starting point on the counts/percents axis. **If it doesn't start at 0, the differences in the bar lengths may appear larger than they really are.** It may appear that you are just taking the top part off the graph and showing the results. It's not always wrong to do that, but you need to watch for it, so you can put the results into proper perspective.

8. **It's certainly okay to include an "other" category in a pie chart. However, you don't want the "other" category to be so large that it eclipses other slices of the pie that are clearly defined.** You want to put all other choices into the "other" category, but you don't want to include so many in the "other" category that it gets larger than the other pie slices.

9. **False.** Three-dimensional pie charts show the pie slices in an inappropriate proportion. The slices closest to the reader appear larger than their actual size, and the slices further from the reader appear smaller than their actual size.

10. You want to avoid having too many slices on your pie chart because **the reader can get confused easily, it's hard to have that many different colors or patterns for each different slice, and some slices could turn out to be so small that you can hardly see them.** Better to keep fewer slices and have them be easy to read and distinguish.

IN THIS CHAPTER

» Making and interpreting histograms and boxplots for numerical data

» Examining time charts for numerical data collected over time

» Gaining strategies for spotting misleading and incorrect graphs

Chapter 7 Going by the Numbers: Graphing Numerical Data

The main purpose of charts and graphs is to summarize data and display the results to make your point clearly, effectively, and correctly. In this chapter, I present data displays used to summarize numerical data — data that represent counts (such as the number of pills a patient with diabetes takes per day, or the number of accidents at an intersection per year) or measurements (the time it takes you to get to work/school each day, or your blood pressure).

You see examples of how to make, interpret, and evaluate the most common data displays for numerical data: time charts, histograms, and boxplots. I also point out many potential problems that can occur in these graphs, including how people often misread what's there. This information will help you develop important detective skills for quickly spotting misleading graphs.

Handling Histograms

A histogram provides a snapshot of all the data broken down into numerically ordered groups, making it a quick way to get the big picture of the data, in particular, its general shape. In this section you find out how to make and interpret histograms, and how to critique them for correctness and fairness.

Making a histogram

A *histogram* is a special graph applied to data broken down into numerically ordered groups, for example, age groups such as 10–20, 21–30, 31–40, and so on. The bars connect to each other in a histogram — as opposed to a bar graph (see Chapter 6) for categorical data, where the bars represent categories that don't have a particular order and are separated. The height of each bar of a histogram represents either the number of individuals (called the *frequency*) in each group or the percentage of individuals (the *relative frequency*) in each group. Each individual in the data set falls into exactly one bar.

REMEMBER

You can make a histogram from any numerical data set; however, you can't determine the actual values of the data set from a histogram because all you know is which group each data value falls into.

An award-winning example

Here's an example of how to create a histogram for all you movie lovers out there (especially those who love old movies). The Academy Awards started in 1928, and one of the most popular categories for this award is Best Actress in a Motion Picture. Table 7-1 shows the winners of the first eight Best Actress Oscars, the years they won (1929–1936), their ages at the time of winning their awards, and the movies they were in. From the table, you see that the ages cover a range from 22 to 63 — much wider than you may have thought it would be.

TABLE 7-1 Ages of Best Actress Oscar Award Winners 1929–1936

Year	Winner	Age	Movie
1929	Laura Gainor	22	*Sunrise*
1930	Mary Pickford	37	*Coquette*
1931	Norma Shearer	28	*The* Divorcee
1932	Marie Dressler	63	*Min and Bill*
1933	Helen Hayes	32	*The Sin of Madelon Claudet*
1934	Katharine Hepburn	26	*Morning Glory*
1935	Collette Colbert	31	*It Happened One Night*
1936	Bette Davis	27	*Dangerous*

To find out more about the ages of the Best Actress winners, I expanded my data set to the period 1929 to 2021. The age variable for this data set is numerical, so you can graph it using a histogram. From there you can answer questions like: What do the ages of these actresses look like? Are they mostly young, old, or in between? Are their ages all spread out, or are they similar? Are most of them in a certain age range, with a few outliers (either very young or very old actresses, compared to the others)? To investigate these questions, a histogram of ages of the Best Actress winners is shown in Figure 7-1.

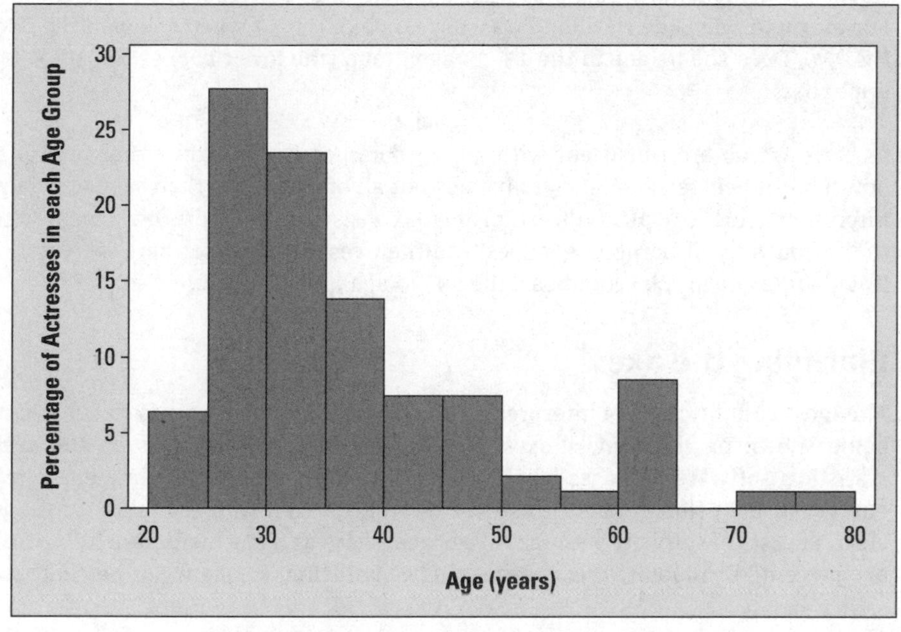

FIGURE 7-1: Histogram of Best Actress Academy Award winners' ages, 1929–2021.

Notice that the age groups are shown on the horizontal (*x*) axis. They go by groups of five years each: 20–25, 25–30, 30–35, ... , 80–85. The percentage (relative frequency) of actresses in each age group appears on the vertical (*y*) axis. For example, about 27 percent of the actresses were between 25 and 30 years of age when they won their Oscars.

Creating appropriate groups

TIP

For Figure 7-1, I used groups of five years each in the preceding example because increments of five create natural breaks for years and because this grouping provides enough bars to look for general patterns. You don't have to use this particular grouping, however; you have a bit of creative license when making a histogram. (However, this freedom also allows others to deceive you, as you see in the later section, "Detecting misleading histograms.") Here are some tips for setting up your histogram:

- Each data set requires different ranges for its groupings, but you want to avoid ranges that are too wide or too narrow.
- If a histogram has really wide ranges for its groups, it places all the data into a very small number of bars that make meaningful comparisons impossible.
- If the histogram has very narrow ranges for its groups, it looks like a big series of tiny bars that cloud the big picture. This can make the data look very choppy with no real pattern.
- Make sure your groups have equal widths. If one bar is wider than the others, it may contain more data than it should.

One idea that may be appropriate for your histogram is to take the range of the data (largest minus smallest) and divide by 10 to get 10 groupings.

Handling borderline values

In the Academy Award example, what happens if an actress's age lies right on a borderline? For example, Olivia de Havilland was 30 years old in 1947 when she won the Oscar for *To Each His Own*. Does she belong in the 25–30 age group (the lower bar) or the 30–35 age group (the upper bar)?

REMEMBER

As long as you are consistent with all the data points, you can either put all the borderline points into their respective lower bars or put all of them into their respective upper bars. The important thing is to pick a direction and be consistent. In Figure 7-1, I went with the convention of putting all borderline values into their respective lower bars — which puts Olivia de Havilland's age in the second bar, the 25–30 age group of Figure 7-1.

Clarifying the axes

The most complex part of interpreting a histogram for the reader is to get a handle on what's being shown on the *x* and *y* axes. Having good descriptive labels on the axes helps. Most statistical software packages label the *x*-axis using the variable name you provide when you enter your data (for example, "age" or "weight"). However, the label for the *y*-axis isn't as clear. Statistical software packages often label the *y*-axis of a histogram by writing "frequency" or "percent" by default. These terms can be confusing: frequency or percentage of what?

TIP

Clarify the *y*-axis label on your histogram by changing "frequency" to "number of" and adding the variable name. To modify a label that simply reads "percent," clarify by writing "percentage of" and the variable. For example, in the histogram of ages of the Best Actress winners shown in Figure 7-1, I labeled the *y*-axis "Percentage of actresses in each age group." In the next section, you see how to interpret the results from a histogram. How old are those actresses anyway?

EXAMPLE

Q. Test scores for a class of 30 students are shown in the following table.

Scores	Frequency
70–79	8
80–89	16
90–99	6

(a) Make a frequency histogram.

(b) Find the relative frequencies for each group.

(c) Without actually drawing it, how would the relative frequency histogram compare to the frequency histogram?

A. Frequency histograms and relative frequency histograms look the same; they're just done using different scales on the y-axis.

(a) The frequency histogram for the scores data is shown in the following figure.

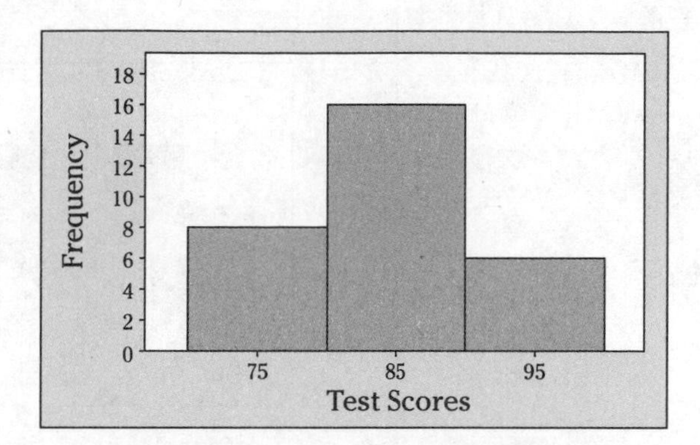

(b) You find the relative frequencies by taking each frequency and dividing by 30 (the total sample size). **The relative frequencies for these three groups are $8 \div 30 = 0.27$ or 27 percent; $16 \div 30 = 0.53$ or 53 percent; and $6 \div 30 = 0.20$ or 20 percent, respectively.**

(c) A histogram based on relative frequencies looks the same as the histogram (of the same data). The only difference is the label on the y-axis.

YOUR TURN

1 You lose information from the data when you create a histogram. What information is lost?

2 Make a histogram from this data set of test scores: 72, 79, 81, 80, 63, 62, 89, 99, 50, 78, 87, 97, 55, 69, 97, 87, 88, 99, 76, 78, 65, 77, 88, 90, and 81. Would a pie chart be appropriate for this data?

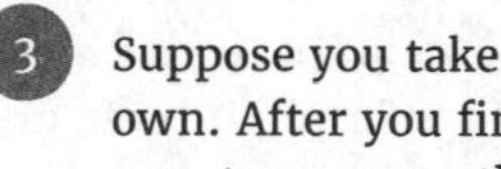

3 Suppose you take a survey of 45 homeowners to find out how many televisions they own. After you finish, you find that 2 people own no TVs, 17 people own one, 22 people own two, 3 own three, and 1 owns four. Make a relative frequency histogram of this data and interpret the results.

4 Suppose you have a loaded die. You roll it several times and record the outcomes, which are shown in the following figure.

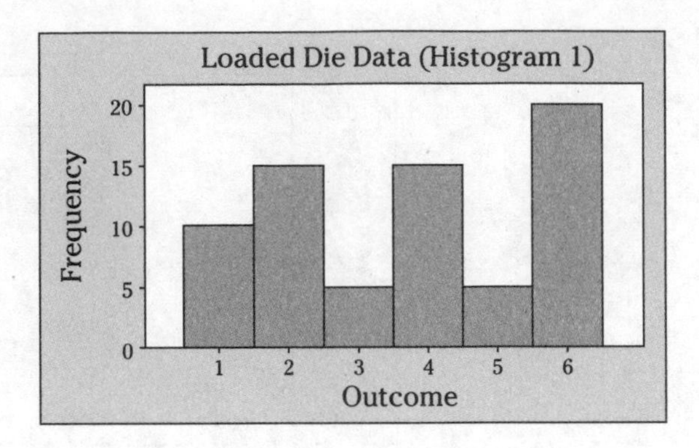

(a) Make a relative frequency histogram of these results.

(b) You can make a relative frequency histogram from a frequency histogram; can you go in the other direction?

Interpreting a histogram

A histogram tells you three main features of numerical data:

- How the data are distributed among the groups (statisticians call this the *shape* of the data)
- The amount of variability in the data (statisticians call this the amount of *spread* in the data)
- Where the center of the data is (statisticians use different measures)

Checking out the shape of the data

One of the features that a histogram can show you is the *shape* of the data — in other words, the manner in which the data fall into the groups. For example, all the data may be exactly the same, in which case the histogram is just one tall bar; or the data may have an equal number in each group, in which case the shape is flat.

Some data sets have a distinct shape. Here are three shapes that stand out.

- **Symmetric:** A histogram is symmetric if you cut it down the middle and the left-hand and right-hand sides resemble mirror images of each other.

  Figure 7-2a shows a symmetric data set; it represents the amount of time each of 50 survey participants took to fill out a certain survey. You see that the histogram is close to symmetric.

- **Skewed right:** A skewed-right histogram looks like a lopsided mound, with a tail going off to the right.

  Figure 7-1 from earlier in the chapter, showing the ages of the Best Actress Award winners, is skewed right. You see on the right side that there are a few actresses whose ages are older than the rest.

- **Skewed left:** If a histogram is skewed left, it looks like a lopsided mound with a tail going off to the left.

  Figure 7-2b shows a histogram of 17 exam scores. The shape is skewed left; you see a few students who scored lower than everyone else.

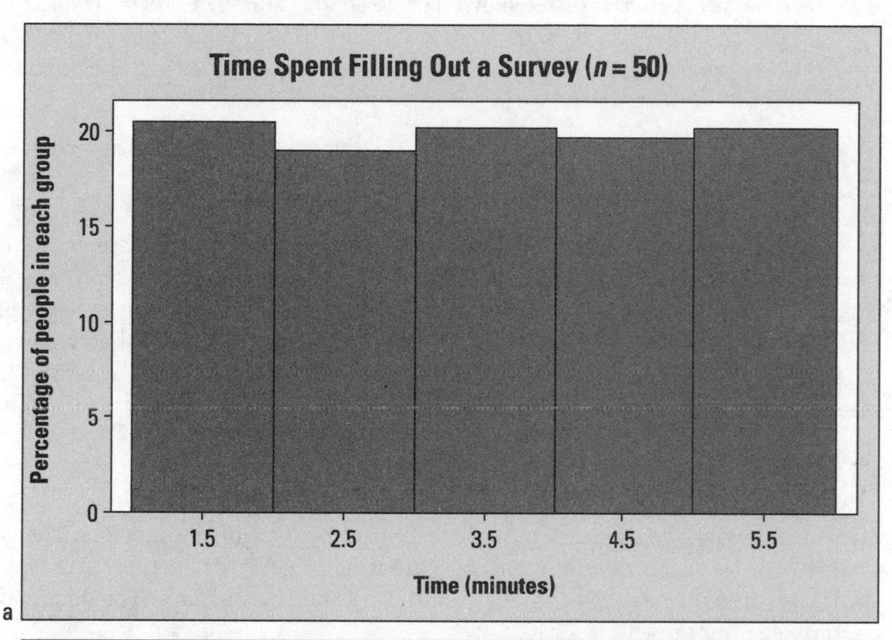

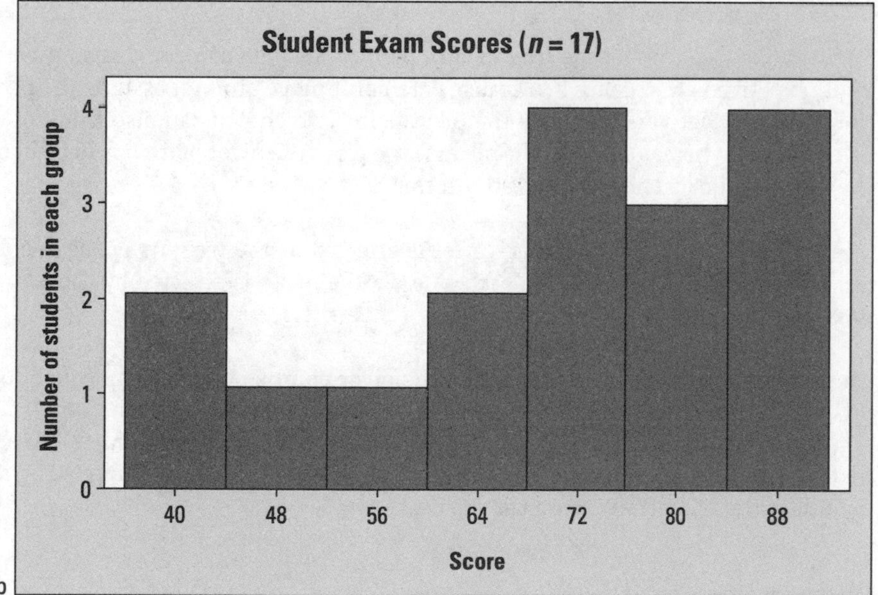

FIGURE 7-2: Comparing the shape of a) a symmetric histogram and b) a skewed-left histogram.

Following are some particulars about classifying the shape of a data set:

- **Don't expect symmetric data to have an exact and perfect shape.** Data hardly ever fall into perfect patterns, so you have to decide whether the data shape is close enough to be called symmetric.

  If the shape is close enough to symmetric that another person would notice it, and the differences aren't enough to write home about, I'd classify it as symmetric or roughly symmetric. Otherwise, you classify the data as nonsymmetric. (More sophisticated statistical procedures exist that actually test data for symmetry, but they're beyond the scope of this book.)

- **Don't assume that data are skewed if the shape is nonsymmetric.** Data sets come in all shapes and sizes, and many of them don't have a distinct shape at all. I include skewness on the list here because it's one of the more common nonsymmetric shapes, and it's one of the shapes included in a standard introductory statistics course.

  If a data set does turn out to be skewed (or close to it), make sure to denote the direction of the skewness (left or right).

As you can see in Figure 7-1, the actresses' ages in the histogram are skewed right. Most of the actresses were between 20 and 50 years of age when they won, with about 27 percent of them between the ages of 25 and 30. A few actresses were older when they won their Oscars; about 8 percent were between 60 and 65 years of age, and not more than 2 percent (total) were over 70 (if you add the percentages from the last two bars in the histogram). The last few bars on the right side are what give the data a shape that is skewed right.

Measuring center: Mean versus median

A histogram gives you a rough idea of where the "center" of the data lies. The word *center* is in quotes because many different statistics are used to designate center. The two most common measures of center are the average (the mean) and the median. (For details on measures of center, see Chapter 5.)

TIP

To visualize the average age (the mean), picture the data as people sitting on a teeter-totter. Your objective is to balance it. Because data don't move around, assume the people stay where they are and that you can move the pivot point (which you can also think of as the hinge or fulcrum) anywhere you want. The mean is the place the pivot point has to be in order to balance the weight on each side of the teeter-totter.

The balancing point of the teeter-totter is affected by the weights of the people on each side, not by the number of people on each side. So the mean is affected by the actual values of the data, rather than the amount of data.

The median is the place where you put the pivot point so you have an equal number of people on each side of the teeter-totter, regardless of their weights. With the same number of people on each side, the teeter-totter wouldn't balance in terms of weight unless the teeter-totter had people with the same total weight on each side. So the median isn't affected by the values of the data, just their location within the data set.

REMEMBER

The mean is affected by *outliers*, values in the data set that are away from the rest of the data, on the high end and/or the low end. The median, being the middle number, is not affected by outliers.

Viewing variability: Amount of spread around the mean

You also get a sense of variability in the data by looking at a histogram. For example, if the data are all the same, they are all placed into a single bar, and there is no variability. If an equal amount of data is in each group, the histogram looks flat, with the bars close to the same height; this shows a fair amount of variability.

WARNING

The idea of a flat histogram indicating some variability may go against your intuition, and if it does, you're not alone. If you're thinking a flat histogram means no variability, you're probably thinking about a time chart, where single numbers are plotted over time (see the section, "Tackling Time Charts," later in this chapter). Remember, though, that a histogram doesn't show data over time — it shows all the data at one point in time.

Equally confusing is the idea that a histogram with a big lump in the middle and tails sloping sharply down on each side actually has less variability than a histogram that's straight across. The curves looking like hills in a histogram represent clumps of data that are close together; a flat histogram shows data equally dispersed, with more variability.

REMEMBER

Variability in a histogram is higher when the taller bars are more spread out around the mean and lower when the taller bars are close to the mean.

For the Best Actress Award winners' ages shown earlier in Figure 7-1, you see many actresses are in the age range from 25 to 30, and most of the ages are between 20 and 50 years, which is quite diverse. Then you have some outliers, those few older actresses (two of them include Jessica Tandy for *Driving Miss Daisy* and Katharine Hepburn for *On Golden Pond*); their ages spread the data out farther, increasing its overall variability.

The most common statistic used to measure variability in a data set is the *standard deviation*, which, in a rough sense, measures the average distance that the data lie from the mean. The standard deviation for the Best Actress Award age data is 12.23 years. (See Chapter 5 for all the details on standard deviation.) A standard deviation of 12.23 years is fairly large in the context of this problem, but the standard deviation is based on average distance from the mean, and the mean is influenced by outliers, so the standard deviation will be as well (see Chapter 5 for more information).

In the later section, "Interpreting a boxplot," I discuss another measure of variability, called the *interquartile range* (IQR), which is a more appropriate measure of variability when you have skewed data.

EXAMPLE

Q. The police checked the speeds of cars after the city painted lines on a certain section of a street where the road narrows. The speeds are organized in the histogram shown in the following figure. Describe what this histogram tells you about the speeds of the cars.

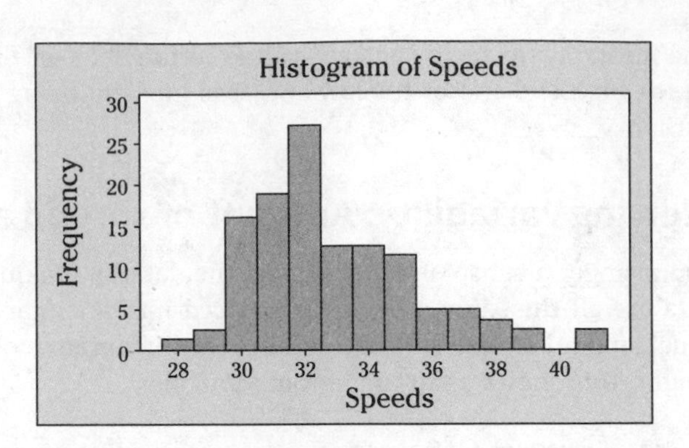

A. The speeds of the cars in this data set range from 28 miles per hour (lowest) to 41 mph (highest). Most of the cars traveled from 30 to 35 mph (you can tell by noting that those bars have the highest frequencies). A few cars drove faster than the rest (noted by the few short bars at the upper end), which indicates a skewed-right shape. The average speed seems to center around 32 mph (but this is hard to tell without doing more analysis).

5 An ATM machine asks customers who use the "fast cash" option to choose an amount in $50 increments from $100 to $500. Results from a recent sample of customer withdrawals are shown in the following figure. Discuss the shape, center, and spread of the data.

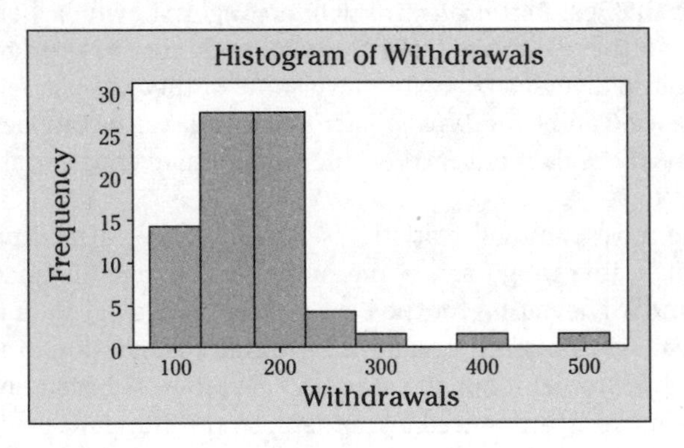

6 A histogram of a sample of rods made by Rowdy Rod's is shown in the following figure. The rods should be 100 inches in length. Discuss the company's accuracy (in terms of meeting the length specification) by interpreting the shape, center, and spread of the data.

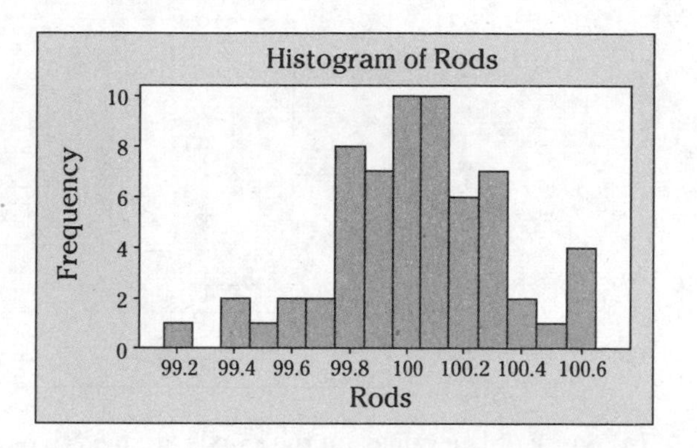

7 A histogram of the amount of money 317 households spent on fruits and vegetables in a year is shown in the following figure (based on a random sample). Discuss the shape, center, and spread. What do these three characteristics say about how much families spend on fruits and vegetables? (Such analysis is called interpreting your results in the context of the problem.)

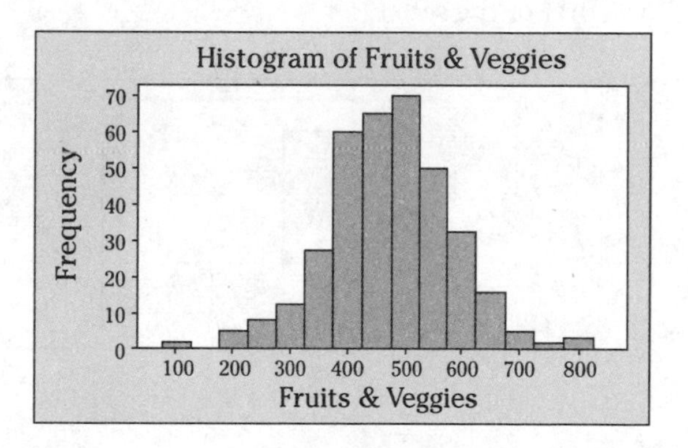

8 An investor monitors the percentage return for a particular group of stocks in Portfolio A over a one-year period. The one-year percentage returns for these stocks are shown in the following figure.

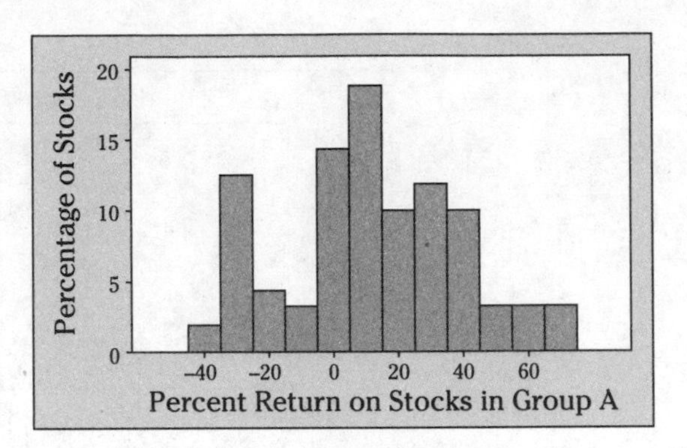

(a) Some of the values on the *x*-axis are negative numbers. What does this mean?

(b) On any histogram in general, when (if ever) can the *x*- and/or *y*-axis contain negative values?

9 You make two histograms from two different data sets (see the following figures), with each one containing 200 observations. Which of the histograms has a smaller spread, the first or the second?

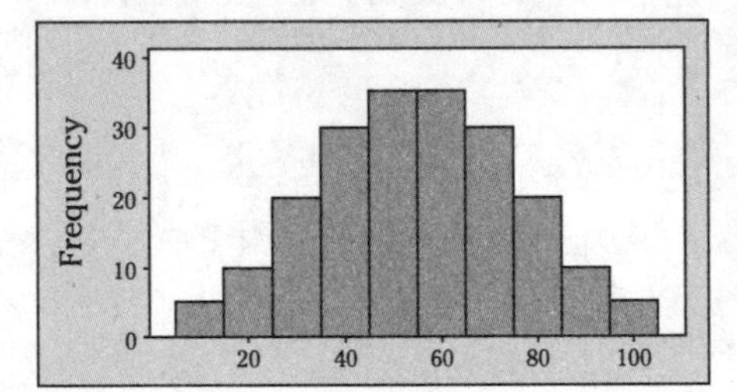

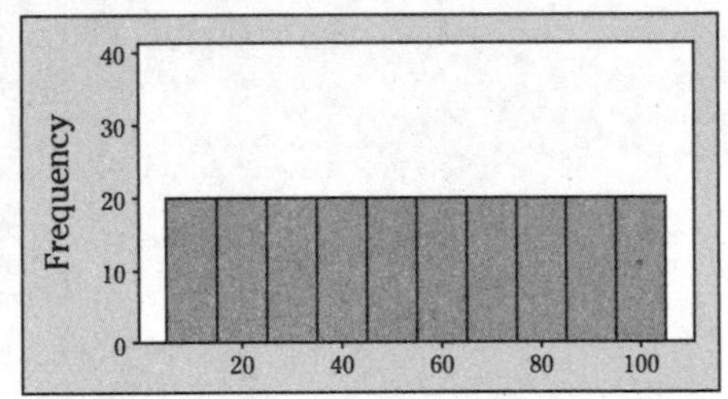

Putting numbers with pictures

You can't actually calculate measures of center and variability from the histogram itself because you don't know the exact data values. To add detail to your findings, you should always calculate the basic statistics of center and variation along with your histogram. (All the descriptive statistics you need, and then some, appear in Chapter 5.)

Figure 7-1 is a histogram for the Best Actress ages; you can see it is skewed right. For Figure 7-3, I calculate some basic (that is, descriptive) statistics from the data set. Examining these numbers, you find the median age is 33.00 years and the mean age is 36.78 years.

FIGURE 7-3: Descriptive statistics for Best Actress ages (1929–2021).

Variable	N	Mean	StDev	Minimum	Q1	Median	Q3	Maximum	IQR
Age	94	36.78	12.23	21.00	28.00	33.00	41.00	80.00	13.00

The mean age is higher than the median age because of a few actresses who were quite a bit older than the rest when they won their awards. For example, Jessica Tandy won for her role in *Driving Miss Daisy* when she was 80, and Katharine Hepburn won the Oscar for *On Golden Pond* when she was 74. The relationship between the median and mean confirms the skewness (to the right) found in Figure 7-1.

Here are some tips for connecting the shape of the histogram (discussed in the previous section) with the mean and median:

» If the histogram is skewed right, the mean is greater than the median.

  This is the case because skewed-right data have a few large values that drive the mean upward but do not affect where the exact middle of the data is (that is, the median). Looking at the histogram of ages of the Best Actress Award winners in Figure 7-1, you see they're skewed right.

» If the histogram is close to symmetric, then the mean and median are close to each other.

  Close to symmetric means it's almost the same on either side; it doesn't need to be exact. *Close* is defined in the context of the data; for example, the numbers 50 and 55 are said to be close if all the values lie between 0 and 1,000, but they are considered to be farther apart if all the values lie between 49 and 56.

  The histogram shown in Figure 7-2a is close to symmetric. Its mean and median are both equal to 3.5.

» If the histogram is skewed left, the mean is less than the median.

  This is the case because skewed-left data have a few small values that drive the mean downward but do not affect where the exact middle of the data is (that is, the median).

Figure 7-2b represents the exam scores of 17 students, and the data are skewed left. I calculated the mean and median of the original data set to be 70.41 and 74.00, respectively. The mean is lower than the median due to a few students who scored quite a bit lower than the others. These findings match the general shape of the histogram shown in Figure 7-2b.

The tips for interpreting histograms found in the previous section can also be used the other way around. If for some reason you don't have a histogram of the data, and you only have the mean and median to go by, you can compare them to each other to get a rough idea as to the shape of the data set.

- If the mean is much larger than the median, the data are generally skewed right; a few values are larger than the rest.
- If the mean is much smaller than the median, the data are generally skewed left; a few smaller values bring the mean down.
- If the mean and median are close, you know the data is fairly balanced, or symmetric, on each side.

Under certain conditions, you can put together the mean and standard deviation to describe a data set in quite a bit of detail. If the data have a normal distribution (a bell-shaped hill in the middle, sloping down at the same rate on each side; see Chapter 5), the Empirical Rule can be applied.

The Empirical Rule (also in Chapter 5) says that if the data have a normal distribution, about 68 percent of the data lie within 1 standard deviation of the mean, about 95 percent of the data lie within 2 standard deviations of the mean, and about 99.7 percent of the data lie within 3 standard deviations of the mean. These percentages are custom-made for the normal distribution (bell-shaped data) only and can't be used for data sets of other shapes.

Q. The police checked the speeds of cars after the city painted lines on a certain section of a street where the road narrows. The speeds are organized in the histogram shown in the following figure. Which is greater: the mean of this data or the median of this data?

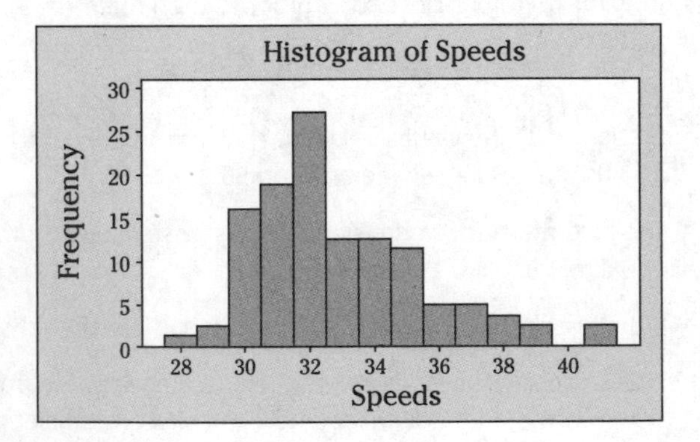

A. **The mean is greater.** The data is skewed to the right because there a few more outlying large values than there are smaller values. So the mean gets pulled out toward the large values to keep that seesaw-type balance.

YOUR TURN

10 The incomes of last year's new graduates of a certain large and very successful program are shown in the following figure.

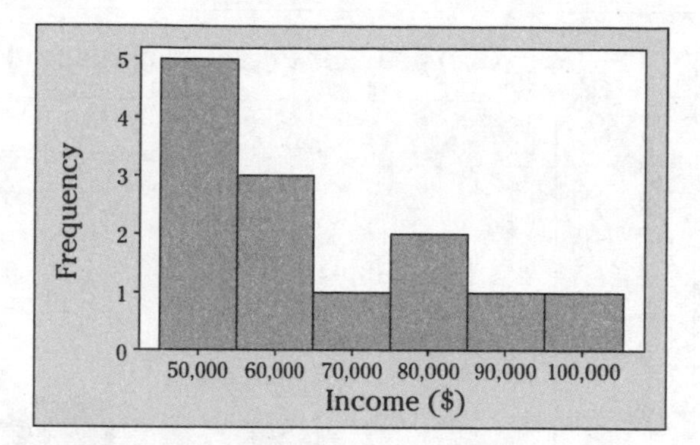

(a) Discuss the implications for graduates of this program.

(b) Estimate where the median salary is in this data set.

(c) Do you see any issues that anyone who tries to interpret this data should take into account?

Detecting misleading histograms

There are no hard and fast rules for how to create a histogram; the person making the graph gets to choose the groupings on the x-axis, as well as the scale and starting and ending points on the y-axis. Just because there is an element of choice, however, doesn't mean every choice is appropriate; in fact, a histogram can be made to be misleading in many ways. In the following sections, you see examples of misleading histograms and how to spot them.

Missing the mark with too few groups

Although the number of groups used for a histogram is up to the discretion of the person making the graph, there is such a thing as going overboard, either by having way too few bars, with everything lumped together, or by having way too many bars, where every little difference is magnified.

TIP

To decide how many bars a histogram should have, take a good look at the groupings used to form the bars on the x-axis and see whether they make sense. For example, it doesn't make sense to talk about exam scores in groups of 2 points; that's too much detail — too many bars. On the other hand, it doesn't make sense to group actresses' ages by intervals of 20 years; that's not descriptive enough.

Figures 7-4 and 7-5 illustrate this point. Each histogram summarizes $n = 222$ observations of the amount of time between eruptions of the Old Faithful geyser in Yellowstone Park. Figure 7-4 uses six bars that group the data by ten-minute intervals. This histogram shows a general skewed-left pattern, but with 222 observations, you are cramming an awful lot of data into only six groups; for example, the bar for 75–85 minutes has more than 90 pieces of data in it. You can break it down further than that.

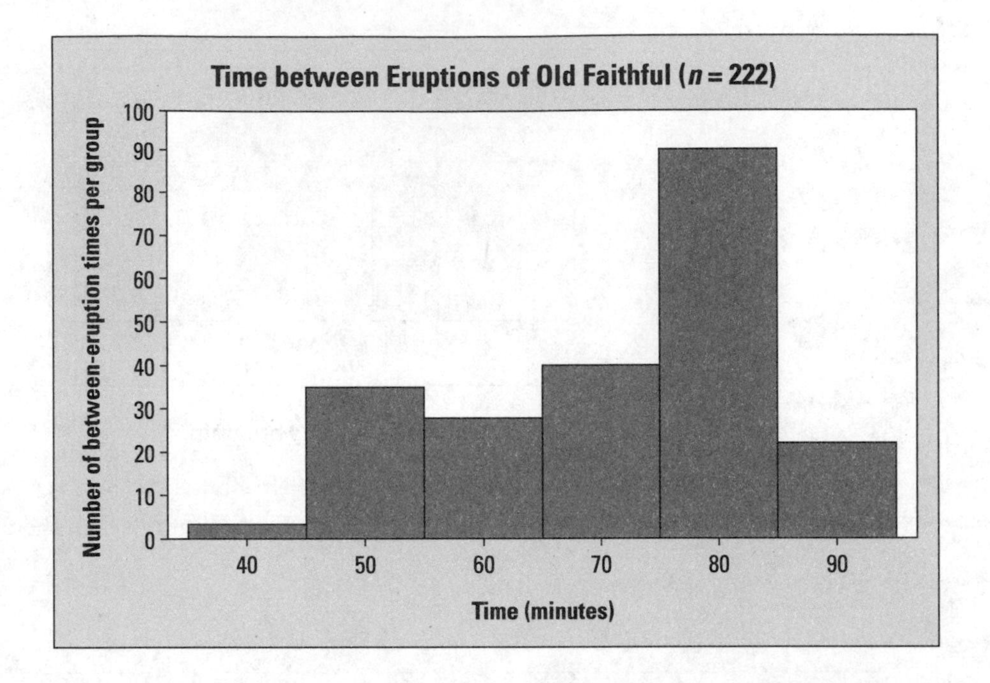

FIGURE 7-4: Histogram #1 showing time between eruptions for Old Faithful geyser ($n = 222$).

Figure 7-5 is a histogram of the same data set, where the time between eruptions is broken into groups of three minutes each, resulting in 19 bars. Notice the distinct pattern in the data that shows up with this histogram, which wasn't uncovered in Figure 7-4. You see two distinct peaks in the data: one peak around the 50-minute mark, and one around the 75-minute mark. A data set with two peaks is called *bimodal*; Figure 7-5 shows a clear example.

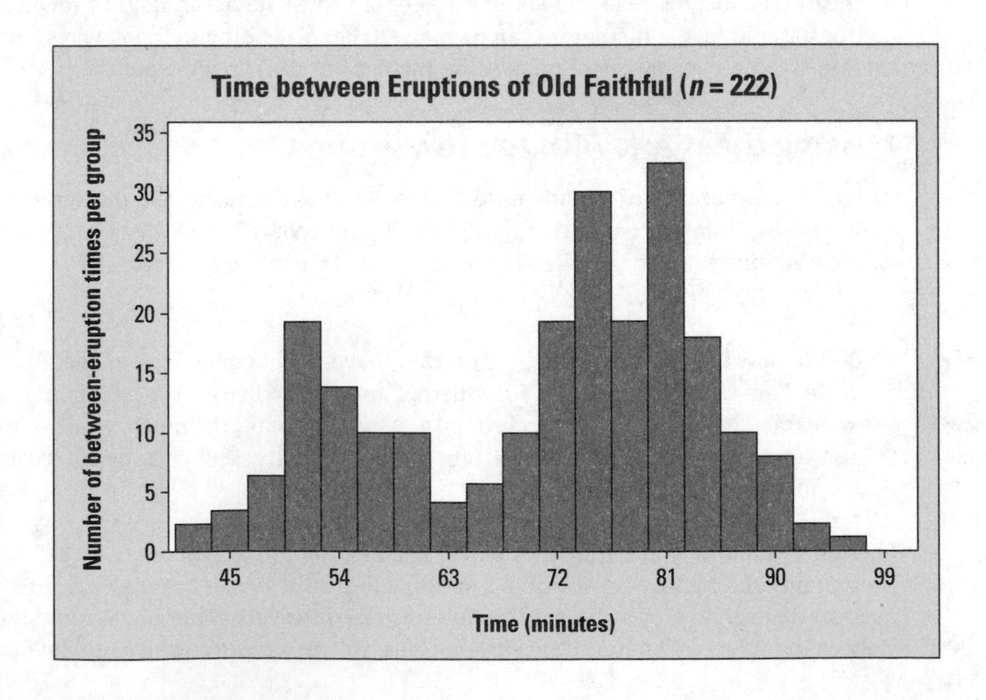

FIGURE 7-5: Histogram #2 showing time between eruptions for Old Faithful geyser ($n = 222$).

Looking at Figure 7-5, you can conclude that the geyser has two categories of eruptions: one group that has a shorter waiting time, and another group that has a longer waiting time. Within each group, you see the data are fairly close to where the peak is located. Looking at Figure 7-4, you couldn't say that.

REMEMBER

If the interval for the groupings of the numerical variable is really small, you see too many bars in the histogram; the data may be hard to interpret because the heights of the bars look more variable than they should. On the other hand, if the ranges are really large, you see too few bars, and you may miss something interesting in the data.

Watching the scale and start/finish lines

The *y*-axis of a histogram shows how many individuals are in each group, using counts or percents. A histogram can be misleading if it has a deceptive scale and/or inappropriate starting and ending points on the *y*-axis.

REMEMBER

Watch the scale on the *y*-axis of a histogram. If it goes by large increments and has an ending point that's much higher than needed, you see a great deal of white space above the histogram. The heights of the bars are squeezed down, making their differences look more uniform than they should. If the scale goes by small increments and ends at the smallest value possible, the bars become stretched vertically, exaggerating the differences in their heights and suggesting a bigger difference than really exists.

An example comparing scales on the vertical (*y*) axes is shown in Figures 7-5 and 7-6. I took the Old Faithful data (time between eruptions) and made a histogram with vertical increments of 20 minutes, from 0 to 100; see Figure 7-6. Compare this to Figure 7-5, with vertical increments of five minutes, from 0 to 35. Figure 7-6 has a lot of white space and gives the appearance that the times are more evenly distributed among the groups than they really are. It also makes the data set look smaller, if you don't pay attention to what's on the *y*-axis. Of the two graphs, Figure 7-5 is more appropriate.

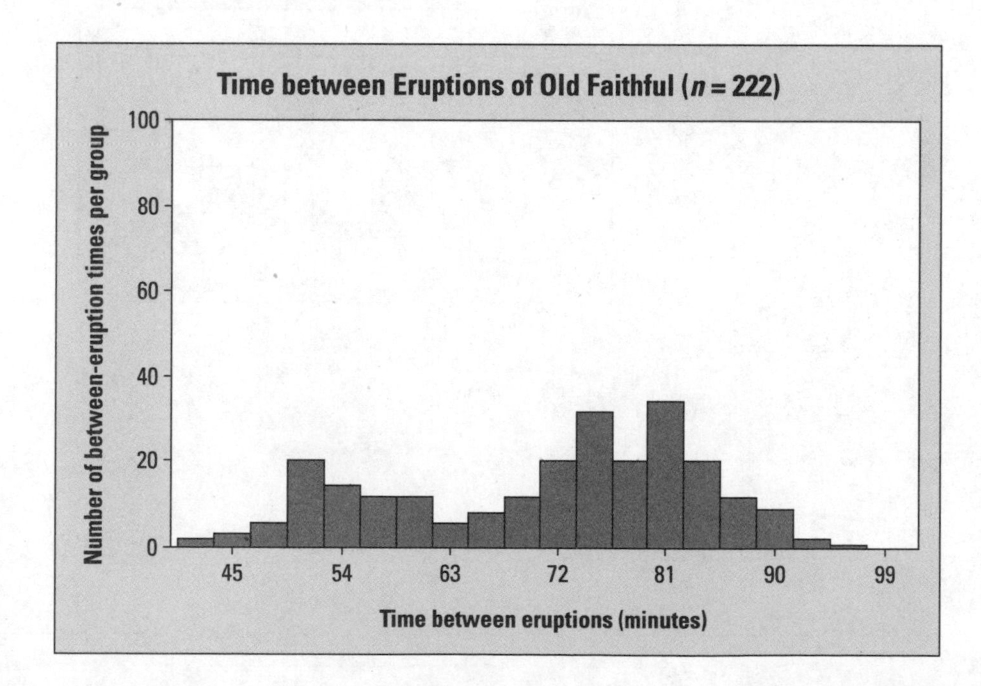

FIGURE 7-6: Histogram #3 of Old Faithful geyser eruption times.

EXAMPLE

Q. In an earlier practice question, you see data from a die that's clearly loaded (not fair). However, someone (the gambler's lawyer, perhaps?) could make that same die appear fair by setting up the histogram a certain way. Explain how the following histogram (made from that same data) makes the die appear to be fair.

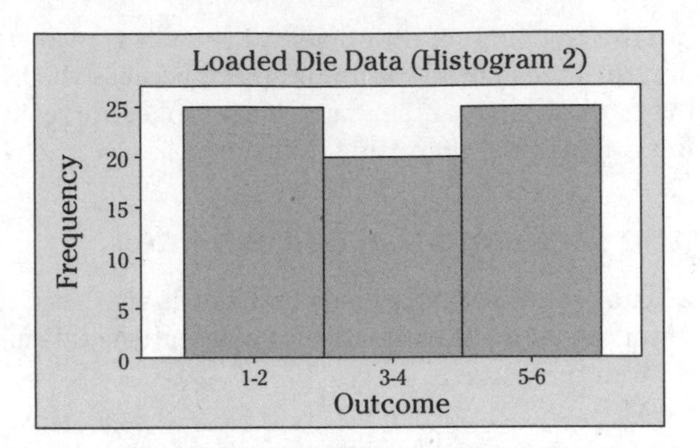

A. This histogram combines the results of rolling a 1 and 2, rolling a 3 and 4, and rolling a 5 and 6, so there are three bars on the histogram now, not six. This histogram is misleading because when you combine the results into three groups of two outcomes each, the differences that make the die loaded don't show up. The lack of precision works to the advantage of the person who created the graph.

YOUR TURN

11 Suppose your friend believes their gambling partner plays with a loaded die (not fair). They show you a graph of the outcomes of the games played with this die (see the following figure). Based on this graph, do you agree with your friend? Why or why not?

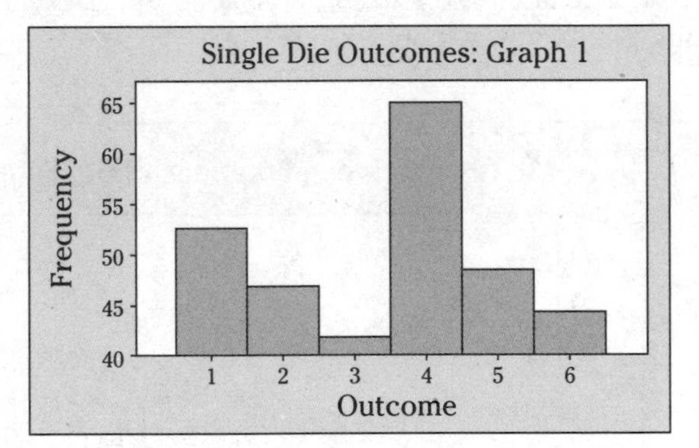

12 The first month's telephone bills for new customers of a certain phone company are shown in the following figure. The histogram showing the bills is misleading, however. Explain why, and suggest a solution.

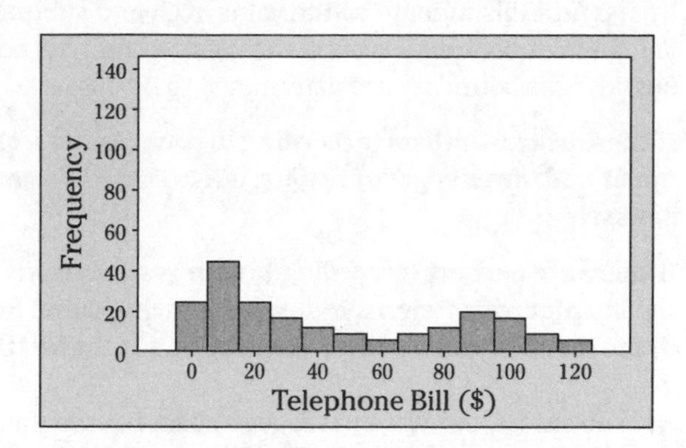

Examining Boxplots

A *boxplot* is a one-dimensional graph of numerical data based on the five-number summary, which includes the minimum value, the 25th percentile (known as Q_1), the median, the 75th percentile (Q_3), and the maximum value. In essence, these five descriptive statistics divide the data set into four parts; each part contains 25 percent of the data. (See Chapter 5 for a full discussion of the five-number summary.)

Making a boxplot

To make a boxplot, follow these steps:

1. **Find the five-number summary of your data set. (Use the steps outlined in Chapter 5.)**
2. **Create a vertical (or horizontal) number line whose scale includes the numbers in the five-number summary and uses appropriate units of equal distance from each other.**
3. **Mark the location of each number in the five-number summary just above the number line (for a horizontal boxplot) or just to the right of the number line (for a vertical boxplot).**
4. **Draw a box around the marks for the 25th percentile and the 75th percentile.**

5. **Draw a line in the box where the median is located.**
6. **Determine whether or not outliers are present.**

   To make this determination, calculate the IQR (by subtracting $Q_3 - Q_1$); then multiply by 1.5. Add this amount to the value of Q_3 and subtract this amount from Q_1. This gives you a wider boundary around the median than the box does. Any data points that fall outside this boundary are determined to be outliers.

7. **If there are no outliers (according to your results from Step 6), draw lines from the upper and lower edges of the box out to the minimum and maximum values in the data set.**
8. **If there are outliers (according to your results from Step 6), indicate their location on the boxplot with * signs. Instead of drawing a line from the edge of the box all the way to the most extreme outlier, stop the line at the last data value that isn't an outlier.**

TIP

Many, if not most, software packages indicate outliers in a data set by using an asterisk (*) or star symbol and use the procedure outlined in Step 6 to identify outliers. However, not all packages use these symbols and procedures; check to see what your package does before analyzing your data with a boxplot.

A vertical boxplot for ages of the Best Actress Academy Award winners from 1929 to 2021 is shown in Figure 7-7. You can see that the numbers separating sections of the boxplot match the five-number summary statistics, as shown previously in Figure 7-3.

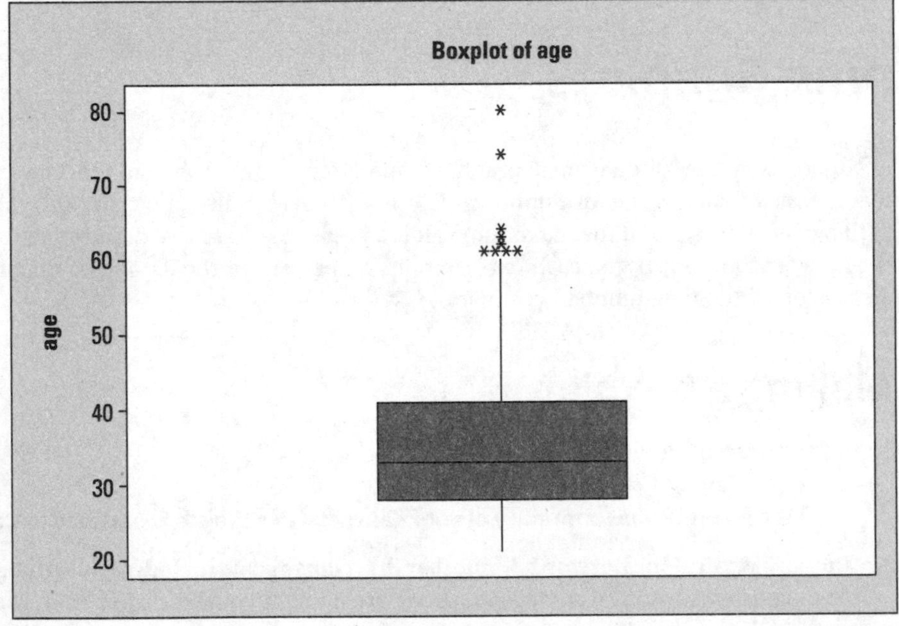

FIGURE 7-7: Boxplot of Best Actress ages (1929–2021; $n = 93$ awards).

REMEMBER

Boxplots can be vertical (straight up and down) with the values on the axis going from bottom (lowest) to top (highest); or they can be horizontal, with the values on the axis going from left (lowest) to right (highest). The next section shows you how to interpret a boxplot.

EXAMPLE

Q. Make a vertical boxplot from this data set of exam scores: 43, 54, 56, 61, 62, 66, 68, 69, 69, 70, 71, 72, 77, 78, 79, 85, 87, 88, 89, 93, 95, 96, 98, 99, 99.

A. From an example question in Chapter 5, you found the five-number summary for this data set to be 43, 68, 77, 89, and 99, which completes Step 1. Steps 2 and 3 say to draw a vertical number line (the y-axis) that spans 43 to 99 and mark the five-number summary values. In Step 4 you draw a box with 68 and 89 marking the edges, and in Step 5 you draw a line for the median at 77.

To start Step 6, you calculate $1.5 \times \text{IQR} = 1.5 \times (89 - 68) = 1.5 \times 21 = 31.5$. Any large outliers would have to be greater than $Q_3 + 31.5 = 89 + 34.5 = 120.5$, and any small outliers would have to be less than $Q_1 - 31.5 = 68 - 31.5 = 36.5$. Because none of the exam scores fall outside of those values, there are no outliers. According to Step 7, you draw the remaining lines from the quartiles to the maximum (99) and minimum (43). See the completed boxplot in the following picture.

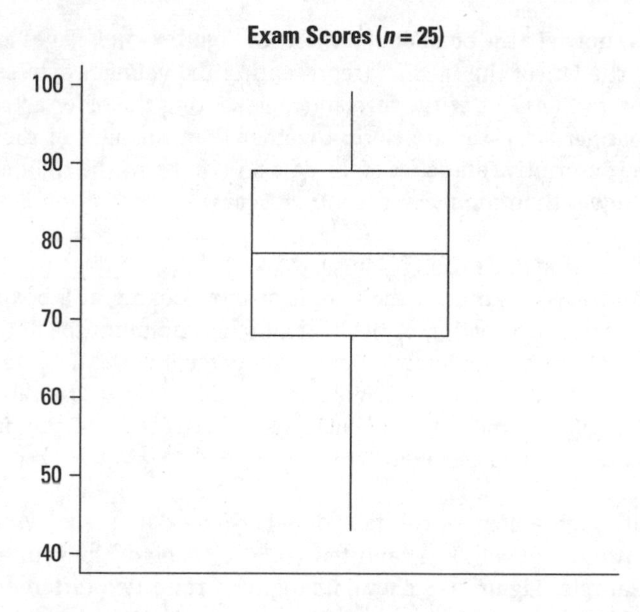

YOUR TURN

13 Suppose you have measured the height in inches of 11 randomly selected adults, and in order from lowest to highest they look like this: 59, 61, 65, 66, 67, 67, 69, 70, 72, 73, 75. Make a horizontal boxplot of these heights.

14 Suppose you have surveyed 14 randomly selected adults for their typical commute times to work, and the responses look like this: 16, 8, 35, 17, 13, 15, 15, 5, 16, 25, 20, 20, 12, 10. Make a vertical boxplot of these times.

Interpreting a boxplot

Similar to a histogram (see the section, "Interpreting a histogram"), a boxplot can give you information regarding the shape, center, and variability of a data set. Boxplots differ from histograms in terms of their strengths and weaknesses, as you see in the upcoming sections, but one of their biggest strengths is how they handle skewed data.

Checking the shape with caution!

A boxplot can show whether a data set is symmetric (roughly the same on each side when cut down the middle) or skewed (lopsided). A symmetric data set shows the median roughly in the middle of the box. Skewed data show a lopsided boxplot, where the median cuts the box into two unequal pieces. If the longer part of the box is to the right of (or above) the median, the data is said to be *skewed right*. If the longer part is to the left of (or below) the median, the data is *skewed left*.

As shown in the boxplot of the data in Figure 7-7, the ages are skewed right. The part of the box to the left of the median (representing the younger actresses) is shorter than the part of the box to the right of the median (representing the older actresses). That means the ages of the younger actresses are closer together than the ages of the older actresses. Figure 7-3 shows the descriptive statistics of the data and confirms the right skewness: the median age (33 years) is lower than the mean age (36.78 years).

REMEMBER

If one side of the box is longer than the other, it does not mean that side contains more data. In fact, you can't tell the sample size by looking at a boxplot; it's based on percentages, not counts. Each section of the boxplot (the minimum to Q_1, Q_1 to the median, the median to Q_3, and Q_3 to the maximum) contains 25 percent of the data no matter what. If one of the sections is longer than another, it indicates a wider range in the values of data in that section (meaning the data are more spread out). A smaller section of the boxplot indicates the data are more condensed (closer together).

WARNING

Although a boxplot can tell you whether a data set is symmetric (when the median is in the center of the box), it can't tell you the shape of the symmetry the way a histogram can. For example, Figure 7-8 shows histograms from two different data sets, each one containing 18 values that vary from 1 to 6. The histogram on the left has an equal number of values in each group, and the one on the right has two peaks at 2 and 5. Both histograms show the data are symmetric, but their shapes are clearly different.

Figure 7-9 shows the corresponding boxplots for these same two data sets; notice they are exactly the same. This is because the data sets both have the same five-number summaries — they're both symmetric with the same amount of distance between Q_1, the median, and Q_3. However, if you just saw the boxplots and not the histograms, you might think the shapes of the two data sets are the same, when indeed they are not.

Despite its weakness in detecting the type of symmetry (you can add a histogram to your analyses to help fill in that gap), a boxplot has a great upside in that you can identify actual measures of spread and center directly from the boxplot, where on a histogram you can't. A boxplot is also good for comparing data sets by showing them on the same graph, side by side.

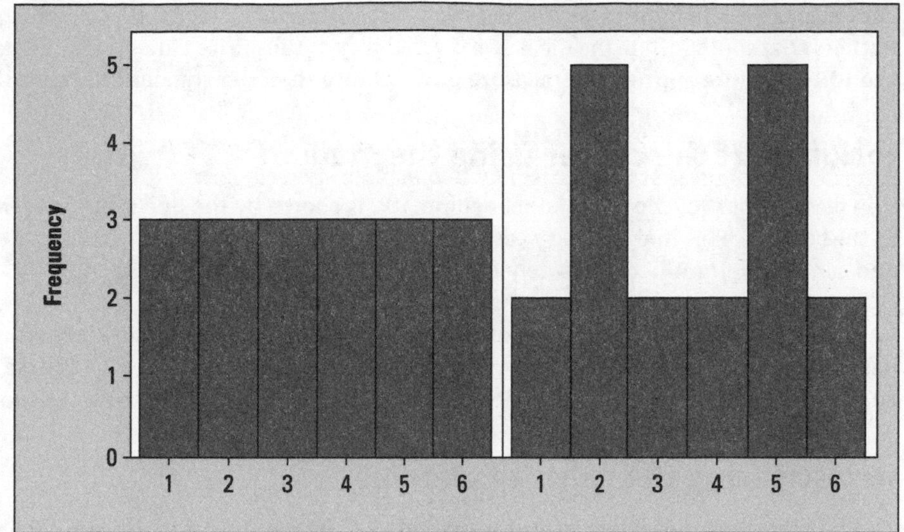

FIGURE 7-8: Histograms of two symmetric data sets.

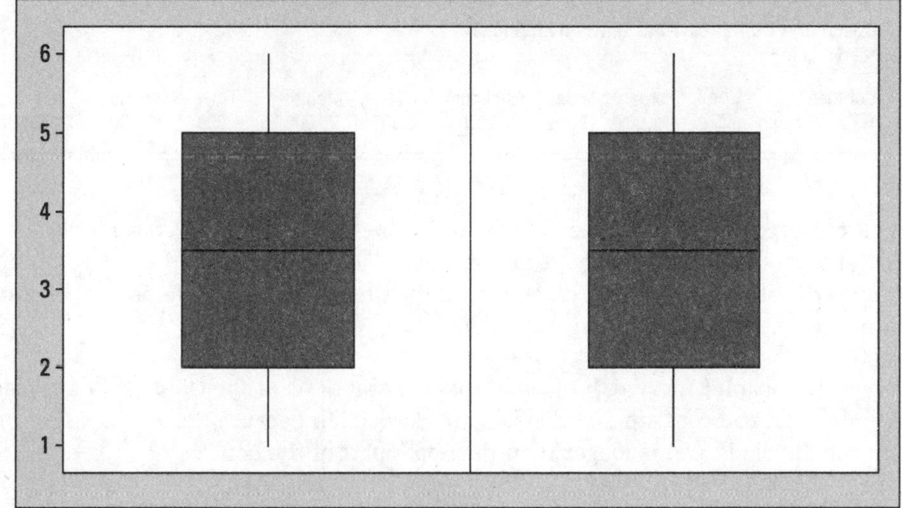

FIGURE 7-9: Boxplots of the two symmetric data sets from Figure 7-8.

TIP

All graphs have strengths and weaknesses; it's always a good idea to show more than one graph of your data for that reason.

Measuring variability with IQR

Variability in a data set that is described by the five-number summary is measured by the interquartile range (IQR). The IQR is equal to $Q_3 - Q_1$, the difference between the 75th percentile and the 25th percentile (the distance covering the middle 50 percent of the data). The larger the IQR, the more variable the data set is.

From Figure 7-3, the variability in age of the Best Actress winners as measured by the IQR is $Q_3 - Q_1 = 41 - 28 = 13$ years. Of the group of actresses whose ages were closest to the median, half of them were within 13 years of each other when they won their awards.

TIP

Notice that the IQR ignores data below the 25th percentile or above the 75th, which may contain outliers that could inflate the measure of variability of the entire data set. So if data are skewed, the IQR is a more appropriate measure of variability than the standard deviation.

Picking out the center using the median

The median, part of the five-number summary, is shown by the line that cuts through the box in the boxplot. This makes it very easy to identify. The mean, however, is not part of the boxplot and can't be determined accurately by just looking at the boxplot.

You don't see the mean on a boxplot because boxplots are based completely on percentiles. If data are skewed, the median is the most appropriate measure of center. Of course, you can calculate the mean separately and add it to your results; it's never a bad idea to show both.

Investigating Old Faithful's boxplot

The relevant descriptive statistics for the Old Faithful geyser data are found in Figure 7-10.

Descriptive Statistics: Time between Eruptions

Variable	Total Count	Mean	StDev	Minimum	Q1	Median	Q3	Maximum	IQR
Time between	222	71.009	12.799	42.000	60.000	75.000	81.000	95.000	21.000

FIGURE 7-10: Descriptive statistics for Old Faithful data.

You can predict from the data set that the shape will be skewed left a bit because the mean is lower than the median by about four minutes. The IQR is $Q_3 - Q_1 = 81 - 60 = 21$ minutes, which shows the amount of overall variability in the time between eruptions; 50 percent of the eruptions are within 21 minutes of each other.

A vertical boxplot for length of time between eruptions of the Old Faithful geyser is shown in Figure 7-11. You confirm that the data are skewed left because the lower part of the box (where the small values are) is longer than the upper part of the box.

You see the values of the boxplot in Figure 7-11 that mark the five-number summary and the information shown in Figure 7-10, including the IQR of 21 minutes to measure variability. The center as marked by the median is 75 minutes; this is a better measure of center than the mean (71 minutes), which is driven down a bit by the left-skewed values (the few that are shorter times than the rest of the data).

Looking at the boxplot (Figure 7-11), you see there are no outliers denoted by asterisks. However, note that the boxplot doesn't pick up on the bimodal shape of the data that you see previously in Figure 7-5. You need a good histogram for that.

Denoting outliers

Looking at the boxplot in Figure 7-7 for the Best Actress ages data, you see a set of what the computer software defines at outliers (ten in all) on the right side of the data set, marked by a group of asterisks (as described in Step 8 in the earlier section, "Making a boxplot"). Four of the asterisks lie on top of one another because four actresses were the same age, 61, when they won their Oscars.

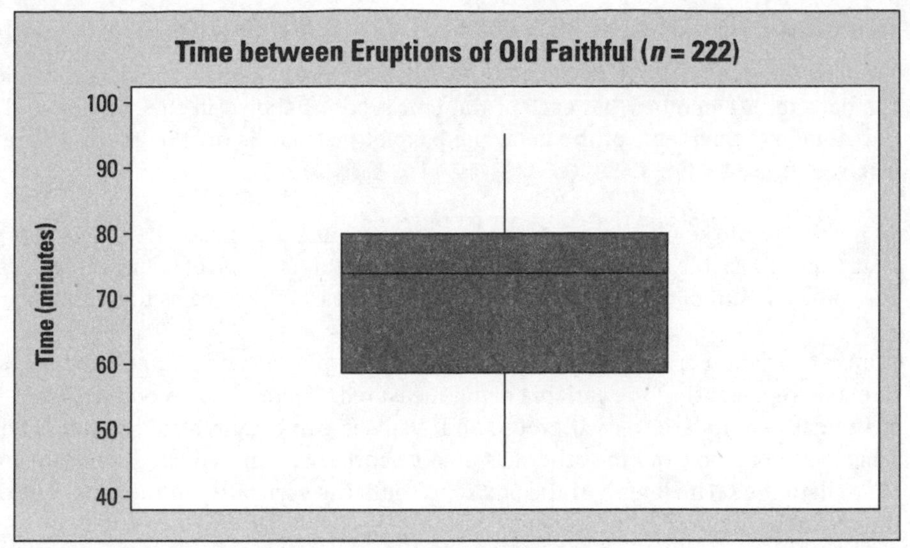

FIGURE 7-11: Boxplot of eruption times for Old Faithful geyser ($n = 222$).

The computer uses certain criteria to determine what it decides is an outlier. You can verify these outliers by applying the rule described in Step 6 of the section, "Making a Boxplot." The IQR is 13 (from Figure 7-3), so you take $13 \times 1.5 = 19.5$ years. Add this amount to Q_3 and you get $39 + 19.5 = 58.5$ years; subtracting this amount from Q_1, you get $28 - 19.5 = 8.5$ years. So an actress whose age was below 8.5 years (that is, 8 years old and under) or above 58.5 years (that is, 59 years old or over) is considered to be an outlier.

Of course, the lower end of this boundary (8 years) isn't relevant because the youngest actress was 21 (Figure 7-3 shows the minimum is 21). So you know there aren't any outliers on the low end of this data set.

However, ten outliers, as defined by Minitab, are on the high end of the data set, where the 59-and-over actresses' ages are. Table 7-2 shows the information on all ten outliers in the Best Actress ages data set.

TABLE 7-2 **Best Actress Winners with Ages Designated as Outliers**

Year	Name	Age	Movie
1968	Katharine Hepburn	60	*Guess Who's Coming to Dinner*
1969	Katharine Hepburn	61	*The Lion in Winter*
1986	Geraldine Page	61	*Trip to Bountiful*
2007	Helen Mirren	61	*The Queen*
2018	Frances McDormand	61	*Three Billboards Outside Ebbing, Missouri*
2012	Meryl Streep	62	*The Iron Lady*
1932	Marie Dressler	63	*Min and Bill*
2021	Frances McDormand	64	*Nomadland*
1982	Katharine Hepburn	74	*On Golden Pond*
1990	Jessica Tandy	80	*Driving Miss Daisy*

Making mistakes when interpreting a boxplot

It's a common mistake to associate the size of the box in a boxplot with the amount of data in the data set. Remember that each of the four sections shown in the boxplot contains an equal percentage (25 percent) of the data; the boxplot just marks off the places in the data set that separate those sections.

REMEMBER

In particular, if the median splits the box into two unequal parts, the larger part contains data that's more variable than the smaller part, in terms of its range of values. However, there is still the same amount of data (25 percent) in the larger part of the box as there is in the smaller part.

Another common error involves sample size. A boxplot is a one-dimensional graph with only one axis representing the variable being measured. There is no second axis that tells you how many data points are in each group. So if you see two boxplots side by side and one of them has a very long box and the other has a very short one, don't conclude that the longer one has more data in it. The length of the box represents the variability in the data, not the number of data values.

REMEMBER

When viewing or making a boxplot, always make sure the sample size (n) is included as part of the title. You can't figure out the sample size otherwise.

EXAMPLE

Q. Here is a boxplot of the most recent test scores from a class of 36 students.

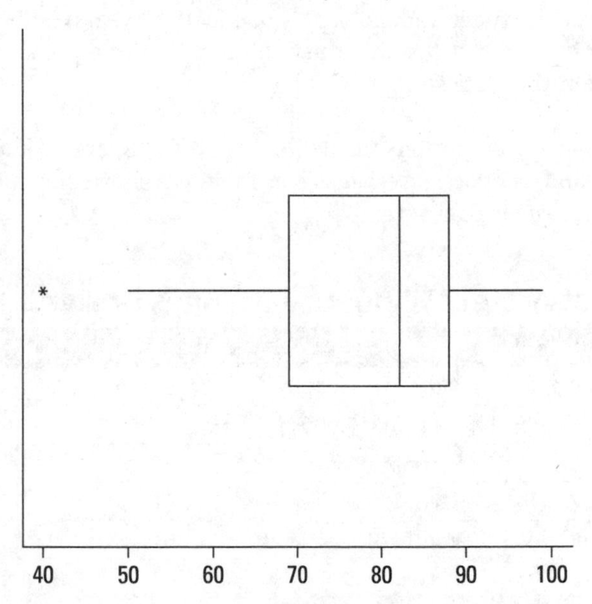

(a) Describe the shape of the distribution. Are there any outliers?

(b) How many students are represented by the box in the middle of the boxplot?

A. Keep in mind that a boxplot is constructed from just the positions of the five values from the five-number summary and not much else. Be careful when making interpretations.

(a) The fact that both the line and the part of the box in the lower half of the boxplot stretch out longer than the same parts in the upper half leads you to believe that the data is also stretched out to smaller values. **This means the data is likely skewed left. Also, it seems there is one outlier** around the value of 40. (Yikes! Maybe that student wasn't studying as hard as you are.)

(b) The middle box of any boxplot will contain 50 percent of the data. That's everybody from Q_1 (the 25th percentile) to Q_3 (the 75th percentile). But that's not enough to know how many students are in there. You also need the sample size. Fortunately, it's given that there are $n = 36$ students in the class. **That means there are $0.50 \times 36 = 18$ students represented by the middle box in the boxplot.**

YOUR TURN

15 Which of the following two boxplots contains more data: the one containing the heights of men or the one containing the heights of women?

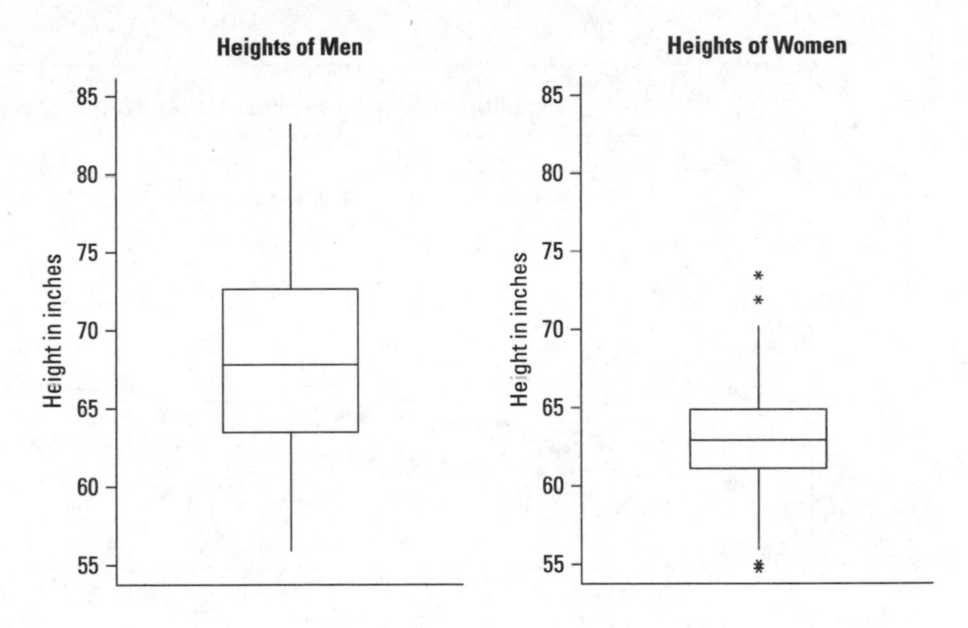

Tackling Time Charts

A *time chart* (also called a *line graph*) is a data display used to examine trends in data over time (also known as *time series data*). Time charts show time on the *x*-axis (for example, by month, year, or day) and the values of the variable being measured on the *y*-axis (like birth rates, total sales, or population size). Each point on the time chart summarizes all the data collected at that particular time, for example, the average of all pepper prices for January or the total revenue for 2010.

Interpreting time charts

To interpret a time chart, look for patterns and trends as you move across the chart from left to right.

The time chart in Figure 7-12 shows the ages of the Best Actress winners, in order of year won, from 1929 to 2021. Each dot indicates the age of a single actress, the one that won the Oscar that year.

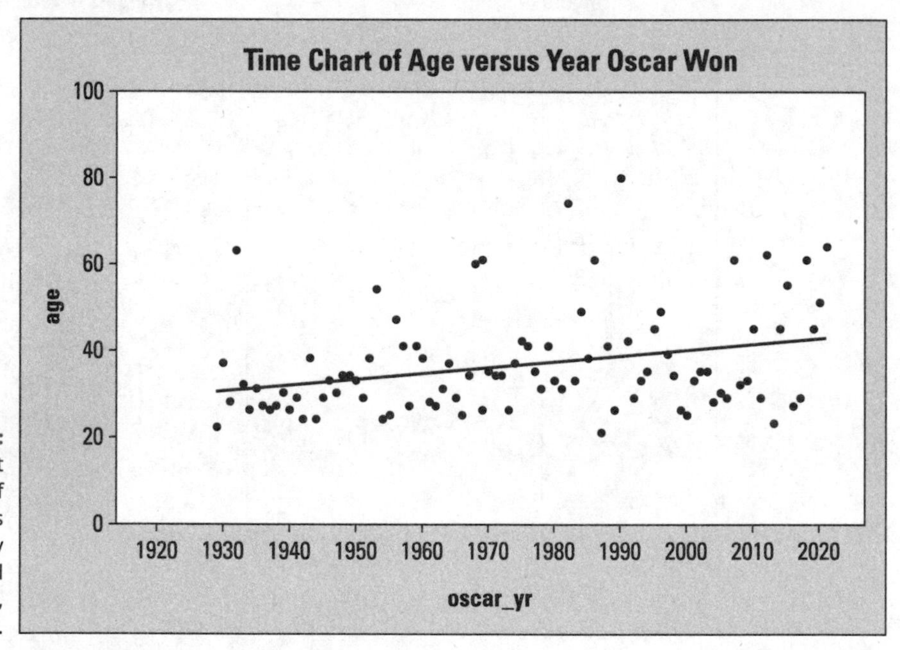

FIGURE 7-12: Time Chart #1 for ages of Best Actress Academy Award winners, 1929–2021.

Figure 7-12 shows a faint trend in age that is tending uphill, indicating that the Best Actress Award winners may be winning their awards increasingly later in life. Again, I wouldn't make too many assumptions from this result because the data have a great deal of variability. It's hard to say what may be going on here; many variables go into determining an Oscar winner, including the type of movie, type of female role, mood of the voters, and so forth, and some of these variables may have a cyclical pattern to them. As far as variability goes, you see that the ages represented by the dots do fluctuate quite a bit on the *y*-axis (representing age); all the dots basically fall between 20 and 80 years, with most of them between 25 and 45 years, I'd say. This goes along with the descriptive statistics found in Figure 7-3.

Understanding variability: Time charts versus histograms

REMEMBER

Variability in a histogram should not be confused with variability in a time chart. If values change over time, they're shown on a time chart as highs and lows, and many changes from high to low (over time) indicate lots of variability. So a flat line on a time chart indicates no change and no variability in the values across time. For example, if the price of a product stays the same for 12 months in a row, the time chart for price is flat.

But when the heights of a histogram's bars appear flat, the data are spread out uniformly across all the groups, indicating a great deal of variability in the data. (For an example, refer to Figure 7-2a.)

Spotting misleading time charts

As with any graph, you have to evaluate the units of the numbers being plotted. For example, it's misleading to chart the *number* of crimes over time, rather than the crime *rate* (crimes per capita) — because the population size of a city changes over time, crime rate is the appropriate measure. Make sure you understand what numbers are being graphed, and examine them for fairness and appropriateness.

Watching the scale and start/end points

The scale on the vertical axis can make a big difference in the way the time chart looks. Refer to Figure 7-12 to see the original time chart of the ages for the Best Actress Academy Award winners from 1929 to 2021 in increments of ten years. You see a fair amount of variability, as discussed previously.

In Figure 7-12, the starting and ending points on the vertical axis are 0 and 100, which creates some extra white space on the top and bottom of the picture. I could have used 10 and 90 as my start/end points, but this graph looks reasonable.

Now, what happens if I change the vertical axis? Figure 7-13 shows the same data, with start/end points of 20 and 80. The increments of 10 years appear longer than the increments of 10 years shown in Figure 7-12. Both of these changes in the graph exaggerate the differences in ages even more.

REMEMBER

How do you decide which graph is the best one for your data? There is no perfect graph, no right or wrong answer; but there are limits. You can quickly spot problems just by zooming in on the scale and the start/end points.

Simplifying excess data

A time chart of the intervals between eruptions for the Old Faithful data is shown in Figure 7-14. You see 222 dots on this graph; each one represents the time between one eruption and the next, for every eruption during a 16-day period.

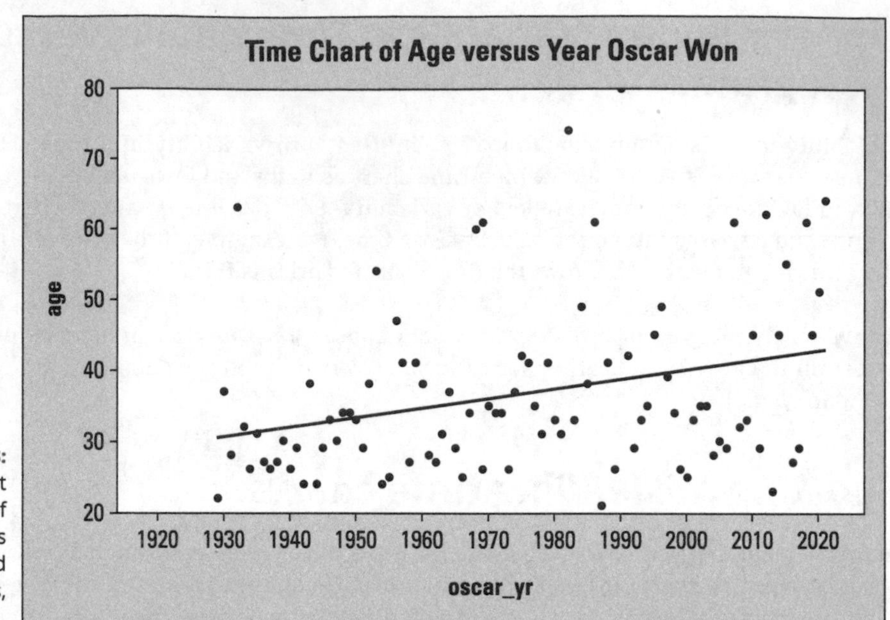

FIGURE 7-13: Time Chart #2 for ages of Best Actress Oscar Award winners, 1929–2021.

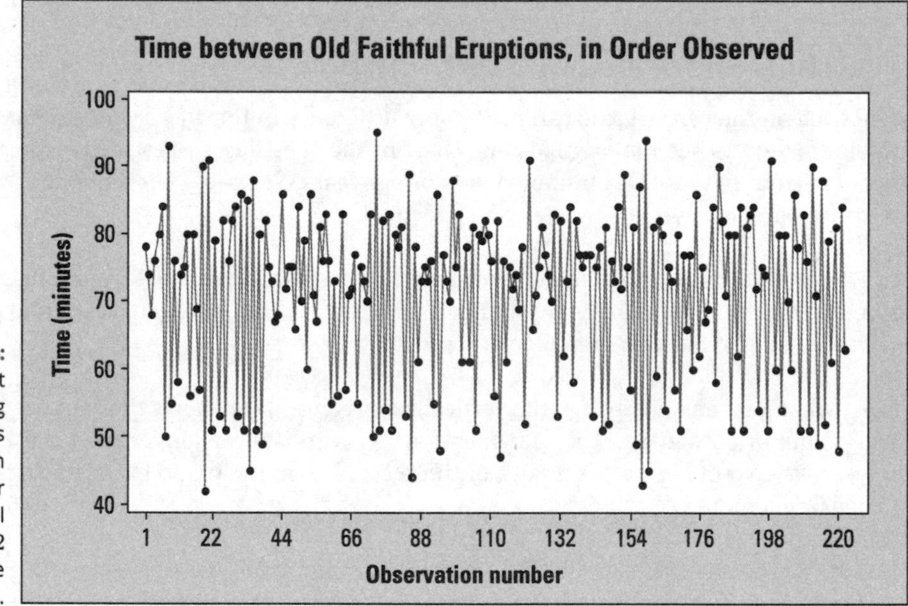

FIGURE 7-14: Time chart showing intervals between eruptions for Old Faithful geyser ($n = 222$ consecutive observations).

This figure looks very complex; data are everywhere, there are too many points to really see anything, and you can't find the forest for the trees. There is such a thing as having too much data, especially nowadays when you can measure data continuously and meticulously using all kinds of advanced technology. I'm betting they didn't have a student standing by the geyser recording eruption times on a clipboard, for example!

To get a clearer picture of the Old Faithful data, I combined all the observations from a single day and found its mean; I did this for all 16 days, and then I plotted all the means on a time chart in order. This reduced the data from 222 points to 16 points. The time chart is shown in Figure 7-15.

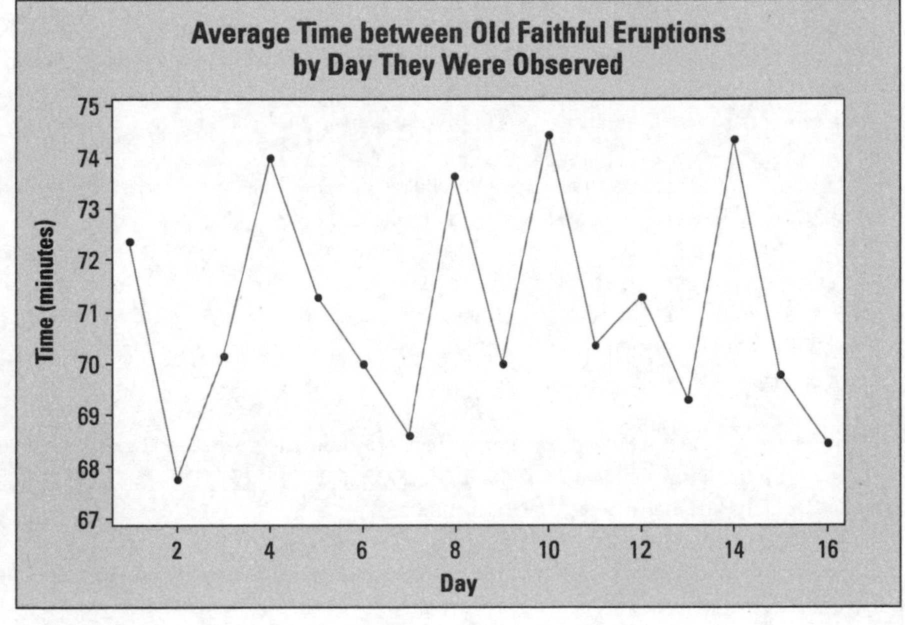

FIGURE 7-15: Time chart showing daily average intervals between eruptions for Old Faithful geyser ($n = 16$ consecutive days).

From this time chart, I see a little bit of a cyclical pattern to the data; every day or two, it appears to shift from short times between eruptions to longer times between eruptions. While these changes are not definitive, it does provide important information for scientists to follow up on when studying the behavior of geysers like Old Faithful.

REMEMBER

A time chart condenses all the data for one unit of time into a single point. By contrast, a histogram displays the entire sample of data that was collected at that one unit of time. For example, Figure 7-15 shows the daily average time between eruptions for 16 days. For any given day, you can make a histogram of all the eruptions observed on that particular day. Displaying a time chart of average times over 16 days accompanied by a histogram summarizing all the eruptions for a particular day would be a great one-two punch.

EXAMPLE

Q. The following figure shows the revenues of a company taken over time. Each dot represents the revenue for that year, in millions of dollars.

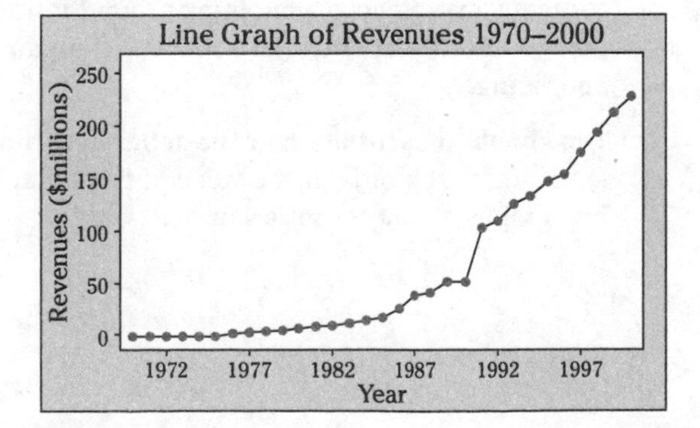

EVALUATING TIME CHARTS

Here is a checklist for evaluating time charts, with a couple more thoughts added in:

Examine the scale and start/end points on the vertical axis (the one showing the values of the data). Large increments and/or lots of white space make differences look less dramatic; small increments and/or a plot that totally fills the page exaggerate differences.

If the amount of data you have is overwhelming, consider boiling it down by finding means/medians for blocks of time and plotting those instead.

Use expert knowledge to think about a meaningful way to combine data rather than just trying different groupings. For example, if sales of milk at the grocery story have a weekly cycle, you wouldn't want to condense it to monthly (or even 10 day) groupings because you might miss the weekly cycle.

Watch for gaps in the timeline on a time chart. For example, it's misleading to show equally spaced points on the horizontal (time) axis for 1990, 2000, 2005, and 2010. This happens when years are just treated like labels, rather than real numbers.

As with any graph, take the units into account; be sure they're appropriate for comparison over time. For example, are dollar amounts adjusted for inflation? Are you looking at the number of crimes or the crime rate?

(a) What's the time period over which this data set was collected?

(b) Describe what the line graph tells you about the revenues of this company over the time period.

(c) What do you need to take into account in order to properly interpret revenues (or any variable reported in dollars) over time?

A. Be aware of the impact of inflation over time on data reported in dollars. Some line graphs adjust for inflation and some don't. Look at the fine print.

(a) **The time period is approximately 1970 to 2000.**

(b) **Revenues increased a little in the 1970s and then began a more steady increase in the '80s until around 1989, when the company broke the $50 million barrier. The company experienced a big jump around 1990–1991, with a very strong and steady increase each year since. In 2000, the company's revenue was up to $225 million and rising.**

(c) **You should take inflation of the dollar over time into account.** The revenues may look larger later on, but the value of the dollar has decreased over time as well. Some line graphs adjust for inflation.

16 Check out the sales of a particular car across the United States over a 60-day period in the following figure.

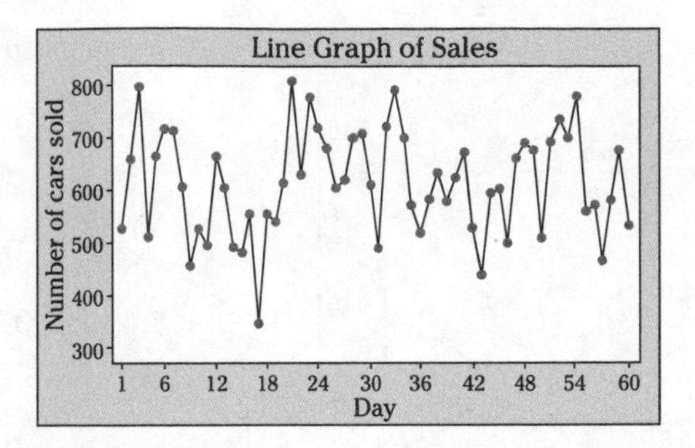

(a) Can you see a pattern to the sales of this car across this time period?

(b) What are the highest and lowest numbers of sales and when did they occur?

(c) Can you estimate the average of all sales over this time period?

17 After a heart attack, Bob decides that it is a good time to get in shape, so he starts exercising each day and plans to increase his exercise time as he goes along. Look at the two line graphs shown in the following figures. One is a good representation of his data, and the other should get as much use as Bob's treadmill before his heart attack.

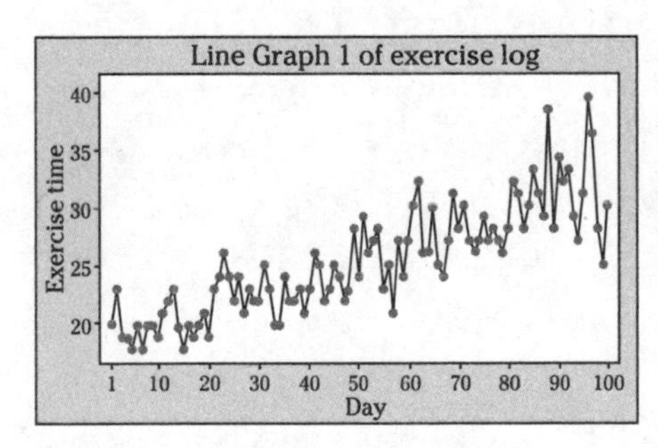

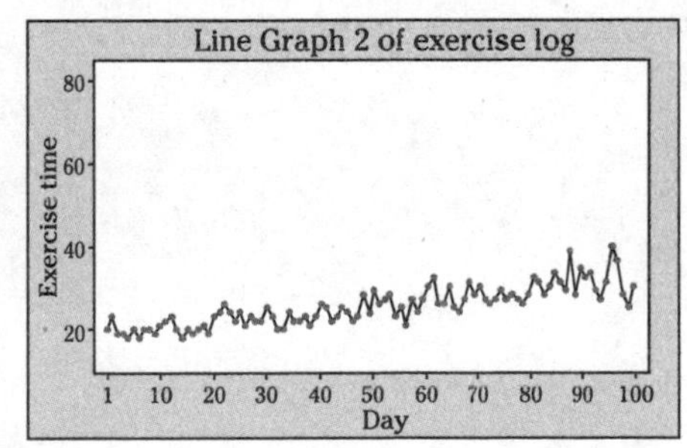

(a) Compare the two graphs. Do they represent the same data set, or do they show totally different data sets?

(b) Assume both graphs are made from the same data. Which graph is more appropriate and why?

18 The line graph in the following figure shows one company's revenues over time. Explain why this graph is misleading and what you can do to fix the problem.

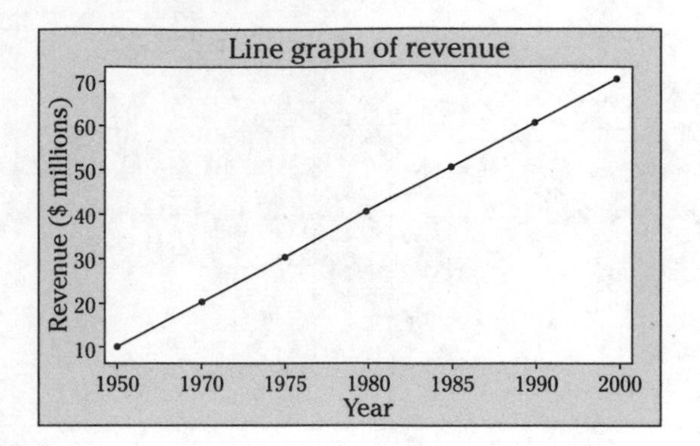

19 Line graphs typically connect the dots that represent the data values over time. If the time increments between the dots are large, explain why the line graph can be somewhat misleading.

Practice Questions Answers and Explanations

1. **You don't know the actual values of the data anymore; you know only what group they fall into.** For example, if eight test scores fall in the group 70–79, they could all be 70, they could all be 79, or they could be some mixture in between.

2. One possible histogram of this data is shown in the following figure. Yours may have different groupings and look slightly different. **The data is quantitative, so a pie chart isn't appropriate.** The groupings I chose for this histogram are as follows: 48–52; 53–57; 58–62; 63–67; 68–72; 73–77; 78–82; 83–87; 88–92; 93–97; and 98 and up.

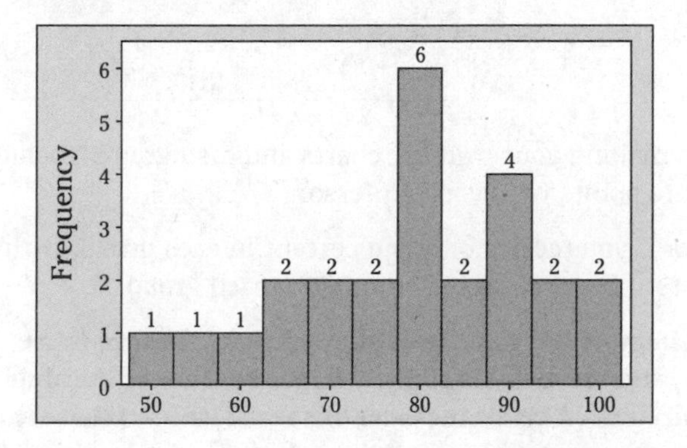

REMEMBER

The scale of the x-axis is continuous on a histogram, so if you have a place for a bar but you have no data, you should leave a space in that spot. If you don't leave a space, you'll have an incorrect histogram, because the gaps help you to distinguish how spread out the data is throughout the bars.

3. The following figure shows the relative frequency histogram for this data. Yours should look similar. **The relative frequencies are $2 \div 45 = 0.044$ or 4.4 percent; $17 \div 45 = 0.38$ or 38 percent; $22 \div 45 = 0.49$ or 49 percent; $3 \div 45 = 0.07$ or 7 percent; and $1 \div 45 = 0.02$ or 2 percent.**

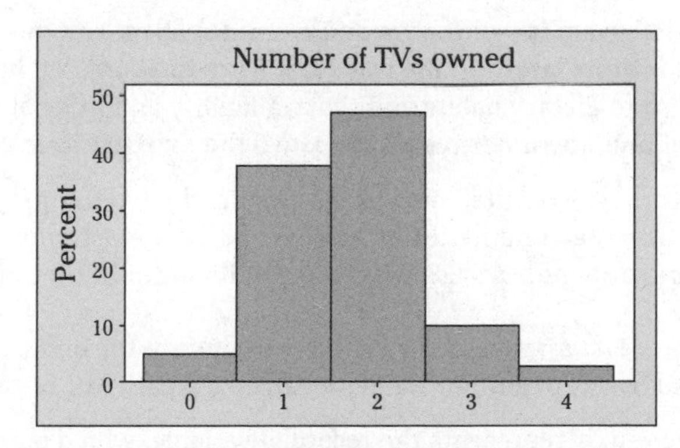

4. You can find the total sample size from a frequency histogram.

   a. The relative frequency histogram is shown in the following figure. Note, the total sample size is missing here, but you can still find it by summing the heights of all the bars. Here, the total frequency is: $10 + 15 + 5 + 15 + 5 + 5 + 20 = 70$. To get the relative frequencies, divide the height of each bar by 70 (the total sample size) to obtain a percentage. For example, the percentage of ones is $10 \div 70 = 14.3$ percent, the height of the first bar.

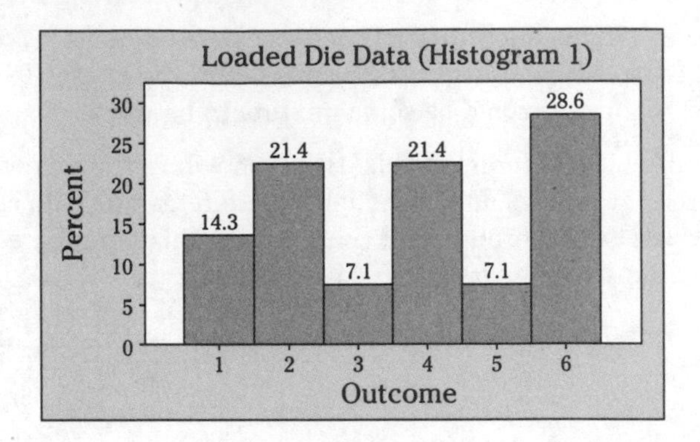

REMEMBER

When making your own bar charts and histograms, include the total sample size if you want to score points with your professor.

b. **No.** If you receive only the percent in each group, without the total sample size, you can't determine the original number in each group.

5. **The shape of this data set is skewed right. The median is somewhere between $150 and $200, and the mean is slightly larger because of the single $300, $400, and $500 withdrawals. A crude measure of spread without these three single values is less than $50 (taking the range without those values divided by six); with these three values, a crude measure of spread is around $80 (the range of all the values divided by six).** You can see a more accurate measure of spread in Chapter 5.

6. **The histogram of rod lengths is skewed left. It appears to be centered very near 100, but the mean drops down a bit because of the one value lower than the rest (99.2). The range of the lengths is very tight, between 100.6 and 99.2 inches (each rod is within 1.4 inches). Most rod lengths are between 99.8 and 100.3 inches, indicating that the company seems to be doing very well in terms of accuracy.** They may want to check into the four values of 100.6 and the one value of 99.2 to see whether they can improve the process somewhere.

7. **The histogram is symmetric and mound shaped. You can see one peak around $500. The range is quite large, all the way from $100 to $800,** due in part to the size of the household and also to dietary habits and the availability and price of fruits and vegetables. **Most households spent between $200 and $700 on these items per year.**

REMEMBER

"Interpret the results" means you need to do two things. First, identify the statistics and what they mean numerically, and second (and most important), apply those statistics to the scenario at hand. Discuss what the results mean in the context of the problem.

8. *Percentage return* means the difference between the ending value minus the beginning value, divided by the beginning value; it can be a positive or negative number or zero for no change.

   a. The *x*-axis represents the recorded variable, which is the one-year return for each stock. **A negative number means that the stock lost money during that year** (the price at the beginning was more than its value at the end).

   b. **The x-axis will contain negative values if some of the data are negative. However, the y-axis of a histogram reports the counts or percentage of data in each grouping, and those values are always greater than or equal to zero.**

9. **The first figure has a smaller spread because the bars are higher in the middle, near the mean.** Therefore, on average, many of the values in this data set are close to the mean. The second figure shows all the data spread out about equally, which indicates that some are close to the mean, but an equal percentage are a medium distance away, and another equal percentage are a long distance away. This results in a larger spread.

WARNING

Don't interpret a flat histogram as meaning the data shows "no change" or no spread. A flat histogram means the data does have quite a bit of spread. Data with a bell shape that's tight around the middle has a much smaller spread, because you generally measure spread as average distance from the middle.

REMEMBER

Note that the values on the x-axis are the same for both graphs. If the values aren't the same, you'll have difficulty comparing the spreads alone; you should also take the scale into account. This may be above and beyond your particular course, but if you want to compare spreads with data sets that have different scales and different means, you can use the *coefficient of variation*, which is the standard deviation divided by the mean. A large coefficient of variation means the spread is large, relative to the mean. A small coefficient of variation means the spread is small, relative to the mean.

10. Be aware that people may be tempted to give an incorrect response to a salary question (in other words, lie). Such a fib is called *response bias*, and it results in data that's systematically over or under the truth (in this case, probably over).

    a. **The graph shows that this group is making plenty of money, although the graph is skewed to the right, with fewer graduates making the large amounts.** The range is quite large, going from $50,000 to $100,000, and the center is probably in the high-$60,000 range.

    b. Because the heights of the bars sum to 13 $(5+3+1+2+1+1)$, you know 13 salaries make up the data set. **The median is the one in the middle, which is the seventh salary.** Because the first bar contains five salaries and the second bar contains three, the seventh number is in the second bar, which is in the $60,000 range.

    c. You have only 13 salaries — from a "large" program — so **the data probably isn't precise, because it represents such a small part of the overall group**. Also, you need to keep in mind that **these values will change over time**.

11. **No. The die isn't loaded; the graph is loaded.** The histogram is misleading because it starts at 40 on the y-axis and goes to only 65. The differences in the heights of the bars are exaggerated because the graph doesn't start at zero. A more fair and balanced-looking graph of the same data is shown here. Notice the frequencies for the outcomes are fairly close, indicating no evidence of the die being loaded.

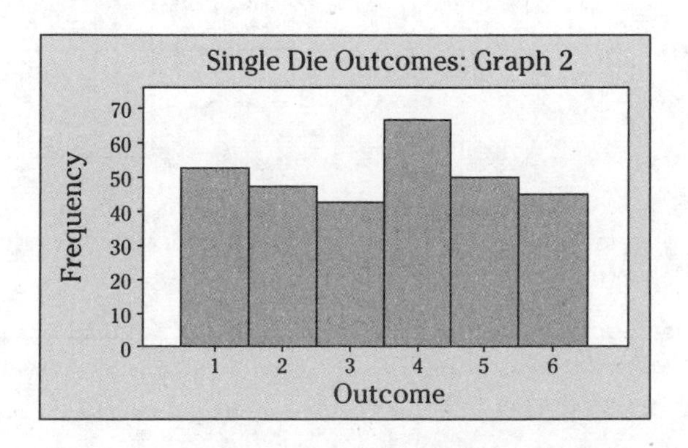

WARNING

Beware of graphs that don't start at zero on the y-axis. They may make the results look more dramatic than necessary, which is misleading to the reader.

12 **This histogram has a great deal of open space at the top, and the bars appear to be very short and close together in height. The scale for the y-axis of this graph uses oversized increments. A better histogram would have smaller increments on the y-axis, and it wouldn't include numbers that go beyond what you need to show the data.** Such a histogram follows.

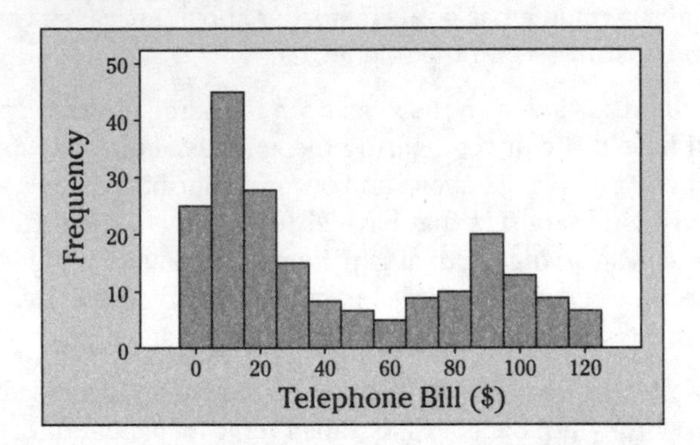

13 First, you use the tricks and tips in Chapter 5 to find the five-number summary, which is 59, 65, 67, 72, and 75. Steps 2 and 3 say to draw a horizontal number line (the x-axis) that spans 59 to 75, and mark the five-number summary values. In Step 4 you draw a box with 65 and 72 marking the edges, and in Step 5 you draw a line for the median at 67.

To start Step 6, you calculate $1.5 \times IQR = 1.5 \times (72-65) = 1.5 \times 7 = 10.5$. Any large outliers would have to be greater than $Q_3 + 10.5 = 72 + 10.5 = 82.5$, and any small outliers would have to be less than $Q_1 - 10.5 = 65 - 10.5 = 54.5$. Because none of the heights fall outside of those values, there are no outliers. According to Step 7, you draw the remaining lines from the quartiles to the maximum (75) and minimum (59). See the completed boxplot in the following figure.

Heights of Randomly Selected Adults ($n = 11$)

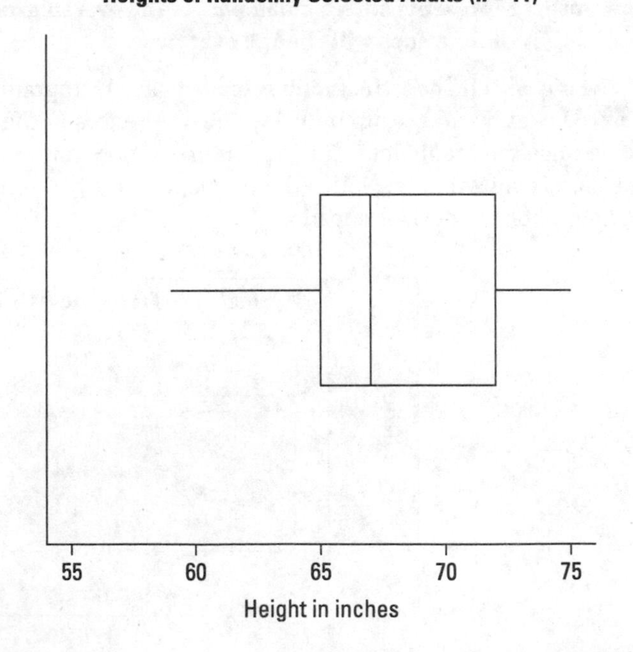

(14) Remember to first order the commute times and then find the five-number summary, which is 5, 12, 15.5, 20, and 35. Next, draw a vertical number line (the y-axis) that spans 5 to 35, and mark the five-number summary values. Then draw a box with 12 and 20 marking the boundaries of the box, and draw a line for the median at 15.5.

To check for outliers, you start by calculating $1.5 \times IQR = 1.5 \times (20 - 12) = 1.5 \times 8 = 12$. Any large outliers would have to be greater than $Q_3 + 12 = 20 + 12 = 32$, and any small outliers would have to be less than $Q_1 - 10.5 = 12 - 12 = 0$. There are no commute times less than 0, but there is one greater than 32: the maximum value, 35. You label the value of 35 with an asterisk in the boxplot and draw the line from the box to the largest value less than the boundary value of 32, namely 25. On the low end, you draw the remaining line from the edge of the box down to the minimum of 5. See the completed boxplot in the following figure.

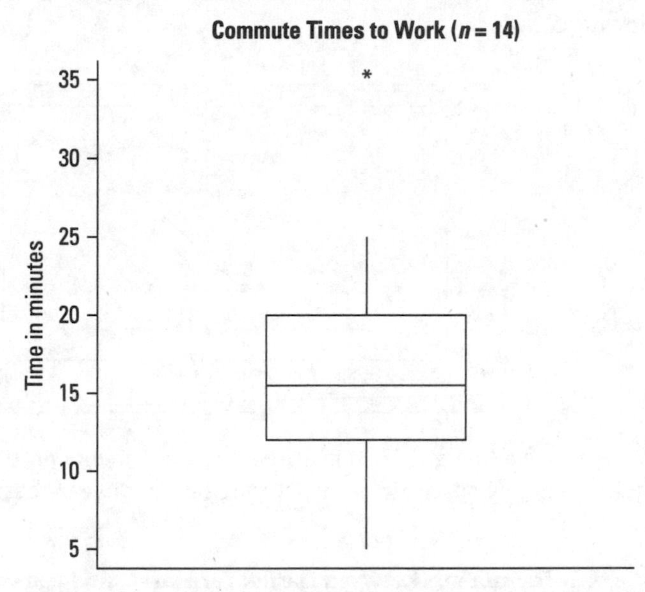

(15) **Without the sample size, there's no way to know for sure.** The heights of the men are more spread out than the heights of the women, but that only tells you about the variation. You can see that the range and IQR of the men's heights are both clearly greater. But for all you know, the boxplot for the men's data is based on 20 males while the boxplot for the women's data is based on 2,000 females. That's why it's so important that plots also include the sample size — so you know how much information your conclusions are based on.

(16) Interpreting a line graph is different than interpreting a histogram; a line graph represents many snapshots of the situation over time, each one summarized by one point, and a histogram shows one single snapshot in detail of all the data at once.

a. **No.** The data seem to fluctuate back and forth from around 350 to 800 cars sold per day.

b. **You can't tell exactly, but the highest sales figure is around 800, occurring on days 1 and 21, and the lowest, at 350, occurred only a few days before, around day 17.** Maybe the customers knew a sale was coming and waited to buy.

c. **The average appears to be around 600 cars sold per day,** looking at what the values on the y-axis seem to center around. (The actual average is 613.)

17 The way data are organized can greatly affect how readers interpret graphs. Look at the way the data are organized before you think about what the graph means.

a. **They do represent the same data set.** The difference is in the scale used on the *y*-axis. The second graph has larger increments on the *y*-axis, so the differences in the data over time are played down.

b. **The first graph is more appropriate. It has a scale that uses most of the space on the graph, and it doesn't understate or overstate Bob's progress over time.**

18 **This line graph is misleading because the time increments on the *x*-axis (time) aren't equally spaced, but the graph presents them as if they are, which incorrectly makes it look like the revenue is increasing at the same rate over time. To fix the problem, you need to space the time periods shown on the *x*-axis properly.** The correct line graph is shown in the following figure.

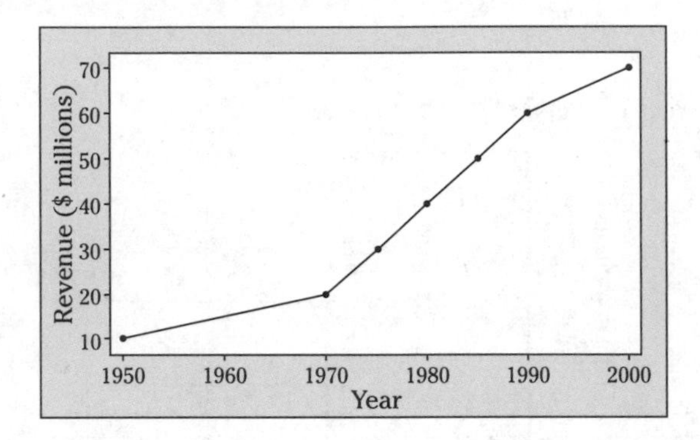

REMEMBER

When you see a line graph, look at the time increments on the *x*-axis and make sure the times are equally spaced in terms of the number of years between them.

19 **If the time increments between the dots are large, and you connect the dots, you assume that the change that took place during the interim period (when data wasn't collected) occurred at a steady rate, represented by the line that connects the dots. That may not always be the case.**

If you're ready to test your skills a bit more, take the following chapter quiz that incorporates all the chapter topics.

Whaddya Know? Chapter 7 Quiz

Quiz time! Complete each problem to test your knowledge on the various topics covered in this chapter. You can then find the solutions and explanations in the next section.

1. You have a data set of exam-taking times, where 60 minutes is the maximum time. Most students finished at between 30 and 40 minutes, but a few of them took longer, and a couple of them turned in their exams at the very end. Is the histogram of exam times symmetric, skewed right, or skewed left?
2. If the mean is much larger than the median, are the data symmetric?
3. Suppose a data set contains the values 10, 11, 12, 13, 12, 10, 11, 12, 10, 16. Is 16 considered an outlier according to the criterion described in this chapter?
4. A larger boxplot means a larger data set. True or false?
5. Twenty-five percent of the data lies between each section of a boxplot. True or false?
6. Changing the scale on a graph can change the way it looks. True or false?
7. You can get the size of the data set from its boxplot. True or false?
8. A flat histogram shows no variability. True or false?
9. Which data set has a higher coefficient of variation: data set 1 (Mean 10, StDev = 2) or data set 2 (Mean 2, StDev = 10)?
10. Which statistics make up the five-number summary?

Answers to Chapter 7 Quiz

1. The answer is **skewed right** because the times tend to be on the left side of the distribution and then trail off to the right, with fewer and fewer of them getting larger and larger.

2. **No.** The data are not symmetric if the mean is much larger than the median. The mean is influenced by skewness and outliers, and the median is not, so some outliers and/or skewness are pushing the mean past the median in this case.

3. **Yes.** The criterion is that if the data value lies above $Q_3 + 1.5 * \text{IQR}$, then it's an outlier on the right side. In this case, Q_3 is 12 because it's the median of the upper half of the ordered data. The value of Q_1 is 10 because it's the median of the lower half of the ordered data. The IQR is $Q_3 - Q_1 = 12 - 10 = 2$, and $1.5 * \text{IQR} = 1.5 * 2 = 3$. So values above $Q_3 + 1.5 * \text{IQR} = 12 + 3 = 15$ are considered outliers by this criterion. This means 16 is considered an outlier in this case.

4. **False.** A larger boxplot means you have more variability in the data between Q_1 and Q_3. The data are more spread out. It does not indicate anything about the size of the data set.

5. **True.** Twenty-five percent of the data lies between the min and Q_1, Q_1 and the median, the median and Q_3, and Q_3 and the max.

6. **True.** If you make the scale include large increments, the graph appears to have smaller changes from group to group. If you make the scale include small increments, the graph appears to have larger changes from group to group.

7. **False.** You cannot get the size of the data set from its boxplot. All you know is that 25 percent of the data lies within each section; you don't know any more than that.

8. **False.** A flat histogram indicates a lot of variability, as you measure variability by distance from the mean. A flat histogram has the same amount of data near the mean as is further from the mean. There is data further from the mean, so there is variability.

9. **Data set 2 has a higher coefficient of variation.** The coefficient of variation is the standard deviation divided by the mean. For data set 1, the coefficient of variation is $2 / 10 = 0.20$. For data set 2, the coefficient of variation is $10 / 2 = 5$.

10. **The five-number summary consists of the minimum value, Q_1, the median, Q_3, and the maximum value.** (Note that the mean is not part of the five-number summary.)

3

Distributions and the Central Limit Theorem

In This Unit. . .

IN THIS CHAPTER

» Nailing down the basic definitions and terms associated with probability

» Examining how probability relates different events

» Solving probability problems with the rules and formulas of probability

» Identifying independent and mutually exclusive events

» Exploring the difference between independence and exclusivity

Chapter 8 Coming to Terms with Probability

Statistics like the ones you calculate in Chapters 4 and 5 are values that describe what you just observed. Probability is calculated as the likelihood or the chance that you would observe some specific event (like your observed statistics) when certain conditions exist. The first step toward probability success is having a clear knowledge of the terms, the notation, and the different types of probabilities you come across. If you use and understand the terms, notation, and types when working on easy problems, you have an edge from the start when the problems get more complex. This chapter sets you on the right track.

A Set Notation Overview

Probability has its own set of notations, symbols, and definitions that provide a shorthand way of expressing what you want to do. *Notation* refers to the symbols that you use as shorthand to talk about probability; for example, *p*(A) means the probability that A will occur. The notation A refers to all the possible ways that an event could happen. Every probability problem starts out by defining the information you have and the quantity you're trying to get, which all comes down to notation and terms.

Noting outcomes: Sample spaces

A *probability* is the chance that a certain outcome, or result, will occur out of all the possible outcomes for the process at hand. The process is called a *random process* because you conduct an experiment, or other form of data collection, and you don't know how the results will come out. Before you can figure out the probability of the result you're interested in, you list all the possible outcomes; this list is called the *sample space* and is typically denoted by S. Any collection of items in probability is called a *set*. Notice that S is a set, so you use set notation to list its outcomes and probabilities for those outcomes (such as using brackets around the list with commas that separate each outcome).

For example, if your random process is rolling a single die, $S = \{1, 2, 3, 4, 5, 6\}$ denotes the sample space. The set S can take on three different types: finite, countably infinite, and uncountably infinite.

Finite sample spaces

If you can write and count all the elements in a set, the set is *finite*. Rolling a single die is an example of a finite random process because you can achieve only six possible outcomes, and you can account for all of them. Another example would be scores on the ACT because possible scores are all integers from 1 to 36. Probability models that you can use for finite sample spaces include the binomial distribution (see Chapter 9).

Countably infinite sample spaces

Countably infinite means that you have a way to show the progression of the values, but they can go on into infinity. For example, if your random process involves the number of phone calls that come in to a switchboard during a week's time, the possible outcomes of S aren't finite, but rather countably infinite. In this case, $S = \{0, 1, 2, 3, 4, \ldots\}$. S goes to infinity because you can't be sure of the maximum number of calls coming in. If you count all the calls, you get a fixed number with a countably infinite sample space S, but to be sure you allow for any maximum, you let S go on to infinity.

Uncountably infinite sample spaces

Uncountably infinite means that you have situations where the possible outcomes are too numerous to write down in a listing, so you include an interval to describe them. An interval is a subset of the number line that falls entirely between two values — [1, 2] is the set of all real numbers between 1 and 2. It may seem weird to have an uncountably infinite set consisting of numbers between 1 and 2, but too many numbers are in this interval to count!

For a real-world example, imagine that you're measuring the lengths of time it could take a computer to complete a task, and the maximum allowable time is 5 seconds. Your measurements of the actual time taken could be anywhere from 0 to 5 seconds and to an infinite number of decimal places. You denote that $S = \{\text{all real numbers } x \text{ such that } 0 \le x \le 5\}$. For this example, S is uncountably infinite.

Noting subsets of sample spaces: Events

Probability problems typically involve figuring the probability of one or more subsets of the sample space, S. A subset of the sample space S is called an *event*, and the notations for events are capital letters: A, B, C, D, and so on. For example, if you roll a single die, $S = \{1, 2, 3, 4, 5, 6\}$. Event A may be the event that you roll an odd number: $A = \{1, 3, 5\}$. Event B may be the event that you roll a number greater than 2: $B = \{3, 4, 5, 6\}$.

If you have to monitor calls that come in to a switchboard for a week's time, you may be interested in the event that at least 10 calls come in (call it event E): $E = \{10, 11, 12\ \}$. If you have to monitor the time it takes a computer to complete tasks in at most 4 seconds, you may want to find the probability that it takes no more than 4 seconds to complete the task (call this event D): $D = \{\text{all real numbers } x \text{ such that } 0 < x <= 4\}$.

TRANSLATING INEQUALITIES

Many probabilities involve a series of outcomes described by phrases such as "at least," "at most," "not more than," "not less than," "more than," or "less than." You need to understand exactly what these phrases mean and be able to translate them into math symbols.

- "At least" means greater than or equal to and is denoted by $\geq$. For example, rolling at least a 3 on a fair die means $x \geq 3$, where x represents the outcomes from S that you're interested in: 3, 4, 5, or 6. Or you may be looking at grade point average (on a 4.00 scale), where at least 3.00 means all possible numbers from 3.00 to 4.00, including 3.00. Whether the statement refers to integers or all real numbers depends on the type of sample space you're dealing with.
- "At most" means you can go up to the number in question but not beyond it, so the notation is $\leq$. To roll a die and get a number that's at most 3 means $x \leq 3: 1,\ 2,$ or 3. If you're looking at grade point average, at most 3.00 means anything from 0.00 to 3.00, including 3.00.
- "Not more than" means you can have the number in question or anything less, but you can't have more. It means the same as "less than or equal to" or "at most."
- "Not less than" has the same meaning as "greater than or equal to" or "at least."
- Strictly "more than" means you don't want to include the number in question, but you include every number beyond it, so you use >. To roll a die and get a number greater than 3 is to roll {4, 5, 6}. A grade point average of greater than 3.00 means every average from 3.00 to 4.00, not including 3.00.
- Strictly "less than" means you don't want to include the number in question itself, so you use <; for example, less than 3 on a single die means {1, 2}. For grade point average, less than 3.00 means everything from 0.00 to 3.00 but not including 3.00.

The easiest way to remember these phrases and their notations is to think of easy-to-remember examples that make sense to you. For example, to remember how "at least" works, you can recall that you need to be at least 21 to go to a bar, which means $x \geq 21$, where x is your age.

You can simplify set notation by using interval notation, which indicates you want an interval of numbers on the number line. In interval notation, you write the left endpoint first, a comma, and then the right endpoint. You indicate whether you want to include the two endpoints by the type of brackets you use:

- If you want to include the endpoint in a set, you use a square bracket.
- If you don't want to include the endpoint in a set, you use a round bracket.

In the previous computer example, you have $S = (0, 4]$. The square bracket indicates that you do include the endpoint 4, and the round bracket indicates that you don't include 0. You include every number in between.

The type of notation that's used to indicate an interval is totally up to the instructor. Both types — the one involving brackets, such as [0, 2], and the one involving inequalities that involve *x*, such as $0 \le x \le 2$ — are used very commonly, so you should get used to both notations. The inequality notation is typically used throughout this book.

Noting a void in the set: Empty sets

The last basic probability definition is the empty set, or *null set*. If an event, or subset of the sample space S, doesn't have any outcomes in it, you have an empty set, or null set. The most common notation used for the empty set is $\varnothing$, although you sometimes see the notation { } (that's the curly brackets with nothing inside). You may see an empty set if you're looking for elements that are common to two sets and you don't find any. For example, let $A = \{1, 2, 3\}$ and $B = \{4, 5, 6\}$. Which outcomes are common to both sets? None. How do you indicate this, using set notation? The set of all outcomes common to sets A and B is the empty set: { }.

Putting sets together: Unions, intersections, and complements

After you identify the sample space, S, and the various events, or subsets, of S that compose the space (see the previous sections), you can put those sets together with unions, intersections, or complements. This action is similar to adding and subtracting; only you're dealing with sets, so you use a different notation.

When you find the union of two sets, you produce a set that's at least as big as the largest of your two sets; it may be as large as the sample space itself. When you find the intersection of two sets, you produce a set that's at most as big as the smaller of the two sets; it can even be as small as the empty set. In general, unions make the sets stay the same or get larger, and intersections make the sets stay the same or get smaller.

Unions

To put two sets together into one possibly larger set is called forming a *union* of those two sets. The notation for a union is $\cup$. $A \cup B$ represents the union of two sets, A and B. ***Note:*** The union of two sets is itself a set.

Say, for example, that you're rolling a single die. Event A is the event that you roll an odd number, and event B is the event that you roll a number greater than 2 (in other words, at least a 3). What would the set $A \cup B$ look like? You know that $A = \{1, 3, 5\}$, and you know that $B = \{3, 4, 5, 6\}$, which means $A \cup B$ is {1, 3, 4, 5, 6}. *Note:* The union is itself a set, so you need to use set notation, or brackets, around the new set. Notice that you represent each distinct outcome in either set only one time in the union set (the number 3 shows up in both sets, but in the union, it appears only one time). Now, suppose you have a third event, C, which represents an even number on the die: $C = \{2, 4, 6\}$. If you want to look at the set $A \cup C$, you get {1, 2, 3, 4, 5, 6}, which is equal to the sample space S.

Intersections

To find only the common outcomes between two sets is to find the *intersection* of those sets. The notation for intersection is $\cap$. $A \cap B$ represents the intersection of two sets, A and B. Note: The intersection of two sets is itself a set.

In the previous die rolling example, you have $A = \{1, 3, 5\}$ and $B = \{3, 4, 5, 6\}$. What are the common outcomes of these two sets? The only common outcomes are the numbers 3 and 5, so $A \cap B$ is {3, 5}. If you look at the intersection of A and C ({2, 4, 6}), what do you find? The sets A and C have no common elements, so you use the notation for the empty set, { }.

Complements

The *complement* of an event A is the set of all outcomes from the sample space, S, that don't reside in A. (Maybe the outcomes used to reside in A but didn't receive enough compliments, so they left.) The notation for the complement of A is A^C. For example, if $S = \{1, 2, 3, 4, 5, 6\}$ and $A = \{1, 3, 5\}$, the complement of A is the set $A^C = \{2, 4, 6\}$. In other words, if you're rolling a die, set A represents the outcomes where the die comes up odd, and set A^C represents the outcomes where the die comes up even.

Probabilities of Events Involving A and/or B

The types of probabilities you want to find will vary according to the question under review, but all probabilities boil down to the big five: marginal probability, union probability, intersection probability, conditional probability, and the probability of a complement. Suppose, for example, that you roll a single fair die. You may want to know the probability that the number you roll is even (marginal probability); the probability that the number is even or less than 4 (union probability); the probability that the number is even and less than 4 (intersection probability); or the probability that the number is 5 if you know it came up odd (conditional probability). Each type of probability is addressed in the pages that follow, along with how probability notation fits into the picture.

Even though it may seem easier to figure probabilities intuitively without the use of formulas, you need to resist the urge and stick to using the definitions and formulas to figure them out. When the problems get more complicated, you'll be glad that you've established a process to make the calculations.

Probability notation

To describe the probability of an outcome (or set of outcomes), you need shorthand notation. But in order to understand the notation, you need to know what a probability really means. You start with a set, such as $A=\{1, 2, 3\}$ outcomes of the roll of a die (see the section, "A Set Notation Overview," earlier in this chapter). To find the probability of A occurring (of the die coming up 1, 2, or 3), you give set A a number between 0 and 1. (In this case, the number is $\frac{3}{6}=\frac{1}{2}$, because a die boasts six possible numbers and three of them are in set A.) The $\frac{1}{2}$ is a probability.

A probability is really a mapping (or a correspondence) that goes from the sample space S to the numbers on the number line between zero and one. Table 8-1 shows this mapping for a single roll of a die. Note that each probability is $\frac{1}{6}$ because you have six possible outcomes and each one has an equal chance of occurring, assuming the die is fair.

Table 8-1 Probability Mappings for a Single Die (S={1, 2, 3, 4, 5, 6})

Outcome from S	Probability
{1}	$\frac{1}{6}$
{2}	$\frac{1}{6}$
{3}	$\frac{1}{6}$
{4}	$\frac{1}{6}$
{5}	$\frac{1}{6}$
{6}	$\frac{1}{6}$

Because the probability of getting a value in the set {1} is $\frac{1}{6}$, you can write $p(1)=\frac{1}{6}$, or 0.167. You say, "The probability of 1 is equal to one-sixth, or 0.167." Similarly, you write $p(2)=\frac{1}{6}$, and so on. If set A is {1, 3, 5}, you write its probability as $p(A)=\frac{3}{6}$, or 0.50, because it contains three elements, each with the probability $\frac{1}{6}$. (You say, "The probability of A is equal to one-half, or 0.50.") The important idea here is that you find the probability of a set of outcomes. In the end, the probability you get is a number between 0 and 1. If you have an empty set, you always give it a probability of zero.

Be clear about the different parts of a probability equation/statement. Consider the example $p(A)=0.50$ from the previous paragraph. The letter A represents the set {1, 3, 5}; it doesn't equal 0.50. The probability of A is what equals 0.50. Also, be careful when using probability notation; $p(1)=\frac{1}{6}$ is correct, but $p(1)=p\left(\frac{1}{6}\right)$ is incorrect.

REMEMBER

Many statisticians will use lowercase p because probability is a function. In algebra, you might define $f(x)$ to equal $x^2 + 3$. Then you plug in 2 to get $f(2) = (2)^2 + 3 = 7$, so you get that 2 maps to 7. It is the same with probabilities; like in the last example and table, $p(1) = 1/6$ because 1 maps to $1/6$. As a heads-up, some statisticians will use capital P for the probability notation instead, for example, $P(1) = 1/6$.

Marginal probabilities

If you're finding the probability of a set A all by itself, the probability you're finding is called the *marginal probability* of A. For example, suppose you roll a fair die twice and want the probability that the second number you roll will be even. The event in question here is $A = \{2, 4, 6\}$. Because three equally likely outcomes make up this set, the probability that the die will come up even on the second roll is $p(A) = 3/6$, or $1/2$.

You aren't concerned with more than one single characteristic of an outcome when you're looking at a marginal probability. For the die example, the die coming up even on the second roll is the only characteristic of concern, and you don't really care about what you get on the first die roll.

Union probabilities

The probability of the union of two events, say A and B, is called *a union probability* and is written $p(A \cup B)$. The language often associated with unions is the word "or"; in other words, $p(A \cup B)$ means the probability of "A or B." For example, suppose you roll a fair die. You let A be the event that the outcome is even and B represent the event that the outcome is less than 4: $A = \{2, 4, 6\}$ and $B = \{1, 2, 3\}$. The union of sets A and B is the set of all numbers present in A or B or both: {1, 2, 3, 4, 6}. This union contains five equally likely elements, so $p(A \cup B) = 5/6$. (In the section, "Understanding and Applying the Rules of Probability," later in this chapter, you find out more about how to calculate union probabilities.)

REMEMBER

In the case of union probabilities, you're concerned about two characteristics of an outcome, and the chance that one or the other (or both) is present. Keep in mind, however, that "or" doesn't mean "either or." It means A or B or both.

Intersection (joint) probabilities

The probability of the intersection of two events, say A and B, is called an *intersection probability*, or a *joint probability*, and is written $p(A \cap B)$. The "joint" part of joint probability means "happening at the same time."

The language often associated with intersection is the word "and" — $p(A \cap B)$ means the probability of "A and B." The joint probability for A and B is the probability of all outcomes jointly located in A and B at the same time. For example, suppose that you roll a fair die. You let A be the event that you roll an even number and B represent the event that the number is less than 4: $A = \{2, 4, 6\}$ and $B = \{1, 2, 3\}$. The intersection of A and B is the set {2}. The intersection contains only one element, which has a probability of $1/6$. That means $p(A \cap B) = 1/6$. (In the section, "Understanding and Applying the Rules of Probability," later in this chapter, you discover a shortcut for calculating joint probabilities under certain circumstances.)

Complement probabilities

The complement of an event A, A^C, is every item in the sample space, S, that isn't in A. For example, suppose you're rolling a single die, and the event $A=\{2, 4\}$. The complement of A is the event $A^C=\{1, 3, 5, 6\}$. Because this set contains the four die outcomes that don't appear in A, the $p(A^C)=4/6$, or $2/3$, because each outcome of the die is equally likely. (In the section, "Understanding and Applying the Rules of Probability," you find out how to get the probability of complements in a general way and how very helpful complements can be in terms of calculating a probability.)

Conditional probabilities

With marginal probabilities, you're just finding the probability of a single action or event regardless of other things going on. Sometimes knowing prior information about an outcome or other related events can change the probability of the outcome. And when you break down outcomes into subgroups (for example, odd or even die outcomes), the probabilities change. Conditional probabilities deal with the change that comes from factoring in prior information. The probability of one event given that another event has already occurred is a *conditional probability.*

Solving conditional probabilities without a formula

Conditional probabilities provide a way to compare groups or to use information that you already know about a situation to your advantage. The notation for the probability of event A given that B has already occurred is $p(A|B)$ — translated as "probability of A given B."

WARNING

Be careful not to confuse conditional notation. The expression $p(A|B)$ is not equivalent to $p(A)$ divided by $p(B)$. The notation separates the event you already know has occurred (B) from the event you want to find the probability for (A). And remember that the event that follows the "|" is the known or given event; the conditional probability of B given A is entirely different from the conditional probability of A given B.

For example, say that you're rolling a single die, and one roll of the die comes up odd. What's the probability that the roll is a 5? In probability notation, you want $p(\text{die is a } 5|\text{die is odd})$, or $p(C|A)$, where A is the event that the die is odd, and C is the event that the die is a 5. After you know that the die is odd, you have only three possibilities — 1, 3, or 5 — and each is equally likely. Therefore, you can say that $p(\text{die is } 5|\text{die is odd})$ is $1/3$, or 0.33.

Solving conditional probabilities with a formula

The definition of $p(A|B)$ in equation form is $p(A|B)=\frac{p(A\cap B)}{p(B)}$. To find an answer, you take the joint probability of A and B and divide it by the probability of B. The numerator is the joint probability because you want the outcomes from B that also appear in A. You divide by the probability of B because B is your new sample space; you know that the item in question is already in set B.

USING CONDITIONAL PROBABILITIES TO EVALUATE DISEASE TESTING

Doctors often want to know how effective a certain test is for detecting a disease. To find out, they have to test it on people who have the disease and on people who don't have the disease. They assign the set A to the event that someone has the disease and set B to the event that someone tests positive for the disease. If the test works properly, the probability of testing positive for people who have the disease should be very high, and the probability for testing negative for people who don't have the disease should also be very high. After all, you don't want to make a mistake and scare people by telling them they have the disease when they actually don't.

The notation for the event that someone who has the disease tests positive is B | A. This means you use the set "B given A," where A means you have the disease, and B means you test positive. You use the word "given" because you know the person belongs to event or group A already (the given part, or the part indicated after the "|" sign). So, of those people who have the disease, what's the probability that they test positive? This probability is represented by $p(B | A)$, and it should be high.

For the test to work properly, you also have to check how often it gives the negative diagnosis to people who don't have the disease. In other words, given that you don't have the disease, what's the probability that you'll test negative? The known part is that you don't have the disease; the unknown part is whether you'll test negative. Participants who don't have the disease are noted by A^C, the complement of set A. Giving the correct result for these people means that the test should come out negative, indicated by B^C, the complement of set B. Therefore, you want to look at $p(B^C | A^C)$ — the conditional probability of B complement, given A complement. You want this probability to be high.

You can't find the conditional probability of A given B if the probability of B is zero — in other words, if B is the empty set. But that's no problem; if B is the empty set, you shouldn't be interested in finding the probability of A given B anyway, because B can't happen.

You can use the formula for conditional probability to find the answer to the example problem in the previous section, where $A = \{1, 3, 5\}$ and $C = \{5\}$.

According to the definition of conditional probability, $p(C | A) = \frac{p(C \cap A)}{p(A)}$. You know that $p(C \cap A)$ equals $1/6$ because the intersection of those two sets is the outcome {5}, and its probability is $1/6$. Now you can say $p(A) = p(\{1, 3, 5\}) = 3/6$. Dividing these two, you get $1/6 \div 3/6$, which is $1/6 \times 6/3 = 1/3$, or 0.33. You get the same answer as when you don't use the formula.

Understanding and Applying the Rules of Probability

You can find the probability of outcomes, events, or combinations of outcomes and/or events by adding, subtracting, multiplying, or dividing the probabilities of the original outcomes and events. You use some combinations so often that they have their own rules and formulas. The better you understand the ideas behind the formulas, the more likely it is that you'll remember them and be able to use them successfully.

Any probability has to follow three basic properties:

- Every probability has to be a number between zero and one. If you ever report that the probability of an event is greater than one or negative, you've made a mistake!
- To find the probability of a set of individual outcomes from S, the sample space, you sum their probabilities. (This isn't necessarily true for combining events, but it is for individual outcomes.)
- If you take the probabilities of all the outcomes in S, they have to sum to one.

The following pages contain the basic rules and formulas of probability that build on these three basic properties.

The complement rule (for opposites, not for flattering a date)

The complement of an event A, A^C, is the set of every item or individual in the sample space, S, that isn't in A. The probability of the complement of event A is the chance that A didn't occur. By definition, if you take the union of A and A complement, you get S, and the probability is $1/1$; therefore, $p(A^C) + p(A) = 1$. Solving for $p(A^C)$, you get what's called the complement rule: $p(A^C) = 1 - p(A)$.

Suppose that you're rolling a single die. The sample space $S = \{1, 2, 3, 4, 5, 6\}$. If you let $A = \{1, 3, 5\}$, the complement of A is the set $A^C = \{2, 4, 6\}$. Say, for example, that you want the probability of rolling a number greater than 1 (in other words, at least 2), which means you create an event D: {2, 3, 4, 5, 6}. You can see that the probability of D, p(D), is $5/6$, but you can also find this probability by using the complement rule. You know that $D^C = \{1\}$ and $p(D^C) = 1/6$; therefore, according to the complement rule, $p(D) = 1 - p(D^C) = 1 - 1/6 = 5/6$.

TIP

Oftentimes, the event you have to deal with is complicated and difficult to get a handle on. In tough cases, before you start pulling your hair out, think of the complement; it may be easier to grasp the outcomes you don't want than the outcomes you do want.

To illustrate this idea, look at the situation where you roll two dice. You have $6 \times 6 = 36$ possible outcomes, from (1, 1) all the way to (6, 6). The following table shows the entire set of outcomes.

(1, 1)	(2, 1)	(3, 1)	(4, 1)	(5, 1)	(6, 1)
(1, 2)	(2, 2)	(3, 2)	(4, 2)	(5, 2)	(6, 2)
(1, 3)	(2, 3)	(3, 3)	(4, 3)	(5, 3)	(6, 3)
(1, 4)	(2, 4)	(3, 4)	(4, 4)	(5, 4)	(6, 4)
(1, 5)	(2, 5)	(3, 5)	(4, 5)	(5, 5)	(6, 5)
(1, 6)	(2, 6)	(3, 6)	(4, 6)	(5, 6)	(6, 6)

Suppose A is the event that at least one of the dice comes up greater than 1 on a roll. To find this probability, you need to add the probabilities of all the outcomes that make up this event. That's a lot of possibilities! All the outcomes in the first column of the preceding table are included, except the first one — (1, 1). All the outcomes in the remaining five columns are also included. In fact, all the outcomes except (1, 1) fit this description. So, you can calculate *p*(A) by finding all the outcomes that meet the description and summing up their probabilities to get $^{35}/_{36}$, but you may find it easier to take a look at the complement. The complement of A is the set of all outcomes in S in which you don't have at least one of the dice greater than 1. The only outcome that meets this criterion is (1, 1); therefore, you have $A^C = (1, 1)$. You know that $p\left(A^C\right) = {}^{1}/_{36}$, and by the complement rule, $p(A) = 1 - p\left(A^C\right) = 1 - {}^{1}/_{36} = {}^{35}/_{36}$. The complement rule makes it much easier to find *p*(A) in this case.

The multiplication rule (for intersections, not for rabbits)

You use the multiplication rule to find the probability of an intersection of two events A and B. It makes sense that the definition of conditional probability involves an intersection (see the section, "Conditional probabilities," earlier in this chapter) because conditional probability is where the multiplication rule comes from. The conditional probability of A given B is defined as $p(A \mid B) = \frac{p(A \cap B)}{p(B)}$. If you cross-multiply this formula, you get $p(A \cap B) = p(B) \times p(A \mid B)$. Here's the translation: The probability that A and B occurred is equal to the probability that B occurred times the probability that A occurred given that B occurred. The multiplication rule splits up the joint probability into two stages; first, B occurs, and then A occurs given that B has occurred.

REMEMBER

If you're given a marginal probability, like *p*(B), and a conditional probability, like *p*(A|B), you can use the multiplication rule to find the joint probability.

Suppose, for example, that a class is made up of 60 percent women, and of these women, 40 percent are married. What's the chance that a person you select at random from the class is a woman and married? To answer this, let the event W = {woman} and M = {married}. What you want is *p*(W and M), which is the joint probability $p(W \cap M)$. You know that 60 percent of the class is made up of women, which means $p(W) = 0.60$. You also know that of the women in the class, 40 percent, 0.40, are married. You have to use a conditional probability to solve the problem because you split up the women and look at the probability that they're married — $p(M \mid W) = 0.40$. By the multiplication rule, to find $p(W \cap M)$, you take $p(W) \times p(M \mid W) = 0.60 \times 0.40 = 0.24$. Of all the people in the class, 24 percent are women and are married, which also means that the chance of you picking a married woman from the class is 24 percent.

A probability, technically, is a number between 0 and 1, but you often see it expressed as a percentage because it's easier to interpret that way. You write a probability as a percentage by multiplying the probability by 100.

Be aware of the difference between a joint probability and a conditional probability. You need a joint probability when you select someone from the entire group who has two characteristics. You need a conditional probability when you pull out a subgroup that has one of the characteristics already, and you want the probability that someone from that subgroup has a second characteristic.

The addition rule (for unions of the nonmarital nature)

The union of two events A and B is the set of all outcomes in the sample space, S, that are in either A or B or both. To find the probability of the union of two events A and B, you do what appears to be the most intuitive calculation — you add the two probabilities together. Only you can't stop there. When you add $p(\text{A})$ and $p(\text{B})$, you double count the outcomes that are both in A and in B; in other words, you double count the outcomes in $\text{A} \cup \text{B}$. If you count those outcomes twice, it makes the probability of the union too large. So how do you fix it? Because you count the outcomes in $\text{A} \cap \text{B}$ twice, you should subtract them out once. The probability of A union B is given by the following formula: $p(\text{A} \cup \text{B}) = p(\text{A}) + p(\text{B}) - p(\text{A} \cap \text{B})$. This formula is called the *addition rule.*

For example, suppose that you have a class made up of 60 percent women, 40 percent of whom are married (for the introduction of this example, see the previous section). Suppose you also know that 50 percent of all the people in the class are married. You want to find the percentage of people in the class who are women or married (or both): $p(\text{M} \cup \text{W})$. Using the addition rule, you have $p(\text{M} \cup \text{W}) = p(\text{M}) + p(\text{W}) - p(\text{M} \cap \text{W})$. You know that 60 percent of the class is made up of women, so $p(\text{W}) = 0.60$. You know that 50 percent of the class is married, so $p(\text{M}) = 0.50$. By using the multiplication rule in the previous section, you already found that $p(\text{M} \cap \text{W}) = 0.24$. Therefore, $p(\text{M} \cup \text{W}) = 0.50 + 0.60 - 0.24 = 0.86$. You calculate that 86 percent of the class is married or female (or both).

Notice that if you don't subtract the intersection in the previous example, you get $0.50 + 0.60 = 1.1$, which is greater than one. You can never have a probability greater than one or less than zero.

Q. Suppose you flip a fair coin three times.

(a) How many outcomes are possible?

(b) What's the probability of each outcome?

(c) What are the possible values for the total number of heads out of three tosses, and what are their probabilities?

A. Before you start flipping your coins, it's good to know exactly what the possibilities and their probabilities are. The table shows you the possibilities and probabilities at a glance for this coin-flipping scenario.

Number of Heads	Possible Outcomes	Probability
3	HHH	$\frac{1}{8}$
2	HHT, HTH, THH	$\frac{3}{8}$
1	HTT, THT, TTH	$\frac{3}{8}$
0	TTT	$\frac{1}{8}$

(a) The sample space contains eight outcomes: $S = \{HHH, HHT, HTH, THH, HTT, THT, TTH, TTT\}$. Notice that each flip has two possible outcomes, so three flips has $2 \times 2 \times 2 = 8$ possible outcomes.

(b) Each outcome has a probability of 1 in 8 (because the total number of outcomes is eight and because you assume the coin to be two-sided and fair).

(c) The total number of heads can be 3, 2, 1, or 0. Three heads happens only one way (HHH), so its probability is 1 in 8. Two heads happens three ways (HHT, HTH, THH), so its probability is 3 in 8. One head happens three ways (HTT, THT, TTH), so its probability is 3 in 8. Zero heads happens in only one way (TTT), so its probability is 1 in 8. All these probabilities sum to one $\left(\frac{1}{8}+\frac{3}{8}+\frac{3}{8}+\frac{1}{8}=1\right)$.

YOUR TURN

1 Suppose M&Ms colors come in the following percentages: 13 percent brown, 14 percent yellow, 13 percent red, 24 percent blue, 20 percent orange, and 16 percent green. Reach into a giant bag of M&Ms without looking.

(a) What's the chance that you pull out a brown or yellow M&M?

(b) What's the chance that you won't pull out a green one?

---

2 Suppose you flip a coin four times, and it comes up heads each time. Does this outcome give you reason to believe that the coin isn't legitimate?

---

3 Consider tossing a fair coin ten times and recording the number of heads that occur.

(a) How many possible outcomes would occur?

(b) What would be the probability of each of the outcomes?

(c) How many of the outcomes would have one head? What is the probability of one head in ten flips?

(d) How many of the outcomes would have zero heads? What is the probability of zero heads in ten flips?

(e) What's the probability of getting one head or less on ten flips of a fair coin?

(f) What's the probability of getting more than one head on ten flips of a fair coin?

Recognizing Independence in Multiple Events

One of the most important assumptions of the basic probability models is independence. Multiple events are *independent* if knowledge that one event has happened doesn't affect the probability of the other event happening. In other words, knowing that A has occurred doesn't change the probability of B occurring given A. If events A and B are independent, you can say bye-bye to conditional probabilities, which makes your life much easier! (See the section, "Conditional probabilities," earlier in this chapter for more information on the topic.)

You have two ways to check for independence (assuming A and B aren't empty):

- Use the definition of independence. Check to see if $p(\text{A} \mid \text{B}) = p(\text{A})$ or if $p(\text{B} \mid \text{A}) = p(\text{B})$.
- Check the multiplication rule for independence. You can also check to see if $p(\text{A} \cap \text{B}) = p(\text{A}) \times p(\text{B})$. If so, A and B are independent.

Checking independence for two events with the definition

Suppose that you're rolling a single die. The sample space, S, equals {1, 2, 3, 4, 5, 6}. If you let event $\text{A} = \{\text{the die comes up odd}\}$ and event $\text{B} = \{\text{the die comes up } 1\}$, are these two events independent? To answer that, first ask the question, "If I know the die is odd, what's the probability that it's a 1?" The answer is $\frac{1}{3}$. Now ask, "What's the probability that the die is a 1 without knowing whether it's odd?" The answer is $\frac{1}{6}$. The probabilities are different, so events A and B are not independent. Knowledge of one event affects the probability of the other event.

Now suppose you add the event $C = \{\text{the die is a 1 or 2}\}$. Are events A and C independent? To find out, you check to see if $p(C)$ equals $p(C|A)$. You could also check to see if $p(A) = p(A \mid C)$. The probability of C is $2/6$, or 0.33. The probability of C given A is the probability of the die being a 1 or 2 given that it's odd. Using the definition of conditional probability (see the section, "Conditional probabilities"), you have $p(C \mid A) = \frac{p(C \cap A)}{p(A)}$. The set $C \cap A$ is the set {1}, whose probability is $1/6$. The probability of A is $3/6$. Dividing these probabilities, you get $1/6 \div 3/6$, which is $1/3$, or 0.33. This is the same as $p(C)$ found earlier. So, events A and C are independent because knowing that the die is odd doesn't change the probability that the die is a 1 or a 2. Some information isn't really worth knowing, because it doesn't affect the chances.

WARNING

One big word of caution when it comes to independent events: If two events are independent, it doesn't mean that they can't happen at the same time. Many people make the mistake of thinking of independent events as being totally separate from each other. In probability, two independent events can happen at the same time, and in essence coexist; they just don't affect each other in terms of their probabilities. For example, consider the events of it raining and you wearing a blue shirt. Those two events don't affect each other; they are independent. But you could certainly be wearing a blue shirt on a rainy day.

Using the multiplication rule for independent events

It's impossible to overemphasize how wonderful life is when events are independent. Suppose you want the probability of five events happening at the same time. If these events aren't independent, you have to find the conditional probabilities at every stage of the process: the second event would depend on the first event; the third event would depend on the first and second events; the fourth event would depend on the first three events; and the fifth event would depend on the first four events. What a complicated mess! If all the events are independent, their five-way joint probability is just the product of the five probabilities of the individual events. Much easier! You can extend the multiplication rule for independent events to any number of events. To find the joint probability of two events, for example, you simply multiply their individual probabilities.

If you know that two events are independent, or if you can show that they're independent, your calculations are much easier for any probabilities affiliated with these events. This is because when you know A and B are independent, you can say that $p(A \mid B) = p(A)$ and $p(B \mid A) = p(B)$. "Well, what's so great about that," you ask? When you go to find the joint probability of A and B, $p(A \cap B)$, it equals $p(A) \times p(B \mid A)$. But because A and B are independent, it equals $p(A) \times p(B)$, because $p(B \mid A) = p(B)$. Therefore, to get the joint probability of events A and B, you multiply the marginal (or individual) probabilities of A and B together if A and B are independent.

WARNING

It may be tempting to take $p(A) \times p(B)$ whenever you need to find $p(A \cap B)$, but you can do this only if A and B are independent. If not, you have to use the formula $p(A) \times p(B \mid A)$ and deal with conditional probabilities.

Suppose, for example, that you're rolling dice. You can safely assume that the outcome on one die doesn't affect the outcome on the other die. If you roll two dice, what's the probability of getting a 1 and a 1? According to the multiplication rule, it's the probability of getting a 1, $\frac{1}{6}$, times the probability of getting a 1, $\frac{1}{6}$. The result is $\frac{1}{36}$. If you roll five dice, the chance of getting all 1s is $\frac{1}{6} \times \frac{1}{6} \times \frac{1}{6} \times \frac{1}{6} \times \frac{1}{6}$, which equals $\left(\frac{1}{6}\right)^5$. In general, if you roll n dice, the probability of getting all 1s is $\left(\frac{1}{6}\right)^n$. Now, if you want to find the probability of tossing a single 1 when you roll five dice five times, that's more difficult, because you can use different ways to get a single 1, and you have to take them all into account.

Including Mutually Exclusive Events

In probability, you often see independent events that can occur at the same time without affecting each other when they do, but you also see the opposite situation, where two events can't occur at the same time, and hence affect each other greatly. Two events A and B are *mutually exclusive* if they can't occur at the same time. In other words, they exclude each other from occurring: $A \cap B = \{\ \}$, or $p(A \cap B) = 0$. If you know A has occurred, you know that B can't occur; and if you know B has occurred, you know that A can't occur.

As with independent events (see the previous section of this chapter), mutually exclusive events can make your calculations much easier, so you should keep an eye out and try to control for them in probability models where possible.

Recognizing mutually exclusive events

REMEMBER

Tagging two events as mutually exclusive doesn't always mean that one event or the other must occur; it only means that if one of the events occurs, the other event can't occur.

Look at the outcomes of a traffic light, where the sample space, S, is {red, green, yellow}. Let A equal the event that the light is green, and let B equal the event that the light is red. If you know the light is red, you know it can't be green, and vice versa. Those two events are mutually exclusive because mutually exclusive events directly affect each other's probabilities. No matter what $p(A)$ is, you know that $p(A|B)$ has to be zero, and vice versa.

If A and B are mutually exclusive, note that $p(A \cap B) = 0$, so $p(A \mid B) = \frac{p(A \cap B)}{p(B)} = \frac{0}{p(B)} = 0$.

You can also use the definition of mutually exclusive events in reverse order to see whether A and B are mutually exclusive, because definitions always go in both directions. You know that if A and B are mutually exclusive, $p(A \cap B) = 0$. Therefore, to check to see whether two events are mutually exclusive, you can check to see if $p(A \cap B) = 0$. If so, the events are mutually exclusive; if not, the events aren't mutually exclusive.

REMEMBER

One special case of mutually exclusive events is events that are complements. Such events are the opposite of each other in terms of the outcomes they contain. So if an outcome in A happens, then no outcome in A^C could have possibly happened.

For example, say you flip a coin twice. The sample space is {HH, HT, TH, and TT} — H denotes getting a head, and T denotes getting a tail. If you let A be the event that the outcome is two heads, the complement, A^C, is the event that you don't get two heads, so $A = \{HH\}$ and $A^C = \{HT, TH, TT\}$. By definition, because A^C contains the outcomes in S that don't appear in A, the events A and A^C can't intersect —$A \cap A^C = \{\,\}$, which means $p(A \cap A^C) = 0$. That makes the events mutually exclusive.

Simplifying the addition rule with mutually exclusive events

If two events are mutually exclusive, they have no intersection, which makes using the addition rule for probability much easier. The addition rule finds the probability of the union of two events A and B (refer to the section, "The addition rule (for unions of the nonmarital nature)," earlier in this chapter); it's the probability of the set of all outcomes in A or B or both. With mutually exclusive events, no outcomes lie in both sets, so the addition rule is simplified to the sum of the two events. Using probability notation, the addition rule looks like this: $p(A \cup B) = p(A) + p(B) - p(A \cap B)$. If A and B are mutually exclusive, $p(A \cap B) = 0$, so the addition rule becomes $p(A \cup B) = p(A) + p(B)$. You don't have to deal with the intersection probabilities, which makes the calculations much easier.

Suppose, for example, that you select a card from a standard 52-card deck, and you want the probability that the card is a 2 or a 3. Let $A = \{\text{card is a 2}\}$ and $B = \{\text{card is a 3}\}$. Note that $p(A) = 4/52$, because you have 52 cards in a standard deck, and 4 of the cards are 2s. Similarly, $p(B) = 4/52$. Because you want the probability that the card is a 2 or a 3, you want a union probability, or $p(A \cup B)$. Events A and B are mutually exclusive because a card can't be a 2 and a 3 at the same time, so $p(A \cap B) = 0$. Therefore, $p(A \cup B) = p(A) + p(B) = 4/52 + 4/52 = 8/52 = 0.154$, or a 15.4 percent probability.

REMEMBER

You may be tempted to take $p(A) + p(B)$ whenever you need to find $p(A \cup B)$, but you can do this only if A and B are mutually exclusive. If not, you have to use the formula $p(A) + p(B) - p(A \cap B)$ and deal with intersection probabilities.

REVIEWING THE STANDARD 52-CARD DECK

A standard card deck contains 52 cards. Each card has one of 13 denominations on it: ace, 2, 3, 4, 5, 6, 7, 8, 9, 10, jack, queen, or king. Sometimes the ace is considered the highest denomination, and sometimes it's considered the lowest denomination; it all depends on what type of game you play with the cards. Each card is also labeled with one of four possible suits. The four suits are diamonds, denoted by ♦; hearts, denoted by ♥; clubs, denoted by ♣; and spades, denoted by ♠. Diamonds and hearts are red cards (their denomination and suit labels are marked in red), and clubs and spades are black (their denomination and suit labels are marked in black). Thirteen cards make up each suit, which makes 26 of the cards red and 26 black. The deck has four cards in each denomination. The cards denoted by J, Q, or K are called face cards or court cards, because on those cards you see the face (and body) of a jack (or prince), a queen, or a king. One deck contains 12 face cards.

Distinguishing Independent from Mutually Exclusive Events

One challenge students of probability face is understanding the difference between independent events and mutually exclusive events. Both events are often easy to understand separately, but when compared to each other, the concepts seem to fall apart. However, if you take the time to study the definitions of each event, you'll see a stunning difference between the two; it all boils down to the intersection probabilities and how they compare.

Comparing and contrasting independence and exclusivity

If events A and B are independent, they can occur at the same time, which means they can intersect. Their intersection (or joint) probability is given by $p(\mathrm{A} \cap \mathrm{B}) = p(\mathrm{A}) \times p(\mathrm{B})$. So, to find the intersection probability of two events, you multiply their marginal probabilities together. If, however, events A and B are mutually exclusive, they can't occur at the same time, which means they can't intersect. Their intersection (or joint) probability is given by $p(\mathrm{A} \cap \mathrm{B}) = 0$.

So now comes the big question. Suppose A and B are nonempty events (which means their probabilities aren't zero) and independent. Can they be mutually exclusive? No, because for them to be mutually exclusive, they must have no intersection, so $p(\mathrm{A} \cap \mathrm{B})$ must be zero. Because $p(\mathrm{A} \cap \mathrm{B}) = p(\mathrm{A}) \times p(\mathrm{B})$ and A and B are independent, the only way to get zero is if $p(\mathrm{A})$ is zero or $p(\mathrm{B})$ is zero, and that isn't the case. So, if two events are nonempty and independent, they can't be mutually exclusive.

Turning this example around, if A and B are mutually exclusive events and nonempty, can they be independent? No. To see this clearly, look at the definition of independence. Suppose events A and B are mutually exclusive. These two events are independent if $p(\mathrm{A} \mid \mathrm{B}) = p(\mathrm{A})$, but you know that $p(\mathrm{A} \mid \mathrm{B}) = \frac{p(\mathrm{A} \cap \mathrm{B})}{p(\mathrm{B})}$, and the numerator is $p(\mathrm{A} \cap \mathrm{B}) = 0$ because A and B are mutually exclusive. This fact forces the entire conditional probability, $p(\mathrm{A}|\mathrm{B})$, to be zero. But for A and B to be independent, $p(\mathrm{A}|\mathrm{B})$ has to be equal to $p(\mathrm{A})$, and it can't be zero unless A is the empty set. So, if events A and B are nonempty and mutually exclusive, they can't be independent.

REMEMBER

Mutually exclusive events can't be independent and independent events can't be mutually exclusive unless one (or both) of the events is the empty set.

Checking for independence or exclusivity in a 52-card deck

Suppose you pull a card from a standard 52-card deck. Let $A = \{\text{the card is a 2}\}$; let $B = \{\text{the card is black}\}$; let $C = \{\text{the card is face card}\}$; and let $D = \{\text{the card isn't a face card}\}$. You find that $p(A) = 4/52$, or $1/13$; $p(B) = 26/52$, or $1/2$; $p(C) = 12/52$, or $3/13$; and $p(D) = 1 - 3/13 = 10/13$ (by the complement rule, because C and D are complementary events; see the earlier section, "The complement rule [for opposites, not for flattering a date]"). Are events A and B mutually exclusive? No, because they have an intersection: Two of the cards in the deck are 2s and black (2♠ and 2♣).

Are events A and B independent? You find that $p(A \cap B) = p(\text{the card is a black } 2) = 2/52$, or $1/26$. You can also find that $p(A) \times p(B) = (1/13)(1/2) = 1/26$. Because these probabilities are equal, the events are independent. This makes sense, because if you know that the card is black, the probability of it being a 2 is $2/26$, which is the same as the probability of the card being a 2 without knowing it's black (which is $4/52 = 2/26$). So, knowing that the card is black doesn't affect the probability of it being a 2.

What about events A and C? Are they independent? Because a card can't be a 2 and a face card at the same time, the intersection is the empty set, so A and C are mutually exclusive — their probabilities directly affect each other. Thus, $p(A) = 4/52$, and $p(A \mid C) = 0$; because these numbers aren't equal, the events aren't independent.

What about events A and D? Their intersection is the set of all cards that are 2s and not face cards, of which there are four (all the 2s), so $p(A \cap D) = 4/52$. Therefore, A and D aren't mutually exclusive. Are the events independent? Because $p(A) = 4/52 = 1/13$ and $p(D) = 40/52 = 10/13$, you find that $p(A) \times p(D) = (1/13)(10/13) = 10/169 = 0.059$. But remember, the intersection of A and D contains four cards (all the 2s), so its probability is $4/52 = 0.077$. These two probabilities aren't equal, so events A and D aren't independent.

PINPOINTING PROBABILITY MODELS: IT'S ALL IN THE WAY YOU SAY IT

The word "or" is a clue that you need to find the probability of a union — for example, when you have to find the probability that someone owns more than one cellphone or more than one land-line phone. The word "and" is a clue that you need to find the probability of an intersection — for example, when you need to find the probability that someone owns at least one cellphone and at least one land-line phone. The word "of" is a good indicator that you're looking at a conditional probability — for example, of those people who own land-line phones, what's the probability that they also own cellphones?

REMEMBER

For two events to be independent, $p(A)$ has to be equal to $p(A|B)$ — not close, but equal. In the previous example, the probabilities 0.059 and 0.077 may seem close, but close doesn't count when it comes to independence. The numbers have to be exact.

EXAMPLE

Q. Suppose you flip a fair coin three times.

(a) Give an example of two events in this scenario that are mutually exclusive.

(b) Give an example of two events in this scenario that are independent.

A. It's important to be able to distinguish these two terms and be able to recognize them when they pop up in examples.

(a) When events are mutually exclusive, it means that they can't happen at the same time. An example here would be flipping exactly three heads ({HHH}) and flipping exactly two heads (one example is {THH}) in the same three tosses.

(b) When events are independent, it means that one can happen or not happen and it won't affect whether or not the other one happens. An example here would be flipping a head on the first toss and flipping a head on the second toss. What you flip on the first toss has no bearing on what you get on the second toss.

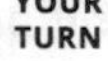

4 Suppose M&Ms' colors come in the following percentages: 13 percent brown, 14 percent yellow, 13 percent red, 24 percent blue, 20 percent orange, and 16 percent green. You're at the M&M factory and get to randomly take two M&Ms without looking, one at a time. Suppose you really want two red M&Ms (your favorite flavor of M&M!).

(a) Consider the two events of first getting a red one (A) and then secondly getting a red one (B). Are A and B mutually exclusive?

(b) Are A and B independent?

(c) What is the chance of getting two red M&Ms?

Suppose you draw two cards from a standard 52-card deck.

(a) What is the probability of drawing two kings?

(b) What is the probability of drawing two kings or two queens?

(c) What is the probability of drawing two cards that have the same rank, also known as a pair?

Avoiding Probability Misconceptions

Probability often goes against your intuition or your desires; people often ignore it or are unaware of its impact. For example, thinking that every situation with two possible outcomes is a 50-50 situation can get you into trouble (yet it seems tempting). And the biggest misconceptions are in the gambling arena, with people thinking they'll hit it big any minute now. The casinos are counting on you having that attitude (and are literally counting your money).

WARNING

Here are some common misconceptions about probability that you want to avoid:

» Believing that outcomes that appear to be "more random" have a higher chance of occurring than outcomes that don't

» Thinking probability works well for predicting short-run behavior

» Claiming you can be "on a roll" or "due for a hit" by the law of averages

» Treating any situation with only two possible outcomes as a "50-50" situation

» Misinterpreting a rare event

The problem with these misconceptions is that they appear to make sense, and you may want them to be true, but they just aren't. With probability practice, you can begin to see through the misconceptions and find out how to avoid them.

EXAMPLE

Q. Suppose an NBA player's free-throw shooting percentage is 70 percent.

(a) Explain what this means as a probability.

(b) What's wrong with thinking that his chances of making his next free throw are 50-50 (because he either makes it or he doesn't, right)?

A. The misconception that random situations having only two possible outcomes are 50-50 situations is a very common one.

(a) The 70 percent shooting clip means that in the long term, over many free throws, this player makes his shots 70 percent of the time, on average.

(b) The 50-50 argument breaks down because the two outcomes aren't equally likely. According to past data, this player hits 70 percent of his free throws and misses only 30 percent of the time. Using these numbers is similar to flipping an unfair coin. You have two sides, yes, but the two outcomes aren't equally likely. If the 50-50 argument worked, then everyone should buy lottery tickets, because either you win or you don't with a 50-50 chance, right? (In your dreams!)

REMEMBER

Just because you have two possible outcomes doesn't mean they each have a 50 percent chance of happening. You have to look at past data and determine what the weight is for each outcome, just like in every other situation.

YOUR TURN

6 Suppose you buy a lottery ticket, and you have to pick six numbers from 1 through 50 (repetitions allowed). Which combination is more likely to win: (13, 48, 17, 22, 6, 39) or (1, 2, 3, 4, 5, 6)?

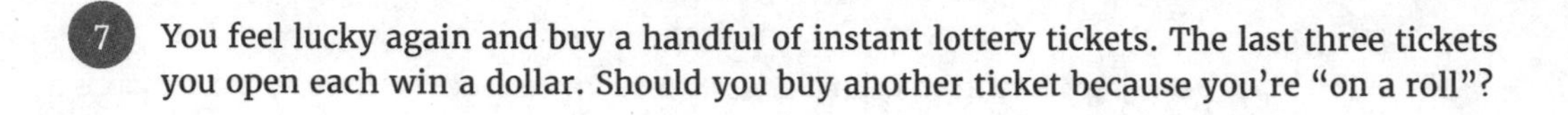

7 You feel lucky again and buy a handful of instant lottery tickets. The last three tickets you open each win a dollar. Should you buy another ticket because you're "on a roll"?

8 Suppose that a small town has five people with a rare form of cancer. Does this automatically mean a huge problem exists that needs to be addressed?

Predictions Using Probability

One of the most important functions of probability used by researchers and the media is its usefulness in making predictions. Forecasts can be very useful, but you have to temper them with the understanding that probability is a long-term predictor, not a short-term fix. You also need to realize that building a model to make a prediction can be a very complicated business.

EXAMPLE

Q. Suppose you know that the chance of winning your money back on an instant lottery ticket is 1 in 10. Does that mean if you buy ten lottery tickets, you know one of them will be an instant winner?

A. No. This misconception of probability is a popular one. Probability is the long-term percentage of instant winners, and it doesn't apply to a sample as small as ten. Results will vary from sample to sample.

YOUR TURN

9 A couple has conceived three girls so far with a fourth baby on the way. Do you predict the newborn will be a girl or a boy? Why?

10 Meteorologists use computer models to predict when and where a hurricane will hit a coast. Suppose they predict that hurricane Stat has a 20 percent chance of hitting the east coast.

(a) On what info are the meteorologists basing this prediction?

(b) Why is this prediction harder to make than your chance of getting a head on your next coin toss?

11 Bob has glued himself to a certain slot machine for four hours in a row now with his bucket of coins and a bad attitude. He doesn't want to leave because he feels the longer he plays, the better chance he has to win eventually. Is poor Bob right?

12 Which situation is more likely to produce exactly 50 percent heads: flipping a coin 10 times or flipping a coin 10,000 times?

Practice Questions Answers and Explanations

1. Probabilities apply to individual selections as well as to long-term frequencies.

   a. To get brown (B) or yellow (Y), you're talking about a union, $B \cup Y$. You use the addition rule to get $p(B \cup Y) = p(B) + p(Y) - p(B \cap Y)$. Because 13 percent of all plain M&Ms are brown, the chance that the one M&M you pick is brown is 0.13. For yellow, the chance is 0.14, or 14 percent. And lastly, because each M&M only has one color, the probability that you'd get one that's brown and yellow at the same time is zero. The probability of getting either a brown or yellow M&M is $p(B \cup Y) = p(B) + p(Y) - p(B \cap Y) = 0.13 + 0.14 - 0 = 0.27$, or **27 percent**.

   b. You could add up the individual probabilities for all colors that aren't green (G) or go the faster route and use the complement rule. The probability of getting a green M&M is 16 percent, so $p(G) = 0.16$. Using the complement rule, you get $p(G^C) = 1 - p(G) = 1 - 0.16 = 0.84$, or **84 percent**.

2. You don't have enough data to determine the probability until you look at the long-run percentage of heads, and four flips isn't a long run of data collecting. But another way to look at this question is to find the probability of flipping four heads on four flips of a coin. Flipping a coin four times gives you $2^4 = 16$ possibilities. You can get four heads in only one way, so the chance of flipping four heads on four flips is $1 \div 16 = 0.06$, or 6 percent. It doesn't happen very often, but it does happen. You may be skeptical of the coin, but you should collect more data before you decide.

3. This problem seems daunting until you realize you have an easy way and a hard way to do this problem (and your professor is banking on you realizing the same thing!).

   a. Flipping a coin ten times results in $2 \times 2 \times 2 \times \quad \times 2 = 2^{10} = \mathbf{1{,}024}$ **possible outcomes**. (Don't try to list them all out!)

   b. Because the coin is fair, each outcome is equally likely and has a probability of **1 in 1,024**.

   c. You have ten ways to get a single head on ten flips. One way is to have HTTTTTTTTT. But the head can come on the second toss, the third toss, or anywhere up to the tenth toss. Counting all the places where the H occurs leaves you with **ten possible outcomes**. The probability is, therefore, 10 in $1{,}024 = \mathbf{0.010}$.

   d. To flip no heads, you have to get all tails, and that happens only **one way out of 1,024**. The probability is $1 \div 1{,}024$, which equals **0.001**.

   e. The event of getting one head or less means that you either get one head or zero heads, so you're talking about a union, $\{1\} \cup \{0\}$. You use the addition rule to get $p(\{1\} \cup \{0\}) = p(1) + p(0) - p(\{1\} \cap \{0\})$. The event of $\{1\} \cap \{0\}$ is the same as saying you'll get one head and zero heads in ten coin flips. No way! So the probability of that is zero. You found the other two probabilities in Parts C and D, so $p(\{1\} \cup \{0\}) = 0.010 + 0.001 - 0 = \mathbf{0.011}$. That means there's a 1.1 percent chance of getting one head or less in ten coin tosses.

f. Solving this problem is much easier if you look at the complement. The complement of flipping more than one head (call that event A) is flipping less than or equal to one head (A^C; a number line can help show you this as well). In Answer (e), you find the probability of less than or equal to one head to be $\frac{11}{1{,}024} = 0.011$. The probability of flipping greater than one head is, therefore, $p(A) = 1 - p(A^C) = 1 - (11 \div 1{,}024) = 1 - 0.011 = \mathbf{0.989}$.

4. It'd be kind of a letdown to go to the M&M factory and only get two M&Ms from the visit.

a. When events are mutually exclusive, it means that they can't happen at the same time. Is it possible to first pull a red M&M and then to secondly pull another red one? Yes! **Because A and B can both happen on the same pull of two M&Ms, these events are not mutually exclusive.**

b. **Here, it's reasonable to assume that they are independent.** When events are independent, whether or not one happens does not affect whether the other one happens. Does the color of the first M&M give you a clue about what the second one will be? No. At the factory, pulling just one M&M out doesn't drastically change the color makeup of all the M&Ms being produced, and you can assume the percentages given previously still hold. So the chance of the second one being red is not influenced by whether or not the first is red.

If you know the specific color makeup in a smaller sample, then the events could be dependent. Suppose you originally have a small bag instead of access to every M&M in the factory. Say there are two reds and one blue left in the bag. If the first is red, then the second has a 50-50 chance of being red. But if the first is blue, then the second has a 100 percent chance of being red. In this case, the color of the first M&M affects the chance that the second is red: That's dependence.

c. The chance of getting two red M&Ms can be split into the two events of first getting a red one (A) and then secondly getting a red one (B). Using notation, that's $p(\text{two red M\&Ms}) = p(A \cap B)$. Then, applying the multiplication rule and the fact that A and B are independent (see Part B), you have $p(A \cap B) = p(A)p(B \mid A) = p(A)p(B)$. Now insert the probability of getting a red M&M, and you get $p(A \cap B) = p(A)p(B) = (0.13)(0.13) = 0.0169$. **There's a 1.69 percent chance of getting two red M&Ms.**

5. It's often a good strategy to assign some notation to the events you're looking at to keep track of everything that's going on.

a. Let K1 = "getting a king on the first draw" and let K2 = "getting a king on the second draw." You're looking for $p(K1 \cap K2)$. Using the multiplication rule, that's the same as $p(K1)$ $p(K2|K1)$. The chance that the first card is a king is $4/52$. Then, given that the first card is a king, you're left with only three kings among the remaining 51 cards in the deck, so $p(K2 \mid K1) = 3/51$. That means the probability of drawing two kings is $p(K1) \times p(K2 \mid K1) = (4/52)(3/51) = \mathbf{0.0045}$.

b. Because you see the reference to "or" in the question, you'll need the addition rule. In this case, you want $p(\text{"draw 2 kings"} \cup \text{"draw 2 queens"})$. Note that drawing two kings and drawing two queens can't both happen, so they are mutually exclusive events. That leaves you with the abbreviated addition rule of $p(\text{"draw 2 kings"} \cup \text{"draw 2 queens"}) = p(\text{"draw 2 kings"}) + p(\text{"draw 2 queens"})$. You already found the probability of drawing two kings in Part A. The result would be exactly the same if you swapped the word *king* with *queen*. So the final answer is $p(\text{"draw 2 kings"} \cup \text{"draw 2 queens"}) = p(\text{"draw 2 kings"}) + p(\text{"draw 2 queens"}) = 0.0045 + 0.0045 = \mathbf{0.0090}$.

c. In Part A, you found the chance of getting one specific pair. In Part B, you found the chance of getting either of two specific pairs. Here, you're asked for the chance of getting any of the 13 possible pairs (because there are 13 ranks). This can be simply solved as $13\times0.0045=0.0585$. But there might be some rounding error from using 0.0045, so use the work from Part A to instead calculate $13\times\left(\frac{4}{52}\right)\left(\frac{3}{51}\right)=13\times\left(\frac{1}{13}\right)\left(\frac{3}{51}\right)=\frac{3}{51}=\mathbf{0.0588}$.

But suppose you hadn't already worked out Part A. How else could you calculate the probability of getting any pair of two matching cards by rank? Basically, you just need the second card to match the first. The first card could be any of the 52 cards. However, of the remaining 51 cards, there are now only three remaining that match the first one you drew. So the probability of getting that second card to match the first is $\frac{3}{51}=0.0588$, exactly the same as before.

6. **Both outcomes are equally likely.** Assuming the lottery process is fair, every single combination of six numbers has an equally likely chance of being selected. The combo 1, 2, 3, 4, 5, 6 seems like it could never happen, but the probability shows you how unlikely any combination is to win. After all, with 50 numbers to choose from and 6 numbers to pick, you have millions of possibilities. But here's a tip: Go ahead and pick 1, 2, 3, 4, 5, 6. If you do win, you won't have to split the money with anyone, because no one else is picking that combination! (Until they read this, at least.)

7. **No.** Probability is a long-term percentage. What recently happened has no impact on what happens in the future. Suppose 5 percent of all instant lottery tickets are winners. This figure tells you nothing about when those winners will come up.

REMEMBER

Probability predicts long-term behavior only; it can't guarantee any kind of short-term outcomes. Numbers may take a long time to "average out," and there's no such thing as being "on a roll" or "due for a hit." The law of averages idea only applies to the long term.

8. **Not necessarily.** The chances of this happening are quite small, but even if the probability is one in a million, you should expect it to happen, on average, once out of every million times. Over a period of years, in a very large country, a situation like this is bound to happen just by chance. The outbreak may point to something else, and the town should investigate, but it doesn't automatically mean that a problem exists.

9. The chance of having a boy this time is the same as it was with the previous births: 1 in 2. Probability has no memory of recent happenings and can't predict short-term behavior.

10. Computer models use a process called *simulation.* You put all your data in, make a mathematical model out of it, and repeatedly run the computer through the scenario to see what happens.

a. The 20 percent comes from the fact that 20 percent of the times the computers repeated the scenario the results pointed to the hurricane hitting the east coast, and 80 percent of the times the results had it veering off into sea.

b. Computer models are based on a great deal of information, but many assumptions fill in the blanks. Some of these assumptions can be wrong, thus throwing the percentage predictions right out the window.

11. **Not really.** In the long run, Bob should expect to lose a small amount with a very high probability every time he pulls the handle on the machine. Yes, he may win big on one pull, but the chance is so tiny that, on average, he doesn't have enough time in his life to sit on that stool and wait to win. Who makes all the money? The person who owns the slot machine, that's who.

12. **Flipping a coin 10,000 times has a higher chance of landing 50 percent heads (exactly),** because when you flip it only 10 times, the results still vary so much that you can get results like three, four, five, six, or seven heads with a fairly high probability. But when you increase the sample size to 10,000, the relative frequency (percentage of heads observed) becomes closer and closer to the true probability you expect (in this case 50 percent). This shows the real law of averages at work.

REMEMBER

The law of averages says that, in the long run, the percentage of occurrences of an event gets closer and closer to the true probability of the event.

If you're ready to test your skills a bit more, take the following chapter quiz that incorporates all the chapter topics.

Whaddya Know? Chapter 8 Quiz

Quiz time! Complete each problem to test your knowledge on the various topics covered in this chapter. You can then find the solutions and explanations in the next section.

1. The chance that a student uses an iPad for school is 70%. You sample ten students at random. What's the chance that at least one of them uses an iPad for school?

2. Which of the following is an uncountably infinite sample space: the number of cars crossing an intersection over a one-week period, or the height of a Sequoia tree measured exactly?

3. If two events A and B are mutually exclusive, what is $p(A \mid B)$?

4. Suppose $p(A) = 0.5$, $p(B) = 0.3$, and $p(B \mid A) = 0.4$. What is $p(A \cap B)$?

5. Suppose two events are complements. Are they independent?

6. Suppose $p(A) = 0.3$, $p(B) = 0.3$. Are A and B independent?

7. Bob and Bill share an apartment. Bob is in the apartment 30% of the time, and Bill is in the apartment 40% of the time. They are in the apartment together 12% of the time. Is Bob being in the apartment independent of Bill being in the apartment?

8. Suppose $p(A) = 0.40$ and $p(B) = 0.30$ and A and B are independent. What is $p(A \cup B)$?

9. If the Super Lotto draws 5 numbers without repeating any of them, you should never bet on a set of numbers like 1, 2, 3, 4, 5 because that combination is too unlikely. True or false?

10. A card being a 2 and a card being red are independent. True or false?

Answers to Chapter 8 Quiz

1. $p(\text{at least one iPad}) = 1 - p(\text{neither has an iPad}) = 1 - p(\text{no iPad}) \times p(\text{no iPad}) = 1 - 0.3 \times 0.3 = 1 - 0.09 = \mathbf{0.81}$. (Note that the probability of no iPad is $1 - p(\text{iPad}) = 1 - 0.7 = 0.3$.)

2. **The height of the Sequoia tree measured exactly is uncountably infinite,** because height can technically be measured to an infinite number of decimal places. The number of cars crossing an intersection over a one-week period is countably infinite.

3. $p(A \mid B) = p(A \cap B) / p(B) = 0 / p(B) = \mathbf{0}$. Also note that if two events are mutually exclusive, they exclude each other. If one happens, the other one cannot happen. That's another reason why $p(A \mid B) = 0$.

4. $p(A \cap B) = p(A) \times p(B \mid A)$ by the multiplication rule, which gives you $0.5 \times 0.4 = \mathbf{0.20}$. Resist the urge to use $p(A) \times p(B)$; A and B are not independent.

5. **If two events are complements, they are mutually exclusive. That means they cannot be independent.** Independent events coexist; the probability of their intersection is the product of their probabilities. Mutually exclusive events exclude each other; the probability of their intersection is zero.

6. **Maybe yes, maybe no.** You don't have enough information to answer this question. You need to be able to check to see if $p(A \mid B) = p(A)$ or if $p(A \cap B) = p(A) \times p(B)$.

7. **Yes,** because $p(\text{Bob and Bill}) = 0.12 = p(\text{Bob}) \times p(\text{Bill}) = 0.30 \times 0.40$.

8. $p(A \cup B) = p(A) + p(B) - p(A \cap B)$ by the Addition Rule. Because A and B are independent, $p(A \cap B) = p(A) \times p(B) = 0.30 \times 0.40 = 0.12$. So $p(A \cup B) = 0.30 + 0.40 - 0.12 = \mathbf{0.58}$.

9. **False.** The combination 1, 2, 3, 4, 5 is just as likely as any other combination of 5 numbers. It just *looks* more unlikely. This shows you how truly unlikely it is to win the Super Lotto!

10. **True.** Given $p(2) = \frac{4}{52}$ and $p(\text{red}) = \frac{26}{52} = \frac{1}{2}$, their product is $\frac{4}{52} \times \frac{1}{2} = \frac{2}{52} = \frac{1}{26}$. You also know that $p(\text{2 and red}) = p(\text{2 of hearts and 2 of diamonds}) = \frac{2}{52} = \frac{1}{26}$.

IN THIS CHAPTER

- » Identifying a binomial random variable
- » Finding probabilities using a formula or table
- » Calculating the mean and variance

Chapter 9

Random Variables and the Binomial Distribution

Scientists and engineers often build models for the phenomena they are studying to make predictions and decisions. For example, where and when is this hurricane going to hit when it makes landfall? How many accidents will occur at this intersection this year if it's not redone? Or, what will the deer population be like in a certain region five years from now?

To answer these questions, scientists (usually working with statisticians) define a characteristic they are measuring or counting (such as number of intersections, location and time when a hurricane hits, population size, and so on) and treat it as a variable that changes in some random way, according to a certain pattern. They cleverly call them — you guessed it — random variables. In this chapter, you find out more about random variables, their types and characteristics, and why they are important. And you look at the details of one of the most common random variables: the binomial.

Defining a Random Variable

A *random variable* is a characteristic, measurement, or count that changes randomly according to a certain set or pattern. Its notation is *X*, *Y*, *Z*, and so on. In this section, you see how different random variables are characterized and how they behave in the long run in terms of their means and standard deviations.

REMEMBER

In math you have variables like *X* and *Y* that take on certain values depending on the problem (for example, the width of a rectangle), but in statistics the variables change in a random way. By *random*, statisticians mean that you don't know exactly what the next outcome will be, but you do know that certain outcomes happen more frequently than others; everything's not 50-50. (Like when my aunt tries to shoot baskets; it's definitely not a 50 percent chance she'll make one and a 50 percent chance she'll miss. It's more like a 5 percent chance of making it and a 95 percent chance of missing it.) You can use that information to better study data and populations and make good decisions. (For example, don't put my aunt in your basketball game to shoot free throws.)

Data have different types: categorical and numerical (see Chapter 3). While both types of data are associated with random variables, only numerical random variables are discussed here (this falls in line with most intro stat courses as well). For information on analyzing categorical variables, see Chapters 4, 6, and 20.

Discrete versus continuous

Numerical random variables represent counts and measurements. They come in two different flavors: discrete and continuous, depending on the type of outcomes that are possible.

- **Discrete random variables:** If the possible outcomes of a random variable can be listed out using whole numbers (for example: 0, 1, 2, 3; or 1, 2, ...), the random variable is *discrete.*
- **Continuous random variables:** If the possible outcomes of a random variable can only be described using an interval of real numbers (for example, all real numbers from zero to infinity), the random variable is *continuous*.

Discrete random variables typically represent counts — for example, the number of people who voted yes for a smoking ban out of a random sample of 100 people (possible values are 0, 1, 2, ... , 100), or the number of accidents at a certain intersection over one year's time (possible values are 0, 1, 2, ...).

REMEMBER

Discrete random variables have two classes: finite and countably infinite. A discrete random variable is *finite* if its list of possible values has a fixed (finite) number of elements in it (for example, the number of smoking ban supporters in a random sample of 100 voters has to be between 0 and 100). One very common finite random variable is the binomial, which is discussed in this chapter in detail.

A discrete random variable is *countably infinite* if its possible values can be specifically listed out but they have no specific end. For example, the number of accidents occurring at a certain intersection over a ten-year period can take on possible values: 0, 1, 2,... . (You know there is a largest value somewhere, but you can't say where exactly, so you list them all. There could be 100 accidents, or 1,000, or 1,001, for example.)

Continuous random variables typically represent measurements, such as time to complete a task (for example, 1 minute 10 seconds, 1 minute 20 seconds, and so on) or the weight of a newborn. What separates continuous random variables from discrete ones is that they are *uncountably infinite;* they have too many possible values to list out or to count and/or they can be measured to a high level of precision (such as the level of smog in the air in Los Angeles on a given day, measured in parts per million).

Examples of commonly used continuous random variables can be found in Chapter 10 (the normal distribution) and Chapter 11 (the *t*-distribution).

TIP

If you can split the measurement in half and it still makes sense as a possible value, then the variable is continuous. For example, there's such a thing as half of a minute (or half of a second) when measuring the time to complete a task. If you can't split the measurement in half, then it's probably a discrete random variable. Talking about "half of a person" or "half of a yes vote" in the smoking ban example and talking about "half of an accident" at an intersection makes no sense.

EXAMPLE

 Which of these is a variable?

(a) The height of the Empire State Building.

(b) The height of a randomly selected building in New York City.

 A variable is a measurement that has to be able to vary.

(a) This is not a variable. The height of the Empire State Building is constant and doesn't change. It is stuck at 1,454 feet.

(b) Because you don't know which building might be selected (maybe the Empire State Building, 30 Rock, or some hotel), the value of the height could vary. This one is a random variable.

YOUR TURN

 Identify whether each of the following is a categorical or numerical variable.

(a) The number of miles on a car's odometer.

(b) The color of a car.

(c) The number of passengers a car holds.

(d) The Zip code of the address that the car is registered at.

 Identify whether each of the following numerical variables is discrete or continuous.

(a) The time it takes you to commute from home to work.

(b) The number of stoplights you pass through during your commute to work.

(c) The number of hours you are at work in a given week, rounded to the nearest hour.

(d) The average annual salary of five randomly selected people at your work.

Probability distributions

A discrete random variable X can take on a certain set of possible outcomes, and each of those outcomes has a certain probability of occurring. The notation used for any specific outcome is a lowercase x. For example, say you roll a die and look at the outcome. The random variable X is the outcome of the die (X takes on possible values of {1, 2, ... , 6}). Now, if you roll the die and get a 1, that's a specific outcome, so you write "$x = 1$."

The probability of any specific outcome occurring is denoted $p(x)$, which you pronounce "p of x." It signifies the probability that the random variable X takes on a specific value, which you call "little x." For example, to denote the probability of getting a 1 on a die, you write $p(1)$.

REMEMBER

Statisticians use an uppercase X when they talk about random variables in their general form; for example, "Let X be the outcome of the roll of a single die." They use lowercase x when they talk about specific outcomes of the random variable, like $x = 1$ or $x = 2$.

A list or function showing all possible values of a discrete random variable, along with their probabilities, is called a *probability distribution*, $p(x)$. For example, when you roll a single die, the possible outcomes are 1, 2, 3, 4, 5, and 6, and each has a probability of 1/6 (if the die is fair). As another example, suppose 40 percent of renters living in an apartment complex own one dog, 7 percent own two dogs, 3 percent own three dogs, and 50 percent own zero dogs. For X = the number of dogs owned, the probability distribution for X is shown in Table 9-1.

TABLE 9-1 Probability Distribution for X = Number of Dogs Owned by Apartment Renters

x	$p(x)$
0	0.50
1	0.40
2	0.07
3	0.03

REMEMBER

Each probability distribution has two important criteria: 1) each probability in the table must be between 0 and 1; and 2) all the probabilities in the table must sum to one.

The mean and variance of a discrete random variable

The *mean* of a random variable is the average of all the outcomes you would expect in the long term (over all possible samples). For example, if you roll a die a billion times and record the outcomes, the average of those outcomes is 3.5. (Each outcome happens with equal chance, so you average the numbers 1 through 6 to get 3.5.) However, if the die is loaded and you roll a 1 more often than anything else, the average outcome from a billion rolls is closer to 1 than to 3.5.

TIP

The notation for the mean of a random variable X is μ_x or μ (pronounced "mu sub x", or just "mu x"). Because you are looking at all the outcomes in the long term, it's the same as looking at the mean of an entire population of values, which is why you denote it μ_x and not $\bar{x}$. (The latter represents the mean of a *sample* of values [see Chapter 5].) You put the X in the subscript to remind you that the variable this mean belongs to is the X variable (as opposed to a Y variable or some other letter).

The *variance* of a random variable is roughly interpreted as the average squared distance from the mean for all the outcomes you would get in the long term, over all possible samples. This is the same as the variance of the population of all possible values. The notation for variance of a random variable X is σ_x^2 or σ^2. You say "sigma sub x, squared" or just "sigma squared."

The standard deviation of a random variable X is the square root of the variance, denoted by σ_x or σ (say "sigma x" or just "sigma"). It roughly represents the average distance from the mean.

Just as for the mean, you use the Greek notation to denote the variance and standard deviation of a random variable. The English notations s^2 and s represent the variance and standard deviation of a *sample* of individuals, not the entire population (see Chapter 5).

REMEMBER

The variance is in square units, so it can't be easily interpreted. You use standard deviation for interpretation because it is in the original units of X. The standard deviation can be roughly interpreted as the average distance away from the mean.

Identifying a Binomial

The most well-known and loved discrete random variable is the binomial. *Binomial* means *two names* and is associated with situations involving two outcomes — for example, yes/no or success/failure (making a basket [in basketball] or missing, hitting a red light or not, developing a side effect or not). This section focuses on the binomial random variable: when you can use it, finding probabilities for it, and finding its mean and variance.

A random variable is binomial (that is, it has a binomial distribution) if the following four conditions are met:

1. There are a fixed number of trials (n).
2. Each trial has two possible outcomes: success or failure.
3. The probability of success (call it p) is the same for each trial.
4. The trials are independent, meaning the outcome of one trial doesn't influence that of any other.

Let X equal the total number of successes in n trials; if all four conditions are met, X has a binomial distribution with probability of success (on each trial) equal to p. The possible values of X are $\{0, 1, 2, \ldots, n\}$.

The lowercase p here stands for the probability of getting a success on one single (individual) trial. It's not the same as $p(x)$, which means the probability of getting x successes in n trials.

Checking binomial conditions step by step

You flip a fair coin ten times and count the number of heads (X). Does X have a binomial distribution? You can check by reviewing your responses to the questions and statements in the list that follows:

1. **Are there a fixed number of trials?**

   You're flipping the coin ten times, which is a fixed number. Condition 1 is met, and $n = 10$.

2. **Does each trial have only two possible outcomes — success or failure?**

   The outcome of each flip is either heads or tails, and you're interested in counting the number of heads. That means success = heads, and failure = tails. Condition 2 is met.

3. **Is the probability of success the same for each trial?**

   Because the coin is fair, the probability of success (getting a head) is $p = \frac{1}{2}$ for each trial. You also know that $1 - \frac{1}{2} = \frac{1}{2}$ is the probability of failure (getting a tail) on each trial. Condition 3 is met.

   Note: The probability of success does not have to be the same as the probability of failure as in this example. The condition would also hold if the coin was biased, and there was a $p = 0.60$ chance of getting heads on any one flip.

4. **Are the trials independent?**

   You assume the coin is being flipped the same way each time, which means the outcome of one flip doesn't affect the outcome of subsequent flips. Condition 4 is met.

Because the random variable X (the number of successes [heads] that occur in ten trials [flips]) meets all four conditions, you conclude it has a binomial distribution with $n = 10$ and $p = \frac{1}{2}$.

But not every situation that appears binomial actually is. Read on to see some examples of this.

No fixed number of trials

Suppose that you're going to flip a fair coin until you get four heads, and you'll count how many flips it takes to get there; in this case, X = number of flips. This certainly sounds like a binomial situation: condition 2 is met because you have success (heads) and failure (tails) on each flip; condition 3 is met with the probability of success (heads) being the same (0.5) on each flip; and the flips are independent, so condition 4 is met.

However, notice that X isn't counting the number of heads (successes); it's counting the number of flips (trials) needed to get four heads. The number of successes (X) is fixed rather than the number of trials (n). Condition 1 is not met, so X does not have a binomial distribution in this case.

More than success or failure

Some situations involve more than two possible outcomes, yet they can appear to be binomial. For example, suppose you roll a fair die ten times and let X be the outcome of each roll {1, 2, 3, ..., 6}. You have a series of $n = 10$ trials, they are independent, and the probability of each outcome is the same for each roll. However, on each roll you're recording the outcome on a six-sided die, a number from 1 to 6. This is not a success/failure situation, so condition 2 is not met.

However, depending on what you're recording, situations originally having more than two outcomes can fall under the binomial distribution. For example, if you roll a fair die ten times and each time you record whether or not you get a 1, then condition 2 is met because your two outcomes of interest are getting a 1 ("success") and not getting a 1 ("failure"). In this case, p (the probability of success) $= \frac{1}{6}$, and $\frac{5}{6}$ is the probability of failure. So if X is counting the number of 1s you get in ten rolls, then X is a binomial random variable.

Trials are not independent

The independence condition is violated when the outcome of one trial affects another trial. Suppose you want to know opinions of adults in your city regarding a proposed casino. Instead of taking a random sample of, say, 100 people, to save time you select 50 married couples and ask each of them what their opinion is. In this case it's reasonable to say couples have a higher chance of agreeing on their opinions than individuals selected at random, so the independence condition 4 is not met.

REMEMBER

If you were random sampling 100 people, it would be reasonable to assume that one person's option or action wouldn't affect another, like in the example of voting in an election. Sometimes it's a little trickier to justify independence between trials when one person is repeating an action again and again. For example, if you take 100 free throws to see how many you make, initially the result of your previous shot might not affect your next one. After a while, though, your arms get tired and you might start to lose focus, perhaps causing your probability of successfully making a shot to go down. If, on the other hand, those 100 free throws are spread throughout a season, then independence between shots is a more reasonable assumption.

Probability of success (p) changes

You have 10 people — 6 women and 4 men — and you want to form a committee of 2 people at random. Let X be the number of women on the committee of 2. The chance of selecting a woman at random on the first try is $\frac{6}{10}$. Because you can't select this same woman again, the chance of selecting another woman is now $\frac{5}{9}$. The value of p has changed, and condition 3 is not met.

REMEMBER

If the population is very large (for example, all U.S. adults), p still changes every time you choose someone, but the change is negligible, so you don't worry about it. You still say the trials are independent with the same probability of success, p. (Life is so much easier that way!)

EXAMPLE

Q. Suppose you'll be given a dollar for every 5 or 6 you get in the next 20 rolls of a die.

(a) Come up with a binomial random variable based on this situation.

(b) Check the four binomial conditions for the random variable chosen in Part A.

A. A binomial experiment basically amounts to asking a question that gets "yes" or "no" as an answer over and over again.

(a) The binomial random variable has to be something that measures whether or not you get "successes" over repeated trials. In this case, you're interested in the number of 5s or 6s in 20 trials.

(b) Condition 1: You will roll the die a fixed number of times, $n = 20$. Condition 2: The sides of the die are {1, 2, 3, 4, 5, 6}, so in terms of your random variable, a success is rolling 5 or 6, and a failure is rolling 1, 2, 3, or 4. Condition 3: The probability of successfully rolling a 5 or 6 is the same for all trials: $p(5 \text{ or } 6) = 2/6 = 1/3$. Condition 4: Die rolls are independent because the current roll is not affected by the previous result. The four conditions are all satisfied.

YOUR TURN

3 Identify whether each of the following is a binomial random variable.

(a) A police officer watches cars go through a lighted intersection for one hour. Each car that passes through has a probability of 0.01 of running a red light. The variable is the number of cars that run a red light within the one hour.

(b) On your commute to work, you pass through seven intersections with stoplights. There's a 0.60 chance that a stoplight will be red when you get near the intersection. Also, there's a stretch of road in which three of the stoplights are timed to go green in sequence so that traffic can flow smoothly for a while. The variable is the number of intersections you need to stop at.

(c) On your commute to work, you pass through four crosswalks. There's a 0.15 chance that a pedestrian will be there to cross the street as you get near and that you'll have to stop for them to pass. The variable is the number of crosswalks you'll need to stop at.

Finding Binomial Probabilities Using a Formula

After you identify that X has a binomial distribution (the four conditions from the section, "Checking binomial conditions step by step," are met), you'll likely want to find probabilities for X. The good news is that you don't have to find them from scratch; you get to use established formulas for finding binomial probabilities, using the values of n and p unique to each problem. Probabilities for a binomial random variable X can be found using the following formula for $p(x)$:

$$\binom{n}{x} p^x (1-p)^{n-x}$$

where

- n is the fixed number of trials.
- x is the specified number of successes.
- $n-x$ is the number of failures.
- p is the probability of success on any given trial.
- $1-p$ is the probability of failure on any given trial. (***Note:*** Some textbooks use the letter q to denote the probability of failure rather than $1-p$.)

These probabilities hold for any value of X between 0 (the lowest number of possible successes in n trials) and n (the highest number of possible successes).

The number of ways to rearrange x successes among n trials is called "n choose x," and the notation is $\binom{n}{x}$. It's important to note that this math expression is not a fraction; it's math shorthand to represent the number of ways to do these types of rearrangements.

In general, to calculate "n choose x," you use the following formula:

$$\binom{n}{x} = \frac{n!}{x!(n-x)!}$$

The notation $n!$ stands for n-*factorial*, the number of ways to rearrange n items. To calculate $n!$, you multiply $n(n-1)(n-2)(2)(1)$. For example, $5!$ is $5(4)(3)(2)(1) = 120$; $2!$ is $2(1) = 2$; and $1!$ is 1. By convention, $0!$ equals 1.

Suppose you have to cross three traffic lights on your way to work. Let X be the number of red lights you hit out of the three. How many ways can you hit two red lights on your way to work? Well, you could hit a green one first, then the other two red; or you could hit the green one in the middle and have red ones for the first and third lights; or you could hit red first, then another red, and then green. Letting G = green and R = red, you can write these three possibilities as: GRR, RGR, RRG. So you can hit two red lights on your way to work in three ways, right?

Check the math. In this example, a "trial" is a traffic light, and a "success" is a red light. (Okay, that seems weird, but a success is whatever you are interested in counting, good or bad.) So you have $n = 3$ total traffic lights, and you're interested in the situation where you get $x = 2$ red ones. Using the fancy notation, $\binom{3}{2}$ means "3 choose 2" and $p(X = 0) =$ stands for the number of ways to rearrange two successes in three trials.

To calculate "3 choose 2," you do the following:

$$\binom{3}{2} = \frac{3!}{2!(3-2)!} = \frac{3(2)(1)}{[(2)(1)](1)} = \frac{6}{2} = 3$$

This confirms the three possibilities listed for getting two red lights.

Now suppose the lights operate independently of each other and each one has a 30 percent chance of being red. Suppose you want to find the probability distribution for X (that is, a list of all possible values of X — $\{0, 1, 2, 3\}$ — and their probabilities).

Before you dive into the calculations, you first check the four conditions (from the section, "Checking binomial conditions step by step") to see whether you have a binomial situation here. You have $n = 3$ trials (traffic lights) — check. Each trial is success (red light) or failure (yellow or green light; in other words, "non-red" light) — check. The lights operate independently, so you have the independent trials taken care of, and because each light is red 30 percent of the time, you know $p = 0.30$ for each light. So X = number of red traffic lights has a binomial distribution. To fill in the nitty-gritty for the formulas, $1 - p$ = probability of a non-red light $= 1 - 0.30 = 0.70$; and the number of non-red lights is $3 - X$.

Using the formula for $p(x)$, you obtain the probabilities for $x = 0, 1, 2,$ and 3 red lights:

$$p(X=0) = \binom{3}{0} 0.3^0(1-0.3)^{3-0} = \frac{3!}{(0)!(3-0)!}(1)(0.7)^3 = 0.343$$

$$p(X=1) = \binom{3}{1} 0.3^1(1-0.3)^{3-1} = \frac{3!}{(1)!(3-1)!}(0.3)^1(0.7)^2 = (3)(0.3)(0.49) = 0.441$$

$$p(X=2) = \binom{3}{2} 0.3^2(1-0.3)^{3-2} = \frac{3!}{(2)!(3-2)!}(0.3)^2(0.7)^1 = (3)(0.09)(0.70) = 0.189$$

$$p(X=3) = \binom{3}{3} 0.3^3(1-0.3)^{3-3} = \frac{3!}{(3)!(3-3)!}(0.3)^3(0.7)^0 = (1)(0.027)(1) = 0.027$$

The final probability distribution for X is shown in Table 9-2. Notice that these probabilities all sum to 1 because every possible value of X is listed and accounted for.

TABLE 9-2 Probability Distribution for X = Number of Red Traffic Lights ($n = 3, p = 0.30$)

X	$p(x)$
0	0.343
1	0.441
2	0.189
3	0.027

EXAMPLE

Q. Suppose you'll be given a dollar for every 5 or 6 you get in the next 20 rolls of a die.

(a) What is the probability that you'll win exactly 5 dollars?

(b) What is the probability that you'll win any money at all?

A. You already checked that this is a binomial experiment. Now, the binomial formula is perfect when you need to calculate the chance that a binomial random variable hits an exact value. It gives you $p(X=x)$.

(a) You're looking at X = the number of 5s and 6s rolled, and you're given that $n=20$, $p=\frac{2}{6}=\frac{1}{3}$, and $x=5$. Plugging into the formula to find $p(X=5)$, you get

$$\begin{aligned}\binom{n}{x}p^x(1-p)^{n-x} &= \binom{20}{5}\left(\frac{1}{3}\right)^5\left(1-\frac{1}{3}\right)^{20-5}\\ &= \left(\frac{20\cdot\ldots\cdot16\cdot15\cdot\ldots\cdot1}{(5\cdot\ldots\cdot1)(15\cdot\ldots\cdot1)}\right)(0.333)^5(0.667)^{15}\\ &= \frac{20\cdot19\cdot18\cdot17\cdot16}{5\cdot4\cdot3\cdot2\cdot1}(0.0041)(0.0023)\\ &= 15504(0.0041)(0.0023)\\ &= 0.1462\end{aligned}$$

Because $p(5)=0.1462$, there is a 14.62 percent chance of winning exactly 5 dollars.

TIP

When solving $\binom{20}{5}=\frac{20!}{5!15!}$, you could multiply out each of the factorials all the way through and then divide. But some of the products might be so big that they'd break your calculator (well, maybe not break it, but maybe put it to sleep for a second). There's a better way that's on display in the solution. You know that the back end of 20! contains $15\times14\times\ldots\times1$, which is the same as 15!. You can cancel off that back end from the numerator with the 15!, and that leaves you with $\frac{20\cdot19\cdot18\cdot17\cdot16}{5\cdot4\cdot3\cdot2\cdot1}$, which isn't so tough for your dusty old Casio. Of course, many calculators now have a $\binom{n}{x}$ feature on them.

(b) You're looking again at X = the number of 5s and 6s rolled. To win any money at all, X could be 1, 2, 3, or any value up to 20 that is $X\geq1$. That's a lot of binomial formulas to crank out. A better option is to solve this by using the complement rule: $p(X\geq1)=1-p(X<1)$ (see Chapter 8). Because the only possible value less than 1 is 0, you're looking for $1-p(X=0)$. Plugging into the formula to find $p(X=0)$, you get

$$\begin{aligned}\binom{n}{x}p^x(1-p)^{n-x} &= \binom{20}{0}\left(\frac{1}{3}\right)^0\left(1-\frac{1}{3}\right)^{20-0}\\ &= \left(\frac{20!}{0!20!}\right)(0.333)^0(0{,}667)^{20}\\ &= 1\cdot1\cdot(0.667)^{20}\\ &= 0.0003\end{aligned}$$

Because $p(0) = 0.0003$, there is a probability of $1 - 0.0003 = 0.9997$, or 99.97 percent that you'll win money. (Unfortunately, you have about a 0 percent chance of finding someone who's going to actually fork over the money for just rolling a die.)

YOUR TURN

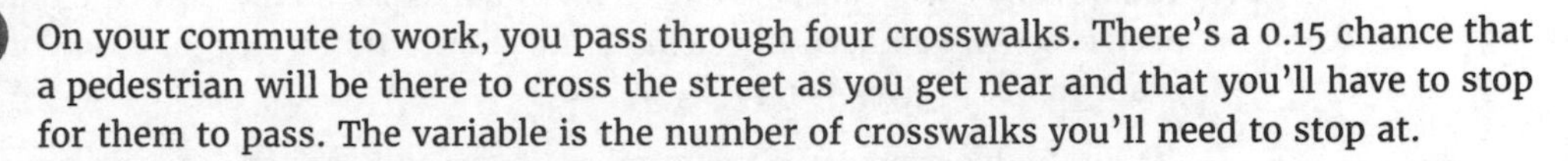

4 On your commute to work, you pass through four crosswalks. There's a 0.15 chance that a pedestrian will be there to cross the street as you get near and that you'll have to stop for them to pass. The variable is the number of crosswalks you'll need to stop at.

(a) What is the probability that you'll have to stop exactly twice?

(b) What is the probability that you'll have to stop for at least one pedestrian?

Finding Probabilities Using the Binomial Table

The previous section deals with values of n that are pretty small, but you may wonder how you are going to handle the formula for calculating binomial probabilities when n gets large. No worries! A large range of binomial probabilities are provided in the binomial table in the Appendix. Here's how to use it:

Within the binomial table you see several mini-tables; each one corresponds with a different n for a binomial ($n = 1$, 2, 3, 15, and 20 are available). Each mini-table has rows and columns. Running down the side of any mini-table, you see all the possible values of X from 0 through n, each with its own row. The columns of the binomial table represent various values of p from 0.10 through 0.90.

Finding probabilities for specific values of *X*

To use the binomial table in the Appendix to find probabilities for X = total number of successes in n trials where p is the probability of success on any individual trial, follow these steps:

1. **Find the mini-table associated with your particular value of n (the number of trials).**
2. **Find the column that represents your particular value of p (or the one closest to it, if appropriate).**
3. **Find the row that represents the number of successes (x) you are interested in.**
4. **Intersect the row and column from Steps 2 and 3.** This gives you the probability for x successes, written as $p(x)$.

For the traffic light example from the section, "Finding Binomial Probabilities Using a Formula," you can use the binomial table (Table A-3 in the Appendix) to verify the results found by the binomial formula as shown in Table 9-2. Go to the mini-table where $n = 3$ and look in the column where $p = 0.30$. You see four probabilities listed for this mini-table: 0.343, 0.441, 0.189, and 0.027; these are the probabilities for $X = 0$, 1, 2, and 3 red lights, respectively, matching those from Table 9-2.

WARNING

Many places in the binomial table (Table A-3 in the Appendix) have probabilities of 0.000, but the 0.000 does not mean that there is zero chance of that value of X occurring. It's just that the probability is so small that it rounds to 0.000 if you're only looking at three decimal places.

Finding probabilities for *X* greater-than, less-than, or between two values

The binomial table (Table A-3 in Appendix) shows probabilities for X being equal to any value from 0 to *n* for a variety of *p*'s. To find probabilities for X being less-than, greater-than, or between two values, just find the corresponding values in the table and add their probabilities. For the traffic light example, you count the number of times (X) that you hit a red light (out of three possible lights). Each light has a 0.30 chance of being red, so you have a binomial distribution with $n = 3$ and $p = 0.30$. If you want the probability that you hit more than one red light, you find $p(X > 1)$ by adding $p(2) + p(3)$ from Table A-3 to get $0.189 + 0.027 = 0.216$.

The probability that you hit between 1 and 3 (inclusive) red lights is $p(1 \leq X \leq 3) = 0.441 + 0.189 + 0.027 = 0.657$.

REMEMBER

You have to distinguish between a *greater-than* (>) and a *greater-than-or-equal-to* (≥) probability when working with discrete random variables. Repackaging the previous two examples, you see $p(X > 1) = 0.216$, but $p(X \geq 1) = 0.657$. This is a non-issue for continuous random variables (see Chapters 10 and 11).

TIP

Other phrases to remember: *at least* means that number or higher, and *at most* means that number or lower. For example, the probability that X is at least 2 is $p(X \geq 2)$, and the probability that X is at most 2 is $p(X \leq 2)$.

EXAMPLE

Q. Change the game a little from earlier in the chapter. Suppose you'll now be given a dollar for every odd number you get in the next 8 rolls of a die.

(a) What is the probability that you'll win exactly 5 dollars?

(b) What is the probability that you'll win any money at all?

Q. A success is rolling a 1, 3, or 5, so the probability of success is $p = 3/6 = 1/2$. Because the sample size ($n = 8$) is small enough and the value $p = 0.50$ appears in Table A-3 in the Appendix, you can use the binomial table. Find the specific mini-table with $n = 8$ and locate the column for $p = 0.50$.

(a) To find the probability that you'll win exactly 5 dollars, find the row for $x = 5$ and then locate where it intersects with the column for $p = 0.50$. The probability is 0.219, or 21.9 percent.

(b) To win any money at all, X could be 1, 2, 3, or any value up to 8 that is $X \geq 1$. You could sum up the values in the column for $p = 0.50$ sitting in rows 1, 2, , all the way to 8. Or use the complement rule: $p(X \geq 1) = 1 - p(X = 0) = 1 - 0.004 = 0.996$. You have a 99.6 percent chance of winning at least a dollar.

YOUR TURN

5 Let X be a binomial random variable with $n = 12$ and $p = 0.25$.

(a) What is the probability that X is 3?

(b) What is the probability that X is greater than 6?

(c) What is the probability that X is at least 6?

6 On your commute to work, you pass through four crosswalks. There's a 0.15 chance that a pedestrian will be there to cross the street as you get near and that you'll have to stop for them to pass. The variable is the number of crosswalks you'll need to stop at. Can you use the binomial table to find the probability that you'll have to stop exactly twice?

Checking Out the Mean and Standard Deviation of the Binomial

Because the binomial distribution is so commonly used, statisticians went ahead and did all the grunt work to figure out nice, easy formulas for finding its mean, variance, and standard deviation. (That is, they've already applied the methods from the section, "Defining a Random Variable," to the binomial distribution formulas, crunched everything out, and presented the results to you on a silver platter — don't you love it when that happens?) The following results are what came out of it.

If X has a binomial distribution with n trials and probability of success p on each trial, then:

1. The mean of X is $\mu = np$.
2. The variance of X is $\sigma^2 = np(1-p)$.
3. The standard deviation of X is $\sigma = \sqrt{np(1-p)}$.

For example, suppose you flip a fair coin 100 times and let X be the number of heads; then X has a binomial distribution with $n = 100$ and $p = 0.50$. Its mean is $\mu = np = 100(0.50) = 50$ heads (which makes sense, because heads and tails are 50-50). The variance of X is $\sigma^2 = np(1-p) = 100(0.50)(1-0.50) = 25$, which is in square units (so you can't interpret it); and the standard deviation is the square root of the variance, which is 5. That means when you flip a coin 100 times, and do that over and over, the average number of heads you'll get is 50, and you can expect that to vary by about 5 heads on average.

TIP

The formula for the mean of a binomial distribution has intuitive meaning. The p in the formula represents the probability of a success, yes, but it also represents the *proportion* of successes you can expect in n trials. Therefore, the total *number* of successes you can expect — that is, the mean of X — is $\mu = np$.

The formula for variance has intuitive meaning as well. The only variability in the outcomes of each trial is between success (with probability p) and failure (with probability $1-p$). Over n trials, the variance of the number of successes/failures is measured by $\sigma^2 = np(1-p)$. The standard deviation is just the square root.

REMEMBER

If the value of n is too large to use the binomial formula or the binomial table to calculate probabilities (see the earlier sections in this chapter), there's an alternative. It turns out that if n is large enough, you can use the normal distribution to get an approximate answer for a binomial probability. The mean and standard deviation of the binomial are involved in this process. All the details are in Chapter 10.

EXAMPLE

Q. Suppose you'll be given a dollar for every 5 or 6 you get in the next 20 rolls of a die. Let X = the number of 5s and 6s rolled. What are the mean and standard deviation of X?

Q. You're given that $n = 20$ and $p = 2/6 = 1/3$. The mean is $\mu = np = (20)(1/3) = 20/3 = 6.67$ dollars. The standard deviation is $\sigma = \sqrt{np(1-p)} = \sqrt{(20)\left(\frac{1}{3}\right)\left(\frac{2}{3}\right)} = \sqrt{4.44} = 2.11$ dollars.

YOUR TURN

7 On your commute to work, you pass through four crosswalks. There's a 0.15 chance that a pedestrian will be there to cross the street as you get near and that you'll have to stop for them to pass. Let X be the number of crosswalks you'll need to stop at. Find the average number of stops you'll have to make as well as the standard deviation.

Practice Questions Answers and Explanations

1. a. **The number of miles on a car's odometer is a numerical variable.** The value is greater than zero and maybe even goes into the hundreds of thousands of miles.

   b. **The color of a car is a categorical variable** with possible values including midnight blue, salsa red, and the ever-unpopular puce.

   c. **The number of passengers a car holds is clearly numerical.**

   d. Even though Zip codes are represented by a number, they don't measure the "amount" of anything. For example, one of Indianapolis's Zip codes is 46256, and of course, in Beverly Hills there's 90210. But does that mean that Beverly Hills is $90210 - 46256 = 43954$ "somethings" better than Indy? No way! **Zip codes are better described as being categorical.**

Just because a variable is measured using a number doesn't mean you can treat the data like a regular numerical variable. Besides Zip codes, other examples include your telephone number and your Social Security number.

2. a. **The time it takes you to commute from home to work is a continuous variable** because any real number from 0 to infinity is fair game.

   b. **The number of stoplights you pass through during your commute to work has to be discrete** because you can count them out while you drive: 0, 1, 2, 3,.

   c. Time spent at work is initially continuous (and some days it seems to be spiraling toward infinity). **But once you're told to round to the nearest hour, it becomes discrete** because you can count out the hours one at a time.

   d. **Salary and other variables related to money are usually treated as continuous.** Typically, statistics generated from numerical variables, like the average, are continuous.

3. Consider all four of the conditions carefully. They all have to be satisfied in order to call what you're counting a binomial variable.

   a. In one hour, there could be 50, 100, or 200 cars. The total number of trials is not fixed. **Because condition 1 is not satisfied, the number of cars running the red light is not a binomial random variable.**

   b. It seems that three of the stoplights are linked. If the first is green, then the other two are also more likely to be green. Because the stoplights' being red is not independent for all seven, condition 4 is not satisfied. **The number of intersections you have to stop at for a red light is not a binomial random variable.**

   c. Condition 1: You drive past a fixed number of crosswalks, $n = 4$. Condition 2: Either a pedestrian is at the crosswalk and you have to stop ("success," but maybe you don't see it that way if you're in a hurry), or there's no one there and you drive on ("failure"). Condition 3: The probability of encountering a pedestrian is fixed at $p = 0.15$. Condition 4: It's reasonable to assume that a person's choice to cross one crosswalk would not affect another person being at another crosswalk. So you can consider the four locations to be independent. **Because the four conditions are met, the number of crosswalks you stop at is a binomial random variable.**

4. This is one of those situations where the "success" is something that you probably aren't hoping for in reality: having to make an extra stop on your way to work.

   a. You're looking at X = the number of crosswalks you'll need to stop at. You're given that $n = 4$, $p = 0.15$, and $x = 2$. Plugging into the formula to find $p(X = 2)$, you get

$$\begin{aligned}\binom{n}{x}p^x(1-p)^{n-x} &= \binom{4}{2}(0.15)^2(1-0.15)^{4-2} \\ &= \frac{4\cdot3\cdot2\cdot1}{(2\cdot1)(2\cdot1)}(0.15)^2(0.85)^2 \\ &= 6(0.0225)(0.7225) \\ &= 0.0975\end{aligned}$$

   Because $p(2) = 0.0975$, **there is a 9.75 percent probability that you'll have to stop for two pedestrians.**

   b. Stopping for at least one pedestrian means you're stopping at 1, 2, 3, or all 4 crosswalks. Basically, you need $p(X \geq 1)$. Instead of calculating $p(x)$ for every value in {1, 2, 3, 4}, you can just solve this by using the complement rule (see Chapter 8): $p(X \geq 1) = 1 - p(X < 1) = 1 - p(X = 0)$. Plugging into the formula to find $p(X = 0)$, you get

$$\begin{aligned}\binom{n}{x}p^x(1-p)^{n-x} &= \binom{4}{0}(0.15)^0(1-0.15)^{4-0} \\ &= \left(\frac{4!}{0!4!}\right)(0.15)^0(0.85)^4 \\ &= 1\cdot1\cdot(0.85)^4 \\ &= 0.5220\end{aligned}$$

   Because $p(0) = 0.5220$, **there is a probability of $1 - 0.5220 = 0.4780$, or 47.8 percent of having to stop at least once.**

5. You can use the binomial table to work through these questions because there's a specific mini-table for $n = 12$ and a column for $p = 0.25$.

   a. Looking at the row for $x = 3$, you find that $p(3) = \mathbf{0.258}$.

   b. If X is greater than 6, possible values include {7, 8, 9, 10, 11, 12}. Look at the rows for those values and sum their probabilities. You get $p(X > 6) = 0.011 + 0.002 + 0 + 0 + 0 + 0 = \mathbf{0.013}$.

   c. Note that the language here is slightly different than the last question. If X is at least 6, possible values include {6, 7, 8, 9, 10, 11, 12}. Again, you can look at the rows for those values and sum their probabilities. You get $p(X \geq 6) = 0.040 + 0.011 + 0.002 + 0 + 0 + 0 + 0 = \mathbf{0.053}$.

6. **No.** There is a mini-table for $n = 4$ in Table A-3, but there is no column for $p = 0.15$. You will have to use the binomial formula.

7. You're given that $n = 4$ and $p = 0.15$. The average number of stops is found by calculating the mean of the binomial random variable X: $\mu = np = (4)(0.15) = \mathbf{0.60}$ **stop**. The standard deviation is $\sigma = \sqrt{np(1-p)} = \sqrt{(4)(0.15)(0.85)} = \sqrt{0.51} = \mathbf{0.71}$ **stop**.

If you're ready to test your skills a bit more, take the following chapter quiz that incorporates all the chapter topics.

Whaddya Know? Chapter 9 Quiz

Quiz time! Complete each problem to test your knowledge on the various topics covered in this chapter. You can then find the solutions and explanations in the next section.

1. Which of the following is a categorical random variable?

   A. The weight of a randomly selected watermelon measured exactly.

   B. The type of a randomly selected watermelon (seedless, seeded).

   C. Both of these are categorical random variables.

   D. Neither of these is a categorical random variable.

5. Do the numbers in the $p(x)$ row of the following table make up a probability distribution?

X	1	2	3
$p(x)$	-0.2	0.8	0.4

6. Which of the following is not a characteristic of a binomial random variable?

   A. Fixed number of observations; n.

   B. Probability of success, p, is the same on each trial.

   C. Observations are independent.

   D. Each observation has the outcome recorded as yes/no.

   E. All of these are characteristics of a binomial distribution.

6. You have 10 people on your work team with 5 women and 5 men, and you have to choose 3 people to be on a committee. You do this at random. Let X be the number of women on the committee. Does X have a binomial distribution?

7. You roll a die until you get a six, and you count the number of rolls it takes to get there. Let X be the number of rolls needed. Does X have a binomial distribution?

8. A multiple-choice test has 4 options on each question, and contains 10 questions. Bob guesses on each question randomly. Let X = the number of correct answers Bob gets on the test. Does X have a binomial distribution?

9. Suppose 10% of Americans own a hybrid fuel car. You sample 100 Americans at random and let X be the number of Americans in the sample who own a hybrid fuel car. What is the mean of X?

10. Suppose 70% of Americans have a social media account. You sample 100 Americans at random and let X be the number of Americans in the sample who have a social media account. What is the standard deviation of X?

11. Suppose X is binomial with $n = 3$ and $p = 0.2$. Find $p(X \leq 1)$ using the probability formula.

12. Suppose X is binomial with $n = 3$ and $p = 0.2$. Find $p(X \leq 1)$ using the binomial table.

Answers to Chapter 9 Quiz

1. **B.** The type of watermelon is categorical. The weight is continuous because it can be measured technically to any decimal place.

2. **No,** the values in the second row of the table are not a probability distribution. You see a value of −0.2, which is impossible for a probability. All probabilities must be between 0 and 1.

3. **E.** All the items on the list are characteristics of a binomial distribution.

4. **No**, X does not have a binomial distribution. Each time you pick someone, they can't be picked again, and the probability p of picking a woman changes each time. For example, the first time, the probability of picking a woman is 5/10. If you pick a woman the first time, the chance of picking one the second time is 4/9 (note that only 9 people are left, so the denominator is 9). If you don't pick a woman, the probability p becomes 5/9. The point is that the probability changes each time, and p is supposed to stay the same on each trial in a binomial distribution.

5. **No,** X does not have a binomial distribution because you don't have a fixed number of rolls (n is not fixed).

6. **Yes.** The number of trials is 10 because you have 10 questions; each question is either right or wrong, so you have 2 options each time; you have independence because Bob is randomly guessing each time; and your probability of "success" = probability of getting the answer right = ¼ because there are 4 possible answers to choose from. This stays constant for each question.

7. The mean of X is np, which here is $100 \times 0.10 = \mathbf{10}$.

8. Here n is 100 and p is 0.70. The standard deviation of X is $\sqrt{np(1-p)} = \sqrt{100(0.70)(1-0.70)} = \mathbf{4.58}$.

9. $p(x \le 1) = p(X=0) + p(X=1) = \binom{3}{0} 0.2^0 (1-0.2)^{3-0} + \binom{3}{1} 0.2^1 (1-0.2)^{3-1}$

$$= \frac{3!}{0!(3-0)!}(1)(0.8)^3 + \frac{3!}{1!(3-1)!}(0.2)(0.8)^2$$
$$= \frac{3*2*1}{1(3*2*1)}0.512 + \frac{3*2*1}{1(2*1)}0.128$$
$$= 0.512 + 0.384$$
$$= \mathbf{0.896}$$

10. If $n=3$ and $p=0.2$, you can then find p: $p(X=0) + p(X=1) = 0.512 + 0.384 = \mathbf{0.896}$.

IN THIS CHAPTER

» Understanding the normal and standard normal distributions

» Going from start to finish when finding normal probabilities

» Working backward to find percentiles

Chapter 10 The Normal Distribution

In your statistical travels you'll come across two major types of random variables: discrete and continuous. Discrete random variables basically count things (number of heads on ten coin flips, number of female Democrats in a sample, and so on). The most well-known discrete random variable is the binomial. (See Chapter 9 for more on discrete random variables and binomials.) A continuous random variable is typically based on measurements; it either takes on an uncountable, infinite number of values (values within an interval on the real line), or it has so many possible values that it may as well be deemed continuous (for example, time to complete a task, exam scores, and so on).

In this chapter, you understand and calculate probabilities for the most famous continuous random variable of all time — the normal distribution. You also find percentiles for the normal distribution, where you are given a probability as a percent and you have to find the value of X that's associated with it. And you can think how funny it would be to see a statistician wearing a T-shirt that said, "I'd rather be normal."

Exploring the Basics of the Normal Distribution

A continuous random variable X has a normal distribution if its values fall into a smooth (continuous) curve with a bell-shaped pattern. Each normal distribution has its own mean, denoted by the Greek letter μ (say "mu" like in the word "music"); and its own standard deviation, denoted by the Greek letter σ (say "sigma"). But no matter what their means and standard deviations are, all normal distributions have the same basic bell shape. Figure 10-1 shows some examples of normal distributions.

Every normal distribution has certain properties. You use these properties to determine the relative standing of any particular result on the distribution, and to find probabilities. The properties of any normal distribution are as follows:

- Its shape is symmetric (that is, when you cut it in half, the two pieces are mirror images of each other).
- Its distribution has a bump in the middle, with tails going down and out to the left and right.
- The mean and the median are the same and lie directly in the middle of the distribution (due to symmetry).
- Its standard deviation measures the distance on the distribution from the mean to the *inflection point* (the place where the curve changes from an upside-down bowl shape to a right-side-up bowl shape).
- Because of its unique bell shape, probabilities for the normal distribution follow the Empirical Rule (full details in Chapter 5), which says the following:
  - About 68 percent of its values lie within one standard deviation of the mean. To find this range, take the value of the standard deviation, then find the mean plus this amount, and the mean minus this amount.
  - About 95 percent of its values lie within two standard deviations of the mean. (Here you take 2 times the standard deviation, then add it to and subtract it from the mean.)
  - Almost all of its values (about 99.7 percent of them) lie within three standard deviations of the mean. (Take 3 times the standard deviation and add it to and subtract it from the mean.)
- Precise probabilities for all possible intervals of values on the normal distribution (not just for those within 1, 2, or 3 standard deviations from the mean) are found using a table with minimal (if any) calculations. (The next section gives you all the info on this table.)

Take a look at Figure 10-1. To compare and contrast the distributions shown in Figure 10-1a, b, and c, you first see they are all symmetric with the signature bell shape. The examples in Figure 10-1a and Figure 10-1b have the same standard deviation, but their means are different; Figure 10-1b is located 30 units to the right of Figure 10-1a because its mean is 120 compared to 90. Figures 10-1a and c have the same mean (90), but Figure 10-1a has more variability than Figure 10-1c due to its higher standard deviation (30 compared to 10). Because of the increased variability, the values in Figure 10-1a stretch from 0 to 180 (approximately), while the values in Figure 10-1c only go from 60 to 120.

Finally, Figures 10-1b and c have different means and different standard deviations entirely; Figure 10-1b has a higher mean, which shifts it to the right, and Figure 10-1c has a smaller standard deviation; its values are the most concentrated around the mean.

REMEMBER

Noting the mean and standard deviation is important so you can properly interpret numbers located on a particular normal distribution. For example, you can compare where the number 120 falls on each of the normal distributions in Figure 10-1. In Figure 10-1a, the number 120 is one standard deviation above the mean (because the standard deviation is 30, you get $90+1\times 30=120$). So on this first distribution, the number 120 is the upper value for the central range where about 68 percent of the data are located, according to the Empirical Rule (see Chapter 5).

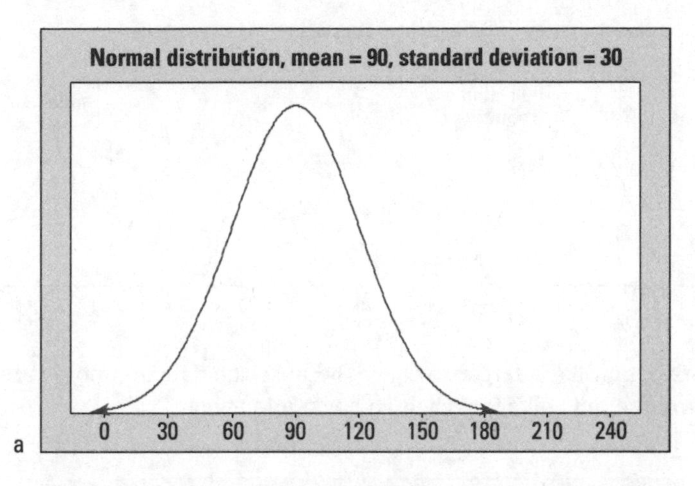

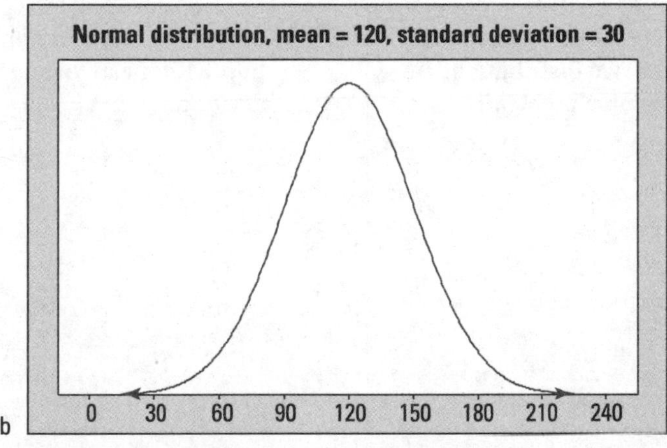

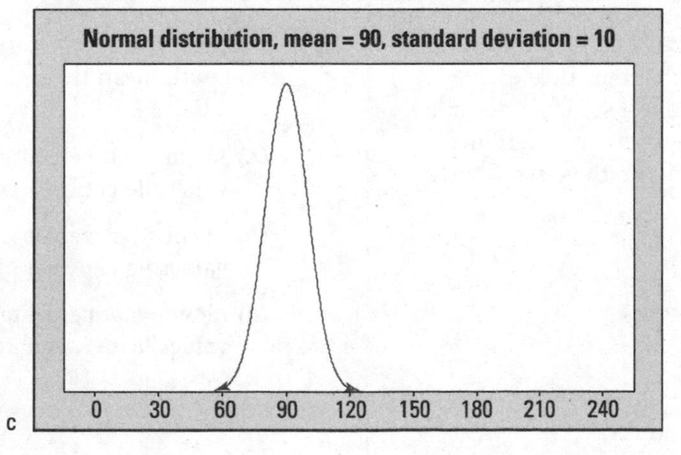

FIGURE 10-1: Three normal distributions, with means and standard deviations of a) 90 and 30; b) 120 and 30; and c) 90 and 10, respectively.

In Figure 10-1b, the number 120 lies directly on the mean, where the values are most concentrated. In Figure 10-1c, the number 120 is way out on the rightmost fringe, 3 standard deviations above the mean (because the standard deviation this time is 10, you get $90+3\times10=120$). In Figure 10-1c, values beyond 120 are very unlikely to occur because they are beyond the central range where about 99.7 percent of the values should be, according to the Empirical Rule.

EXAMPLE

Q. Which of these two normal distributions has a larger variance?

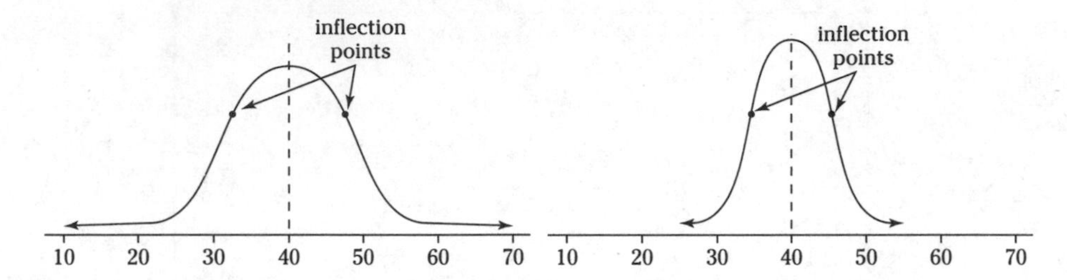

A. The first distribution has a larger variance, because the data are more spread out from the center, and the tails take longer to go down and away.

YOUR TURN

1 What do you guess are the standard deviations of the two distributions in the previous example problem?

2 Draw a picture of a normal distribution with mean 70 and standard deviation 5.

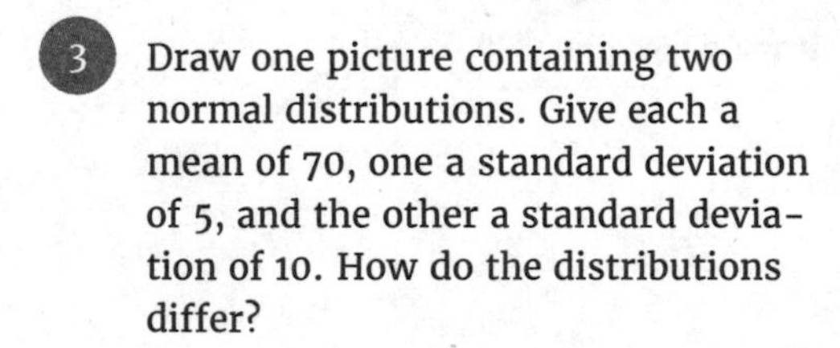

3 Draw one picture containing two normal distributions. Give each a mean of 70, one a standard deviation of 5, and the other a standard deviation of 10. How do the distributions differ?

4 Suppose you have a normal distribution with mean 110 and standard deviation 15.

(a) About what percentage of the values lie between 110 and 125?

(b) About what percentage of the values lie between 95 and 140?

(c) About what percentage of the values lie between 80 and 95?

Meeting the Standard Normal (Z-) Distribution

One very special member of the normal distribution family is called the standard normal distribution, or Z-distribution. The *Z-distribution* is used to help find probabilities and percentiles for regular normal distributions (X). It serves as the standard by which all other normal distributions are measured.

Checking out Z

The Z-distribution is a normal distribution with mean zero and standard deviation 1; its graph is shown in Figure 10-2. Almost all (about 99.7 percent) of its values lie between –3 and 3 according to the Empirical Rule. Values on the Z-distribution are called z-values, z-scores, or standard scores. A *z-value* represents the number of standard deviations that a particular value lies above or below the mean. For example, $z = 1$ on the Z-distribution represents a value that is 1 standard deviation above the mean. Similarly, $z = -1$ represents a value that is one standard deviation below the mean (indicated by the minus sign on the z-value). And a z-value of 0 is — you guessed it — right on the mean. All z-values are universally understood.

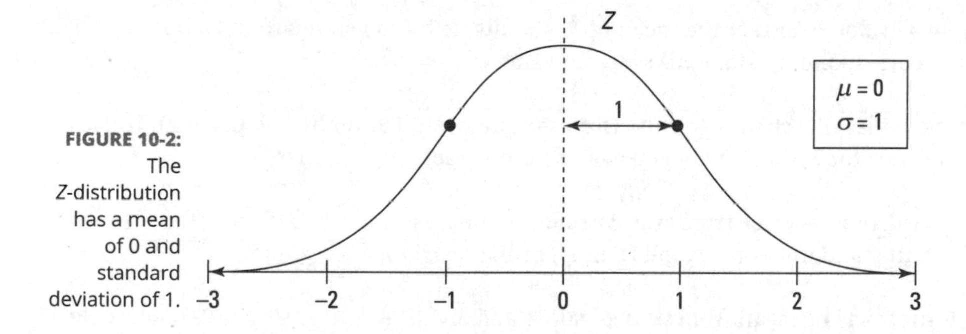

FIGURE 10-2: The Z-distribution has a mean of 0 and standard deviation of 1.

If you refer to Figure 10-1 and the discussion regarding where the number 120 lies on each normal distribution in the section, "Exploring the Basics of the Normal Distribution," you can now calculate z-values to get a much clearer picture. In Figure 10-1a, the number 120 is located one standard deviation above the mean, so its z-value is 1. In Figure 10-1b, 120 is equal to the mean, so its z-value is 0. Figure 10-1c shows that 120 is 3 standard deviations above the mean, so its z-value is 3.

REMEMBER

High standard scores (z-values) aren't always the best. For example, if you're measuring the amount of time needed to run around the block, a standard score of 2 is a bad thing because your time was two standard deviations above (more than) the overall average time. In this case, a standard score of –2 would be much better, indicating your time was two standard deviations below (less than) the overall average time.

Standardizing from X to Z

Probabilities for any continuous distribution are found by finding the area under a curve (if you're into calculus, you know that means integration; if you're not into calculus, don't worry about it). Although the bell-shaped curve for the normal distribution looks easy to work with, calculating areas under its curve turns out to be a nightmare requiring high-level math procedures (believe me, I won't be going there in this book!). Plus, every normal distribution is different, causing you to repeat this process over and over each time you have to find a new probability.

To help get over this obstacle, statisticians worked out all the math gymnastics for one particular normal distribution, made a table of its probabilities, and told the rest of us to knock ourselves out. Can you guess which normal distribution they chose to crank out the table for?

Yes, all the basic results you need to find probabilities for any normal distribution (X) can be boiled down into one table based on the standard normal (Z-) distribution. This table is called the Z-table and is found in the Appendix. Now all you need is one formula that transforms values from your normal distribution (X) to the Z-distribution; from there you can use the Z-table to find any probability you need.

Changing an x-value to a z-value is called *standardizing.* The so-called "z-formula" for standardizing an x-value to a z-value is:

$$z = \frac{x - \mu}{\sigma}$$

You take your x-value, subtract the mean of X, and divide by the standard deviation of X. This gives you the corresponding standard score (z-value or z-score).

Standardizing is just like changing units (for example, from Fahrenheit to Celsius). It doesn't affect probabilities for X; that's why you can use the Z-table to find them!

You can standardize an x-value from any distribution (not just the normal) using the z-formula. Similarly, not all standard scores come from a normal distribution.

Because you subtract the mean from your x-values and divide everything by the standard deviation when you standardize, you are literally taking the mean and standard deviation of X out of the equation. This is what allows you to compare everything on the scale from -3 to 3 (the Z-distribution), where negative values indicate being below the mean, positive values indicate being above the mean, and a value of 0 indicates you're right on the mean.

Standardizing also allows you to compare numbers from different distributions. For example, suppose Bob scores 80 on both his math exam (which has a mean of 70 and a standard deviation of 10) and his English exam (which has a mean of 85 and a standard deviation of 5). On which exam did Bob do better, in terms of his relative standing in the class?

Bob's math exam score of 80 standardizes to a z-value of $\frac{80-70}{10} = \frac{10}{10} = 1$. That tells you his math score is one standard deviation above the class average. His English exam score of 80 standardizes to a z-value of $\frac{80-85}{5} = \frac{-5}{5} = -1$, putting him one standard deviation below the class average. Even though Bob scored 80 on both exams, he actually did better on the math exam than the English exam, relatively speaking.

REMEMBER

To interpret a standard score, you don't need to know the original score, the mean, or the standard deviation. The standard score gives you the relative standing of a value, which in most cases is what matters most. In fact, on most national achievement tests, they won't even tell you what the mean and standard deviation were when they report your results; they just tell you where you stand on the distribution by giving you your z-score.

Finding probabilities for *Z* with the *Z*-table

A full set of less-than probabilities for a wide range of z-values is in the Z-table (Table A-1 in the Appendix). To use the Z-table to find probabilities for the standard normal (Z-) distribution, do the following:

1. **Go to the row that represents the first digit of your z-value and the first digit after the decimal point.**
2. **Go to the column that represents the second digit after the decimal point of your z-value.**
3. **Intersect the row and column.**

   This result represents $p(Z \le z)$, the probability that the random variable Z is less than or equal to the number z (also known as the percentage of z-values that are less than yours).

For example, suppose you want to find $p(Z \le 2.13)$. Using the Z-table, find the row for 2.1 and the column for 0.03. Intersect that row and column to find the probability: 0.9834. You find that $p(Z \le 2.13) = 0.9834$.

Suppose you want to look for $p(Z \le -2.13)$. You find the row for −2.1 and the column for 0.03. Intersect the row and column and you find 0.0166; that means $p(Z \le -2.13)$ equals 0.0166. (This happens to be one minus the probability that Z is less than 2.13 because $p(Z \le 2.13)$ equals 0.9834. That's true because the normal distribution is symmetric; more on that in the following section.)

EXAMPLE

Q. Suppose you play a round of golf and want to compare your score to the other members of your club population. You find that your score is below the mean.

(a) What does this tell you about your standard score?

(b) Is this a good thing or a bad thing?

A. To interpret the standard score, the sign is the first part to look at.

(a) Falling below the mean indicates a negative standard score.

(b) Often, being below the mean is a bad thing, but in golf you have an advantage because golf scores measure the number of swings you need to get around the course, and you want to have a low golf score. So being below the mean in this case is a good thing.

YOUR TURN

5 Exam scores have a normal distribution with a mean of 70 and a standard deviation of 10. Bob's score is 80. Find and interpret his standard score.

6 Bob scores 80 on both his math exam (which has a mean of 70 and a standard deviation of 10) and his English exam (which has a mean of 85 and a standard deviation of 5). Find and interpret Bob's z-scores on both exams to let him know which exam (if either) he did better on. Don't, however, let his parents know; let them think he's just as good at both subjects.

7 Sue's math class's exam has a mean of 70 with a standard deviation of 5. Her standard score is –2. What's her original exam score?

8 Suppose your score on an exam is directly at the mean. What's your standard score?

9 Suppose the weights of cereal boxes have a normal distribution with a mean of 20 ounces and a standard deviation of half an ounce. A box that has a standard score of zero weighs how much?

10 Suppose you want to put fat Fido on a weight-loss program. Before the program, his weight had a standard score of 2 compared to dogs of his breed and age, and after the program, his weight has a standard score of –2. His weight before the program was 150 pounds, and the standard deviation for the breed is 5 pounds.

(a) What's the mean weight for Fido's breed and age?

(b) What's his weight after the weight-loss program?

Finding Probabilities for a Normal Distribution

Here are the steps for finding a probability when X has any normal distribution:

1. **Draw a picture of the distribution.**
2. **Translate the problem into one of the following: $p(X \le a)$, $p(X \ge b)$, or $p(a \le X \le b)$. Shade in the area on your picture.**
3. **Standardize a (and/or b) to a z-score using the z-formula:**

   $$z = \frac{x - \mu}{\sigma}$$

4. **Look up the z-score on the Z-table (Table A-1 in the Appendix) and find its corresponding probability.**

   (See the section, "Standardizing from X to Z," for more on the Z-table).

5a. **If you need a "less-than" probability — that is, $p(X \le a)$ — you're done.**

5b. **If you want a "greater-than" probability — that is, $p(X \ge b)$ — take one minus the result from Step 4.**

5c. **If you need a "between-two-values" probability — that is, $p(a \le X \le b)$ — do Steps 1–4 for b (the larger of the two values) and again for a (the smaller of the two values), and subtract the results.**

REMEMBER

The probability that X is equal to any single value is 0 for any continuous random variable (like the normal). That's because continuous random variables consider probability as being area under the curve, and there's no area under a curve at one single point. So, $p(X \le a) = p(X < a)$, and also $p(X \ge b) = p(X > b)$. This isn't true of discrete random variables.

Suppose, for example, that you enter a fishing contest. The contest takes place in a pond where the fish lengths have a normal distribution with mean $\mu = 16$ inches and standard deviation $\sigma = 4$ inches.

- Problem 1: What's the chance of catching a small fish — say, less than or equal to 8 inches?
- Problem 2: Suppose a prize is offered for any fish over 24 inches. What's the chance of winning a prize?
- Problem 3: What's the chance of catching a fish between 16 and 24 inches?

To solve these problems using the steps that I just listed, first draw a picture of the normal distribution at hand. Figure 10-3 shows a picture of X's distribution for fish lengths. You can see where the numbers of interest (8, 16, and 24) fall.

Next, translate each problem into probability notation. Problem 1 is really asking you to find $p(X \le 8)$. For Problem 2, you want $p(X > 24)$. And Problem 3 is looking for $p(16 < X < 24)$.

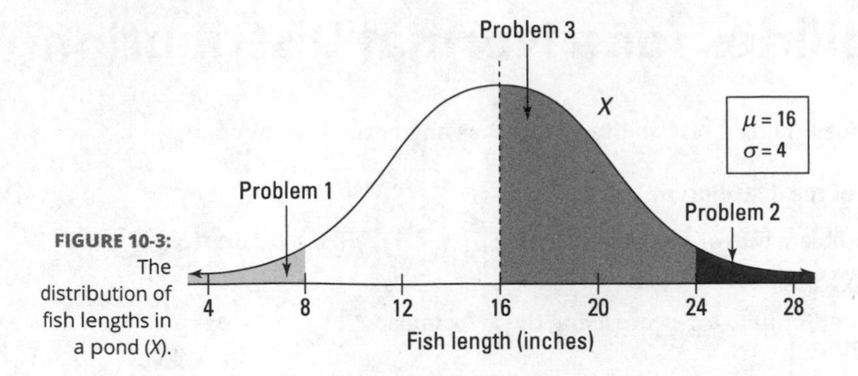

FIGURE 10-3: The distribution of fish lengths in a pond (X).

Step 3 says to change the x-values to z-values using the z-formula:

$$z = \frac{x - \mu}{\sigma}$$

For Problem 1 of the fish example, you have the following:

$$p(X \le 8) = p\left(Z \le \frac{8 - 16}{4}\right) = p(Z \le -2)$$

Similarly for Problem 2, $p(X > 24)$ becomes

$$p(X > 24) = p\left(Z > \frac{24 - 16}{4}\right) = p(Z > 2)$$

And Problem 3 translates from $p(16 < X < 24)$ to

$$p(16 < X < 24) = p\left(\frac{16 - 16}{4} < Z < \frac{24 - 16}{4}\right) = p(0 < Z < 2)$$

Figure 10-4 shows a comparison of the X-distribution and Z-distribution for the values $x = 8$, 16, and 24, which standardize to $z = -2$, 0, and 2, respectively.

Now that you have changed x-values to z-values, you move to Step 4 and find (or calculate) probabilities for those z-values using the Z-table (in the Appendix). In Problem 1 of the fish example, you want $p(Z \le -2)$; go to the Z-table and look at the row for −2.0 and the column for 0.00, intersect them, and you find 0.0228 — according to Step 5a, you're done. The chance of a fish being less than or equal to 8 inches is equal to 0.0228.

For Problem 2, find $p(Z > 2.00)$. Because it's a "greater-than" problem, this calls for Step 5b. To be able to use the Z-table, you need to rewrite this in terms of a "less-than" statement. Because the entire probability for the Z-distribution equals 1, you know $p(Z > 2.00) = 1 - p(Z < 2.00) = 1 - 0.9772 = 0.0228$ (using the Z-table). So, the chance that a fish is greater than 24 inches is also 0.0228. (Note: The answers to Problems 1 and 2 are the same because the Z-distribution is symmetric; refer to Figure 10-3.)

In Problem 3, you find $p(0 < Z < 2.00)$; this requires Step 5c. First, find $p(Z < 2.00)$, which is 0.9772 from the Z-table. Then find $p(Z < 0)$, which is 0.5000 from the Z-table. Subtract them to get $0.9772 - 0.5000 = 0.4772$. The chance of a fish being between 16 and 24 inches is 0.4772.

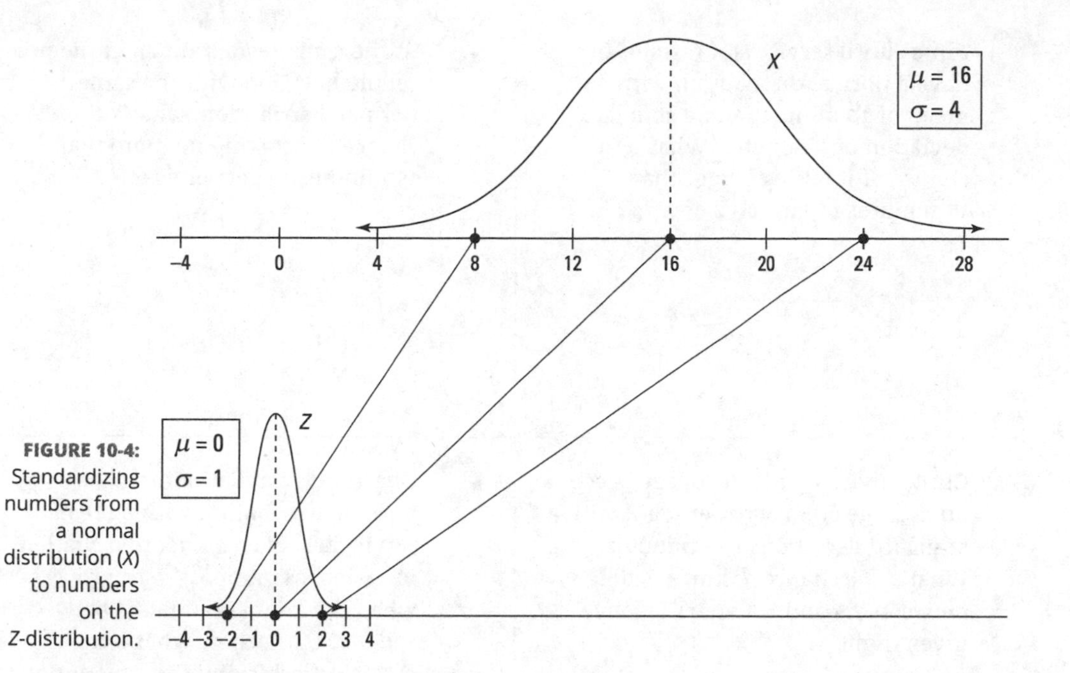

FIGURE 10-4: Standardizing numbers from a normal distribution (*X*) to numbers on the *Z*-distribution.

REMEMBER

The *Z*-table does not list every possible value of *Z*; it just carries them out to two digits after the decimal point. Use the one closest to the one you need. And just like in an airplane where the closest exit may be behind you, the closest *z*-value may be the one that is lower than the one you need.

EXAMPLE

Q. The weights of single-dip ice cream cones at Bob's ice cream parlor have a normal distribution with a mean of 8 ounces and a standard deviation of one-half ounce (0.5 ounce). What's the chance that an ice cream cone weighs between 7 and 9 ounces?

A. In this case, you want $p(7 < X < 9)$, the area between two values. Convert both of the values to *z*-scores, find their probabilities on the *Z*-table (see Table A-1), and subtract them taking the largest one minus the smallest one. Nine ounces becomes $z = \frac{9-8}{0.5} = 2$. The corresponding probability is $p(X < 9) = p(Z < 2.00) = 0.9772$. Seven ounces becomes $z = \frac{7-8}{0.5} = -2$, which has a corresponding probability of $p(X < 7) = p(Z < -2.00) = 0.0228$. Subtracting these probabilities gives you the area between them: $0.9772 - 0.0228 = 0.9544$. Note that 0.9544 is equivalent to $0.9544 \times 100\% = 95.44\%$.

YOUR TURN

11 Bob's commuting times to work have a normal distribution with a mean of 45 minutes and a standard deviation of 10 minutes. How often does Bob get to work in 30 to 45 minutes?

12 The times taken to complete a statistics exam have a normal distribution with a mean of 40 minutes and a standard deviation of 6 minutes. What's the chance of Deshawn completing the exam in 30 to 35 minutes?

13 Times until service at a restaurant have a normal distribution with a mean of 10 minutes and a standard deviation of 3 minutes. What's the chance of it taking longer than 15 minutes to get service?

14 At the same restaurant as in the preceding question with the same normal distribution, what's the chance of it taking no more than 15 minutes to get service?

15 Clint, obviously not in college, sleeps an average of 8 hours per night with a standard deviation of 15 minutes. What's the chance of him sleeping between 7.5 and 8.5 hours on any given night?

16 One state's annual rainfall has a normal distribution with a mean of 100 inches and a standard deviation of 25 inches. Suppose corn grows best when the annual rainfall is between 100 and 140 inches. What's the chance of achieving this amount of rainfall?

Knowing Where You Stand with Percentiles

Percentiles are a way to measure where you stand in a data set. Do you remember the last time you took a standardized test? Not only did the testing company give you your raw score, but they probably also gave you a percentile. If you come in at the 90th percentile, for example, 90 percent of the test scores of all students are the same as or below yours (and 10 percent are above yours). Pediatricians track the growth of newborn babies by the percentile of their length, weight, and head circumference. In general, being at the kth percentile means k percent of the data lie at or below that point and $(100-k)$ percent lie above it.

You saw how to calculate the 25th and 75th percentiles for getting a five-number summary in Chapter 5. To calculate a percentile when the data has a normal distribution is essentially like finding a probability for a given value:

1. **Convert the original score to a standard score by taking the original score minus the mean and dividing by the standard deviation (in other words, use the z-formula).**
2. **Use the Z-table to find the corresponding percentile for the standard score.**

REMEMBER

The percentage is the probability you find in the Z-table. The percentile is the value of X (a test score, your height, and so forth) with a specific percentage of values less than its value.

EXAMPLE

Q. Weights for single-dip ice cream cones at Adrian's have a normal distribution with a mean of 8 ounces and a standard deviation of one-quarter ounce. Suppose your ice cream cone weighs 8.5 ounces. What percentage of cones is smaller than yours?

A. Your cone weighs 8.5 ounces, and you want the corresponding percentile. Before you can use the Z-table to find that percentile, you need to standardize the 8.5 — in other words, run it through the z-formula. This gives you $z = \frac{8.5 - 8}{0.25} = \frac{0.5}{0.25} = 2$, so the z-score for your ice cream cone is 2. Now you use the Z-table to find that a z-value of 2.00 corresponds with 0.9772, which equals 97.72 percent. Your ice cream cone is at the 97.72th percentile, and 97.72 percent of the other single-dip cones at Adrian's are smaller than yours. Lucky you!

YOUR TURN

17 Bob's commuting times to work have a normal distribution with a mean of 45 minutes and a standard deviation of 10 minutes.

(a) What percentile does Bob's commute time represent if he gets to work in 30 minutes or less?

(b) Bob's workday starts at 9 a.m. If he leaves at 8 a.m., how often is he late?

 Suppose your exam score has a standard score of 0.90. Does this mean that 90 percent of the other exam scores are lower than yours?

 If a baby's weight is at the median, what's their percentile?

 Suppose you know that Layton's test score is above the mean in a normal distribution of scores, but he doesn't remember by how much. At least how many students must score lower than Layton?

Finding *X* When You Know the Percent

Another popular normal distribution problem involves finding percentiles for X (see Chapter 5 for a detailed rundown on percentiles). That is, you are given the percentage or probability of being at or below a certain x-value, and you have to find the x-value that corresponds to it. For example, if you know that the people whose golf scores were in the lowest 10 percent got to go to the tournament, you may wonder what the cutoff score was; that score would represent the 10th percentile.

WARNING

A percentile isn't a percent. A percent is a number between 0 and 100; a percentile is a value of X (a height, an IQ, a test score, and so on).

Figuring out a percentile for a normal distribution

Certain percentiles are so popular that they have their own names and their own notation. The three "named" percentiles are Q_1 — the first quartile, or the 25th percentile; Q_2 — the second quartile (also known as the *median* or the 50th percentile); and Q_3 — the third quartile or the 75th percentile. (See Chapter 5 for more information on quartiles.)

Here are the steps for finding any percentile for a normal distribution *X*:

1a. If you're given the probability (percent) less than x and you need to find x, you translate this as follows: Find *a* where $p(X \le a) = p$ (and p is the given probability). That is, find the *p*th percentile for X. Go to Step 2.

1b. If you're given the probability (percent) greater than x and you need to find x, you translate this as follows: Find *b* where $p(X > b) = p$ (and p is given). Rewrite this as a percentile (less-than) problem: Find *b* where $p(X \le b) = 1 - p$. This means find the (1 – *p*)th percentile for X.

2. Find the corresponding percentile for Z by looking in the body of the Z-table (Table A-1 in the Appendix) and finding the probability that is closest to *p* (from Step 1a) or 1 – *p* (from Step 1b). Find the row and column this probability is in (using the table backwards). This is the desired z-value.

3. Change the z-value back into an x-value (original units) by using $x = \mu + z\sigma$. You've (finally!) found the desired percentile for X.

The formula in this step is just a rewriting of the z-formula, $z = \frac{x - \mu}{\sigma}$, so it's solved for x.

Doing a low percentile problem

Look at the fish example used previously in "Finding Probabilities for a Normal Distribution," where the lengths (*X*) of fish in a pond have a normal distribution with a mean of 16 inches and a standard deviation of 4 inches. Suppose you want to know what length marks the bottom 10 percent of all the fish lengths in the pond. What percentile are you looking for?

TIP

Being at the bottom 10 percent means you have a "less-than" probability that's equal to 10 percent, and you are at the 10th percentile.

Now go to Step 1a in the preceding section and translate the problem. In this case, because you're dealing with a "less-than" situation, you want to find x such that $p(X \leq x) = 0.10$. This represents the 10th percentile for X. Figure 10-5 shows a picture of this situation.

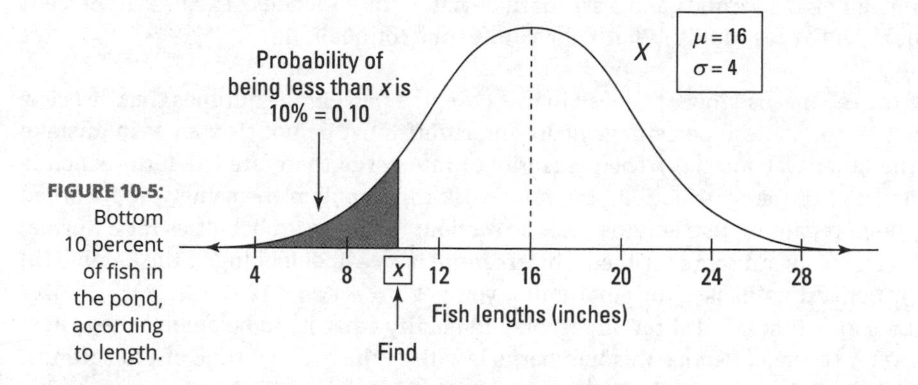

FIGURE 10-5: Bottom 10 percent of fish in the pond, according to length.

Now go to Step 2, which says to find the 10th percentile for Z. Looking in the body of the Z-table (in the Appendix), the probability closest to 0.10 is 0.1003, which falls in the row for $z = -1.2$ and the column for 0.08. That means the 10th percentile for Z is −1.28; so a fish whose length is 1.28 standard deviations below the mean marks the bottom 10 percent of all fish lengths in the pond.

But exactly how long is that fish, in inches? In Step 3, you change the z-value back to an x-value (fish length in inches) using the z-formula solved for x; you get $x = 16 + (-1.28)(4) = 10.88$ inches. So 10.88 inches marks the lowest 10 percent of fish lengths. Ten percent of the fish are as short as or shorter than that.

Working with a higher percentile

Now suppose you want to find the length that marks the *top* 25 percent of all the fish in the pond. This problem calls for Step 1b (in the section, "Finding a percentile for a normal distribution") because being in the top part of the distribution means you're dealing with a greater-than probability. The number you are looking for is somewhere in the right tail (upper area) of the X-distribution, with $p = 25\%$ of the probability to its right and $1 - p = 75\%$ to its left. Thinking in terms of the Z-table and how it only uses less-than probabilities, you need to find the 75th percentile for Z, and then change it to an x-value.

Step 2: The 75th percentile of Z is the z-value where $p(Z < z) = 0.75$. Using the Z-table (in the Appendix), you find the probability closest to 0.7500 is 0.7486, and its corresponding z-value is in the row for 0.6 and the column for 0.07. Put these together and you get a z-value of 0.67. This is the 75th percentile for Z. In Step 3, change the z-value back to an x-value (length in inches) using the z-formula solved for x to get $x = 16 + (0.67)(4) = 18.68$ inches. So, 75 percent of the fish are at least as short as 18.68 inches. And to answer the original question, the top 25 percent of the fish in the pond are longer than 18.68 inches.

Translating tricky wording in percentile problems

REMEMBER

Some percentile problems are especially challenging to translate. For example, suppose the amount of time for a racehorse to run around a track in a qualifying round has a normal distribution with a mean of 120 seconds and a standard deviation of 5 seconds. The best 10 percent of the times qualify; the rest don't. What's the cutoff time for qualifying?

Because "best times" means "lowest times" in this case, the percentage of times that lie *below* the cutoff must be 10, and the percentage *above* the cutoff must be 90. (It's an easy mistake to think it's the other way around.) The percentile of interest is therefore the 10th, which is down on the left tail of the distribution. You now work this problem the same way I worked the earlier problem regarding fish lengths (see the section, "Finding Probabilities for a Normal Distribution"). The standard score for the 10th percentile is $z = -1.28$ looking at the Z-table (in the Appendix). Converting back to original units, you get $x = \mu + z\sigma = 120 + (-1.28)(5) = 113.6$ seconds. So the cutoff time needed for a racehorse to qualify (that is, to be among the fastest 10 percent) is 113.6 seconds. (Notice this number is less than the average time of 120 seconds, which makes sense; a negative z-value is what makes this happen.)

TIP

The 50th percentile for the normal distribution is the mean (because of symmetry) and its z-score is zero. Smaller percentiles, like the 10th, lie below the mean and have negative z-scores. Larger percentiles, like the 75th, lie above the mean and have positive z-scores.

Here's another style of wording that has a bit of a twist: Suppose times to complete a statistics exam have a normal distribution with a mean of 40 minutes and a standard deviation of 6 minutes. Deshawn's time comes in at the 90th percentile. What percentage of the students are still working on their exams when Deshawn leaves? Because Deshawn is at the 90th percentile, 90 percent of the students have exam times lower than hers. That means 90 percent of the students left before Deshawn, so $100 - 90 = 10$ percent of the students are still working when Deshawn leaves.

TIP

To be able to decipher the language used to imply a percentile problem, look for clues like *the bottom 10 percent* (also known as the 10th percentile) and *the top 10 percent* (also known as the 90th percentile). For *the best 10 percent*, you must determine whether low or high numbers qualify as "best."

EXAMPLE

Q. Racehorses race around the track in a qualifying round according to a normal distribution with a mean of 120 seconds and a standard deviation of 5 seconds. Now suppose the bottom 10 percent of the times get eliminated from the next race. What's the cutoff time for being eliminated?

A. You may think you need to find the 10th percentile, but no. The smaller values for times don't get eliminated. The largest 10 percent of the times do get eliminated, so in terms of data on a normal distribution, you know that the percentile of interest is the 90th. Remember that the percentage of times below the cutoff is 90, and the percentage above the cutoff is 10 (that's how percentiles work). The standard score for the 90th percentile is found by seeking out 0.90 in the body of the Z-table. The closest value to this is 0.8997, which matches up with a z-value of +1.28. Converting this back to the original units, you get $x = 120 + (1.28)(5) = 120 + 6.4 = 126.4$. So the cutoff time for being eliminated from the next race is 126.4 seconds.

YOUR TURN

21 Weights have a normal distribution with a mean of 100 and a standard deviation of 10. What weight has 60 percent of the values lying below it?

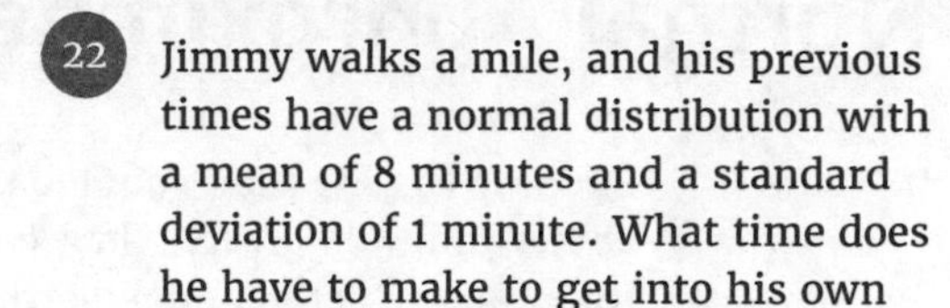

22 Jimmy walks a mile, and his previous times have a normal distribution with a mean of 8 minutes and a standard deviation of 1 minute. What time does he have to make to get into his own top 10 percent of his fastest times?

23 The times it takes to complete a statistics exam have a normal distribution with a mean of 40 minutes and a standard deviation of 6 minutes. Deshawn's time falls at the 42nd percentile. How long does Deshawn take to finish her exam?

24 Exam scores for a particular test have a normal distribution with a mean of 75 and a standard deviation of 5. The instructor wants to give the top 20 percent of the scores an A. What's the cutoff for an A?

25 Service call times for one company have a normal distribution with a mean of 10 minutes and a standard deviation of 3 minutes. Researchers study the longest 25 percent of the calls to make improvements. How long do the longest 25 percent last?

26 Statcars are new vehicles whose mileage has a normal distribution with a mean of 75 miles per gallon. Twenty percent of the vehicles get more than 100 miles per gallon. What's the standard deviation?

Normal Approximation to the Binomial

Suppose you flip a fair coin 100 times and you let X equal the number of heads. What's the probability that X is greater than 60? In Chapter 9, you solve problems like this (involving fewer flips) using the binomial distribution. For binomial problems where n (the number of trials) is small, you can use the direct formula (found in Chapter 9), the binomial table (found in the Appendix), or technology if it is available (such as a graphing calculator or Microsoft Excel).

However, if n is large, the calculations get unwieldy and the binomial table runs out of numbers. If there's no technology available (like when taking an exam), what can you do to find a binomial probability? Turns out, if n is large enough, you can use the normal distribution to find a very close approximate answer with a lot less work.

But what do I mean by n being "large enough"? To determine whether n is large enough to use what statisticians call the *normal approximation to the binomial*, both of the following conditions must hold:

- $n \times p \geq 10$ (at least 10), where p is the probability of success
- $n \times (1-p) \geq 10$ (at least 10), where $1-p$ is the probability of failure

To find the normal approximation to the binomial distribution when n is large, use the following steps:

1. **Verify whether n is large enough to use the normal approximation by checking the two appropriate conditions.**

   For the coin-flipping question, the conditions are met because $n \times p = 100 \times 0.50 = 50$, and $n \times (1-p) = 100 \times (1-0.50) = 50$, both of which are at least 10. So go ahead with the normal approximation.

2. **Translate the problem into a probability statement about X.**

   For the coin-flipping example, you need to find $p(X > 60)$.

3. **Standardize the x-value to a z-value, using the z-formula:**

   $$z = \frac{x - \mu}{\sigma}$$

   For the mean of the normal distribution, use $\mu = np$ (the mean of the binomial), and for the standard deviation σ, use $\sqrt{np(1-p)}$ (the standard deviation of the binomial; see Chapter 9).

   For the coin-flipping example, $\mu - np = (100)(0.50) = 50$ and $\sigma \sqrt{np(1-p)} = \sqrt{100(0.50)(1-0.50)} = 5$. Now put these values into the z-formula to get $z = \frac{x-\mu}{\sigma} = \frac{60-50}{5} = 2$. To solve the problem, you need to find $p(Z > 2)$.

REMEMBER

On an exam, you won't see μ and σ in the problem when you have a binomial distribution. However, you know the formulas that allow you to calculate both of them using n and p (values which will be given in the problem). Just remember you have to do that extra step to calculate the μ and σ needed for the z-formula.

4. **Proceed as you usually would for any normal distribution. That is, do Steps 4 and 5 described in the earlier section, "Finding Probabilities for a Normal Distribution."**

   Continuing the example, $p(Z > 2.00) = 1 - 0.9772 = 0.0228$ from the Z-table (found in the Appendix). So the chance of getting more than 60 heads in 100 flips of a coin is only about 2.28 percent. (I wouldn't bet on it.)

REMEMBER

When using the normal approximation to find a binomial probability, your answer is an *approximation* (not exact) — be sure to state that. Also show that you checked both necessary conditions for using the normal approximation.

EXAMPLE

Q. Suppose that 60 percent of the crowd at a music festival is female. If 40 people will be randomly selected to win an attendance prize, what is the chance that at least 20 of the winners will be female?

A. You have a fixed number of trials $(n = 40)$ with a defined probability of success $(p = 0.60)$ of a selected person being female. Hey, this is talking about a binomial random variable! Unfortunately, Table A-3 only goes up to $n = 20$. Instead, check to see whether you can use the normal approximation to the binomial. You find that $np = (40)(0.60) = 24$ and $n(1-p) = (40)(0.40) = 16$ are both greater than 10, so you can go ahead with the approximation.

For Step 2, identify the probability by writing it as $p(X \geq 20)$. Next, you need to standardize the x-value of 20 into its proper z-value. Remember to first find the mean $\mu = np = (40)(0.60) = 24$ and the standard deviation $\sigma = \sqrt{np(1-p)} = \sqrt{40(0.60)(0.40)} = \sqrt{9.6} = 3.10$. Putting these into the z-formula, you get $z = \frac{20-24}{3.10} = \frac{-4}{3.10} = -1.29$. To solve the problem, you need to find $p(Z \geq -1.29)$. From the Z-table, you can look up that $p(Z \leq -1.29) = 0.0985$, which means that $p(Z \geq -1.29) = 1 - p(Z \leq -1.29) = 1 - 0.0985 = 0.9015$. So the chance that at least 20 of the 40 winners will be female is *approximately* 90.15 percent.

YOUR TURN

27 At a certain major airport, 15 percent of all travelers are headed to an international destination. If you ask 500 random people where they're headed, what's the probability that less than 80 of them are flying internationally?

You are in charge of one of the rides at StatWorld, the greatest amusement park on earth! Historically, 95 percent of park guests who get on your ride are children. Your manager wants to know the chance that less than 140 of the next 150 riders will be children. Can you use a normal approximation to the binomial to answer the manager's question?

29 As a basketball player, you successfully make 50 percent of your shots. (Wow! You should be in the NBA.) Assuming that the likelihood of you scoring a basket is independent from one shot to the next, you're interested in guessing how many of the next 20 shots you will make.

(a) Calculate the probability that you will make at least 10 of your next 20 shots using the binomial table.

(b) Calculate the probability that you will make at least 10 of your next 20 shots using a normal approximation.

(c) How do the two calculations compare?

Practice Questions Answers and Explanations

1. In the first normal distribution, most of the data falls between 10 and 70, each within 3 times 10 units of the mean (40), so you can assume the standard deviation is around 10. You can also see that the saddle point (the place where the picture changes from an upside-down bowl shape to a right-side-up bowl shape) occurs about 10 units away from the mean of 40, which means the standard deviation is around 10. In the second normal distribution, the saddle point occurs about 5 units from the mean, so the standard deviation is at about 5. Notice also that most all the data lies within 15 units of the mean, which is 3 standard deviations.

2. See the following figure. Check that you have saddle points at 65 and 75.

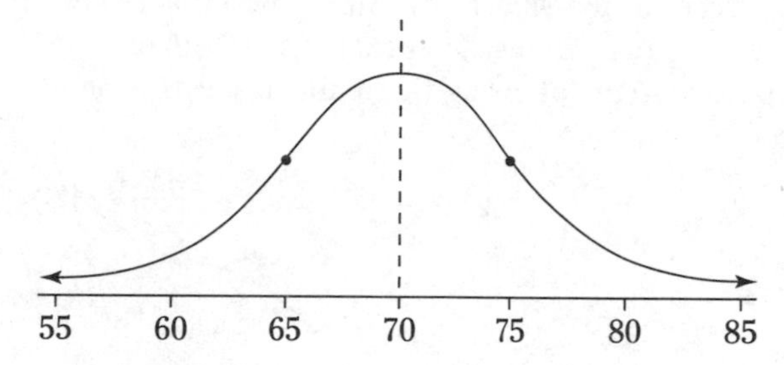

3. See the following figure. A graph with a standard deviation of 5 is taller and thinner than a graph whose standard deviation is 10. The normal distribution with a standard deviation of 10 is more spread out and flatter looking than the normal distribution with a standard deviation of 5, which looks more squeezed together close to the mean (which it is).

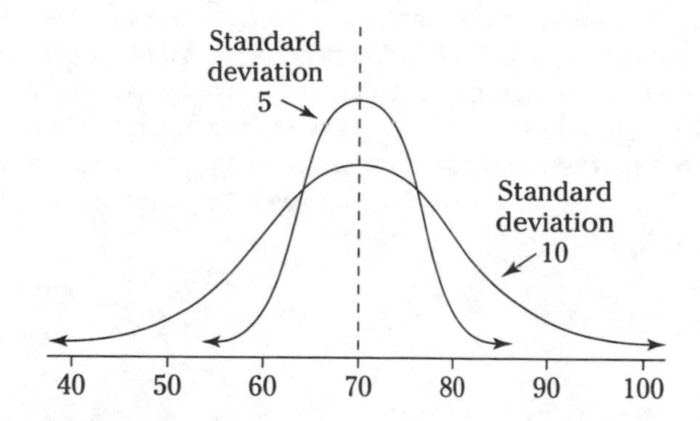

A normal distribution that looks flatter actually has more variability than one that goes from low to high to low as you look from left to right, because you measure variability by how far away the values are from the middle; more data close to the middle means low variability, and more data farther away means less variability. This characteristic differs from what you see on a graph that shows data over time (time series or line graphs). On the time graphs, a flat line means no change over time, and going from low to high to low means great variability.

(4) A picture is worth a thousand points here (along with the Empirical Rule). See the following figure.

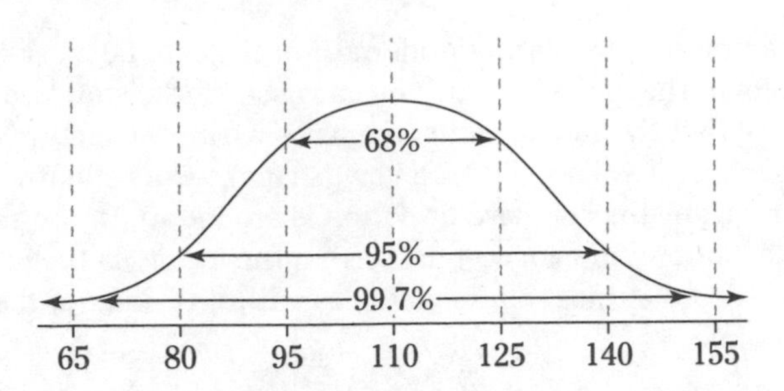

a. For the two values 110 and 125, 125 is one standard deviation above the mean, and about 68 percent of the values lie within one standard deviation of the mean (on both sides of it). So, the percentage of values between 110 and 125 is half of 68 percent, which is 34 percent. The following figure illustrates this point.

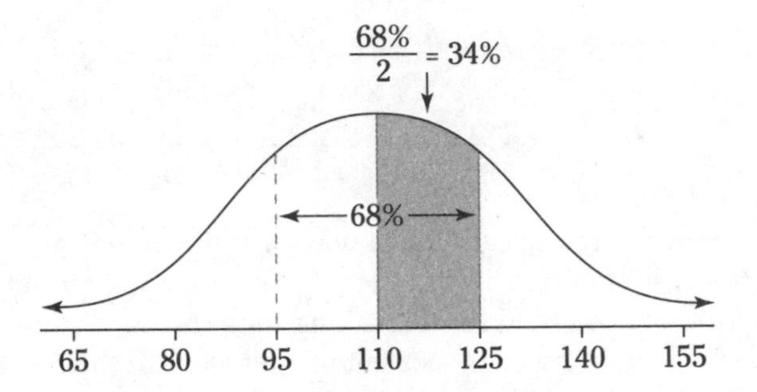

b. For the two values 95 and 140, 95 is one standard deviation below the mean (110), so that distance covers half of the 68 percent again, or 34 percent. To get from 110 to 140, you need to go two standard deviations above the mean. Because about 95 percent of the data lie within two standard deviations of the mean (on both sides of it), from 110 to 140 covers about half of the 95 percent, which is 47.5 percent. Add the 34 and the 47.5 to get 81.5 percent for your approximate answer. See the following figure for an illustration.

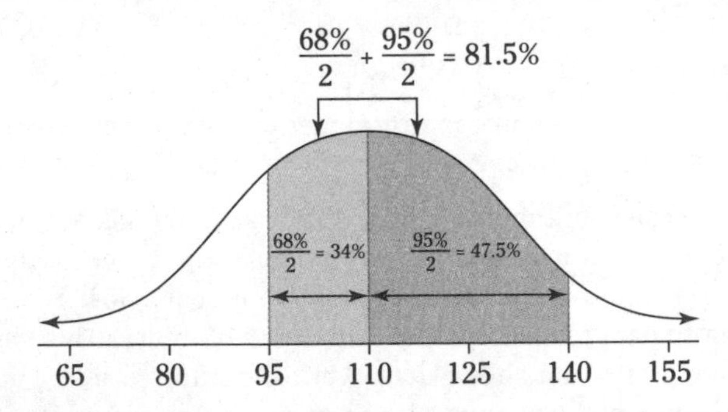

c. For the two values 80 and 95, 80 is two standard deviations below the mean of 110, which represents $95 \div 2$ or 47.5 percent of the data. And 95 is one standard deviation below 110, which represents $68 \div 2$ or 34 percent of the data. You want the area between these two values, so subtract the percentages: $47.5 - 34 = 13.5$ percent. See the following figure for a visual.

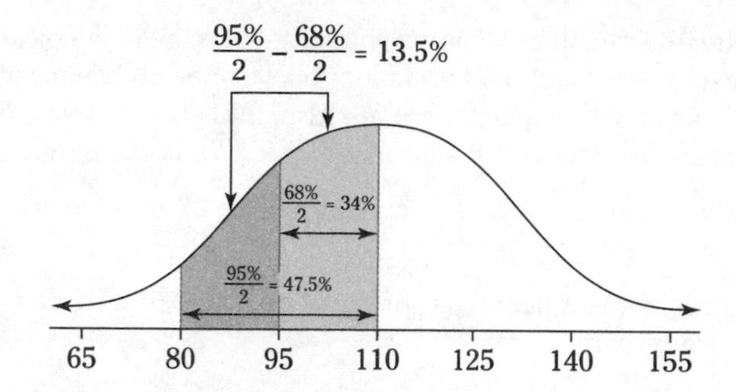

TIP

When using the Empirical Rule to find the percentage between two values, you may have to use a different approach, depending on whether both values are on the same side of the mean or one is above the mean and one below the mean. First, find the percentages that lie between the mean and each value separately. If the values fall on different sides of the mean, add the percentages together. When they appear on the same side of the mean, subtract their percentages (largest minus smallest, so the answer isn't negative). Or, better yet, draw a picture to see what you need to do.

5. The following figure shows a picture of this distribution. Take $80 - 70 = 10$ and divide by 10 to get 1. Bob's score is one standard deviation above the mean. You can see this on a picture as well, because Bob's score falls one "tick mark" above the mean on the picture of the original normal distribution.

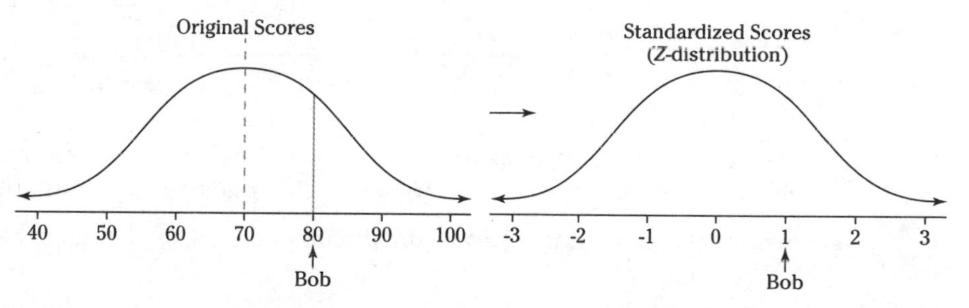

6. The 80 Bob scores on his math exam converts to a standard score of 1 (see the preceding answer). The 80 Bob scores on his English exam converts to a standard score of $(80 - 85) \div 5 = -5 \div 5 = -1$. His score on the English exam is one standard deviation below the mean, so his math score is better.

REMEMBER

Your actual scores don't matter; what matters is how you compare to the mean, in terms of number of standard deviations.

7. Here you can use the same formula, $z = \frac{x - \mu}{\sigma}$, but you need to plug in different values. You know the mean is 70 and the standard deviation is 5. You know $z = -2$, but you don't know *x*, the original score. So what you have looks like $-2 = \frac{x - 70}{5}$. Solving for *x*, you get $x - 70 = -2(5)$, so $x - 70 = -10$, or $x = 60$. The answer makes sense because each standard deviation is worth five, and you start at 70 and go down two of these standard deviations: $70 - (2)(5) = 60$. So Sue scored a 60 on the test.

8. A standard score of zero means your original score is the mean itself, because the standard score is the number of standard deviations above or below the mean. When your score is on the mean, you don't move away from it at all. Also, in the z-formula, after you take the value (which is at the mean) and subtract the mean, you get zero in the numerator, so the answer is zero.

9. Exactly 20 ounces, because a standard score of zero means the observation is right on the mean.

10. This problem is much easier to calculate if you first draw a picture of what you know and work from there.

    a. The following figure shows a picture of the situation before and after Fido's weight-loss program. You know that 150 has a z-score of 2 and the standard deviation is 5, so you have $2 = \frac{150-\mu}{5}$. And solving for the mean (μ), you calculate $\mu = 150 - 2(5) = 140$. The mean weight for his breed and age group is 140 pounds.

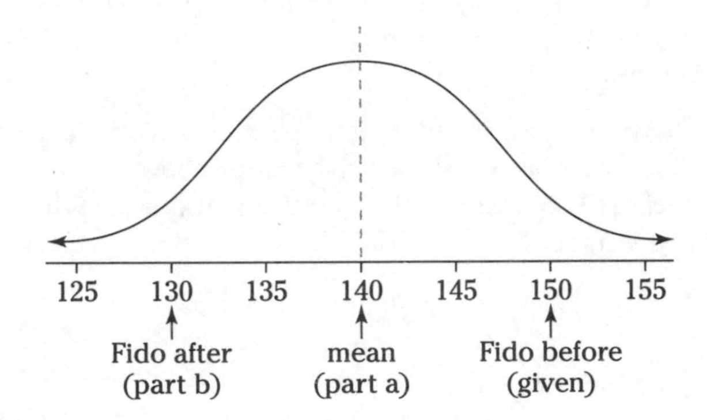

    b. A z-score of –2 corresponds to a weight 2 standard deviations below the mean of 150, which brings you down to $140 - 2(5) = 130$. Fido weighs 130 pounds after the program.

TIP

For any problem involving a normal distribution, drawing a picture is the key to success.

11. The following figure shows a picture of the situation. You want $p(30 < X < 45)$, the probability that X is between 30 and 45, where X is the commuting time. Converting the 30 to a standard score with the z-formula, you get $z = \frac{30-45}{10} = \frac{-15}{10} = -1.5$. The value 45 converts to $z = 0$ because it's right at the mean. (Or, note that $45 - 45 = 0$, and 0 divided by 10 is still 0.) Because you have to find the probability of being between two numbers, look up each of the probabilities associated with their standard scores on the Z-table (see Table A-1) and subtract their values (largest minus smallest, to avoid a negative answer). Using the Z-table, $p(Z < 2.00) = 0.0668$ and $p(Z < 0) = 0.5$. Subtracting those values gives you $0.5000 - 0.0668 = 0.4332$, which is equivalent to 43.32 percent.

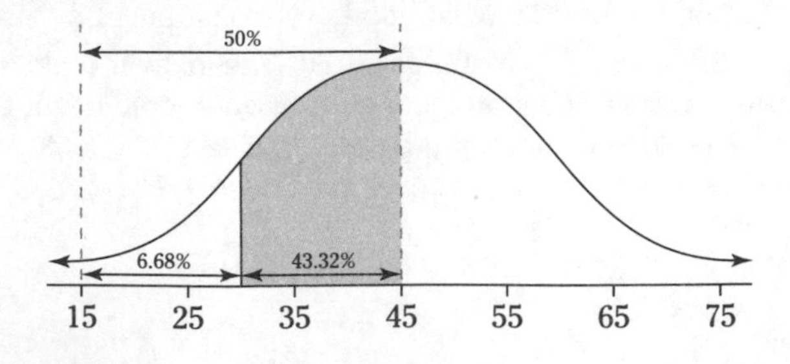

REMEMBER

The reason you subtract the two probabilities when finding the chance of being between two numbers is because the probability includes all the area less than or equal to a certain value. You want the probability of being less than or equal to the larger number, but you don't want the probability of being less than or equal to the smaller number. Subtracting the percentiles allows you to keep the part you want and throw away the part you don't want.

12. The following figure shows a picture of this situation. You want the probability that X (exam time) is between two values, 30 and 35, on the normal distribution. First, convert each of the values to standard scores, using the z-formula. The 30 converts to $z = \frac{30-40}{6} = \frac{-10}{6} = -1.67$, and $p(Z \le -1.67) = 0.0475$. The 35 converts to $z = \frac{35-40}{6} = \frac{-5}{6} = -0.83$, and $p(Z \le -0.83) = 0.2033$. To get the probability, or area, between the two values, subtract each of their percentiles — the larger one minus the smaller one — to get $0.2033 - 0.0475 = 0.1558$, which is equivalent to 15.58 percent.

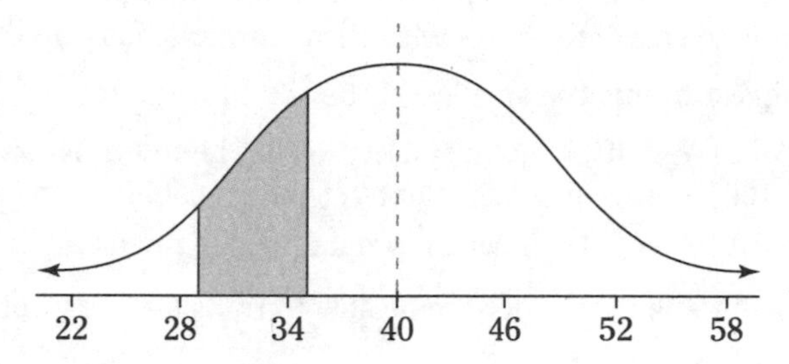

13. You are looking for the probability that X (service time) is more than 15 minutes (see the following figure). In this case, you convert the value to a standard score, find its probability, and take 100 minus that probability because you want the percentage that falls above it. Substituting the values, you have $z = \frac{15-10}{3} = \frac{5}{3} = 1.67$, and using Table A-1, you find that $p(Z \le 1.67) = 0.9525$. The probability you want is $p(X > 15) = 1 - p(X \le 15) = 1 - p(Z \le 1.67) = 1 - 0.9525 = 0.0475$, which is equivalent to 4.75 percent.

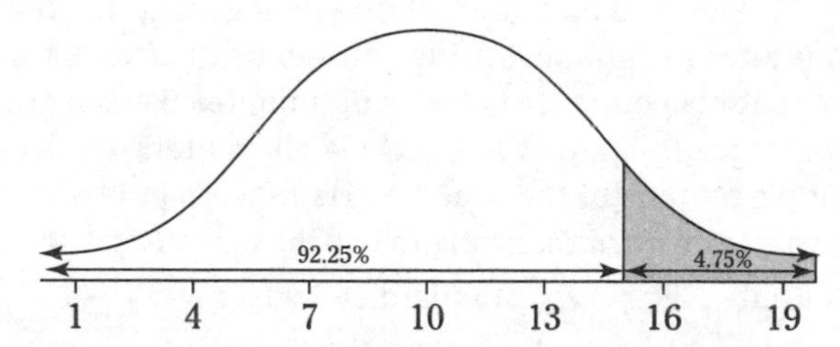

14. This problem is the exact opposite (or complement) of the preceding problem. Here you want the probability that X is no more than 15, which means the probability X is less than or equal to 15, $p(X \le 15)$. If 4.75 percent (.0475) is the probability of being more than 15 in the last problem, then the probability of being less than or equal to 15 is $1 - 0.0475$, which is 0.9525 or 95.25 percent.

TIP

The chance of X being exactly equal to a certain value on a normal distribution is zero, so it doesn't matter whether you take the probability that X is less than 15, $p(X < 15)$, or the probability that X is less than or equal to 15, $p(X \le 15)$. You get the same answer, because the probability that X equals the one specific value of exactly 15 is zero.

15. For this problem, you need to convert 7.5 and 8.5 to standard scores, look up their percentiles on the Z-table (Table A-1), and subtract them by taking the largest one minus the smallest one. Note that the standard deviation given in the problem is stated in minutes. Convert this to hours first by taking $15 \div 60 = 0.25$. For this problem, 7.5 converts to a standard score of $z = \frac{7.5-8}{0.25} = \frac{-0.5}{0.25} = -2$. And 8.5 converts to a standard score of $z = \frac{8.5-8}{0.25} = \frac{0.5}{0.25} = 2$. The corresponding probabilities for $z = -2$ and $z = 2$ are 0.0228 and 0.9772, respectively. Subtracting the largest percentile minus the smallest, you get $0.9772 - 0.0228 = 0.9544$. Clint has a high chance (95.44 percent) of sleeping between 7.5 and 8.5 hours on a given night.

TIP

Practice wording problems in different ways, and also pay attention to homework and in-class examples that are worded differently. You don't want to be thrown off in an exam situation. Your instructor may have their own way to word questions. This comes through in their examples in class and on homework questions.

16. Here you want the probability that X (annual rainfall) is between 100 and 140, so convert both numbers to standard scores with the z-formula, look up their percentiles, and subtract (take the biggest minus the smallest). The 140 converts to $z = \frac{140-100}{25} = \frac{40}{25} = 1.6$ and has a probability of $p(X < 140) = p(Z < 1.60) = 0.9452$. The 100 converts to zero and has a probability of $p(X < 100) = p(Z < 0) = 0.50$. Subtract the probabilities to get the area between them: $0.9452 - 0.5000 = 0.4452$, which is equal to 44.52 percent.

17. For this problem, you need to be able to translate the information into the right statistical task.

    a. Because you want a percentage of time that falls below a certain value (in this case 30), you need to look for a percentile that corresponds with that value (30). You standardize 30 to get a z-score of $z = \frac{30-45}{10} = \frac{-15}{10} = -1.5$. This means a 30-minute commuting time is well below the mean (so it shouldn't happen very often). The probability that corresponds to –1.5, according to the Z-table, is 0.0668, which means Bob gets to work in 30 minutes or less only 6.68 percent of the time. So, a commute time of 30 minutes represents the 6.68th percentile.

    b. If Bob leaves at 8 a.m. and is still late for work, his commuting time must be more than 60 minutes (a time that brings him to work after 9 a.m.), so you want the percentage of time that his commute is above 60 minutes. Percentiles don't automatically give you the percentage of data that lies above a given number, but you can still find it. If you know what percentage of the time he gets to work in less than 60 minutes, you can take 100 percent minus that time to get the percentage of time he gets to work in over 60 minutes. First, you standardize 60 to $z = \frac{60-45}{10} = \frac{15}{10} = 1.5$ and find the probability that goes with that, 0.9332. Then take $1 - 0.9332 = 0.0668$. If he leaves at 8 a.m., he will be late 6.68 percent of the time.

    The answer to Part B is the same as the answer to Part A by symmetry of the normal distribution. The values 30 and 60 are both 1.5 standard deviations away from the mean, so the percentage below 30 and the percentage above 60 will be the same.

18. **No.** It means your exam score is 0.90 standard deviation above the mean, or in other words, $z = 0.90$. You have to look up 0.90 on the Z-table to find its corresponding percentile, which is 0.8159; therefore, 81.59 percent of the exam scores are lower than yours.

REMEMBER

Make sure you keep your units straight. A z-score between zero and one looks a lot like a percentile; however, it isn't!

19. The median is the value in the middle of the data set; it cuts the data set in half. Half the data fall below the median, and half rise above it. So, the median is at the 50th percentile. (See Chapter 5 for more on the median.)

20. If his score is right at the mean, Layton scores in the 50th percentile. So if Layton's score is above the mean, his standard score is positive, and the percentage of values below his score is greater than the 50th percentile.

21. This problem essentially asks for the score that corresponds to the 60th percentile (the tricky part is recognizing this). First, you look up the standard score for the 60th percentile. The closest value to 0.60 in the Z-table is 0.5987, which matches with a z-value of 0.25. Using the reworked z-formula to solve for x, you get $x = 100 + (0.25)(10) = 102.5$.

22. The wording "10 percent of his fastest times" indicates that only 10 percent of his times are *less* than his current one; therefore, the percentile is the 10th (not the 90th). The standard score for the 10th percentile is found by seeking out 0.10 in the Z-table. The closest value to this is 0.1003, which matches up with a z-value of –1.28. Converting this back to the original units, you get $x = 8 + (-1.28)(1) = 8 - 1.28 = 6.72$ minutes. Jimmy has to walk a mile in 6.72 minutes to get into his top 10 percent.

23. You know the percentile, and you want the original score. The middle step is to take the percentile, find the standard score that goes with it, and then convert to the original score (x) with the z-formula solved for x. In this case, you have a percentile of 42. The standard score for the 42nd percentile is found by seeking out 0.42 in the Z-table. The closest value to this is 0.4207, which matches up with a z-value of –0.20. Converting this back to the original units, you get $x = 40 + (-0.20)(6) = 40 - 1.2 = 38.8$. It takes Deshawn about 38.8 minutes to finish the exam.

24. Because the top 20 percent of scores get As, the percentage below the cutoff for an A is $100 - 20 = 80\%$. The standard score for the 80th percentile is found by seeking out 0.80 in the Z-table. The closest value to this is 0.7995, which matches up with a z-value of 0.84. Converting 0.84 to original units (x) with the z-formula solved for x, you have $x = 75 + (0.84)(5) = 75 + 4.2 = 79.2$. The cutoff for an A is 79.2.

REMEMBER

Whatever Z-table you use, make sure you understand how to use it. The Z-table in this book doesn't list every possible percentile; you need to choose the one that's closest to the one you need. And just like an airplane where the closest exit may be behind you, the closest percentile may be the one that's lower than the one you need.

25. The longest 25 percent are the 25 calls with the biggest values. The lengths of calls below the cutoff are 75 minutes, which is the percentile (percentile is the area below the value). The standard score for the 75th percentile is found by seeking out 0.75 in the Z-table. The closest value to this is 0.7486, which matches up with a z-value of 0.67. Converting 0.67 to original units (x) with the z-formula solved for x, you have $x = 10 + (0.67)(3) = 10 + 2.01 = 12.01$. So the cutoff for the longest 10 percent of customer service calls is roughly 12 minutes.

WARNING

You may have thought that the z-score here would be –0.67 because it corresponds to the 10th percentile. But remember, the longer the phone call, the larger the number will be, and you're looking for the longest phone calls, which means the top 10 percent of the values. Be careful in testing situations; these kinds of problems are used very often to make sure you can decide whether you need the upper tail of the distribution or the lower tail of the distribution.

26. I saved the best for last! I consider this problem a bit of an advanced, extra-credit type exercise. It gives you a percent, but it asks you for the standard deviation. When in doubt, work it like all the other problems and see what materializes. You know the percentage of these vehicles getting more than 100 mpg is 20 percent, so the percentage getting less than 100 mpg is 80 (that's the percentile). The standard score for the 80th percentile is found by seeking out 0.80 in the Z-table. The closest value to this is 0.7995, which matches up with a z-value of 0.84. You know the standard score, the mean, and the value for x in original units. Put all this into the z-formula solved for x to get $100 = 75 + 0.84\sigma$. Solving this for σ (the standard deviation) gives you $\sigma = \frac{100-75}{0.84} = \frac{25}{0.84} = 29.76$. The standard deviation is roughly 30 miles per gallon.

27. First, you should check that X, the number of international travelers, is a binomial random variable. You have a fixed number of trials ($n = 500$), which will each result in a success (international destination) or failure (not international). Because they're chosen randomly, you can assume their travels are independent. Lastly, the probability of success is a constant $p = 0.15$. So this is a binomial random variable you're working with. Now check to see whether you can use the normal approximation. You find that $np = (500)(0.15) = 75$ and $n(1-p) = (500)(0.85) = 425$ are both greater than 10, so you're good to continue on.

    You're looking for $p(X < 80)$, so you need to standardize the x-value of 80 into its proper z-value. Remember to first find the mean $\mu = np = (500)(0.15) = 75$ and the standard deviation $\sigma = \sqrt{np(1-p)} = \sqrt{500(0.15)(0.85)} = \sqrt{63.75} = 7.98$. Putting these into the z-formula, you get $z = \frac{80-75}{7.98} = \frac{5}{7.98} = 0.63$. To solve the problem, you need to find $p(Z < 0.63)$. From the Z-table, you can look up that $p(Z < 0.63) = 0.7357$. So, the chance that less than 80 of the next 500 people will be flying internationally is *approximately* 73.57 percent. (This is a really good approximation because the exact value found with some fancy statistical software is 71.68 percent; the approximation is off by only 1.89 percent.)

28. **No.** This is asking about a binomial random variable where X is the number of children who get on the ride out of the next $n = 150$ riders. When you check to see whether you can use the normal approximation to the binomial, you find that $np = (150)(0.95) = 142.5$, which is greater than 10. But $n(1-p) = (150)(0.05) = 7.5$, which is not greater than 10, so you should not use the normal approximation.

    And why not? Well, when you calculate the exact binomial probability and the normal approximation in that fancy statistical software, the approximation is actually about 4.5 percent lower than the exact answer. That's not so good.

29. This exercise should help demonstrate to you why going with the approximation isn't always best, even if it's faster.

    a. First, you should check that X, the number of shots made, is a binomial random variable. You have a fixed number of trials ($n = 20$), which will each result in a success (shot made) or failure (shot missed). The trials are independent, and the probability of success is a constant $p = 0.50$. So you can use the binomial table (Table A-3) to find $p(X \geq 10)$. Look at the portion of the table where $n = 20$ and focus on the column where $p = 0.5$. You can rewrite the probability as $p(X \geq 10) = p(X = 10) + p(X = 11) + p(X = 12) + \ldots + p(X = 20) = 0.176 + 0.160 + 0.120 + \ldots + 0.000 = 0.588$. So there is a 58.8 percent chance that you'll make at least 10 out of your next 20 shots.

    b. You know that X is a binomial, so check to see whether you can use the normal approximation. You find that $np = (20)(0.50) = 10$ and $n(1-p) = (20)(0.50) = 10$ are both greater than or equal to 10. This just barely meets the criteria, so you can go ahead with the approximation.

You're looking for $p(X \geq 10)$, so you need to standardize the x-value of 15 into its proper z-value. Remember to first find the mean $\mu = np = (20)(0.50) = 10$ and the standard deviation $\sigma = \sqrt{np(1-p)} = \sqrt{10(0.50)(0.50)} = \sqrt{2.5} = 1.58$.

Putting these into the z-formula, you get $z = \frac{10-10}{1.58} = \frac{0}{1.58} = 0$. To solve the problem, you need to find $p(Z \geq 0)$. You don't need the Z-table for this one, because you know that $z = 0$ splits the standard normal in half. So by symmetry, $p(Z \geq 0) = 0.50$. According to this, the chance that you'll make at least 10 of the next 20 shots is *approximately* 50 percent.

c. The approximation is 8.8 percentage points lower than the exact calculation $(58.8 - 50 = 8.8)$. This is partly due to the fact that n is so low. As I cautioned earlier, the approximation works best when n is large.

If n isn't large enough, you may not be able to rely too strongly on your approximations. Recall from Chapter 9 that the binomial distribution is discrete, and in this chapter, you see that the normal distribution is continuous. Some statisticians use a method known as "continuity correction" to work out the change. But better than that, as the value of n increases in binomial distributions, at a distance, it begins to look and behave more like a continuous one. You see much more on this useful idea through the Central Limit Theorem in Chapter 12.

If you're ready to test your skills a bit more, take the following chapter quiz that incorporates all the chapter topics.

Whaddya Know? Chapter 10 Quiz

Quiz time! Complete each problem to test your knowledge on the various topics covered in this chapter. You can then find the solutions and explanations in the next section.

1. The normal random variable is a discrete random variable. True or false?

2. The Z-distribution has mean ____________________ and standard deviation ____________________.

3. Suppose test scores have a normal distribution with mean 70 and standard deviation 5. According to the Empirical Rule, about what percentage of test scores passed the test with a grade of more than 60?

4. Suppose test scores have a normal distribution with mean 70 and standard deviation 5. According to the Z-table, exactly what percentage of test scores passed the test with a grade of more than 60?

5. Suppose the miles you get on a certain brand of tire have a normal distribution with mean 40,000 and standard deviation 5,000. What's the chance that a randomly chosen tire gets more than 50,000 miles on it?

6. Suppose the miles you get on a certain brand of tire have a normal distribution with mean 40,000 and standard deviation 5,000. Ten percent of the tires get more mileage than what value?

7. The weights of Nigerian Dwarf baby goats (aka kids) have a normal distribution with mean 3 pounds and standard deviation 0.5 pound. What is the chance that a randomly chosen baby goat weighs more than 4.5 pounds?

8. My baby Nigerian Dwarf goat weighed 1.75 pounds at birth. What percentile was she at, in terms of weight?

9. Suppose 90% of basketball players graduate from College *X*. What is the chance that out of 100 randomly chosen basketball players from College *X*, more than 80 of them graduate?

10. Suppose it is reported that 70% of all college graduates worked at a restaurant sometime during their college experience. You sample 100 college graduates at random. What is the chance that at most 65 of them worked at a restaurant sometime during their college experience?

Answers to Chapter 10 Quiz

1. **False.** The normal random variable has a continuous distribution (bell-shaped).

2. **The Z-distribution has a mean of 0 and a standard deviation of 1.**

3. If you passed the test, you scored more than 60. You divide this problem into two parts: the probability of getting 60 to 70 (the mean) and the probability of getting higher than 70. Because half of the probability in a Z-distribution is above the mean, the probability for scoring higher than 70 is $\frac{1}{2}$. Because 60 is 2 standard deviations below the mean of 70 (every standard deviation is worth 5 points), the probability for scoring between 60 and 70 is half of 95%. This is because the Empirical Rule tells you that about 95% of the values in a normal distribution are within 2 standard deviations on either side of the mean. You only need one side of the mean, so you take $0.95/2 = 0.475$. Sum these two probabilities to get $0.475 + 0.50 = 0.975$, **or 97.5 percent**. This is your approximate answer to the question. Most of the students passed!

4. You start out with $P(X > 60)$. Then you standardize 60 to a z-value by taking $Z = \frac{X - \mu}{\sigma} = \frac{60 - 70}{5} = -2$. You want $p(Z > -2))$. On the Z-table you go to the row for –2.0 and the column for .00 and you find 0.0228. This is $p(Z < -2)$. To get $p(Z > -2)$, you take $1 - 0.0228 = 0.9772$, or **97.72 percent**. This is your exact answer to the question.

5. Here you want $p(X > 50{,}000)$ where X is the number of miles you get on the tire. Change to a z-value by taking $Z = \frac{X - \mu}{\sigma} = \frac{50{,}000 - 40{,}000}{5{,}000} = 2$. Now you find $p(Z > 2)$. Look up 2.00 in the Z-table by finding the row for 2.0 and the column for .00 and intersecting them to get 0.9772. This is $p(Z < 2)$. To calculate $p(Z > 2)$, take $1 - 0.9772 = \mathbf{0.0228}$. It's highly unlikely that your randomly chosen tire will last over 50,000 miles.

6. Because you know 10% of the tires last more than a certain value, that value is standing at the 90th percentile. If 10% lie above it, then 90% lie below it, and percentiles reflect area below a certain value. To find the 90th percentile for X (the tire mileage), you first find the 90th percentile for Z, then change it to X. The 90th percentile of Z is that value of Z that goes with a value of .90 in the body of the table. Find the value closest to .90, which is 0.8997; then follow it across to determine which row it's in, and go up to determine which column it's in. Here you see 0.8997 is in row 1.2 and column .08, so the 90th percentile of Z is 1.28. Now change to X by using the z-formula, and solve for X:

$$Z = \frac{X - \mu}{\sigma} \rightarrow 1.28 = \frac{X - 40{,}000}{5{,}000} \rightarrow X = 1.28(5{,}000) + 40{,}000 \rightarrow X = 46{,}400 \text{ miles}$$

   So 90 percent of the tires get at least **46,400 miles** on them in this case.

7. Here you are looking for $p(X > 4.5)$. Change to Z and get the following:

   $p\left(Z > \frac{4.5 - 3}{0.5}\right) = p(Z > 3.0) = 1 - p(Z < 3.0) = 1 - 0.9987 = \mathbf{0.0013}$ according to the Z-table (row for 3.0 and column for .00).

   So it's really unlikely that a baby goat will be that large in this scenario.

8. This is a true example; I have a goat that was born this small. Her name is Half-Pint and she's doing great! To find her percentile, you find $P(X<1.75)$. You change to Z and get $p\left(Z<\frac{1.75-3}{0.5}\right)=p(Z<-2.5)$. Looking at the Z-table you find the row for –2.5 and the column for .00 and you get .0082. **So Half-Pint was at the .82th percentile** (not the 82nd percentile — this is less than the 1st percentile range.) She was small. But now she's the herd queen, so go figure.

9. Let X = number of graduating basketball players at College X out of 100 randomly chosen ones. X has a binomial distribution because the players either graduate or not (yes/no); you have a fixed number of trials ($n=100$); $p=.90$ is the same for each randomly chosen player; the results are independent because you chose them randomly. You are looking for $p(X>80)$, which is a tall order for a binomial distribution, because you would have to find $p(X=81)+...+p(X=100)$ and sum them all up. However, you may be able to use the normal approximation if your conditions are met — let's check. You have $np=100(.90)=90$, which is at least 10; you have $n(1-p)=100(0.10)=10$, which is at least 10 — whew! The conditions are met. You want $p(X>80)$ so you change to Z, which means $p\left(Z>\frac{80-\mu}{\sigma}\right)$. For μ you use np, the mean of the binomial distribution, and for σ you use $\sqrt{np(1-p)}=\sqrt{100(0.9)(1-0.9)}=3$. Going back to Z, you have $p(X>80)=p\left(Z>\frac{80-90}{3}\right)=p(Z>-3.33)=1-p(Z<-3.33)$. Go to the row for –3.3 on the table and the column for 0.30 and intersect them. You get $1-p(Z<-3.33)=1-0.004=$ **0.9996**. This is an approximate answer. You're very likely to get at least 80 players to graduate out of 100 with these circumstances.

10. Here X = number of college graduates out of 100 who have worked at a restaurant at some point in their college careers. You want P(X at most 65) where $n=100$ and $p=0.70$. "At most" means less-than-or-equal-to, so you are looking for $P(X\leq 65)$. In this case, n is large enough to use the normal approximation to help you out — otherwise, you'd have to find $p(X=0)+...+p(X=65)$, which is quite arduous. To change to Z, you need to find μ and σ, which are equal to $np=100(.70)=70$ and $\sqrt{np(1-p)}=\sqrt{100(.70)(1-.70)}=4.58$, respectively. You have $p(X\leq 65)=p\left(Z\leq\frac{65-70}{4.58}\right)=p(Z\leq -1.09)=0.1379$. This is an approximate answer, contingent on the conditions being met: 1) $np=100(0.70)=70$, which is at least 10; and $n(1-p)=100(0.30)=30$, which is at least 10.

IN THIS CHAPTER

» Discovering characteristics of the *t*-distribution

» Seeing the relationship between the *Z*- and *t*-distributions

» Understanding and using the *t*-table

Chapter 11

The *t*-Distribution

The *t*-distribution is one of the mainstays of data analysis. You may have heard of the "*t*-test," for example, which is often used to compare two groups in medical studies and scientific experiments.

This short chapter covers the basic characteristics and uses of the *t*-distribution. You find out how it compares to the normal distribution (more on that in Chapter 10) and how to use the *t*-table to find probabilities and percentiles.

Basics of the *t*-Distribution

In this section, you get an overview of the *t*-distribution, its main characteristics, when it's used, and how it's related to the *Z*-distribution (see Chapter 10).

Comparing the *t*- and *Z*-distributions

The normal distribution is that well-known bell-shaped distribution whose mean is μ and whose standard deviation is σ (see Chapter 10 for more on the normal distribution). The most commonly used normal distribution is the standard normal (also called the *Z*-distribution), whose mean is 0 and standard deviation is 1.

The *t*-distribution can be thought of as a cousin of the standard normal distribution — it looks similar in that it's centered at zero and has a basic bell shape, but it's shorter and flatter than the *Z*-distribution. Its standard deviation is proportionally larger compared to the *Z*, which is why you see the fatter tails on each side.

Figure 11-1 compares the t- and standard normal (Z-) distributions in their most general forms.

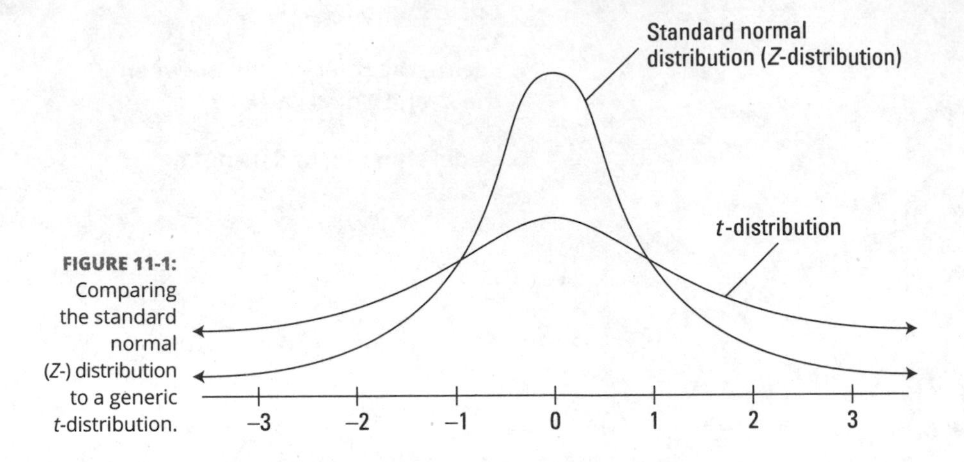

FIGURE 11-1: Comparing the standard normal (Z-) distribution to a generic t-distribution.

The t-distribution is typically used to study the mean of a population rather than to study the individuals within a population. In particular, it is used in many cases when you use data to estimate the population mean — for example, to estimate the average price of all the new homes in California. Or, when you use data to test someone's claim about the population mean — for example, is it true that the mean price of all the new homes in California is $500,000?

TIP

These procedures are called *confidence intervals* and *hypothesis tests* and are discussed in Chapters 14 and 15, respectively.

The connection between the normal distribution and the t-distribution is that the t-distribution is often used for analyzing the mean of a population if the population itself has a normal distribution (or fairly close to it). Its role is especially important if your data set is small or if you don't know the standard deviation of the population (which is often the case).

When statisticians use the term *t-distribution,* they aren't talking about just one individual distribution. There is an entire family of specific t-distributions, depending on what sample size is being used to study the population mean. Each t-distribution is distinguished by what statisticians call its *degrees of freedom* (*df*). In situations where you have one population and your sample size is n, the degrees of freedom for the corresponding t-distribution is $n - 1$. For example, a sample of size 10 uses a t-distribution with $10 - 1$, or 9, degrees of freedom, denoted t_9 (pronounced *tee sub-nine*). Situations involving two populations use different degrees of freedom and are discussed in Chapter 16.

Discovering the effect of variability on *t*-distributions

When t-distributions are based on smaller sample sizes, they have larger standard deviations than those based on larger sample sizes. Their shapes are flatter; their values are more spread out. That's because results based on smaller data sets are more variable than results based on large data sets.

REMEMBER

The larger the sample size is, the larger the degrees of freedom will be, and the more the t-distributions will look like the standard normal distribution (Z-distribution). A rough cutoff point where the t- and Z-distributions become similar enough that either could be used (at least similar enough for jazz or government work) is around $n = 30$.

Figure 11-2 shows what different t-distributions look like for different sample sizes and how they all compare to the standard normal (Z-) distribution.

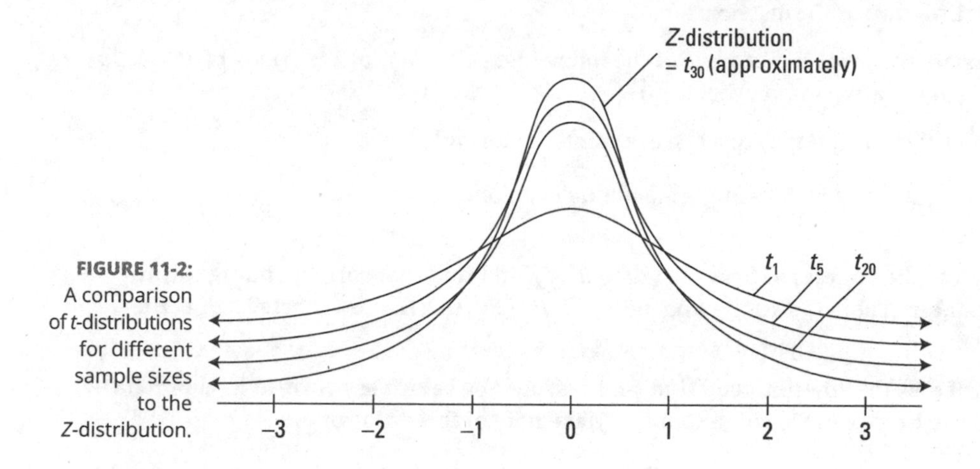

FIGURE 11-2: A comparison of *t*-distributions for different sample sizes to the *Z*-distribution.

Using the *t*-Table

Each normal distribution has its own mean and standard deviation that classify it, so finding probabilities for each normal distribution on its own is not the way to go. Thankfully, you can standardize the values of any normal distribution to become values on a standard normal (Z-) distribution (whose mean is 0 and standard deviation is 1) and use a Z-table (found in the Appendix) to find probabilities. (Chapter 10 has info on normal distributions.)

In contrast, a t-distribution is not classified by its mean and standard deviation, but by the sample size of the data set being used (n). Unfortunately, there is no single "standard t-distribution" that you can use to transform the numbers and find probabilities on a table. Because it wouldn't be humanly possible to create a table of probabilities and corresponding t-values for every possible t-distribution, statisticians created one table showing certain values of t-distributions for a selection of degrees of freedom and a selection of probabilities. This table is called the *t-table* (it appears in the Appendix). In this section, you find out how to determine probabilities, percentiles, and critical values (for confidence intervals) using the t-table.

Finding probabilities with the *t*-table

Each row of the t-table (Table A-2 in the Appendix) represents a different t-distribution, classified by its degrees of freedom. The columns represent various common greater-than probabilities, such as 0.40, 0.25, 0.10, and 0.05. The numbers across a row indicate the values on the t-distribution (the t-values) corresponding to the greater-than probabilities shown at the top of the columns. Rows are arranged by degrees of freedom.

REMEMBER

Another term for greater-than probability is *right-tail probability,* which indicates that such probabilities represent areas on the right-most end (tail) of the *t*-distribution.

The steps for finding a probability related to a sample mean *X* when you need to use the *t*-distribution are very similar to those when using the normal distribution. The only difference comes at the end because of the different way that the *t*-table is arranged:

1. **Draw a picture of the distribution.**
2. **Translate the problem into one of the following: $p(X \geq a)$, $p(X \leq b)$, or $p(a \leq X \leq b)$. Shade in the area on your picture.**
3. **Standardize *a* (and/or *b*) to a *t*-score using the formula:**

   $t = \frac{\bar{x} - \mu}{s/\sqrt{n}}$, where *s* is the sample standard deviation.

4. **Calculate the degrees of freedom ($df = n - 1$), and find that corresponding row in the *t*-table (Table A-2 in the Appendix). If $df > 30$, use the row labeled "z" at the bottom.**
5. **Scan across the row until you find your *t*-score or a value very close to it. Now look up at the header in the same column. Make note of that value of *p*.**

   Note: If your *t*-score isn't really close to any of the values in the *df* row, say within 0.1, then find which two values it falls between. Now look up at the headers and make note of which two values of *p* the probability must fall between. Though you won't be able to find a precise single answer, you'll at least be able to give a range that you know the probability lies within.

6a. **If you need a "tail" probability — that is, $p(X \leq -a)$ or $p(X \geq a)$— you're done. The value of *p* you found is the probability.**

6b. **If you need a probability that's everything except a tail (one that cuts through the middle) — that is, $p(X \leq a)$ or $p(X \geq a)$— take 1 minus the result from Step 5.**

   Note: If you had to make note of two values of *p*, subtract each from 1. The probability is in that range.

7. **If you need a "between-two-values" probability — that is, $p(a \leq X \leq b)$— do Steps 1–6 for *b* (the larger of the two values) and again for *a* (the smaller of the two values), and subtract the results.**

REMEMBER

The probability that *X* is equal to any single value is 0 for any continuous random variable (just like the normal and *t*). So, $p(X \leq a) = p(X < a)$, and also $p(X \geq b) = p(X > b)$. This isn't true of discrete random variables.

For example, the second row of the *t*-table is for the t_2 distribution. With 2 degrees of freedom, you see that the second number, 0.816, is the value on the t_2 distribution whose area to its right (its right-tail probability) is 0.25 (see the heading for the second column). In other words, the probability that t_2 is greater than 0.816 equals 0.25. In probability notation, that means $p(t_2 \geq 0.816) = 0.25$.

The next number in row two of the t-table is 1.886, which lies in the 0.10 column. This means the probability of being greater than 1.886 on the t_2 distribution is 0.10. Because 1.886 falls to the right of 0.816, there is less area under the curve to the right of 1.886, so its right-tail probability is lower.

EXAMPLE

Q. The weights of single-dip ice cream cones at Bob's ice cream parlor have a normal distribution with a mean of 8 ounces. If you order six single-dip ice cream cones for you and your friends, what's the chance that they weigh an average of at least 9 ounces with a standard deviation of 1 ounce?

A. You want $p(\bar{X} \geq 9)$, the probability that the average weight of the cones is 9 ounces or more. Converting the 9 to a standard score, you get $t = \frac{\bar{x} - \mu}{s/\sqrt{n}} = \frac{9-8}{1/\sqrt{6}} = \frac{1}{0.408} = 2.449$.

Calculate $df = 6 - 1 = 5$ and look at that row in the t-table (Table A-2). Scanning across that row, you find that none of the t-values are that close, but $t = 2.449$ does fall between 2.015 and 2.571 in that row. By the note in Step 6, move up to the column header and see that this corresponds to probability values of 0.05 and 0.025, respectively. You were looking for a tail probability, $p(\bar{X} \geq 8)$, and that is exactly what the t-table provides, so you are done! Because your t-value falls between the two in the row, your probability has to fall between the ones given at the top. The probability that the average ice cream cone weight is greater than 9 ounces is between 0.025 and 0.05, or between 2.5 and 5 percent. (Not really great odds of getting more ice cream than you paid for.)

YOUR TURN

1 Suppose the average length of a stay in Europe for American tourists is 17 days. You choose a random sample of 16 American tourists. The sample of 16 tourists stays an average of 18.5 days or more with a standard deviation of 4.5 days. What's the chance of that happening?

2 Suppose a class's test scores have a mean of 80. You randomly choose 25 students from the class. What's the chance that the group's average test score is less than 82 with a standard deviation of 5?

Figuring percentiles for the *t*-distribution

You can also use the *t*-table (in the Appendix) to find percentiles for a *t*-distribution. A *percentile* is a number on a distribution whose less-than probability is the given percentage; for example, the 95th percentile of the *t*-distribution with $n-1$ degrees of freedom is that value of t_{n-1} whose left-tail (less-than) probability is 0.95 (and whose right-tail probability is 0.05). (See Chapter 5 for particulars on percentiles.)

REMEMBER

The normal table shows "less than" probabilities, so when you want to find a percentile, the normal table is ready to go. However, the *t*-table shows "greater than" probabilities. If you want to calculate a percentile using the *t*-table, you need to subtract the value of *p* you find (based on the right tail) from 1 to get the desired percentile's probability (the left tail).

Suppose you have a sample of size 10 and you want to find the 95th percentile of its corresponding *t*-distribution. You have $n-1=9$ degrees of freedom, so you look at the row for $df=9$. The 95th percentile is the number where 95 percent of the values lie below it and 5 percent lie above it, so you want the right-tail area to be 0.05. Move across the column headers until you find the column for 0.05. Matching the column for 0.05 with the row where $df=9$ gives you the cell with $t_9=1.833$. This is the 95th percentile of the *t*-distribution with 9 degrees of freedom.

Now, if you increase the sample size to $n=20$, the value of the 95th percentile decreases; look at the row for $20-1=19$ degrees of freedom, and in the column for 0.05 (a right-tail probability of 0.05), you find $t_{19}=1.729$. Notice that the 95th percentile for the t_{19} distribution is less than the 95th percentile for the t_9 distribution (1.833). This is because larger degrees of freedom indicate a smaller standard deviation and the *t*-values are more concentrated about the mean, so you reach the 95th percentile with a smaller value of *t*. (See the section, "Discovering the effect of variability on *t*-distributions," earlier in this chapter.)

WARNING

You might look at the very bottom of the *t*-table and see percentages like 90 percent, 95 percent, and 99 percent already listed for you. But these are not to be used for calculating any of the percentile problems you're working on here. Move your eyes to the left and you'll see that the row is titled "CI," short for "confidence interval." This last row is only to be used if constructing confidence intervals as described in the section, "Picking out *t**-values for confidence intervals," in this chapter and Chapter 14.

EXAMPLE

Q. Suppose you find the mean of ten quiz scores, convert it to a standard score, and check the table to find out it's equal to the 99th percentile.

(a) What's the standard score?

(b) Compare the result to the standard score you have to get to be at the 99th percentile on the Z-distribution.

A. The *t*-distributions push you farther out to get to the same percentile that the Z-distribution would.

(a) Your sample size is $n=10$, so you need the *t*-distribution with $10-1=9$ degrees of freedom, also known as the t_9 distribution. If you want a 99th percentile, that means 99 percent is in the lower (left-end) tail, and 1 percent is in the upper (right-end) tail. Using the *t*-table (Table A-2 in the Appendix), you find the column header with the right-tail probability of 0.01. Matching the row for $df=9$ and the column for 0.01, you find the *t*-value to be $t_9=2.821$.

(b) Using the Z-distribution (Table A-1) and looking for the left-tail probability closest to 0.99, you find that 0.9901 corresponds to a z-value of 2.33. You could also get a more precise value using the z row in the t-table. That gives the exact z-value for a 99th percentile, which has 0.01 in the right tail, to be 2.326. The standard score associated with the 99th percentile is around 2.33, which is much smaller than the 2.821 from Part A of this question. The number from Part B is smaller because the t-distribution is flatter than the Z-distribution, with more area or probability out in the tails. So to get all the way out to the 99th percentile, you have to go farther out on the t-distribution than on the normal curve.

YOUR TURN

3 Suppose you collect data on ten products and check their weights. The average should be 10 ounces, but your sample mean is 9 ounces with a standard deviation of 2 ounces.

(a) Find the standard score.

(b) What percentile is the standard score found in Part A closest to?

(c) Suppose the mean really is 10 ounces. Do you find these results unusual? Use probabilities to explain.

Picking out t^*-values for confidence intervals

Confidence intervals estimate population parameters, such as the population mean, by using a statistic (for example, the sample mean) plus or minus a margin of error. (See Chapter 14 for all the information you need on confidence intervals and more.) To compute the margin of error for a confidence interval, you need a *critical value* (the number of standard errors you add and subtract to get the margin of error you want; see Chapter 14). When the sample size is large (at least 30), you use critical values on the Z-distribution (shown in Chapter 14) to build the margin of error. When the sample size is small (less than 30) and/or the population standard deviation is unknown, you use the t-distribution to find critical values.

To help you find critical values for the t-distribution, you can use the last row of the t-table, which lists common confidence levels, such as 80 percent, 90 percent, and 95 percent. To find a critical value, look up your confidence level in the bottom row of the table; this tells you which column of the t-table you need. Intersect this column with the row for your df (see Chapter 14 for degrees of freedom formulas). The number you see is the critical value (or the t^*-value) for your confidence interval. For example, if you want a t^*-value for a 90 percent confidence interval when you have 9 degrees of freedom, go to the bottom of the table, find the column for 90 percent, and intersect it with the row for $df = 9$. This gives you a t^*-value of 1.833 (rounded).

REMEMBER

Across the top row of the t-table, you see right-tail probabilities for the t-distribution. But confidence intervals involve both left- and right-tail probabilities (because you add and subtract the margin of error). So half of the probability left from the confidence interval goes into each tail. You need to take that into account. For example, a t^*-value for a 90 percent confidence interval has 5 percent for its greater-than probability and 5 percent for its less-than probability (taking 100 percent minus 90 percent and dividing by 2). Using the top row of the t-table, you would have to look for 0.05 (rather than 10 percent, as you might be inclined to do). But using the bottom row of the table, you just look for 90 percent. (The result you get using either method ends up being in the same column.)

TIP

When looking for t^*-values for confidence intervals, use the bottom row of the t-table as your guide rather than the headings at the top of the table.

Studying Behavior Using the *t*-Table

You can use computer software to calculate any probabilities, percentiles, or critical values you need for any t-distribution (or any other distribution) if it's available to you. (On exams it may not be available.) However, one of the nice things about using a table (rather than computer software) to find probabilities is that the table can tell you information about the behavior of the distribution itself — that is, it can give you the big picture. Here are some nuggets of big-picture information about the t-distribution you can glean by scanning the t-table (in the Appendix).

As the degrees of freedom increase, the values on each t-distribution become more concentrated around the mean, eventually resembling the Z-distribution (see Chapter 10). The t-table confirms this pattern as well. Because of the way the t-table is set up, if you choose any column and move down through the numbers in the column, you're increasing the degrees of freedom (and sample size) and keeping the right-tail probability the same. As you do this, you see the t-values getting smaller and smaller, indicating the t-values are becoming closer to (hence, more concentrated around) the mean.

I labeled the second-to-last row of the t-table with a z in the *df* column. This indicates the "limit" of the t-values as the sample size (n) goes to infinity. The t-values in this row are approximately the same as the z-values on the Z-table (in the Appendix) that correspond to the same greater-than probabilities. This confirms what you already know: As the sample size increases, the t- and the Z-distributions look more and more alike. For example, the t-value in row 30 of the t-table corresponding to a right-tail probability of 0.05 (column 0.05) is 1.697. This lies close to $z = 1.645$, the value corresponding to a right-tail area of 0.05 on the Z-distribution. (See row Z of the t-table.)

REMEMBER

It doesn't take a super-large sample size for the values on the t-distribution to get close to the values on a Z-distribution. For example, when $n = 31$ and $df = 30$, the values in the t-table are already quite close to the corresponding values on the Z-table.

Practice Questions Answers and Explanations

1. You want $p(\bar{X} \geq 18.5)$, the probability that the average is 18.5 days or more. Converting the 18.5 to a standard score, you get $t = \frac{\bar{x} - \mu}{s / \sqrt{n}} = \frac{18.5 - 17}{4.5 / \sqrt{16}} = \frac{1.5}{1.125} = 1.33$. Calculate $df = 16 - 1 = 15$ and look at that row in the t-table (Table A-2). Scanning across that row, you find the value 1.345 in the third column. This is close enough to your $t = 1.33$ to use. Moving up to the column header, you see that this corresponds to a probability value of 0.10. Because you were looking for a tail probability, $p(\bar{X} \geq 18.5)$, and that is exactly what the t-table provides, you are done! The probability that the average is 18.5 days or more is **0.10, or 10 percent**.

2. You want $p(\bar{X} < 82)$, the probability that the average is less than 82. Converting the 82 to a standard score, you get $t = \frac{\bar{x} - \mu}{s / \sqrt{n}} = \frac{82 - 80}{5 / \sqrt{25}} = \frac{2}{1} = 2.00$. Calculate $df = 25 - 1 = 24$ and look at that row in the t-table (Table A-2). Scanning across that row, you find the value 2.06 in the third column. This is close enough to your $t = 2.00$ to use. Moving up to the column header, you see that this corresponds to a probability value of 0.025. Because you were looking for a probability that cuts through the middle ($\mu = 80$), using Step 6b you get $p(\bar{X} < 82) = 1 - p(\bar{X} < 82) = 1 - 0.025 = 0.975$. The probability that the student average is less than 82 is **0.975, or 97.5 percent**.

TIP

If the area under the curve that you want cuts through the middle and contains everything but a little tail, then the final probability has to be at least 50 percent. If you would have finished this last problem and said that the answer was 0.025, that should have set off alarms because the picture of $p(\bar{X} < 82)$ covers over half of the curve.

3. Here you compare what you expect to see with what you actually get (which comes up in hypothesis testing; see Chapter 15). The basic information here is that you have $\bar{x} = 9$, $s = 2$, $\mu = 10$, and $n = 10$. Anytime you have the sample standard deviation (s) and not the population standard deviation (σ), you should use a t-distribution.

   a. The standard score is $t = \frac{\bar{x} - \mu}{s / \sqrt{n}} = \frac{9 - 10}{2 / \sqrt{10}} = \frac{-1}{0.632} = \mathbf{-1.58}$.

   b. The standard score is negative, meaning you're 1.58 standard deviations below the mean on the t_9 distribution ($n - 1 = 10 - 1 = 9$). Because the t-distribution is symmetric, the area below the negative standard score is equal to the area above the positive version of the standard score. In the row with $df = 9$, the value 1.58 lies between 1.38 and 1.83, which correspond to 0.10 and 0.05 in the "greater-than" tails, respectively. So the value –1.58 lies between –1.38 and –1.83, which correspond to 0.10 and 0.05 in the "less-than" tails, respectively. **Thus, –1.58 lies between the 10th percentile and the 5th percentile on the t_9 distribution.**

   c. **The results aren't entirely unusual because, according to Part B, they happen between 5 percent and 10 percent of the time.**

If you're ready to test your skills a bit more, take the following chapter quiz that incorporates all the chapter topics.

Whaddya Know? Chapter 11 Quiz

Quiz time! Complete each problem to test your knowledge on the various topics covered in this chapter. You can then find the solutions and explanations in the next section.

1. The t-distribution looks exactly like the Z-distribution. True or false?
2. What is the mean of the t-distribution?
3. Each t-distribution in the family of all t-distributions is characterized by its ____________________.
4. As you increase the sample size, the degrees of freedom increase, and the t-distribution looks more and more like ____________________.
5. The standard deviation of the t-distribution is ____________________ than the standard deviation of the Z-distribution.
6. Suppose your sample size is 10. What are the degrees of freedom for the corresponding t-distribution?
7. Suppose $n = 20$. What's the probability that t is greater than 2.09?
8. Suppose $n = 6$. What's the probability of being less than 6.87?
9. What is the 90th percentile of the t_{22} distribution?
10. What does the next-to-last row of the t-table represent and why is it on the t-table?

Answers to Chapter 11 Quiz

1. **False.** The t-distribution is fatter and flatter than the Z-distribution. Although it has a mound in the middle like the Z-distribution, its tails are thicker. It's kind of like you sat on top of the Z-distribution and it spread itself out — seriously!

2. **Zero**

3. **degrees of freedom**

4. **the Z-distribution**

5. **larger**

6. $\mathbf{10 - 1 = 9}$

7. **0.025**, according to the t-table with 19 degrees of freedom. Follow to the top of the column where 2.09 is found (the row is 19), and you find 0.025.

8. $p(t_{6-1} < 6.87) = 1 - 0.0005 = 0.9995$. According to the t-table intersecting row 5 (5 degrees of freedom) and the column where 6.87 resides, you find $p(t_{6-1} > 6.87) = 0.0005$. You want $p(t_{6-1} < 6.87)$, so take one minus this amount.

If your particular value of t is not on the t-table, you can locate it between two other t-values in the appropriate row, and list its probability as being between their two corresponding probabilities at the top of their columns. Or you may use technology to find the exact value.

9. **1.32.** Look in row 22 because the degrees of freedom in this case are 22; then go across the top and find the column that has 0.10 probability in the upper part of the distribution (column 0.10). Intersect this row and column to find 1.32. This number has 10% of the values above it, but it also has 90% of the values below it, making it the 90th percentile of the t_{22} distribution.

10. The next-to-last row of the t-table represents certain values on the Z-distribution whose right-tail areas (greater-than probabilities) are listed at the top of their respective columns. It's on the t-table because as the *df* of the t-distribution increase as you move down a particular column, you can see the values in the column (values on those t-distributions) get closer and closer to the values on the Z-distribution. If n is large, you can see that using the z-value is about the same as using the t-value with what would have been a large *df*.

IN THIS CHAPTER

» Understanding the concept of a sampling distribution

» Putting the Central Limit Theorem to work

» Determining the factors that affect precision

Chapter 12

Sampling Distributions and the Central Limit Theorem

When you take a sample of data, it's important to realize the results will vary from sample to sample. Statistical results based on samples should include a measure of how much those results are expected to vary. When the media reports statistics like the average price of a gallon of gas in the U.S. or the percentage of homes on the market that were sold over the last month, you know they didn't sample every possible gas station or every possible home sold. The question is, how much would their results change if another sample were selected?

This chapter addresses this question by studying the behavior of means for all possible samples and the behavior of proportions from all possible samples. By studying the behavior of all possible samples, you can gauge where your sample results fall and understand what it means when your sample results fall outside of certain expectations.

Defining a Sampling Distribution

A *random variable* is a characteristic of interest that takes on certain values in a random manner. For example, the number of red lights you hit on the way to work or school is a random variable; the number of children a randomly selected family has is a random variable. You use capital letters such as X or Y to denote random variables, and you use lowercase letters such as x or y to denote actual outcomes of random variables. A *distribution* is a listing, graph, or function of all possible outcomes of a random variable (such as X) and how often each actual outcome (x), or set of outcomes, occurs. (See Chapter 9 for more details on random variables and distributions.)

For example, suppose a million of your closest friends each roll a single die and you record each actual outcome (x). A table or graph of all these possible outcomes (one through six) and how often they occurred represents the distribution of the random variable X. A graph of the distribution of X in this case is shown in Figure 12-1a. It shows the numbers 1 through 6 appearing with equal frequency (each one occurring $\frac{1}{6}$ of the time), which is what you expect over many rolls if the die is fair.

Now suppose each of your friends rolls this single die 50 times ($n = 50$) and you record the average, $\bar{x}$. The graph of all their averages of all their samples represents the distribution of the random variable $\bar{X}$. Because this distribution is based on sample averages rather than individual outcomes, this distribution has a special name. It's called the *sampling distribution* of the sample mean, $\bar{X}$. Figure 12-1b shows the sampling distribution of $\bar{X}$, the average of 50 rolls of a die.

Figure 12-1b (average of 50 rolls) shows the same range (1 through 6) of outcomes as Figure 12-1a (individual rolls), but Figure 12-1b has more possible outcomes. You could get an average of 3.3 or 2.8 or 3.9 for 50 rolls, for example, whereas someone rolling a single die can only get whole numbers from 1 to 6. Also, the shapes of the graphs are different; Figure 12-1a shows a flat shape, where each outcome is equally likely, and Figure 12-1b has a mound shape; that is, outcomes near the center (3.5) occur with high frequency and outcomes near the edges (1 and 6) occur with extremely low frequency. A detailed look at the differences and similarities in shape, center, and spread for individuals versus averages, and the reasons behind them, is the topic of the following sections. (See Chapters 5, 7, and 9 if you need background info on shape, center, and spread of random variables before diving in.)

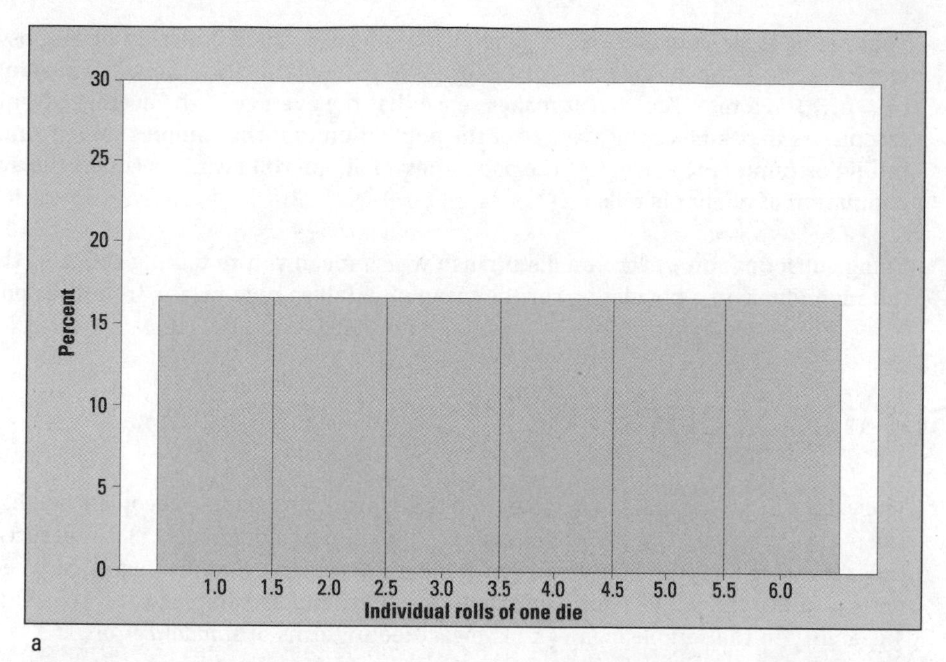

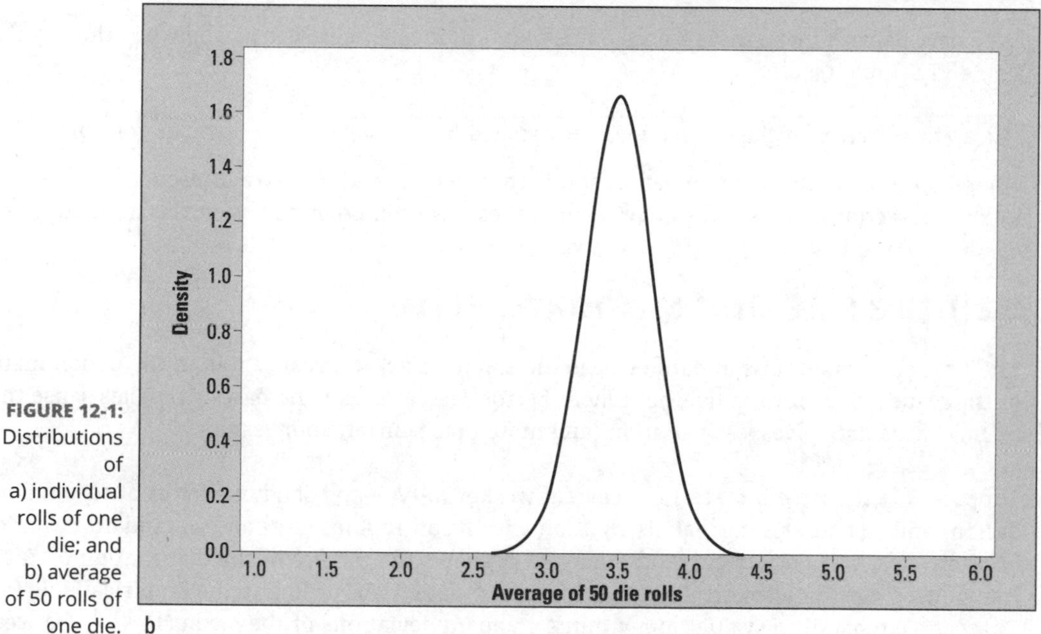

FIGURE 12-1: Distributions of a) individual rolls of one die; and b) average of 50 rolls of one die.

The Mean of a Sampling Distribution

Using the die-rolling example from the preceding section, X is a random variable denoting the outcome you can get from a single die (assuming the die is fair). The mean of X (over all possible outcomes) is denoted by μ_x (pronounced *mu sub-x*); in this case its value is 3.5 (as shown in Figure 12-1a). If you roll a die 50 times and take the average, the random variable $\bar{X}$ represents any outcome you could get. The mean of $\bar{X}$, denoted $\mu_{\bar{x}}$ (pronounced *mu sub-x-bar*), equals 3.5 as well. (You can see this result in Figure 12-1b.)

This result is no coincidence! In general, the mean of the population of all possible sample means is the same as the mean of the original population. (Notationally speaking, you write $\mu_{\bar{x}} = \mu_x$.) It's a mouthful, but it makes sense that the average of the averages from all possible samples is the same as the average of the population that the samples came from. In the die-rolling example, the average of the population of all 50-roll averages equals the average of the population of all single rolls (3.5).

TIP

Using subscripts on μ, you can distinguish which mean you're talking about — the mean of X (all individuals in a population) or the mean of $\bar{X}$ (all sample means from the population).

Measuring Standard Error

The values in any population deviate from their mean; for instance, people's heights differ from the overall average height. Variability in a population of individuals (X) is measured in *standard deviations* (see Chapter 5 for details on standard deviation). Sample means vary because you're not sampling the whole population, only a subset; and as samples vary, so will their means. Variability in the sample mean ($\bar{X}$) is measured in terms of *standard errors.*

REMEMBER

Error here doesn't mean there's been a mistake — it means there is a gap between the population and sample results.

The standard error of the sample mean is denoted by $\sigma_{\bar{x}}$ *(sigma sub-x-bar)*. Its formula is $\frac{\sigma_x}{\sqrt{n}}$, where σ_x *(sigma sub-x)* is population standard deviation and *n* is size of each sample. In the following sections, you see the effect each of these two components has on the standard error.

Sample size and standard error

The first component of standard error is the sample size, *n*. Because *n* is in the denominator of the standard error formula, the standard error decreases as *n* increases. It makes sense that having more data gives less variation (and more precision) in your results.

Suppose *X* is the time it takes for a clerical worker to type and send one letter of recommendation, and say X has a normal distribution with mean 10.5 minutes and a standard deviation of 3 minutes. The bottom curve in Figure 12-2 shows the picture of the distribution of X, the individual times for all clerical workers in the population. According to the Empirical Rule (see Chapter 5), most of the values are within 3 standard deviations of the mean (10.5) — between 1.5 and 19.5.

Now take a random sample of ten clerical workers, measure their times, and find the average, $\bar{x}$, each time. Repeat this process over and over, and graph all the possible results for all possible samples. The middle curve in Figure 12-2 shows the picture of the sampling distribution of $\bar{X}$. Notice that it's still centered at 10.5 (which you expected) but its variability is smaller; the standard error in this case is $\frac{\sigma_x}{\sqrt{n}} = \frac{3}{\sqrt{10}} = 0.95$ minute (quite a bit less than 3 minutes, the standard deviation of the individual times).

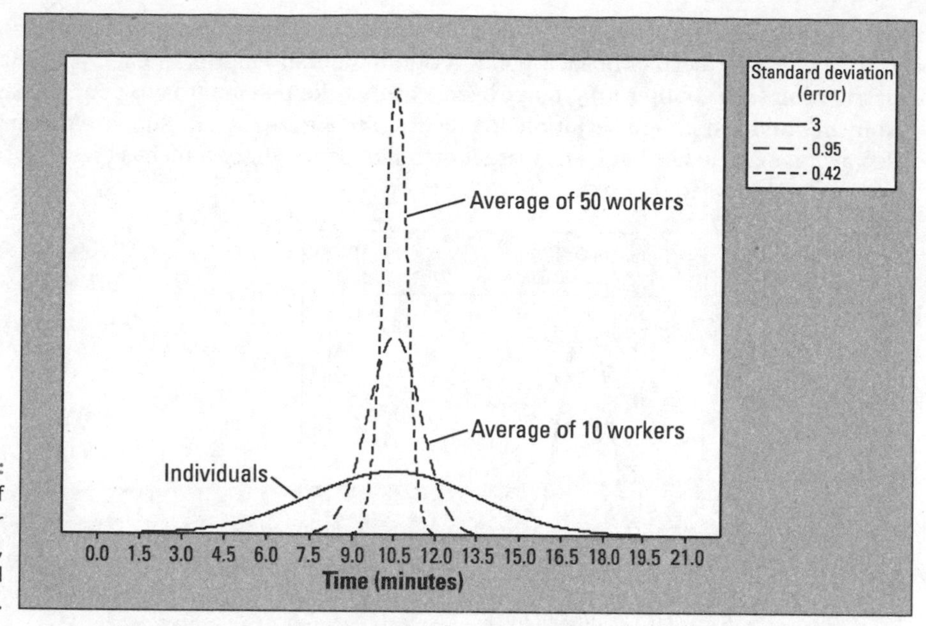

FIGURE 12-2: Distributions of times for 1 worker, 10 workers, and 50 workers.

Looking at Figure 12-2, the average times for samples of 10 clerical workers are closer to the mean (10.5) than the individual times are. That's because average times don't change as much from sample to sample as individual times change from person to person.

Now take all possible random samples of 50 clerical workers and find their means; the sampling distribution is shown in the tallest curve in Figure 12-2. The standard error of $\bar{X}$ goes down to $\frac{\sigma_x}{\sqrt{n}} = \frac{3}{\sqrt{50}} = 0.42$ minute. You can see the average times for 50 clerical workers are even closer to 10.5 than the ones for 10 clerical workers. By the Empirical Rule, most of the values fall between $10.5 - 3(.42) = 9.24$ and $10.5 + 3(.42) = 11.76$. Larger samples give even more precision around the mean because they change even less from sample to sample.

REMEMBER

Why is having more precision around the mean important? Because sometimes you don't know the mean but want to determine what it is, or at least get as close to it as possible. How can you do that? By taking a large random sample from the population and finding its mean. You know that your sample mean will be close to the actual population mean if your sample is large, as Figure 12-2 shows (assuming your data are collected correctly; see Chapter 17 for details on collecting good data).

Population standard deviation and standard error

The second component of standard error involves the amount of diversity in the population (measured by standard deviation). In the standard error formula, $\frac{\sigma_x}{\sqrt{n}}$, for $\bar{X}$, you see the population standard deviation, σ_x, is in the numerator. That means as the population standard deviation increases, the standard error of the sample means also increases. Mathematically this makes sense; how about statistically?

Suppose you have two ponds full of fish (call them pond #1 and pond #2), and you're interested in the length of the fish in each pond. Assume the fish lengths in each pond have a normal distribution (see Chapter 10). You've been told that the fish lengths in pond #1 have a mean of 20 inches and a standard deviation of 2 inches (see Figure 12-3a). Suppose the fish in pond #2 also average 20 inches but have a larger standard deviation of 5 inches (see Figure 12-3b).

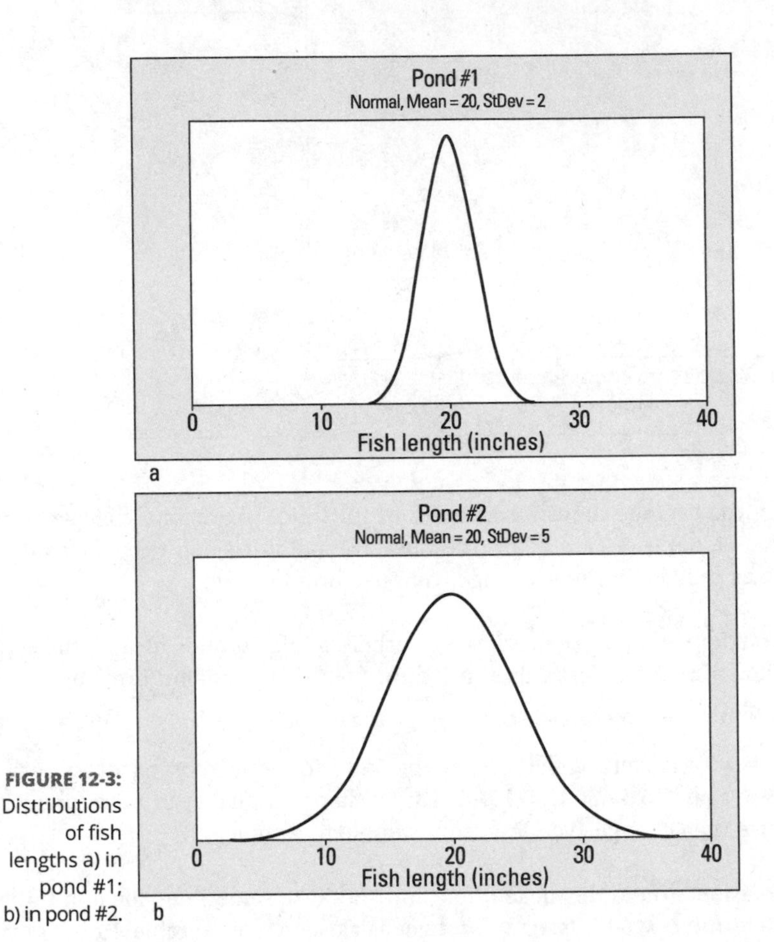

FIGURE 12-3: Distributions of fish lengths a) in pond #1; b) in pond #2.

Comparing Figures 12-3a and 12-3b, you see the lengths for the two populations of fish have the same shape and mean, but the distribution in Figure 12-3b (for pond #2) has more spread, or variability, than the distribution shown in Figure 12-3a (for pond #1). This spread confirms that the fish in pond #2 vary more in length than those in pond #1.

Now suppose you take a random sample of 100 fish from pond #1, find the mean length of the fish, and repeat this process over and over. Then you do the same with pond #2. Because the lengths of individual fish in pond #2 have more variability than the lengths of individual fish in pond #1, you know the average lengths of samples from pond #2 will have more variability than the average lengths of samples from pond #1 as well. (In fact, you can calculate their standard errors using the formula earlier in this section to be 0.20 and 0.50, respectively.)

Estimating the population average is harder when the population varies a lot to begin with; estimating the population average is much easier when the population values are more consistent. The bottom line is that the standard error of the sample mean is larger when the population standard deviation is larger.

Looking at the Shape of a Sampling Distribution

Now that you know about the mean and standard error of $\bar{X}$, the next step is to determine the shape of the sampling distribution of $\bar{X}$; that is, the shape of the distribution of all possible sample means (all possible values of $\bar{x}$) from all possible samples. You proceed differently for different conditions, which I divide into two cases: 1) the original distribution for X (the population) is normal, or has a normal distribution; and 2) the original distribution for X (the population) is *not* normal, or is unknown.

Case 1: The distribution of *X* is normal

If X has a normal distribution, then $\bar{X}$ does too, no matter what the sample size n is. In the example regarding the amount of time (X) for a clerical worker to complete a task (refer to the section, "Sample size and standard error"), you knew X had a normal distribution (refer to the lowest curve in Figure 12-2). If you refer to the other curves in Figure 12-2, you see the average times for samples of $n = 10$ and $n = 50$ clerical workers, respectively, also have normal distributions.

When X has a normal distribution, the sample means also always have a normal distribution, no matter what size samples you take, even if you take samples of only two clerical workers at a time.

The difference between the curves in Figure 12-2 is not their means or their shapes, but rather their amount of variability (how close the values in the distribution are to the mean). Results based on large samples vary less and will be more concentrated around the mean than results from small samples or results from the individuals in the population.

Case 2: The distribution of *X* is not normal — Enter the Central Limit Theorem

If X has any distribution that is *not* normal, or if its distribution is unknown, you can't automatically say the sample mean ($\bar{X}$) has a normal distribution. But incredibly, you can use a normal distribution to *approximate* the distribution of $\bar{X}$ — if the sample size is large enough. This momentous result is due to what statisticians know and love as the Central Limit Theorem.

The *Central Limit Theorem* (abbreviated CLT) says that if X does *not* have a normal distribution (or its distribution is unknown and hence can't be deemed to be normal), the shape of the sampling distribution of $\bar{X}$ is *approximately* normal, as long as the sample size, n, is large enough. That is, you get an *approximate* normal distribution for the means of large samples, even if the distribution of the original values (X) is *not* normal.

TIP

Most statisticians agree that if n is at least 30, this approximation will be reasonably close in most cases, although different distribution shapes for X have different values of n that are needed. The larger the sample size (n), the closer the distribution of the sample means will be to a normal distribution.

Averaging a fair die is approximately normal

Consider the die-rolling example from the earlier section, "Defining a Sampling Distribution." Notice in Figure 12-1a, the distribution of X (the population of outcomes based on millions of single rolls) is flat; the individual outcomes of each roll go from 1 to 6, and each outcome is equally likely.

Things change when you look at averages. When you roll a die a large number of times (say a sample of 50 times) and look at your outcomes, you'll probably find about the same number of 6s as 1s (note that 6 and 1 average out to 3.5); 5s as 2s (5 and 2 also average out to 3.5); and 4s as 3s (which also average out to 3.5 — do you see a pattern here?). So if you roll a die 50 times, you have a high probability of getting an overall average that's close to 3.5. Sometimes just by chance things won't even out as well, but that won't happen very often with 50 rolls.

Getting an average at the extremes with 50 rolls is a very rare event. To get an average of 1 on 50 rolls, you need all 50 rolls to be 1. How likely is that? (If it happens to you, buy a lottery ticket right away; it's the luckiest day of your life!) The same is true for getting an average near 6.

So the chance that your average of 50 rolls is close to the middle (3.5) is highest, and the chance of it being at or close to the extremes (1 or 6) is extremely low. As for averages between 1 and 6, the probabilities get smaller as you move farther from 3.5, and the probabilities get larger as you move closer to 3.5; in particular, statisticians show that the shape of the sampling distribution of sample means in Figure 12-1b is *approximately* normal as long as the sample size is large enough. (See Chapter 10 for particulars on the shape of the normal distribution.)

Note that if you roll the die even more times, the chance of the average being close to 3.5 increases, and the sampling distribution of the sample means looks more and more like a normal distribution.

Averaging an unfair die is still approximately normal

However, sometimes the values of X don't occur with equal probability like they do when you roll a fair die. What happens then? For example, say the die isn't fair, and the average value for many individual rolls turns out to be 2 instead of 3.5. This means the distribution of X is skewed right (more low values like 1, 2, and 3, and fewer high values like 4, 5, and 6). But if the distribution of X (millions of individual rolls of this unfair die) is skewed right, how does the distribution of $\bar{X}$ (average of 50 rolls of this unfair die) end up with an approximate normal distribution?

Say that one person, Bob, is doing 50 rolls. What will the distribution of Bob's outcomes look like? Bob is more likely to get low outcomes (like 1 and 2) and less likely to get high outcomes (like 5 and 6) — the distribution of Bob's outcomes will be skewed right as well.

In fact, because Bob rolled his die a large number of times (50), the distribution of his individual outcomes has a good chance of matching the distribution of X (the outcomes from millions of rolls). However, if Bob had only rolled his die a few times (say, six times), he would be unlikely to even get the higher numbers like 5 and 6, and hence his distribution wouldn't look as much like the distribution of X.

If you run through the results of each of a million people like Bob who rolled this unfair die 50 times, each of their million distributions will look very similar to each other and very similar to the distribution of X. The more rolls they make each time, the closer their distributions get to the distribution of X and to each other. And here is the key: If their distributions of outcomes have a similar shape, no matter what that similar shape is, then their averages will be similar as well. Some people will get higher averages than 2 by chance, and some will get lower averages by chance, but these types of averages get less and less likely the farther you get from 2. This means you're getting an *approximate* normal distribution centered at 2.

The big deal is, it doesn't matter if you started out with a skewed distribution or some totally wacky distribution for X. Because each of them had a large sample size (number of rolls), the distributions of each person's sample results end up looking similar, so their averages will be similar, close together, and close to a normal distribution. In fancy lingo, the distribution of $\bar{X}$ is *approximately* normal as long as n is large enough. This is all due to the Central Limit Theorem.

In order for the CLT to work when X does *not* have a normal distribution, each person needs to roll their die enough times (that is, n must be large enough) to have a good chance of getting all possible values of X, especially those outcomes that won't occur as often. If n is too small, some folks will not get the outcomes that have low probabilities, and their means will differ from the rest by more than they should. As a result, when you put all the means together, they may not congregate around a single value. In the end, the approximate normal distribution may not show up.

Clarifying three major points about the Central Limit Theorem

I want to alert you to a few sources of confusion about the Central Limit Theorem (CLT) before they happen to you:

- The CLT is needed only when the distribution of X is not a normal distribution or is unknown. It is *not* needed if X started out with a normal distribution.
- The formulas for the mean and standard error of $\bar{X}$ are *not* due to the CLT. These are just mathematical results that are always true. To see these formulas, check out the sections, "The Mean of a Sampling Distribution" and "Measuring Standard Error," earlier in this chapter.
- The n stated in the CLT refers to the size of the sample you take each time, *not* the number of samples you take. Bob rolling a die 50 times is one sample of size 50, so $n = 50$. If ten people do it, you have 10 samples, each of size 50, and n is still 50.

EXAMPLE

Q. Suppose you take a sample of 100 from a population that has a normal distribution with a mean of 50 and a standard deviation of 15.

(a) What sample size condition do you need to check here (if any)?

(b) Where's the center of the sampling distribution for $\bar{X}$?

(c) What's the standard error?

A. Remember to check the original distribution to see whether it's normal or not before talking about the sampling distribution of the mean.

(a) This sample distribution already has a normal distribution, so you don't need approximations. The distribution of $\bar{X}$ has an exact normal distribution for any sample size. (Don't get so caught up in the $n > 30$ condition that you forget situations where the data has a normal distribution to begin with. In these cases, you don't need to meet any sample size conditions; they hold true for any n.)

(b) The center is equal to the mean of the population, which is 50 in this case.

(c) The standard error is the standard deviation of the population (15) divided by the square root of the sample size (100); in this case, 1.5.

YOUR TURN

1 How do you recognize that a statistical problem requires you to use the Central Limit Theorem? Think of one or two clues you can look for. (Assume quantitative data.)

2 Suppose you take a sample of 100 from a skewed population with a mean of 50 and a standard deviation of 15.

(a) What sample size condition do you need to check here (if any)?

(b) What's the shape and center of the sampling distribution for $\bar{X}$?

(c) What's the standard error?

Finding Probabilities for the Sample Mean

After you've established through the conditions addressed in Case 1 or Case 2 (see the previous sections) that $\bar{X}$ has a normal or *approximately* normal distribution, you're in luck. The normal distribution is a very friendly distribution that has a table for finding probabilities and anything else you need. For example, you can find probabilities for $\bar{X}$ by converting the $\bar{x}$-value to a z-value and finding probabilities using the Z-table (provided in the Appendix). (See Chapter 10 for all the details on the normal and Z-distributions.)

The general conversion formula from $\bar{x}$-values to z-values is

$$z = \frac{\bar{x} - \mu_{\bar{x}}}{\sigma_{\bar{x}}}$$

Substituting the appropriate values of the mean and standard error of $\bar{X}$, the conversion formula becomes

$$z = \frac{\bar{x} - \mu_x}{\sigma_x / \sqrt{n}}$$

TIP

Don't forget to divide by the square root of n in the denominator of z. Always divide by the square root of n when the question refers to the *average* of the x-values.

Revisiting the clerical worker example from the previous section, "Sample size and standard error," suppose X is the time it takes a randomly chosen clerical worker to type and send a standard letter of recommendation. Suppose X has a normal distribution, and assume the mean is 10.5 minutes and the standard deviation is 3 minutes. You take a random sample of 50 clerical workers and measure their times. What is the chance that their average time is less than 9.5 minutes?

This question translates to finding $p(\bar{X} < 9.5)$. As X has a normal distribution to start with, you know $\bar{X}$ also has an exact (not approximate) normal distribution. Converting to z, you get:

$$z = \frac{\bar{x} - \mu_{\bar{x}}}{\sigma_x / \sqrt{n}} = \frac{9.5 - 10.5}{3 / \sqrt{50}} = -2.36$$

So you want $p(Z < -2.36)$, which equals 0.0091 (from the Z-table in the Appendix). So the chance that a random sample of 50 clerical workers averages less than 9.5 minutes to complete this task is 0.91 percent (very small).

How do you find probabilities for $\bar{X}$ if X is *not* normal, or unknown? As a result of the CLT, the distribution of X can be non-normal or even unknown, and as long as n is large enough, you can still find *approximate* probabilities for $\bar{X}$ using the standard normal (Z-) distribution and the process described earlier. That is, convert to a z-value and find approximate probabilities using the Z-table (in the Appendix).

REMEMBER

When you use the CLT to find a probability for $\bar{X}$ (that is, when the distribution of X is *not* normal or is unknown), be sure to say that your answer is an *approximation.* You also want to say the approximate answer should be close because you have a large enough n to use the CLT. (If n is not large enough for the CLT, you can use the t-distribution in many cases — see Chapter 11.)

Beyond actual calculations, probabilities about $\bar{X}$ can help you decide whether an assumption or a claim about a population mean is on target, based on your data. In the clerical workers example, it was assumed that the average time for all workers to type up a recommendation letter was 10.5 minutes. Your sample averaged 9.5 minutes. Because the probability that they would average less than 9.5 minutes was found to be tiny (0.0091), you either got an unusually high number of fast workers in your sample just by chance, or the assumption that the average time for all workers is 10.5 minutes was simply too high. (I'm betting on the latter.) The process of checking assumptions or challenging claims about a population is called hypothesis testing; details are in Chapter 15.

EXAMPLE

Q. Suppose you have a population with a mean of 50 and a standard deviation of 10. Select a random sample of 40. What's the chance the mean will be less than 55?

A. To find this probability, you take the sample mean, 55, and convert it to a standard score, using $z = \frac{\bar{x} - \mu}{\sigma / \sqrt{n}} = \frac{55 - 50}{10 / \sqrt{40}} = 3.16$. Using the Z-table, the probability of being less than or equal to 3.16 is 0.9992. So, the probability that the sample mean is less than 55 is equal to the probability that Z is less than 3.16, which is 0.9992, or 99.92 percent.

YOUR TURN

3 Suppose you have a normal population of quiz scores with a mean of 40 and a standard deviation of 10.

(a) Select a random sample of 40. What's the chance that the mean of the quiz scores won't exceed 45?

(b) Select one individual from the population. What's the chance that a quiz score won't exceed 45?

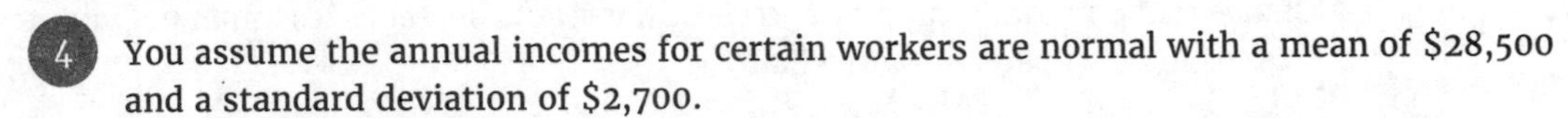

4 You assume the annual incomes for certain workers are normal with a mean of \$28,500 and a standard deviation of \$2,700.

(a) What's the chance that a randomly selected employee makes more than \$30,000?

(b) What's the chance that 36 randomly selected employees make more than \$30,000, on average?

The Sampling Distribution of the Sample Proportion

The Central Limit Theorem doesn't apply only to sample means for numerical data. You can also use it with other statistics, including sample proportions for categorical data (see Chapter 6). The *population proportion, p,* is the proportion of individuals in the population who have a certain characteristic of interest (for example, the proportion of all Americans who are registered voters, or the proportion of all teenagers who own cellphones). The *sample proportion,* denoted $\hat{p}$ (pronounced *p-hat*), is the proportion of individuals in the sample who have that particular characteristic; in other words, it's the number of individuals in the sample who have that characteristic of interest divided by the total sample size (n).

For example, if you take a sample of 100 teens and find 60 of them own cellphones, the sample proportion of cellphone-owning teens is $\hat{p} = \frac{60}{100} = 0.60$. This section examines the sampling distribution of all possible sample proportions, $\hat{p}$, from samples of size n from a population.

The sampling distribution of $\hat{p}$ has the following properties:

- Its mean, denoted by $\mu_{\hat{p}}$ (pronounced *mu sub-p-hat*), equals the population proportion, p.
- Its standard error, denoted by $\sigma_{\hat{p}}$ (say *sigma sub-p-hat*), equals:

  $$\sqrt{\frac{p(1-p)}{n}}$$

- (Note that because n is in the denominator, the standard error decreases as n increases.)
- Due to the CLT, its shape is *approximately* normal, provided that the sample size is large enough. Therefore you can use the normal distribution to find approximate probabilities for $\hat{p}$.
- The larger the sample size (n), the closer the distribution of the sample proportion is to a normal distribution.

TIP

If you are interested in the number (rather than the proportion) of individuals in your sample with the characteristic of interest, you use the binomial distribution to find probabilities for your results (see Chapter 9).

REMEMBER

How large is large enough for the CLT to work for sample proportions? Most statisticians agree that both np and $n(1-p)$ should be greater than or equal to 10. That is, the average number of successes (np) and the average number of failures $n(1-p)$ needs to be at least 10.

To help illustrate the sampling distribution of the sample proportion, consider a student survey that accompanies the ACT test each year asking whether the student would like some help with math skills. Assume (through past research) that 38 percent of all the students taking the ACT respond yes. That means p, the population proportion, equals 0.38 in this case. The distribution of responses (yes, no) for this population is shown in Figure 12-4 as a bar graph (see Chapter 6 for information on bar graphs).

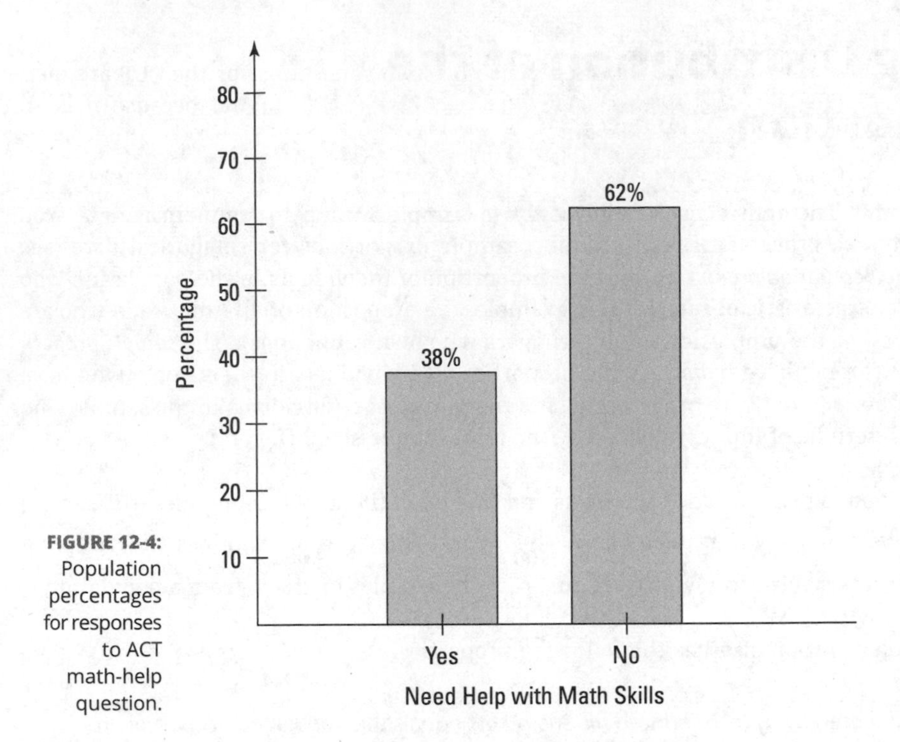

FIGURE 12-4: Population percentages for responses to ACT math-help question.

Because 38 percent applies to all students taking the exam, you use p to denote the population proportion, rather than $\hat{p}$, which denotes sample proportions. Typically p is unknown, but you're giving it a value here to point out how the sample proportions from samples taken from the population behave in relation to the population proportion.

Now take all possible samples of $n = 1{,}000$ students from this population and find the proportion in each sample who said they need math help. The distribution of these sample proportions is shown in Figure 12-5. It has an *approximate* normal distribution with mean $p = 0.38$ and a standard error equal to:

$$\sqrt{\frac{p(1-p)}{n}} = \sqrt{\frac{0.38(1-0.38)}{1{,}000}} = 0.015$$

(or about 1.5 percent).

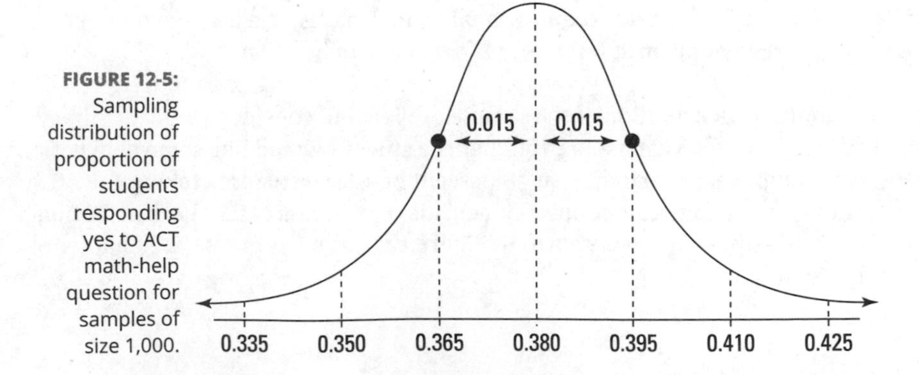

FIGURE 12-5: Sampling distribution of proportion of students responding yes to ACT math-help question for samples of size 1,000.

REMEMBER

The *approximate* normal distribution works because the two conditions for the CLT are met: $np = 1{,}000(0.38) = 380\ (\geq 10)$; and $n(1-p) = 1{,}000(0.62) = 620\ (\text{also} \geq 10)$. And because n is so large (1,000), the approximation is excellent.

EXAMPLE

Q. Suppose you want to find p where p = the proportion of college students in the United States who make over a million dollars a year. Obviously, p is very small (most likely around 0.1 percent = 0.001), meaning the data is very skewed.

(a) How big of a sample size do you need to take in order to use the Central Limit Theorem to talk about your results?

(b) Suppose you live in a dream world where p equals 0.5. Now what sample size do you need to meet the conditions for the Central Limit Theorem?

(c) Explain why you think that skewed data may require a larger sample size than a symmetric data set for the Central Limit Theorem to kick in.

A. Check to be sure the "is it large enough" condition is met before applying the Central Limit Theorem.

(a) You need np to be at least 10; so, $n \times 0.001 \geq 10$, which says $n \geq 10 \div 0.001 = 10{,}000$. And checking $n(1-p)$ gives you $10{,}000(1-0.001) = 9{,}990$, which is greater than or equal to 10, which is fine. But wow, the skewness creates a great need for size. You need a larger sample size to meet the conditions with skewed data (because p is far from $\frac{1}{2}$, the case where the distribution is symmetric).

(b) You need np to be at least 10; so, $n \times 0.5 \geq 10$, which says $n \geq 10 \div 0.5 = 20$. And checking $n(1-p)$ gives you $20(1-0.5) = 10$ also, so both conditions check out if n is at least 20. You don't need a very large sample size to meet the conditions with symmetric data $\left(p = \frac{1}{2}\right)$.

(c) With a symmetric data set, the data averages out to the mean fairly quickly because the data is balanced on each side of the mean. Skewed data takes longer to average out to the middle, because when sample sizes are small, you're more likely to choose values that fall in the big lump of data, not in the tails.

YOUR TURN

5 Suppose you take a sample of 100 from a population that contains 45 percent Democrats.

(a) What sample size condition do you need to check here (if any)?

(b) What's the standard error of $\hat{p}$?

(c) Compare the standard errors for $n = 100$, $n = 1{,}000$, and $n = 10{,}000$, and comment.

---

6 Suppose a coin is fair (so $p = \frac{1}{2}$ for heads or tails).

(a) Guess how many times you have to flip the coin to get the standard error down to only 1 percent (don't do any calculations).

(b) Now use the standard error formula to figure it out.

Finding Probabilities for the Sample Proportion

You can find probabilities for $\hat{p}$, the sample proportion, by using the normal approximation as long as the conditions are met (see the previous section for those conditions). For the ACT test example, you assume that 0.38 or 38 percent of all the students taking the ACT test would like math help. Suppose you take a random sample of 100 students. What is the chance that more than 45 of them say they need math help? In terms of proportions, this is equivalent to the chance that more than $45 \div 100 = 0.45$ of them say they need help; that is, $p(\hat{p} > 0.45)$.

To answer this question, you first check the conditions: First, is np at least 10? Yes, because $100 \times 0.38 = 38$. Next, is $n(1-p)$ at least 10? Again yes, because $100 \times (1-0.38) = 62$ checks out. So you can go ahead and use the normal approximation.

You make the conversion of the $\hat{p}$-value to a z-value using the following general equation:

$$z = \frac{\hat{p} - \mu_{\hat{p}}}{\sigma_{\hat{p}}} = \frac{\hat{p} - p}{\sqrt{\frac{p(1-p)}{n}}}$$

When you plug in the numbers for this example, you get

$$z = \frac{0.45 - 0.38}{\sqrt{\frac{0.38(1-0.38)}{100}}} = 1.44$$

And then you find $p(Z > 1.44) = 1 - 0.9251 = 0.0749$ using Table A-1 in the Appendix. So if it's true that 0.38 percent of all students taking the exam want math help, the chance of taking a random sample of 100 students and finding more than 45 needing math help is *approximately* 0.0749 (by the CLT).

As noted in the previous section on sample means, you can use sample proportions to check out a claim about a population proportion. (This procedure is a hypothesis test for a population proportion; all the details are found in Chapter 16.) In the ACT example, the probability that more than 45 percent of the students in a sample of 100 need math help (when you assumed 38 percent of the population needed math help) was found to be 0.0749. Because this probability is higher than 0.05 (the typical cutoff for blowing the whistle on a claim about a population value), you can't dispute their claim that the percentage in the population needing math help is only 38 percent. Our sample result is just not a rare enough event. (See Chapter 16 for more on hypothesis testing for a population proportion.)

EXAMPLE

Q. What's the chance that a fair coin comes up heads more than 60 times when you toss it 100 times?

A. A fair coin means $p = 0.5$, where p is the proportion of heads in the population of all possible tosses. In this case, you want the probability that your sample proportion is beyond $\hat{p} = 60 \div 100 = 0.6$. So for the z-score, you have $z = \frac{\hat{p} - \mu_{\hat{p}}}{\sigma_{\hat{p}}} = \frac{\hat{p} - p}{\sqrt{\frac{p(1-p)}{n}}} = \frac{0.6 - 0.5}{\sqrt{\frac{0.5(1-0.5)}{100}}} = \frac{0.1}{0.05} = 2.00$. Using the Z-table, you find $p(Z > 2.00) = 1 - p(Z \le 2.00) = 1 - 0.9772 = 0.0228$, or 2.28 percent.

YOUR TURN

7 Suppose studies claim that 40 percent of cellphone owners use their phones in the car while driving. What's the chance that more than 425 out of a random sample of 1,000 cellphone owners say they use their phones while driving?

Practice Questions Answers and Explanations

1 The main clue is when you have to find a probability about an average. Also, the problem gives you a value for μ, the population mean.

TIP

Watch for little words or phrases that can really help you lock on to a problem and know how to work it. It may sound cheesy, but there's no better (statistical) feeling than the feeling that comes over you when you recognize how to do a problem. Practicing that skill while the points are free is always better than sweating over the skill when the points cost you something (like on an exam).

2 The facts that you have a skewed population to start with and that you end up using a normal distribution in the end are important parts of the Central Limit Theorem.

a. The condition is $n > 30$, which you meet.

b. The shape is approximately normal by the Central Limit Theorem. The center is the population mean, 50.

c. The standard error is $\frac{\sigma}{\sqrt{n}} = \frac{15}{\sqrt{100}} = 1.5$. Notice that when the problem gives you the population standard deviation, σ, you use it in the formula for standard error. If not, you use the sample standard deviation.

REMEMBER

Checking conditions is becoming more and more of an "in" thing in statistics classes, so be sure you put it on your radar screen. Check conditions before you proceed.

3 Problems such as this require quite a bit of calculation compared to other types of statistical problems. Hang in there; show all your steps and you can make it.

a. "Not exceeding" 45 means "less than or equal to" 45, so you want $p(\bar{X} \leq 45)$. First, convert to a standard score: $z = \frac{\bar{x} - \mu}{\sigma / \sqrt{n}} = \frac{45 - 40}{10 / \sqrt{40}} = 3.16$. So, $p(\bar{X} \leq 45) = p(Z \leq 3.16)$. Finally, looking up this left-hand probability in the Z-table gives $p(Z \leq 3.16) = 0.9993$.

b. In this case, you focus on one individual, not the average, so you don't use sampling distributions to answer it; you use the population standard deviation (without dividing by $\sqrt{n}$) and the old z-formula (see Chapter 10). That means you want $p(X \leq 45)$, not $\bar{X}$. Take 45 and convert it to a z-score to get $z = \frac{x - \mu}{\sigma} = \frac{45 - 40}{10} = 0.5$. The probability of being less than 0.5 using the Z-table is 0.6915, or 69.15 percent.

WARNING

Be on the lookout for two-part problems where, in one part, you find a probability about a sample mean (Part A), and in the other part, you find the probability about a single individual (Part B). Both convert to a z-score and use the Z-table, but the difference is that the first part requires you to divide by the standard error, and the second part requires you to divide by the standard deviation. Instructors really want you to understand these ideas, and they put them on exams almost without exception.

(4) Here's another problem that dedicates one part to a probability about one individual (Part A) and another part to a sample of individuals (Part B).

a. In this case, you focus on one individual, not the average, so you don't use sampling distributions to answer it. This asks for $p(X > 30{,}000)$, so first, convert 30,000 to a z-score to get $z = \frac{x-\mu}{\sigma} = \frac{30{,}000-28{,}500}{2{,}700} = 0.56$. Then, using the Z-table, you get $p(X > 30{,}000) = p(Z > 0.56) = 1 - p(Z \le 0.56) = 1 - 0.7123 = 0.2877$.

b. Here, you're dealing with the sample mean, so you want $p(\bar{X} > 30{,}000)$. First, convert to a standard score: $z = \frac{\bar{x}-\mu}{\sigma/\sqrt{n}} = \frac{30{,}000-28{,}500}{2{,}700/\sqrt{36}} = 3.33$. So, $p(\bar{X} > 30{,}000) = p(Z > 3.33) = 1 - p(Z \le 3.33)$. Finally, looking up this left-hand probability in the Z-table gives you $1 - p(Z \le 3.33) = 1 - 0.9996 = 0.0004$.

(3) Here's a situation where you deal with percents and want a probability. The CLT is your route, provided your conditions check out.

a. You need to check $np \ge 10$ and $n(1-p) \ge 10$. In this case, you have $(100)(0.45) = 45$, which is fine, and $(100)(1-0.45) = 55$, which is also fine.

b. The standard error is $\sqrt{\frac{p(1-p)}{n}} = \sqrt{\frac{0.45(1-0.45)}{100}} = 0.050$.

c. The standard errors for $n = 100$, $n = 1{,}000$, and $n = 10{,}000$, respectively, are 0.050 (from Part B), $\sqrt{\frac{0.45(1-0.45)}{1{,}000}} = 0.016$, and $\sqrt{\frac{0.45(1-0.45)}{10{,}000}} = 0.0050$. The standard errors get smaller as n increases, meaning you get more and more precise with your sample proportions as the sample size goes up.

(6) Based on guesses from other students before they learn the CLT, you're going to predict that you guessed too high in your answer to Part A.

a. Whatever you guessed is fine. After all, it's just a guess.

b. The actual answer is 2,500. You can use trial and error (not a great idea) and plug different values for n into the standard error formula to see which n gets you a standard error of 1 percent (or 0.01). Or, if you don't have that kind of time, you can take the standard error formula (labeled here as SE), plug in the parts you know, and use algebra to solve for n. Here's how:

$$SE = \sqrt{\frac{p(1-p)}{n}}$$

$$0.01 = \sqrt{\frac{0.5(1-0.5)}{n}}$$

$$0.01 = \sqrt{\frac{0.25}{n}}$$

$$(0.01)^2 = \frac{0.25}{n}$$

$$n = \frac{0.25}{(0.01)^2} = 2{,}500$$

I know what you're thinking: Will my instructor really ask a question like Part B? Probably not. But if you train at a little higher setting of the bar, you're more likely to jump over the basic stuff in a test situation. If you solved Problem 2, great. If not, no big deal.

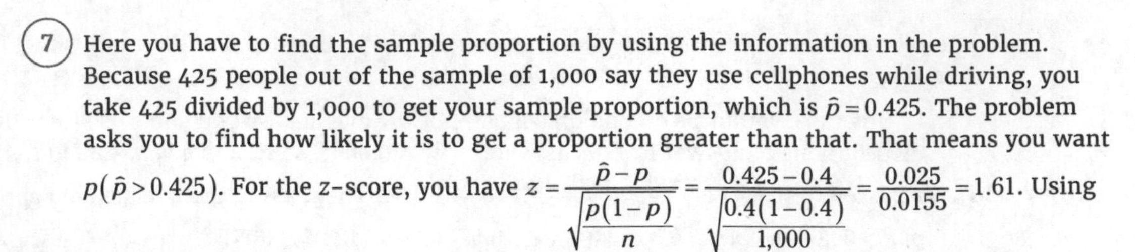

7. Here you have to find the sample proportion by using the information in the problem. Because 425 people out of the sample of 1,000 say they use cellphones while driving, you take 425 divided by 1,000 to get your sample proportion, which is $\hat{p}=0.425$. The problem asks you to find how likely it is to get a proportion greater than that. That means you want $p(\hat{p}>0.425)$. For the *z*-score, you have $z=\dfrac{\hat{p}-p}{\sqrt{\dfrac{p(1-p)}{n}}}=\dfrac{0.425-0.4}{\sqrt{\dfrac{0.4(1-0.4)}{1,000}}}=\dfrac{0.025}{0.0155}=1.61$. Using the Z-table, you find $p(Z>1.61)=1-p(Z\leq 1.61)=1-0.9463=0.0537$, or 5.37 percent.

TIP

If you're given the sample size and the number of individuals in the group you're interested in, divide those to get your sample proportion.

If you're ready to test your skills a bit more, take the following chapter quiz that incorporates all the chapter topics.

Whaddya Know? Chapter 12 Quiz

Quiz time! Complete each problem to test your knowledge on the various topics covered in this chapter. You can then find the solutions and explanations in the next section.

1. What is the mean of the sampling distribution of $\bar{X}$?
2. What is the standard error of the sampling distribution of $\bar{X}$?
3. What is the shape of the sampling distribution of $\bar{X}$ when the population has a normal distribution?
4. What is the shape of the sampling distribution of $\bar{X}$ when the population does not have a normal distribution?
5. Suppose exam scores have a normal distribution with mean 80 and standard deviation 10. You take a random sample of 36 exam scores and find the mean. What is the chance that the mean is over 85?
6. Suppose exam scores have a skewed distribution with mean 80 and standard deviation 5. You take a random sample of 36 exam scores and find the mean. What is the chance that the mean is over 85?
7. What is the difference between Problems 5 and 6, and how does that affect the answers?
8. The Central Theorem is needed whenever the sample size is more than 30. True or false?
9. The Central Limit Theorem applies to what characteristics of $\bar{X}$: the mean, the standard error, or the shape? (Note all that apply.)
10. What is the mean of the sampling distribution of $\hat{p}$?
11. What is the standard error of the sampling distribution of $\hat{p}$?
12. What are the conditions by which the sampling distribution of $\hat{p}$ has an approximate normal distribution?
13. Suppose 90% of students in a large introductory statistics course pass the final exam. You take a random sample of 100 students and calculate their average final exam score. What's the chance that less than 85% of these students passed?

Answers to Chapter 12 Quiz

1. The mean of $\bar{X}$ equals the mean of X. In symbols, you have $\mu_{\bar{X}} = \mu_X$.

2. The standard error of $\bar{X}$ equals the standard deviation of X over the square root of n.

   In symbols, you have $\sigma_{\bar{X}} = \frac{\sigma_X}{\sqrt{n}}$.

3. The shape of the sampling distribution of $\bar{X}$ is exactly normal if the population is normal (for any n).

4. The shape of the sampling distribution of $\bar{X}$ is approximately normal if the population is non-normal (for $n > 30$).

5. You want the probability that the mean test score is greater than 85:

$$p(\bar{X} > 85) \doteq p\left(Z > \frac{85-80}{10/\sqrt{36}}\right) = p(Z > 3.0) = 1 - p(Z < 3.0) = 1 - 0.9987 = 0.0013$$

6. $p(\bar{X} > 85) \doteq p\left(Z > \frac{85-80}{5/\sqrt{36}}\right) = p(Z < 6.0) = 1 - 0.9999999 = 0.0000$ (approximately 0 since 6 is off the chart).

7. The answer to Problem 5 is an exact answer because the scores have a normal distribution. The answer to Problem 6 is an approximate answer because the original scores do not have a normal distribution; they are skewed. Because $n > 30$, the Central Limit Theorem lets you use Z to solve the problem.

8. **False.** If the original population (X) has a normal distribution, you do not need the Central Limit Theorem.

9. **The Central Limit Theorem only applies to the shape of the distribution.** The mean and standard error formulas for $\bar{X}$ are not due to the Central Limit Theorem.

10. **The mean of $\hat{p}$ is p.**

11. **The standard error of $\hat{p}$ is $\sqrt{\frac{p(1-p)}{n}}$.**

12. You need np and $n(1-p)$ to be ≥ 10.

13. $p(\hat{p} < 0.85) \doteq p\left(Z < \frac{0.85 - 0.90}{\sqrt{\frac{0.90(1-0.90)}{100}}}\right) = p(Z < -1.67) = 0.0475$. You can use Z here

    $np = 100(0.90) = 90$ and $n(1-p) = 100(1-0.90) = 10$ are both ≥ 10.

4

Guesstimating and Hypothesizing with Confidence

In This Unit . . .

IN THIS CHAPTER

» Understanding and calculating margin of error

» Exploring the effect of sample size

» Finding out what margin of error doesn't measure

Chapter 13
Leaving Room for a Margin of Error

Good survey and experiment researchers always include some measure of how accurate their results are so consumers of the information can put the results into perspective. This measure is called the *margin of error* (MOE) — it's a measure of how close the sample statistic (one number that summarizes the sample) is expected to be to the population parameter being studied. (A population parameter is one number that summarizes the population. Find out more about statistics and parameters in Chapter 3.) Thankfully, many journalists are also realizing the importance of the MOE in assessing information, so reports that include the margin of error are beginning to appear in the media. But what does the margin of error really mean, and does it tell the whole story?

This chapter looks at the margin of error and what it can and can't do to help you assess the accuracy of statistical information. It also examines the issue of sample size; you may be surprised at how small a sample can be used to get a good handle on the pulse of America — or the world — if the research is done correctly.

Seeing the Importance of that Plus or Minus

Margin of error is probably not a new term to you. You've probably heard of it before, most likely in the context of survey results. For example, you may have heard someone report, "This survey had a margin of error of plus or minus three percentage points." And you may have wondered what you're supposed to do with that information and how important it really is. The truth is, the survey results themselves (with no MOE) are only a measure of how the *sample* of selected individuals felt about the issue; they don't reflect how the *entire population* may have felt, had they *all* been asked. The margin of error helps you estimate how close you are to the truth about the population based on your sample data.

REMEMBER

Results based on a sample won't be exactly the same as what you would've found for the entire population, because when you take a sample, you don't get information from everyone in the population. However, if the study is done right (see Chapters 17 and 18 for more about designing good studies), the results from the sample should be close to and representative of the actual values for the entire population, with a high level of confidence.

The MOE doesn't mean someone made a mistake; all it means is that you didn't get to sample everybody in the population, so you expect your sample results to vary from that population by a certain amount. In other words, you acknowledge that your results will change with subsequent samples and are only accurate to within a certain range — which can be calculated using the margin of error.

Consider one example of the type of survey conducted by some of the leading polling organizations, such as the Gallup Organization. Suppose its latest poll sampled 1,000 people from the United States, and the results show that 520 people (52 percent) think the president is doing a good job, compared to 48 percent who don't think so. Suppose Gallup reports that this survey had a margin of error of plus or minus 3 percent. Now, you know that the majority (more than 50 percent) of the people in this *sample* approve of the president, but can you say that the majority of *all Americans* approve of the president? In this case, you can't. Why not?

You need to include the margin of error (in this case, 3 percent) in your results. If 52 percent of *those sampled* approve of the president, you can expect that the percent of the *population of all Americans* who approve of the president will be 52 percent, plus or minus 3 percent. Therefore, between 49 percent and 55 percent of all Americans approve of the president. That's as close as you can get with your sample of 1,000. But notice that 49 percent, the lower end of this range, represents a minority, because it's less than 50 percent. So you really can't say that a majority of the American people support the president, based on this sample. You can only say you're confident that between 49 percent and 55 percent of all Americans support the president, which may or may not be a majority.

Think about the sample size for a moment. Isn't it interesting that a sample of only 1,000 Americans out of a population of well over 332,000,000 can lead you to be within plus or minus only 3 percent on your survey results? That's incredible! That means for large populations you only need to sample a tiny portion of the total to get close to the true value (assuming, as always, that you have good data). Statistics is indeed a powerful tool for finding out how people feel about issues, which is probably why so many people conduct surveys and why you're so often bothered to respond to them as well.

TIP

When you are working with categorical variables (those that record certain characteristics that don't involve measurements or counts; see Chapter 4), a quick-and-dirty way to get a rough idea of the margin of error for proportions, for any given sample size (n), is simply to find 1 divided by the square root of n. For the Gallup poll example, $n = 1{,}000$, and its square root is roughly 31.62, so the margin of error is roughly 1 divided by 31.62, or about 0.03, which is equivalent to 3 percent. In the remainder of this chapter, you see how to get a more accurate measure of the margin of error.

Finding the Margin of Error: A General Formula

The margin of error is the amount of "plus or minus" that is attached to your sample result when you move from discussing the sample itself to discussing the whole population that it represents. Therefore, you know that the general formula for the margin of error contains a "±" in front of it. So, how do you come up with that plus or minus amount (other than taking a rough estimate, as shown earlier)? This section shows you how.

Measuring sample variability

Sample results vary, but by how much? According to the Central Limit Theorem (see Chapter 12), when sample sizes are large enough, the so-called sampling distribution of the sample proportions (or the sample means) follows a bell-shaped curve (or approximate normal distribution — see Chapter 10). Some of the sample proportions (or sample means) overestimate the population value and some underestimate it, but most are close to the middle.

And what's in the middle of this sampling distribution? If you average out the results from all the possible samples you could take, the average is the actual *population proportion*, in the case of categorical data, or the actual *population average*, in the case of numerical data. Normally, you don't know all the values of the population, so you can't look at all the possible sample results and average them out — but knowing something about all the other sample possibilities does help you to measure the amount by which you expect your own sample proportion (or average) to vary. (See Chapter 12 for more on sample means and proportions.)

REMEMBER

Standard errors are the basic building blocks of the margin of error. The *standard error* of a statistic is basically equal to the standard deviation of the population divided by the square root of n (the sample size). This reflects the fact that the sample size greatly affects how much that sample statistic is going to vary from sample to sample. (See Chapter 12 for more about standard errors.)

REMEMBER

The number of standard errors you have to add or subtract to get the MOE depends on how confident you want to be in your results (this is called your *confidence level*). Typically, you want to be about 95 percent confident, so the basic rule is to add or subtract about 2 standard errors (1.96, to be exact) to get the MOE (you get this from the Empirical Rule; see Chapter 5). This allows you to account for about 95 percent of all possible results that may have occurred with repeated sampling. To be 99 percent confident, you add and subtract 2.58 standard errors. (This assumes a normal distribution on large n where the standard deviation is known; see Chapter 12.)

You can be more precise about the number of standard errors you have to add or subtract in order to calculate the MOE for any confidence level; if the conditions are right, you can use values on the standard normal (Z-) distribution. (See Chapter 14 for details.) For any given confidence level, a corresponding value on the standard normal distribution (called a *z^*-value*) represents the number of standard errors to add and subtract to account for that confidence level. For 95 percent confidence, a more precise z^*-value is 1.96 (which is "about" 2), and for 99 percent confidence, the exact z^*-value is 2.58. Some of the more commonly used confidence levels (also known as percentage confidence), along with their corresponding z^*-values, are given in Table 13-1.

TABLE 13-1 z^*-values for Selected Confidence Levels

Confidence Level	z^*-value
80%	1.28
90%	1.645
95%	1.96
98%	2.33
99%	2.58

REMEMBER

To find a z^*-value like those in Table 13-1, add to the confidence level to make it a less-than probability and find its corresponding z-value on the Z-table (Table A-1 in the Appendix). For example, a 95 percent confidence level means the "between" probability is 95 percent, so the "less-than" probability is 95 percent plus 2.5 percent (half of what's left), or 97.5 percent. Look up 0.975 in the body of the Z-table, and you find $z^* = 1.96$ for a 95 percent confidence level. That is, $p(-1.96 \le Z \le 1.96) = 0.95 = 95\%$. The z^*-values summarized in Table 13-1 can also be found using the z and CI (confidence interval) rows at the bottom of the t-table (Table A-2 in the Appendix).

REMEMBER

A z-value and a z^*-value are effectively the same thing. A z-value (from Chapter 10) represents how far a specific value lies from the sample mean. A z^*-value represents how far the boundaries of a confidence interval lie from the sample mean (or proportion) resting in the center. Statisticians often use the * notation to indicate a different calculation. In Chapter 10, you use the z-value to find the probability that a certain event might happen (like the probability that Bob gets a score of 80 or higher on his exam). Here, you use the z^*-value when given a percentage (*not* a probability) of confidence to find an interval of plausible values for the population parameter (like a confidence interval for the true mean of all students nationwide who take a standardized exam).

EXAMPLE

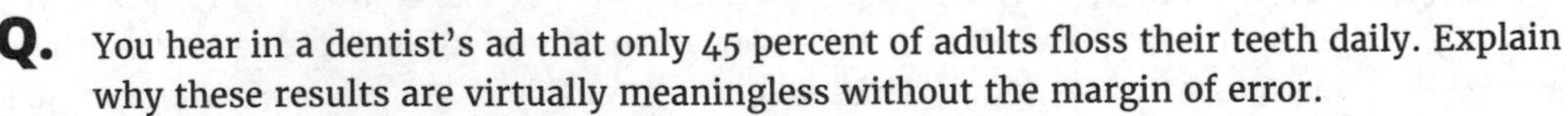

Q. You hear in a dentist's ad that only 45 percent of adults floss their teeth daily. Explain why these results are virtually meaningless without the margin of error.

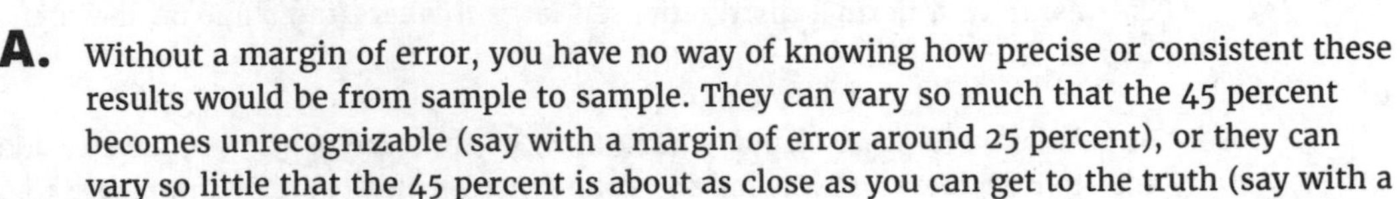

A. Without a margin of error, you have no way of knowing how precise or consistent these results would be from sample to sample. They can vary so much that the 45 percent becomes unrecognizable (say with a margin of error around 25 percent), or they can vary so little that the 45 percent is about as close as you can get to the truth (say with a margin of error closer to 1 percent).

YOUR TURN

1. Suppose you want to be 99 percent confident in your results, and you plan on having a large sample size. What's your z^*-value?

2. Suppose polling for the upcoming election says the results are "too close to call," because they fall within the margin of error. Explain what the pollsters mean.

3. Explain why it makes sense that the margin of error should depend on the standard deviation of the population.

4. What margin of error should you require in order to have 100 percent confidence in your sample result?

Calculating margin of error for a sample proportion

When a polling question asks people to choose from a range of answers (for example, "Do you approve or disapprove of the president's performance?"), the statistic used to report the results is the proportion of people from the sample who fell into a certain group (for example, the "approve" group). This is known as the *sample proportion.* You find this number by taking the number of people in the sample that fell into the group of interest, divided by the sample size, *n*.

Along with the sample proportion, you need to report a margin of error. The general formula for margin of error for the sample proportion (if certain conditions are met) is $z^*\sqrt{\frac{\hat{p}(1-\hat{p})}{n}}$, where $\hat{p}$ is the sample proportion, n is the sample size, and z^* is the appropriate z^*-value for your desired level of confidence (from Table 13-1). Here are the steps for calculating the margin of error for a sample proportion:

1. **Find the sample size, *n*, and the sample proportion, $\hat{p}$.**

   The sample proportion is the number in the sample with the characteristic of interest, divided by *n*.

2. **Multiply the sample proportion by $(1-\hat{p})$.**

3. **Divide the result by *n*.**

4. **Take the square root of the calculated value.**

   You now have the standard error, $\sqrt{\frac{\hat{p}(1-\hat{p})}{n}}$.

5. **Multiply the result by the appropriate *z**-value for the confidence level desired.**

   Refer to Table 13-1 for the appropriate *z**-value. If the confidence level is 95 percent, the *z**-value is 1.96.

Looking at the example involving whether Americans approve of the president, you can find the actual margin of error. First, assume you want a 95 percent level of confidence, so $z^* = 1.96$. The number of Americans in the sample who said they approve of the president was found to be 520. This means that the sample proportion, $\hat{p}$, is $520 \div 1{,}000 = 0.52$. (The sample size, *n*, was 1,000.) The margin of error for this polling question is calculated in the following way:

$$z^*\sqrt{\frac{\hat{p}(1-\hat{p})}{n}} = 1.96\sqrt{\frac{0.52(0.48)}{1{,}000}} = 1.96(0.0158) = 0.0310$$

According to this data, you conclude with 95 percent confidence that 52 percent of all Americans approve of the president, plus or minus 3.1 percent.

REMEMBER

Two conditions need to be met in order to use a *z**-value in the formula for margin of error for a sample proportion:

- You need to be sure that $n\hat{p}$ is at least 10.
- You need to make sure that $n(1-\hat{p})$ is at least 10.

In the preceding example of a poll on the president, $n = 1{,}000$, $\hat{p} = 0.52$, and $1-\hat{p}$ is $1-0.52 = 0.48$. Now check the conditions: $n\hat{p} = 1{,}000 \times 0.52 = 520$, and $n(1-\hat{p}) = 1{,}000 \times 0.48 = 480$. Both of these numbers are at least 10, so everything is okay.

Most surveys you come across are based on hundreds or even thousands of people, so meeting these two conditions is usually a piece of cake (unless the sample proportion is very large or very small, requiring a larger sample size to make the conditions work).

TIP

A sample proportion is the decimal version of the sample percentage. In other words, if you have a sample percentage of 5 percent, you must use 0.05 in the formula, not 5. To change a percentage into decimal form, simply divide by 100. After all your calculations are finished, you can change back to a percentage by multiplying your final answer by 100 percent.

Reporting results

Including the margin of error allows you to make conclusions beyond your sample to the population. After you calculate and interpret the margin of error, report it along with your survey results. To report the results from the president approval poll in the previous section, you say, "Based on my sample, 52 percent of all Americans approve of the president, plus or minus a margin of error of 3.1 percent. I am 95 percent confident in these results."

How does a real-life polling organization report its results? Here's an example from Gallup:

> Based on the total random sample of 1,000 adults in (this) survey, we are 95 percent confident that the margin of error for our sampling procedure and its results is no more than ±3.1 percentage points.

It sounds sort of like that long list of disclaimers that comes at the end of a car-leasing advertisement. But now you can understand the fine print!

REMEMBER

Never accept the results of a survey or study without the margin of error for the study. The MOE is the only way to estimate how close the sample statistics are to the actual population parameters you're interested in. Sample results vary, and if a different sample had been chosen, a different sample result would have been obtained; the MOE measures that amount of difference.

The next time you hear a media story about a survey or poll that was conducted, take a closer look to see whether the margin of error is given; if it's not, you should ask why. Some news outlets are getting better about reporting the margin of error for surveys, but what about other studies?

Calculating margin of error for a sample mean

When a research question asks you to estimate a parameter based on a numerical variable (for example, "What's the average age of teachers?"), the statistic used to help estimate the results is the average of all the responses provided by people in the sample. This is known as the *sample mean* (or average — see Chapter 5). And just as for sample proportions, you need to report an MOE for sample means.

The general formula for margin of error for the sample mean (assuming a certain condition is met) is $z^* \frac{\sigma}{\sqrt{n}}$, where σ is the population standard deviation, n is the sample size, and z^* is the appropriate z^*-value for your desired level of confidence (which you can find in Table 13-1).

Here are the steps for calculating the margin of error for a sample mean:

1. **Find the population standard deviation, σ, and the sample size, n.**

   The population standard deviation will be given in the problem.

2. **Divide the population standard deviation by the square root of the sample size.**

   $\frac{\sigma}{\sqrt{n}}$ gives you the standard error.

3. **Multiply by the appropriate z^*-value (refer to Table 13-1).**

   For example, the z^*-value is 1.96 if you want to be about 95 percent confident.

TIP

The condition you need to meet in order to use a z^*-value in the margin of error formula for a sample mean is either: 1) The original population has a normal distribution to start with, or 2) The sample size is large enough so the normal distribution can be used (that is, the Central Limit Theorem kicks in; see Chapter 12). In general, the sample size, n, should be above about 30 for the Central Limit Theorem. Now, if it's 29, don't panic — 30 is not a magic number, it's just a general rule of thumb. (The population standard deviation must be known either way.)

Suppose you're the manager of an ice cream shop, and you're training new employees to be able to fill the large-size cones with the proper amount of ice cream (10 ounces each). You want to estimate the average weight of the cones they make over a one-day period, including a margin of error. Instead of weighing every single cone made, you ask each of your new employees to randomly spot-check the weights of a random sample of the large cones they make and record those weights on a notepad. For $n = 50$ cones sampled, the sample mean is found to be 10.3 ounces. Suppose the population standard deviation of $\sigma = 0.6$ ounces is known.

What's the margin of error (assuming you want a 95 percent level of confidence)? It's calculated this way:

$$z^* \frac{\sigma}{\sqrt{n}} = 1.96 \frac{0.6}{\sqrt{50}} = (1.96)(0.0849) = 0.17$$

So to report these results, you say that based on the sample of 50 cones, you estimate that the average weight of all large cones made by the new employees over a one-day period is 10.3 ounces, with a margin of error of plus or minus 0.17 ounce. In other words, the range of likely values for the average weight of all large cones made for the day is estimated (with 95 percent confidence) to be between $10.30 - 0.17 = 10.13$ ounces and $10.30 + 0.17 = 10.47$ ounces. The new employees appear to be giving out too much ice cream (but I have a feeling the customers aren't offended).

WARNING

Notice that in the ice-cream-cone example, the units are ounces, not percentages! When working with and reporting results about data, always remember what the units are. Also, be sure that statistics are reported with their correct units of measure, and if they're not, ask what the units are.

REMEMBER

In cases where n is too small (in general, less than 30) for the Central Limit Theorem to be used, but the data at least resembles a normal distribution, and/or you don't know the population standard deviation but you know the sample standard deviation, you can use a t^*-value instead of a z^*-value in your formulas. A t^*-value is one that comes from a t-distribution with $n-1$ degrees of freedom. (Chapter 11 gives you all the in-depth details on the t-distribution.) In fact, many statisticians go ahead and use t^*-values instead of z^*-values consistently, because if the sample size is large, t^*-values and z^*-values are approximately equal anyway. In addition, for cases where you don't know the population standard deviation, σ, you can substitute it with s, the sample standard deviation; from there, you use a t^*-value instead of a z^*-value in your formulas as well.

Being confident you're right

If you want to be *more* than 95 percent confident about your results, you need to add and subtract more than 1.96 standard errors (see Table 13-1). For example, to be 99 percent confident, you add and subtract 2.58 standard errors to obtain your margin of error. More confidence means a larger margin of error, though (assuming the sample size stays the same); so you have to ask yourself whether it's worth it. When going from 95 percent to 99 percent confidence, the z^*-value increases by $2.58 - 1.96 = 0.62$ (see Table 13-1). Most people don't think adding and subtracting this much more of an MOE is worthwhile, just to be 4 percent more confident (99 percent versus 95 percent) in the results obtained.

You can never be completely certain that your sample results do reflect the population, even with the margin of error included. Even if you're 95 percent confident in your results, that actually means that if you repeat the sampling process over and over, 5 percent of the time the sample won't represent the population well, simply due to chance (not because of problems with the sampling process or anything else). In these cases, you would miss the mark. So all results need to be viewed with that in mind.

Suppose you want to estimate the percentage of cellphone users at a university. Which formula for margin of error do you need?

You should use the formula for sample proportions because you have to deal with categorical (yes/no) data: The people either use cellphones or they don't. Also, recall that proportions are the decimal representation for percentages (for example, a proportion of 0.40 is 40 percent).

5 A survey of 1,000 dental patients produces 450 people who floss their teeth adequately. What's the margin of error for this result? Assume 90 percent confidence.

A survey of 1,000 dental patients shows that the average cost of a regular six-month cleaning/checkup is $150.00 with a standard deviation of $80. What's the margin of error for this result? Assume 95 percent confidence.

Out of a sample of 200 babysitters, 70 percent are girls and 30 percent are guys.

(a) What's the margin of error for the percentage of female babysitters? Assume 95 percent confidence.

(b) What's the margin of error for the percentage of male babysitters? Assume 95 percent confidence.

You sample 100 fish in Pond A at the fish hatchery and find that they average 5.5 inches with a standard deviation of 1 inch. Your sample of 100 fish from Pond B has the same mean, but the standard deviation is 2 inches. How do the margins of error compare? (Assume the confidence levels are the same.)

9 Suppose you conduct a study twice, and the second time you use four times as many people as you did the first time. How does the change affect your margin of error? (Assume the other components remain constant.)

10 Suppose Sue and Bill each make a confidence interval out of the same data set, but Sue wants a confidence level of 80 percent compared to Bill's 90 percent. How do their margins of error compare?

11 Suppose you find the margin of error for a sample proportion. What unit of measurement is it in?

12 Suppose you find the margin of error for a sample mean. What unit of measurement is it in?

Determining the Impact of Sample Size

The two most important ideas regarding sample size and margin of error are the following:

- Sample size and margin of error have an inverse relationship.
- After a point, increasing *n* beyond what you already have gives you a diminished return.

This section illustrates both concepts.

Sample size and margin of error

The relationship between margin of error and sample size is simple: As the sample size increases, the margin of error decreases. This relationship is called an inverse because the two move in opposite directions. If you think about it, it makes sense that the more information you have, the more accurate your results are going to get (in other words, the smaller your margin of error will get). (That assumes, of course, that the data were collected and handled properly.)

TIP

In the previous section, you see that the impact of a larger confidence level is a larger MOE. But if you increase the sample size, you can offset the larger MOE and bring it down to a reasonable size! Find out more about this concept in Chapter 14.

Bigger isn't always (that much) better!

In the example of the poll involving the approval rating of the president (see the earlier section, "Calculating margin of error for a sample proportion"), the results of a sample of only 1,000 people from well over 310 million residents in the United States could get to within about 3 percent of what the whole population would have said, if they had all been asked.

Using the formula for margin of error for a sample proportion, you can look at how the margin of error changes dramatically for samples of different sizes. Suppose in the presidential approval poll that n was 500 instead of 1,000. (Recall that $\hat{p} = 0.52$ for this example.) Therefore, the margin of error for 95 percent confidence is $1.96\sqrt{\frac{(0.52)(0.48)}{500}} = (1.96)(0.0223) = 0.0438$, which is equivalent to 4.38 percent. When $n = 1{,}000$ in the same example, the margin of error (for 95 percent confidence) is $1.96\sqrt{\frac{(0.52)(0.48)}{1{,}000}} = (1.96)(0.0158) = 0.0310$, which is equal to 3.10 percent. If n is increased to 1,500, the margin of error (with the same level of confidence) becomes $1.96\sqrt{\frac{(0.52)(0.48)}{1{,}500}} = (1.96)(0.0129) = 0.0253$, or 2.53 percent. Finally, when $n = 2{,}000$, the margin of error is $1.96\sqrt{\frac{(0.52)(0.48)}{2{,}000}} = (1.96)(0.0112) = 0.0219$, or 2.19 percent.

Looking at these different results, you can see that larger sample sizes decrease the MOE, but after a certain point, you have a diminished return. Each time you survey one more person, the cost of your survey increases, and going from a sample size of, say, 1,500 to a sample size of 2,000 decreases your margin of error by only 0.34 percent (one third of one percent!) — from 0.0253 to 0.0219. The extra cost and trouble to get that small decrease in the MOE may not be worthwhile. Bigger isn't always that much better!

But what may really surprise you is that bigger can actually be worse! I explain this surprising fact in the following section.

Keeping margin of error in perspective

The margin of error is a measure of how closely you expect your sample results to represent the entire population being studied. (Or at least it gives an upper limit for the amount of error you should have.) Because you're basing your conclusions about the population on your one sample, you have to account for how much those sample results could vary just due to chance.

Another view of margin of error is that it represents the maximum expected distance between the sample results and the actual population results (if you'd been able to obtain them through a census). Of course, if you had the absolute truth about the population, you wouldn't be trying to do a survey, would you?

Just as important as knowing what the margin of error measures is realizing what the margin of error does *not* measure. The margin of error does not measure anything other than chance variation. That is, it doesn't measure any bias or errors that happen during the selection of the participants, the preparation or conduct of the survey, the data collection and entry process, or the analysis of the data and the drawing of the final conclusions.

A good slogan to remember when examining statistical results is "garbage in equals garbage out." No matter how nice and scientific the margin of error may look, remember that the formula that was used to calculate it doesn't have any idea of the quality of the data that the margin of error is based on. If the sample proportion or sample mean was based on a *biased sample* (one that favored certain people over others), a bad design, bad data-collection procedures, biased questions, or systematic errors in recording, then calculating the margin of error is pointless because it won't mean a thing.

For example, 50,000 people surveyed sounds great, but if they were all visitors to a certain website, the margin of error for this result is bogus because the calculation is all based on biased results! In fact, many extremely large samples are the result of biased sampling procedures. Of course, some people go ahead and report them anyway, so you have to find out what went into the formula: good information or garbage? If it turns out to be garbage, you know what to do about the margin of error. Ignore it. (For more information on errors that can take place during a survey or experiment, see Chapters 17 and 18, respectively.)

The Gallup Organization addresses the issue of what margin of error does and doesn't measure in a disclaimer that it uses to report its survey results. Gallup tells you that besides sampling error, surveys can have additional errors or bias due to question wording and some of the logistical issues involved in conducting surveys (such as missing data due to phone numbers that are no longer current).

This means that even with the best of intentions and the most meticulous attention to details and process control, stuff happens. Nothing is ever perfect. But what you need to know is that the margin of error can't measure the extent of those other types of errors. And if a highly credible polling organization like Gallup admits to possible bias, imagine what's really going on with other people's studies that aren't nearly as well designed or conducted.

Q. Suppose you increase a sample size and keep everything else the same. What happens to the margin of error?

A. The margin of error decreases, because n is in the denominator of a fraction, and increasing the denominator decreases the fraction. This also makes sense because including more data should increase the level of precision in your results (provided you include good data).

13 What happens to the margin of error if you increase your confidence level (but keep all other elements fixed)?

14 Suppose you have two ponds of fish in a fish hatchery, and the first pond has twice as much variability in fish lengths as the second. You take a sample of 100 fish from each pond and calculate 95 percent confidence intervals for the average fish lengths for each pond. Which confidence interval has a larger margin of error?

15 How can you increase your confidence level and keep the margin of error small? (Assume you can do anything you want with the components of the confidence interval.)

16 Suppose you conduct a pilot study and find that your target population has a very large amount of variability. What can you do (if anything) to ensure a small margin of error in your full-blown study?

Practice Questions Answers and Explanations

1. **The z^*-value for a 99 percent confidence interval is 2.58**, because the area on the Z-distribution between –2.58 and +2.58 is about 0.99, or 99 percent.

2. They mean that after you add and subtract the margin of error from one of the percentages in the poll, the other percentage is included in that interval, so they can't be statistically different. Here's another way to look at this. If 49 percent of respondents say they want to vote for Candidate A and 51 percent back Candidate B, with a margin of error of plus or minus three points, then Candidate A may get anywhere from 46 percent to 52 percent of the vote in the population. Candidate B may get anywhere from 48 percent to 54 percent of the vote. The intervals overlap, which means the results are too close to call.

3. If a population has a wide amount of variety in its values, then its average value becomes harder to pinpoint, which makes the margin of error larger. You can offset this, however, by sampling more data, because that increases n and offsets the larger standard deviation in the margin of error formula.

4. **You need an infinite margin of error in the case of the sample mean and a 50 percent margin in the case of the sample proportion** (because you want to cover all possible values from 0 to 100 percent). Having 100 percent confidence is meaningless, however. Who wants to say that they know the percentage of people owning cellphones is likely somewhere from 0 to 100 percent? Duh!

5. This problem concerns the percentage of people in a certain category (those who floss their teeth daily). The formula to use is the margin of error for the sample proportion. Note that z^* is 1.645 because the confidence level is 90 percent, and that $\hat{p}$, the sample proportion, is $450 \div 1{,}000 = 0.45$. The margin of error is

$$\pm z^* \sqrt{\frac{\hat{p}(1-\hat{p})}{n}} = \pm 1.645 \cdot \sqrt{\frac{0.45(1-0.45)}{1{,}000}} = \pm 1.645(0.0157) = \mathbf{\pm 0.026, \text{ or } 2.6\%}$$

REMEMBER

Any formula involving the sample proportion, $\hat{p}$, requires that you use the decimal version of the percentage (also known as the proportion), or you get inaccurate results. For example, if the sample proportion is 70 percent, you must use 0.70 for $\hat{p}$ in the formula.

6. This problem concerns the average of a quantitative variable (cost of dental cleaning and exam). The formula to use is the margin of error for the sample mean. You are given the sample standard deviation (s) instead of the population standard deviation (σ), but the sample size is so large that it's okay to use a z^*-value instead of a t^*-value. Note that z^* is 1.96, because the confidence level is 95 percent. The margin of error is

$$\pm z^* \frac{s}{\sqrt{n}} = \pm 1.96 \cdot \frac{80}{\sqrt{1{,}000}} = \pm 1.96(2.53) = \mathbf{\pm 4.96 \text{ (dollars)}}$$

TIP

When making margin of error calculations, keep at least two significant digits after the decimal point throughout the calculations, rounding only at the very end to avoid accumulated round-off errors.

(7) This problem concerns the percentage of babysitters in a certain category, so you have to use the formula for the margin of error for the sample proportion.

a. Here you concentrate on the percentage of female babysitters. Note that z^* is 1.96, because the confidence level is 95 percent, and that $\hat{p}$, the sample proportion, is 0.70. The margin of error is

$$\pm z^* \sqrt{\frac{\hat{p}(1-\hat{p})}{n}} = \pm 1.96 \cdot \sqrt{\frac{0.70(1-0.70)}{200}} = \pm 1.96(0.0324) = \mathbf{\pm 0.0635}$$

b. Here you concentrate on the percentage of male babysitters. Note that z^* is 1.96, because the confidence level is 95 percent, and that $\hat{p}$, the sample proportion, is 0.30. The margin of error is

$$\pm z^* \sqrt{\frac{\hat{p}(1-\hat{p})}{n}} = \pm 1.96 \cdot \sqrt{\frac{0.30(1-0.30)}{200}} = \pm 1.96(0.0324) = \mathbf{\pm 0.0635}$$

Notice the similarities to Part A, because 0.30 equals $1-0.70$, and 0.70 equals $1-0.30$, so you plug the exact same numbers into the equation.

(8) **The sample from Pond B has a larger margin of error** because the standard deviation is larger, and standard deviation is involved in the numerator of the fraction for margin of error.

(9) **It reduces the margin of error by a factor of 2, because 2 is the square root of 4.** (You substitute n in the margin of error equation with $4n$, and the square root of $4n$ is 2 times the square root of n.) Notice, the change doesn't reduce margin of error by a factor of 4, because n is under a square root sign.

To cut the margin of error in half, quadruple the sample size.

(10) This problem looks at the effect of confidence level on the size of the margin of error. **Sue's margin of error is smaller than Bill's**, because Sue's z^*-value is 1.28 and Bill's is 1.645.

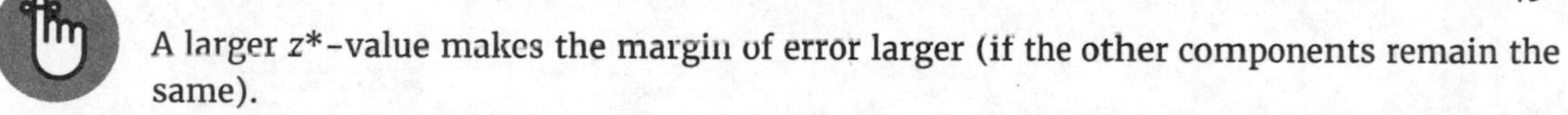

REMEMBER

A larger z^*-value makes the margin of error larger (if the other components remain the same).

(11) **The margin of error for a sample proportion is in the same units as the proportion** — a number between zero and one. You can rewrite it in the end as a percentage if you want, after the calculations are done.

(12) **The margin of error for a sample mean is in the same units as the original data.** For example, if you want to calculate the margin of error for average fish length, and the fish are measured in inches, the margin of error is in inches as well.

TIP

Keep track of the units you work with when you do your calculations. This helps you recognize when something just doesn't appear right, and it scores you big points with your professor if you write down the (correct) units — trust me!

13. **Margin of error increases** if you increase the confidence level and keep all the other components the same, because as your confidence level increases, the z^*-value increases.

14. **The first confidence interval has the largest margin of error.** The first pond has more variability in the population, so the value of the standard deviation is larger. Because standard deviation appears in the top part of the fraction in margin of error, increasing the standard deviation increases the margin of error. This makes good sense, because more variability in the population makes it harder to pin down the actual population average with precision.

TIP

The population standard deviation for quantitative data is denoted by σ. You typically don't know it, so you substitute s, the standard deviation of the sample, in the formulas, but only if the sample size is large enough (typically over 30). If not, you use a t-distribution value (see Chapter 11) in place of the z^*-value.

15. If you increase the confidence level, you increase the z^*-value and thereby increase the margin of error. **You can offset this increase by also increasing the sample size,** and because the sample size is in the denominator of the margin of error, increasing it has the opposite effect — lowering the margin of error. This makes good sense, because having more data allows you to be more precise.

REMEMBER

You can offset an increase in margin of error by increasing the sample size.

16. **Increase the sample size in your actual (full-blown) study** to offset the anticipated increase in margin of error that stems from the larger value of s.

If you're ready to test your skills a bit more, take the following chapter quiz that incorporates all the chapter topics.

Whaddya Know? Chapter 13 Quiz

Quiz time! Complete each problem to test your knowledge on the various topics covered in this chapter. You can then find the solutions and explanations in the next section.

1. Why is margin of error important?

2. What is the margin of error when you are trying to estimate the population mean?

3. What 3 things affect the margin of error for the sample mean?

4. What happens to the MOE for the mean as the sample size increases?

5. What happens to the MOE for the mean as the population standard deviation increases?

6. What happens to the MOE for the mean as the confidence level increases?

7. Can you have a small MOE yet have a high confidence level?

8. What is the margin of error when you try to estimate the population proportion?

9. What happens to the margin of error for the proportion as the sample size increases?

10. What is the difference between margin of error and standard error?

Answers to Chapter 13 Quiz

1. Without it, you only have one number as your estimate of the population mean or proportion. If you were to take another sample, you would have a different estimate. You need a plus or minus to go with your estimate to cover most of the other estimates you would have gotten with other samples.

2. $z^* \frac{\sigma}{\sqrt{n}}$

3. **The population standard deviation, σ, the confidence level (which affects Z), and the sample size, n.**

4. **The MOE decreases as the sample size increases.**

5. **The MOE increases as the population standard deviation increases.**

6. **The MOE increases as the confidence level increases** (because this increases the z-value).

7. **Yes.** If you increase n, you can offset increases due to standard deviation or confidence level.

8. $z^* \sqrt{\frac{\hat{p}(1-\hat{p})}{n}}$

9. **The MOE decreases.**

10. The standard error is like the standard deviation for the sample mean or sample proportion. The margin of error is the number of standard errors to add and subtract when you are trying to estimate the population mean or population proportion.

IN THIS CHAPTER

- » Understanding confidence interval pieces, parts, and interpretation
- » Calculating with confidence
- » Examining factors that influence the width of a confidence interval
- » Detecting misleading results

Chapter 14
Confidence Intervals: Making Your Best Guesstimate

Most statistics are used to estimate some characteristic about a population of interest, such as average household income, the percentage of people who buy birthday gifts online, or the average amount of ice cream consumed in the United States every year (and the resulting average weight gain — nah!). Such characteristics of a population are called *parameters*. Typically, people want to estimate (take a good guess at) the value of a parameter by taking a sample from the population and using statistics from the sample that will give them a good estimate. The question is: How do you define "good estimate"?

As long as the process is done correctly (and in the media, it often isn't!), an estimate can often get very close to the parameter. This chapter gives you an overview of confidence intervals (the type of estimates used and recommended by statisticians); why they should be used (as opposed to just a one-number estimate); how to set up, calculate, and interpret the most commonly used confidence intervals; and how to spot misleading estimates.

Not All Estimates Are Created Equal

Read any magazine or newspaper or listen to any newscast, and you hear a number of statistics, many of which are estimates of some quantity or another. You may wonder how they came up with those statistics. In some cases, the numbers are well researched; in other cases, they're just a shot in the dark. Here are some examples of estimates that I came across in one single issue of a leading business magazine. They come from a variety of sources:

- Even though some jobs are harder to get these days, employers in some areas are really looking for recruits: Over the next eight years, 13,000 nurse anesthetists will be needed. Pay starts from $80,000 to $95,000.
- The average number of bats used by a major league baseball player per season is 90.
- The Lamborghini Murciélago can go from 0 to 60 mph in 3.7 seconds with a top speed of nearly 205 mph.

Some of these estimates are easier to obtain than others. Here are some observations I was able to make about those estimates:

- How do you estimate how many nurse anesthetists are needed over the next eight years? You can start by looking at how many will be retiring in that time; but that won't account for growth. A prediction of the need in the next year or two would be close, but eight years into the future is much harder to predict.
- The average number of bats used per major league baseball player in a season could be found by surveying the players themselves, the people who take care of their equipment, or the bat companies that supply the bats.
- Determining car speed is more difficult but could be conducted as a test with a stopwatch. And they should find the average speed of many different cars (not just one) of the same make and model, under the same driving conditions each time.

REMEMBER

Not all statistics are created equal. To determine whether a statistic is reliable and credible, don't just take it at face value. Think about whether it makes sense and how you would go about formulating an estimate. If the statistic is really important to you, find out what process was used to come up with it. (Chapter 17 handles all the elements involving surveys, and Chapter 18 gives you the lowdown on experiments.)

Linking a Statistic to a Parameter

A *parameter* is a single number that describes a population, such as the median household income for all households in the U.S. A *statistic* is a single number that describes a sample, such as the median household income of a sample of, say, 1,200 households. You typically don't know the values of parameters of populations, so you take samples and use statistics to give your best estimates.

Suppose you want to know the percentage of vehicles in the U.S. that are pickup trucks (that's the parameter, in this case). You can't look at every single vehicle, so you take a random sample of 1,000 vehicles over a range of highways at different times of the day. You find that 7 percent of the vehicles in your sample are pickup trucks. Now, you don't want to say that *exactly* 7 percent of all vehicles on U.S. roads are pickup trucks, because you know this is only based on the 1,000 vehicles you sampled. Though you hope 7 percent is close to the true percentage, you can't be sure because you based your results on a sample of vehicles, not on all the vehicles in the U.S.

So what to do? You take your sample result and add and subtract some number to indicate that you are giving a range of possible values for the population parameter, rather than just assuming the sample statistic equals the population parameter (which would not be good, although it's done in the media all the time). This number that is added to and subtracted from a statistic is called the *margin of error* (MOE). This plus or minus (denoted by ±) that's added to any estimate helps put the results into perspective. When you know the margin of error, you have an idea of how much the sample results could change if you took another sample.

REMEMBER

The word *error* in *margin of error* doesn't mean a mistake was made or the quality of the data was bad. It just means the results from a sample are not exactly equal to what you would have gotten if you had used the entire population. This gap measures error due to random chance, the luck of the draw — not due to bias. (That's why minimizing bias is so important when you select your sample and collect your data; see Chapters 17 and 18.)

Getting with the Jargon

A statistic plus or minus a margin of error is called a *confidence interval:*

- The word *interval* is used because your result becomes an interval. For example, say the percentage of kids who like baseball is 40 percent, plus or minus 3.5 percent. That means the percentage of kids who like baseball is somewhere between $40\% - 3.5\% = 36.5\%$ and $40\% + 3.5\% = 43.5\%$. The lower end of the interval is your statistic minus the margin of error, and the upper end is your statistic plus the margin of error.
- With all confidence intervals, you have a certain amount of confidence in being correct (guessing the parameter) with your sample in the long run. Expressed as a percent, the amount of confidence is called the *confidence level.*

You can find formulas and examples for the most commonly used confidence intervals later in this chapter.

Following are the general steps for estimating a parameter with a confidence interval. Details on Steps 1, 4, and 6 are included throughout the remainder of this chapter. Steps 2 and 3 involve sampling and data collection, which are detailed in Chapter 17 (sampling and survey data collection) and Chapter 18 (data collection from experiments).

1. **Choose your confidence level and your sample size.**
2. **Select a random sample of individuals from the population.**

3. **Collect reliable and relevant data from the individuals in the sample.**
4. **Summarize the data into a statistic, such as a mean or proportion.**
5. **Calculate the margin of error.**
6. **Take the statistic plus or minus the margin of error to get your final estimate of the parameter.**

   This step calculates the *confidence interval* for that parameter.

Interpreting Results with Confidence

Suppose you, a research biologist, are trying to catch a fish using a hand net, and the size of your net represents the margin of error of a confidence interval. Now say your confidence level is 95 percent. What does this really mean? It means that if you scoop this particular net into the water over and over again, you'll catch a fish 95 percent of the time. Catching a fish here means your confidence interval was correct and contains the true parameter (in this case, the parameter is represented by the fish itself).

But does this mean that on any given try you have a 95 percent chance of catching a fish after the fact? No. Is this confusing? It certainly is. Here's the scoop (no pun intended): On a single try, say you close your eyes before you scoop your net into the water. At this point, your chances of catching a fish are 95 percent. But then go ahead and scoop your net through the water with your eyes still closed. *After* that's done, however, you open your eyes and see one of only two possible outcomes: You either caught a fish or you didn't; probability isn't involved anymore.

Likewise, *after* data have been collected, and the confidence interval has been calculated, you either captured the true population parameter or you didn't. So you're not saying you're 95 percent confident that the parameter is in your particular interval. What you are 95 percent confident about is the process by which random samples are selected and confidence intervals are created. (That is, 95 percent of the time in the long run, you'll catch a fish.)

You know that this process will result in intervals that capture the population mean 95 percent of the time. The other 5 percent of the time, the data collected in the sample just by random chance has abnormally high or low values in it and doesn't represent the population. This 5 percent measures errors due to random chance only and doesn't include bias.

WARNING

The margin of error is meaningless if the data that went into the study were biased and/or unreliable. However, you can't tell that by looking at anyone's statistical results. My best advice is to look at how the data were collected before accepting a reported margin of error as the truth (see Chapters 17 and 18 for details on data collection issues). That means asking questions before you believe a study.

EXAMPLE

Q. Suppose you are interested in the opinions of people in the U.S. An Internet survey has 50,000 respondents, and the margin of error is plus or minus 0.1 percent. Do you believe the results are that precise? Explain.

A. **No.** The results are based on an Internet survey where people select themselves to participate. Therefore, the reported margin of error isn't meaningful, because the error in the sample results is off by way more than that amount. The survey is biased because you find it on the Internet. So even though the numbers you plug into the margin of error formula seem nice and small, they're based on garbage and should be ignored.

YOUR TURN

1. You find out that the dietary scale you use each day is off by a factor of 2 ounces (over; at least that's what you say!). The margin of error for your scale was plus or minus 0.5 ounce before you found this out. What's the margin of error now?

2. You're fed up with keeping Fido locked inside, so you conduct a mail survey to find out people's opinions on the new dog barking ordinance in a certain city. Of the 10,000 people who receive surveys, 1,000 respond, and 80 percent are in favor of it. You calculate the margin of error to be 1.2 percent. Explain why this reported margin of error is misleading.

3. Does the margin of error measure the amount of error that goes into collecting and recording data?

4. If you add and subtract the margin of error to/from the sample mean, do you guarantee the population mean to be in your resulting interval?

5. A television news channel samples 25 gas stations from its local area and uses the results to estimate the average gas price for the state. What's wrong with its margin of error?

6. If you control for all possible confounding variables in an experiment, can you reduce the margin of error in your results?

Zooming In on Width

The *width* of your confidence interval is two times the margin of error. For example, suppose the margin of error is ±5 percent. A confidence interval of 7 percent, plus or minus 5 percent, goes from $7\% - 5\% = 2\%$ all the way up to $7\% + 5\% = 12\%$. So the confidence interval has a width of $12\% - 2\% = 10\%$. A simpler way to calculate this is to say that the width of the confidence interval is two times the margin of error. In this case, the width of the confidence interval is $2 \times 5\% = 10\%$.

REMEMBER

The width of a confidence interval is the distance from the lower end of the interval (statistic minus margin of error) to the upper end of the interval (statistic plus margin of error). You can always calculate the width of a confidence interval quickly by taking two times the margin of error.

The ultimate goal when making an estimate using a confidence interval is to have a narrow width, because that means you're zooming in on what the parameter is. Having to add and subtract a large margin of error only makes your result much less accurate.

REMEMBER

So, if a small margin of error is good, is smaller even better? Not always. A narrow confidence interval is a good thing — to a point. To get an extremely narrow confidence interval, you have to conduct a much larger — and more expensive — study, so a point comes where the increase in price doesn't justify the marginal difference in accuracy. Most people are pretty comfortable with a margin of error of 2 to 3 percent when the estimate itself is a percentage (like the percentage of women, Republicans, or smokers).

How do you go about ensuring that your confidence interval will be narrow enough? You certainly want to think about this issue before collecting your data; after the data are collected, the width of the confidence interval is set.

Three factors affect the width of a confidence interval:

- Confidence level
- Sample size
- Amount of variability in the population

Each of these three factors plays an important role in influencing the width of a confidence interval. In the following sections, you explore details of each element and how they affect width.

Choosing a Confidence Level

Every confidence interval (and every margin of error, for that matter) has a percentage associated with it that represents how confident you are that the results will capture the true population parameter, depending on the luck of the draw with your random sample. This percentage is called a *confidence level.*

A confidence level helps you account for the other possible sample results you could have gotten, when you're making an estimate of a parameter using the data from only one sample. If you want to account for 95 percent of the other possible results, your confidence level would be 95 percent.

TIP

What level of confidence is typically used by researchers? I've seen confidence levels ranging from 80 percent to 99 percent. The most common confidence level is 95 percent. In fact, statisticians have a saying that goes, "Why do statisticians like their jobs? Because they have to be correct only 95 percent of the time." (Sort of catchy, isn't it? And let's see weather forecasters beat that.)

Variability in sample results is measured in terms of number of standard errors. A *standard error* is similar to the standard deviation of a data set, only a standard error applies to sample means or sample percentages that you could have gotten if different samples were taken. (See Chapter 12 for information on standard errors.)

REMEMBER

Standard errors are the building blocks of confidence intervals. A confidence interval is a statistic plus or minus a margin of error, and the margin of error is the number of standard errors you need to get the confidence level you want.

Every confidence level has a corresponding number of standard errors that have to be added or subtracted. This number of standard errors is a called a *critical value.* In a situation where you use a Z-distribution to find the number of standard errors (as described later in this chapter), you call the critical value the *z*-value* (pronounced *z-star value*). See Table 14-1 for a list of z*-values for some of the most common confidence levels.

TABLE 14-1 ***z*-values for Selected Confidence Levels***

Confidence Level	z*-value
80%	1.28
90%	1.645
95%	1.96
98%	2.33
99%	2.58

REMEMBER

As the confidence level increases, the number of standard errors increases, so the margin of error increases.

If you want to be more than 95 percent confident about your results, you need to add and subtract more than about two standard errors. For example, to be 99 percent confident, you would add and subtract about two and a half standard errors to obtain your margin of error (2.58 to be exact). The higher the confidence level, the larger the z*-value, the larger the margin of error, and the wider the confidence interval (assuming everything else stays the same). You have to pay a certain price for more confidence.

Note that I said, "assuming everything else stays the same." You can offset an increase in the margin of error by increasing the sample size. See the following section for more on this.

Factoring In the Sample Size

The relationship between margin of error and sample size is simple: As the sample size increases, the margin of error decreases, and the confidence interval gets narrower. This relationship confirms what you hope is true: The more information (data) you have, the more accurate your results are going to be. (That, of course, assumes that you are working with good, credible information. See Chapter 2 for how statistics can go wrong.)

The margin of error formulas for the confidence intervals in this chapter all involve the sample size (n) in the denominator. For example, the formula for margin of error for the sample mean, $\pm z^* \frac{\sigma}{\sqrt{n}}$ (which you see in great detail later in this chapter), has an n in the denominator of the fraction (this is the case for most margin-of-error formulas). As n increases, the denominator of this fraction increases, which makes the overall fraction get smaller. That makes the margin of error smaller and results in a narrower confidence interval.

REMEMBER

When you need a high level of confidence, you have to increase the z^*-value and, hence, margin of error, resulting in a wider confidence interval, which isn't good. (See the previous section.) But you can offset this wider confidence interval by increasing the sample size and bringing the margin of error back down, thus narrowing the confidence interval.

The increase in sample size allows you to still have the confidence level you want, but also ensures that the width of your confidence interval will be small (which is what you ultimately want). You can even determine the sample size you need before you start a study: If you know the margin of error you want to get, you can set your sample size accordingly. (See the later section, "Figuring Out What Sample Size You Need," for more info.)

TIP

When your statistic is going to be a percentage (such as the percentage of people who prefer to wear sandals during summer), a rough way to figure margin of error for a 95 percent confidence interval is to take 1 divided by the square root of n (the sample size). You can try different values of n, and you can see how the margin of error is affected. For example, a survey of 100 people from a large population will have a margin of error of about $\frac{1}{\sqrt{100}} = 0.10$ or plus or minus 10 percent (meaning the width of the confidence interval is 20 percent, which is pretty large).

However, if you survey 1,000 people, your margin of error decreases dramatically, to plus or minus about 3 percent; the width now becomes only 6 percent. A survey of 2,500 people results in a margin of error of plus or minus 2 percent (so the width is down to 4 percent). That's quite a small sample size to get so accurate, when you think about how large the population is (the U.S. population, for example, is over 310 million!).

Keep in mind, however, that you don't want to go *too* high with your sample size, because a point comes where you have a diminished return. For example, moving from a sample size of 2,500 to 5,000 narrows the width of the confidence interval to about $2 \times 1.4 = 2.8\%$, down from 4 percent. Each time you survey one more person, the cost of your survey increases, so adding another 2,500 people to the survey just to narrow the interval by a little more than 1 percent may not be worthwhile.

The first step in any data analysis problem (and when critiquing another person's results) is to make sure you have good data. Statistical results are only as good as the data that went into them, so real accuracy depends on the quality of the data as well as on the sample size. A large sample size that has a great deal of bias (see Chapter 17) may appear to have a narrow confidence interval — but means nothing. That's like competing in an archery match and shooting your arrows consistently, but finding out that the whole time you're shooting at the next person's target; that's how far off you are. With the field of statistics, though, it's really difficult to measure bias after the fact; it's best to try to minimize it by designing good samples and studies (see Chapters 17 and 18).

Counting On Population Variability

One of the factors influencing variability in sample results is the fact that the population itself contains variability. For example, in a population of houses in a fairly large city like Columbus, Ohio, you see a great deal of variety in not only the types of houses, but also the sizes and the prices. And the variability in prices of houses in Columbus should be more than the variability in prices of houses in a selected housing development in Columbus.

That means if you take a sample of houses from the entire city of Columbus and find the average price, the margin of error should be larger than if you take a sample from that single housing development in Columbus, even if you have the same confidence level and the same sample size.

Why? Because the houses in the entire city have more variability in price, and your sample average would change more from sample to sample than it would if you took the sample only from that single housing development, where the prices tend to be very similar because houses tend to be comparable in a single housing development. So you need to sample more houses if you're sampling from the entire city of Columbus in order to have the same amount of accuracy that you would get from that single housing development.

The standard deviation of the population is denoted σ. Notice that σ appears in the numerator of the standard error in the formula for margin of error for the sample mean: $\pm z^* \frac{\sigma}{\sqrt{n}}$. Therefore, as the standard deviation (the numerator) increases, the standard error (the entire fraction) also increases. This results in a larger margin of error and a wider confidence interval. (Refer to Chapter 12 for more info on the standard error.)

More variability in the original population increases the margin of error, making the confidence interval wider. This increase can be offset by increasing the sample size.

EXAMPLE

Q. Suppose you're trying to estimate the population mean, and your sample size of 100 gives you a mean of 9, and your margin of error is 3 for a 95 percent confidence interval.

(a) What is the resulting confidence interval?

(b) What is the confidence level?

(c) What is the sampling error?

A. You need to be able to pick out all the various parts of a confidence interval, even before you can calculate one for yourself.

(a) The resulting confidence interval here is 9 ± 3 because 3 is the margin of error. Another way to write this interval is (6, 12) because $9 + 3 = 12$ and $9 - 3 = 6$. *Interpretation:* You're 95 percent confident that the population mean is in the interval (6, 12).

(b) The confidence level is 95 percent (given in the problem).

(c) The sampling error is 100 percent − 95 percent = 5 percent, or 0.05.

YOUR TURN

7 Suppose a 95 percent confidence interval for the average number of minutes a regular (obviously not teenage) customer uses with a cellphone in a month is 110 plus or minus 35 minutes.

(a) What's the margin of error?

(b) What are the lower and upper boundaries for this confidence interval?

8 Suppose you make two confidence intervals with the same data set — one with a 95 percent confidence level and the other with a 99 percent confidence level.

(a) Which interval is wider?

(b) Is a wide confidence interval a good thing?

9 Is it true that a 95 percent confidence interval means you're 95 percent confident that the sample statistic is in the interval?

10 Is it true that a 95 percent confidence interval means there's a 95 percent chance that the population parameter is in the interval?

Calculating a Confidence Interval for a Population Mean

When the characteristic that's being measured (such as income, IQ, price, height, quantity, or weight) is *numerical*, most people want to estimate the mean (average) value for the population. You estimate the population mean, μ, by using a sample mean, $\bar{x}$, plus or minus a margin of error. The result is called a *confidence interval for the population mean*, μ. Its formula depends on whether certain conditions are met. You split the conditions into two cases, illustrated in the following sections.

Case 1: Population standard deviation is known

In Case 1, the population standard deviation is known. The formula for a confidence interval (CI) for a population mean in this case is $\bar{x} \pm z^* \frac{\sigma}{\sqrt{n}}$, where $\bar{x}$ is the sample mean, σ is the population standard deviation, n is the sample size, and z^* represents the appropriate z^*-value from the standard normal distribution for your desired confidence level. (Refer to Table 14-1 for values of z^* for the given confidence levels.)

REMEMBER

In this case, the data either have to come from a normal distribution, or if not, then n has to be large enough (at least 30 or so) for the Central Limit Theorem to kick in (see Chapter 12), allowing you to use z^*-values in the formula.

To calculate a CI for the population mean (average), under the conditions for Case 1, do the following:

1. **Determine the confidence level and find the appropriate z^*-value.**

   Refer to Table 14-1.

2. **Find the sample mean ($\bar{x}$) for the sample size (n).**

   Note: The population standard deviation is assumed to be a known value, σ.

3. **Multiply z^* times σ and divide that by the square root of n.**

   This calculation gives you the margin of error.

4. **Take $\bar{x}$ plus or minus the margin of error to obtain the CI.**

   The lower end of the CI is $\bar{x}$ minus the margin of error, whereas the upper end of the CI is $\bar{x}$ plus the margin of error.

For example, suppose you work for the Department of Natural Resources and you want to estimate, with 95 percent confidence, the mean (average) length of walleye fingerlings in a fish hatchery pond. Follow these steps:

1. **Because you want a 95 percent confidence interval, your z^*-value is 1.96.**
2. Suppose you take a random sample of 100 fingerlings and determine that the average length is 7.5 inches; assume the population standard deviation is 2.3 inches. This means $\bar{x} = 7.5$, $\sigma = 2.3$, and $n = 100$.

3. Multiply 1.96 times 2.3 divided by the square root of 100 (which is 10). The margin of error is, therefore, $\pm 1.96 \times (2.3 \div 10) = (1.96)(0.23) = 0.45$ inch.

4. Your 95 percent confidence interval for the mean length of walleye fingerlings in this fish hatchery pond is 7.5 inches ± 0.45 inch. (The lower end of the interval is $7.5 - 0.45 = 7.05$ inches; the upper end is $7.5 + 0.45 = 7.95$ inches.)

REMEMBER

After you calculate a confidence interval, make sure you always interpret it in words a non-statistician would understand. That is, talk about the results in terms of what the person in the problem is trying to find out — statisticians call this interpreting the results "in the context of the problem." In this example you can say: "With 95 percent confidence, the average length of walleye fingerlings in this entire fish hatchery pond is between 7.05 and 7.95 inches, based on my sample data." (Always be sure to include appropriate units.)

Case 2: Population standard deviation is unknown and/or n is small

In many situations, you don't know σ, so you estimate it with the sample standard deviation, s; and/or the sample size is small (less than 30), and you can't be sure your data came from a normal distribution. (In the latter case, the Central Limit Theorem can't be used; see Chapter 12.) In either situation, you can't use a z^*-value from the standard normal (Z-) distribution as your critical value anymore; you have to use a larger critical value than that, because of not knowing what σ is and/or having less data.

The formula for a confidence interval for one population mean in Case 2 is $\bar{x} \pm t^*_{n-1} \frac{s}{\sqrt{n}}$, where t^*_{n-1} is the critical t^*-value from the t-distribution with $n-1$ degrees of freedom (where n is the sample size). The t^*-values for common confidence levels are found using the CI row at the bottom of the t-table (Table A-2 in the Appendix). Chapter 11 gives you the full details on the t-distribution and how to use the t-table.

REMEMBER

The t-distribution has a similar shape to the Z-distribution except it's flatter and more spread out. For small values of n and a specific confidence level, the critical values on the t-distribution are larger than those on the Z-distribution, so when you use the critical values from the t-distribution, the margin of error for your confidence interval will be wider. As the values of n get larger, the t^*-values are closer to z^*-values. (Chapter 11 gives you the full details on the t-distribution and its relationships to the Z-distribution.)

In the fish hatchery example from Case 1, suppose your sample size was 10 instead of 100, and everything else was the same. The t^*-value in this case comes from a t-distribution with $10 - 1 = 9$ degrees of freedom (or df). This t^*-value is found by looking at the t-table (in the Appendix). Look in the last row where the confidence levels are located, and find the confidence level of 95 percent; this marks the column you need. Then find the row corresponding to $df = 9$. Intersect the row and column, and you find $t^* = 2.262$. This is the t^*-value for a 95 percent confidence interval for the mean with a sample size of 10. (Notice this is larger than the z^*-value of 1.96 found in Table 14-1.) Calculating the confidence interval, you get $7.5 \pm 2.262 \frac{2.3}{\sqrt{10}} = 7.50 \pm 1.645$, or 5.86 to 9.15 inches. (Chapter 11 gives you the full details on the t-distribution and how to use the t-table.)

Notice this confidence interval is wider than the one found when $n = 100$. In addition to having a larger critical value (t^* versus z^*), the sample size is much smaller, which increases the margin of error, because n is in its denominator.

REMEMBER

In a case where you need to use s because you don't know σ, the confidence interval will be wider as well. It is also often the case that σ is unknown and the sample size is small, in which case the confidence interval is also wider.

EXAMPLE

Q. Suppose 100 randomly selected used cars on a lot have an average of 30,250 miles on them, with a standard deviation of 500 miles. Find a 95 percent confidence interval for the average miles on all the cars in this lot.

A. You need to find a 95 percent confidence interval for the population mean, and the formula you need is $\bar{x} \pm t^*_{n-1} \frac{s}{\sqrt{n}}$. The sample mean, $\bar{x}$, is 30,250; the standard deviation, s, is 500; and because n is rather large at 100, the t^*-value is getting very close to the z^*-value, so from Table 14-1 or the t-table in the Appendix, you'll use $t^* = 1.96$ because the confidence level is 95 percent. So, your confidence interval is

$$\bar{x} \pm t^*_{n-1} \frac{s}{\sqrt{n}} = 30{,}250 \pm 1.96 \cdot \frac{500}{\sqrt{100}} = 30{,}250 \pm 98 = (30{,}152; 30{,}348) \text{ miles}$$

Interpretation: You're 95 percent confident that the average mileage for all cars in this lot is between 30,152 and 30,348 miles.

YOUR TURN

11 You want to estimate the average number of tines on the antlers of male deer in a nearby metro park. A sample of 30 deer has an average of 5 tines, with a standard deviation of 3 tines.

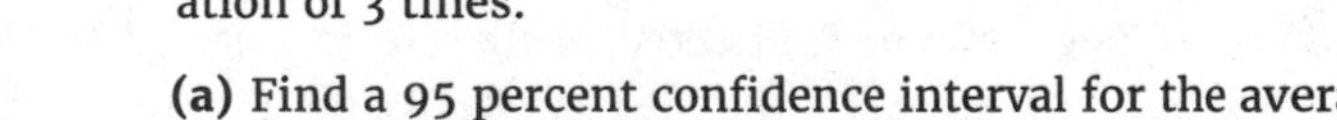

(a) Find a 95 percent confidence interval for the average number of tines for all male deer in this metro park.

(b) Find a 98 percent confidence interval for the average number of tines for all male deer in this metro park.

12 Based on a sample of 100 participants, a 95 percent confidence interval for the average weight loss the first month under a new weight loss plan is 11.4 pounds, plus or minus 0.51.

(a) Explain what this confidence interval means.

(b) What's the margin of error for this confidence interval?

13 A 95 percent confidence interval for the average miles per gallon for all cars of a certain type is 32.1, plus or minus 1.8. The interval is based on a sample of 40 randomly selected cars.

(a) What units represent the margin of error?

(b) Suppose you want to decrease the margin of error, but you want to keep 95 percent confidence. What should you do?

14 Suppose you want to increase the confidence level of a particular confidence interval from 80 percent to 95 percent without changing the width of the confidence interval. Can you do it?

Figuring Out What Sample Size You Need

The margin of error of a confidence interval is affected by size (see the earlier section, "Factoring In the Sample Size"); as size increases, margin of error decreases. Looking at this the other way around, if you want a smaller margin of error (and doesn't everyone?), you need a larger sample size. Suppose you are getting ready to do your own survey to estimate a population mean; wouldn't it be nice to see ahead of time what sample size you need to get the margin of error you want? Thinking ahead will save you money and time and it will give you results you can live with in terms of the margin of error — you won't have any surprises later.

REMEMBER

The formula for the sample size required to get a desired margin of error (MOE) when you are doing a confidence interval for μ is $n=\left(\frac{z^{*}\sigma}{\text{MOE}}\right)^{2}$; always round up the sample size no matter what decimal value you get. (For example, if your calculations give you 126.2 people, you can't just have 0.2 of a person — you need the whole person, so include them by rounding up to 127.)

In this formula, MOE is the number representing the margin of error you want, and z^* is the z^*-value corresponding to your desired confidence level (from Table 14-1; most people use 1.96 for a 95 percent confidence interval). If the population standard deviation, σ, is unknown, you can put in a worst-case scenario guess for it or run a pilot study (a small trial study) ahead of time, find the standard deviation of the sample data (s), and use that number. This can be risky if the sample size is very small because it's less likely to reflect the whole population; try to get the largest trial study that you can and/or make a conservative estimate for σ.

TIP

Often a small trial study is worth the time and effort. Not only will you get an estimate of σ to help you determine a good sample size, but you may also learn about possible problems in your data collection.

REMEMBER

I only include one formula for calculating sample size in this chapter: the one that pertains to a confidence interval for a population mean. (You can, however, use the quick and dirty formula in the earlier section, "Factoring In the Sample Size," for handling proportions.)

Here's an example where you need to calculate n to estimate a population mean. Suppose you want to estimate the average number of songs college students store on their portable devices. You want the margin of error to be *no more than* plus or minus 20 songs. You want a 95 percent confidence interval. How many students should you sample?

Because you want a 95 percent CI, z^* is 1.96 (found in Table 14-1); you know your desired MOE is 20. Now you need a number for the population standard deviation, σ. This number is not known, so you do a pilot study of 35 students and find the standard deviation (s) for the sample is 148 songs — use this number as a substitute for σ. Using the sample size formula, you calculate that the sample size you need is $n \geq \left(\frac{1.96(148)}{20} \right)^2 = (14.504)^2 = 210.37$, which you round *up* to 211 students (you always round up when calculating n). So you need to take a random sample of *at least* 211 college students in order to have a margin of error in the number of stored songs of *no more than* 20. That's why you see a greater-than-or-equal-to sign in the formula here.

REMEMBER

You always round up to the nearest integer when calculating sample size, no matter what the decimal value of your result is (for example, 0.37). That's because you want the margin of error to be *no more than* what you stated. If you round down when the decimal value is under 0.50 (as you normally do in other math calculations), your MOE will be a little larger than you wanted.

If you are wondering where this formula for sample size comes from, it's actually created with just a little math gymnastics. Take the margin of error formula (which contains n), fill in the remaining variables in the formula with numbers you glean from the problem, set it equal to the desired MOE, and solve for n.

Determining the Confidence Interval for One Population Proportion

When a characteristic being measured is categorical — for example, opinion on an issue (support, oppose, or are neutral), gender, political party, or type of behavior (do/don't wear a seatbelt while driving) — most people want to estimate the proportion (or percentage) of people in the population that fall into a certain category of interest. For example, consider the percentage of people in favor of a four-day work week, the percentage of Republicans who voted in the last election, or the proportion of drivers who don't wear seat belts. In each of these cases, the object is to estimate a population proportion, p, using a sample proportion, $\hat{p}$, plus or minus a margin of error. The result is called a *confidence interval for the population proportion, p.*

The formula for a CI for a population proportion is $\hat{p} \pm z^* \sqrt{\frac{\hat{p}(1-\hat{p})}{n}}$, where $\hat{p}$ is the sample proportion, n is the sample size, and z^* is the appropriate value from the standard normal distribution for your desired confidence level. Refer to Table 14-1 for values of z^* for certain confidence levels.

To calculate a CI for the population proportion:

1. **Determine the confidence level and find the appropriate z^*-value.**

   Refer to Table 14-1 for z^*-values.

2. **Find the sample proportion, $\hat{p}$, by dividing the number of people in the sample having the characteristic of interest by the sample size (n).**

   Note: This result should be a decimal value between 0 and 1.

3. **Multiply $\hat{p}(1-\hat{p})$ and then divide that amount by n.**
4. **Take the square root of the result from Step 3.**
5. **Multiply your answer by z^*.**

   This step gives you the margin of error.

6. **Take $\hat{p}$ plus or minus the margin of error to obtain the CI; the lower end of the CI is $\hat{p}$ minus the margin of error, and the upper end of the CI is $\hat{p}$ plus the margin of error.**

The formula shown in the preceding example for a CI for p is used under the condition that the sample size is large enough for the Central Limit Theorem to kick in and allow you to use a z^*-value (see Chapter 12), which happens in cases when you are estimating proportions based on large-scale surveys (see Chapter 10). For small sample sizes, confidence intervals for the proportion are typically beyond the scope of an intro statistics course.

For example, suppose you want to estimate the percentage of the time you're expected to get a red light at a certain intersection. Follow these steps:

1. **Because you want a 95 percent confidence interval, your z^*-value is 1.96.**
2. You take a random sample of 100 different trips through this intersection and find that you hit a red light 53 times, so $\hat{p} = 53 \div 100 = 0.53$.
3. Find $\frac{\hat{p}(1-\hat{p})}{n} = 0.53(1-0.53) \div 100 = 0.2491 \div 100 = 0.002491$.
4. Take the square root to get 0.0499.

   The margin of error is, therefore, plus or minus $1.96(0.0499) = 0.0978$, or 9.78 percent.

5. Your 95 percent confidence interval for the percentage of times you will ever hit a red light at that particular intersection is 0.53 (or 53 percent), plus or minus 0.0978 (rounded to 0.10 or 10 percent). (The lower end of the interval is $0.53 - 0.10 = 0.43$ or 43 percent; the upper end is $0.53 + 0.10 = 0.63$ or 63 percent.)

   To interpret these results within the context of the problem, you can say that with 95 percent confidence the percentage of the times you should expect to hit a red light at this intersection is somewhere between 43 percent and 63 percent, based on your sample.

TIP

While performing any calculations involving sample percentages, use the decimal form. After the calculations are finished, convert to percentages by multiplying by 100. To avoid round-off error, keep at least 2 decimal places throughout.

EXAMPLE

Q. A random sample of 1,000 U.S. college students finds that 28 percent watch the Super Bowl every year. Find a 95 percent confidence interval for the proportion of all U.S. college students who watch the Super Bowl every year.

A. The sample proportion is 0.28, $n = 1{,}000$, and $z^* = 1.96$ (from Table 14-1), so the 95 percent confidence interval is

$$0.28 \pm 1.96 \cdot \sqrt{\frac{0.28(1-0.28)}{1{,}000}} = 0.28 \pm 1.96 \cdot \sqrt{0.0002} = 0.28 \pm 0.028$$

Interpretation: You're 95 percent confident that the proportion of all U.S. college students who watch the Super Bowl every year is between 0.252 and 0.308, or 25.2 percent and 30.8 percent.

YOUR TURN

15 A random sample of 1,117 U.S. college students finds that 729 go home at least once each term. Find a 98 percent confidence interval for the proportion of all U.S. college students who go home at least once each term.

16 A poll of 2,500 people shows that 50 percent approve of a smoking ban in bars and restaurants. What's the margin of error for this confidence interval? (Assume 95 percent confidence.)

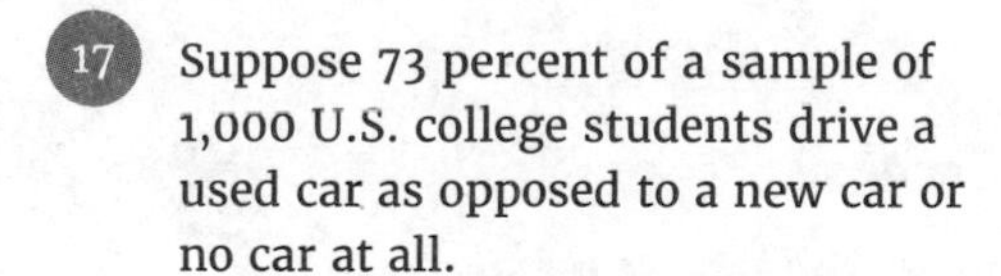

17 Suppose 73 percent of a sample of 1,000 U.S. college students drive a used car as opposed to a new car or no car at all.

(a) Find an 80 percent confidence interval for the percentage of all U.S. college students who drive a used car.

(b) What sample size would cut this margin of error in half?

18 A special interest group reports a tiny margin of error (plus or minus 0.04 percent) for its online survey based on 50,000 responses. Is the margin of error legitimate? (Assume the group's math is correct.)

Creating a Confidence Interval for the Difference of Two Means

The goal of many surveys and studies is to compare two populations, such as men versus women, low- versus high-income families, and Republicans versus Democrats. When the characteristic being compared is numerical (for example, height, weight, or income), the object of interest is the amount of difference in the means (averages) for the two populations.

For example, you may want to compare the difference in average age of Republicans versus Democrats, or the difference in average incomes of men versus women. You estimate the difference between two population means, $\mu_1 - \mu_2$, by taking a sample from each population (say, Sample 1 and Sample 2) and using the difference of the two sample means $\bar{x}_1 - \bar{x}_2$, plus or minus a margin of error. The result is a *confidence interval for the difference of two population means,* $\mu_1 - \mu_2$. The formula for the CI is different depending on certain conditions, as shown in the following sections; I call them Case 1 and Case 2.

Case 1: Population standard deviations are known

Case 1 assumes that both of the population standard deviations are known. The formula for a CI for the difference between two population means (averages) is $\bar{x}_1 - \bar{x}_2 \pm z^* \sqrt{\frac{\sigma_1^2}{n_1} + \frac{\sigma_2^2}{n_2}}$, where $\bar{x}_1$ and n_1 are the mean and size of the first sample, and the first population's standard deviation, σ_1, is given (known); $\bar{x}_2$ and n_2 are the mean and size of the second sample, and the second population's standard deviation, σ_1, is given (known). Here, z^* is the appropriate value from the standard normal distribution for your desired confidence level. (Refer to Table 14-1 for values of z^* for certain confidence levels.)

To calculate a CI for the difference between two population means, do the following:

1. **Determine the confidence level and find the appropriate z^*-value.**

   Refer to Table 14-1.

2. **Identify $\bar{x}_1$, n_1, and σ_2; find $\bar{x}_2$, n_2, and σ_2.**
3. **Find the difference, $(\bar{x}_1 - \bar{x}_2)$, between the sample means.**
4. **Square σ_1 and divide it by n_1; square σ_2 and divide it by n_2. Add the results together and take the square root.**
5. **Multiply your answer from Step 4 by z^*.**

   This answer is the margin of error.

6. **Take $\bar{x}_1 - \bar{x}_2$ plus or minus the margin of error to obtain the CI.**

   The lower end of the CI is $\bar{x}_1 - \bar{x}_2$ *minus* the margin of error, whereas the upper end of the CI is $\bar{x}_1 - \bar{x}_2$ *plus* the margin of error.

Suppose you want to estimate with 95 percent confidence the difference between the mean (average) length of the cobs of two varieties of sweet corn (allowing them to grow the same number of days under the same conditions). Call the two varieties Corn-e-stats and Stats-o-sweet. Assume from prior research that the population standard deviations for Corn-e-stats and Stats-o-sweet are 0.35 inch and 0.45 inch, respectively.

1. **Because you want a 95 percent confidence interval, your z^* is 1.96.**
2. Suppose your random sample of 100 cobs of the Corn-e-stats variety averages 8.5 inches, and your random sample of 110 cobs of Stats-o-sweet averages 7.5 inches. So the information you have is: $\bar{x}_1 = 8.5, \sigma_1 = 0.35, n_1 = 100, \bar{x}_2 = 7.5, \sigma_2 = 0.45$, and $n_2 = 110$.
3. The difference between the sample means, $\bar{x}_1 - \bar{x}_2$, from Step 3, is $8.5 - 7.5 = +1$ inch. This means the average for Corn-e-stats minus the average for Stats-o-sweet is positive, making Corn-e-stats the larger of the two varieties, in terms of this sample. Is that difference enough to generalize to the entire population, though? That's what this confidence interval is going to help you decide.
4. Square σ_1 (0.35) to get 0.1225; divide by 100 to get 0.0012. Square σ_2 (0.45) and divide by 110 to get $0.2025 \div 110 = 0.0018$. The sum is $0.0012 + 0.0018 = 0.0030$; the square root is 0.0554 inch (if no rounding was done).
5. Multiply 1.96 times 0.0554 to get 0.1085 inch, the margin of error.
6. Your 95 percent confidence interval for the difference between the average lengths for these two varieties of sweet corn is 1 inch, plus or minus 0.1085 inch. (The lower end of the interval is $1 - 0.1085 = 0.8915$ inch; the upper end is $1 + 0.1085 = 1.1085$ inches.) Notice all the values in this interval are positive. That means Corn-e-stats is estimated to be longer than Stats-o-sweet, based on your data.

   To interpret these results in the context of the problem, you can say with 95 percent confidence that the Corn-e-stats variety is longer, on average, than the Stats-o-sweet variety, by somewhere between 0.8915 and 1.1085 inches, based on your sample.

REMEMBER

Notice that you could get a negative value for $\bar{x}_1 - \bar{x}_2$. For example, if you had switched the two varieties of corn, you would have gotten -1 for this difference. You would say that Stats-o-sweet averaged one inch shorter than Corn-e-stats in the sample (the same conclusion, but stated differently).

TIP

If you want to avoid negative values for the difference in sample means, always make the group with the larger sample mean your first group — all your differences will be positive (that's what I do).

Case 2: Population standard deviations are unknown and/or sample sizes are small

In many situations, you don't know σ_1 and σ_2, and you estimate them with the sample standard deviations, s_1, and s_2; and/or the sample sizes are small (less than 30) and you can't be sure whether your data came from a normal distribution.

A confidence interval for the difference in two population means under Case 2 is $\left(\bar{x}_1 - \bar{x}_2\right) \pm t^*_{n_1+n_2-2}\sqrt{\frac{(n_1-1)s_1^2+(n_2-1)s_2^2}{n_1+n_2-2}}\sqrt{\frac{1}{n_1}+\frac{1}{n_2}}$, where t^* is the critical value from the t-distribution with $n_1 + n_2 - 2$ degrees of freedom; n_1 *and* n_2 are the two sample sizes, respectively; and s_1 and s_2 are the two sample standard deviations. This t^*-value is found on the t-table (in the Appendix) by intersecting the row for $df = n_1 + n_2 - 2$ with the column for the confidence level you need, as indicated by looking at the last row of the table (see Chapter 11). Here I assume the population standard deviations are similar; if not, modify them by using the standard error and degrees of freedom. See Chapter 16 for more info. In the corn example from Case 1, suppose the mean cob lengths of the two brands of corn, Corn-e-stats (Group 1) and Stats-o-sweet (Group 2), are the same as they were before: $\bar{x}_1 = 8.5$ and $\bar{x}_2 = 7.5$ inches. But this time you don't know the population standard deviations, so you use the sample standard deviations instead — suppose they turn out to be $s_1 = 0.40$ and $s_2 = 0.50$ inch, respectively. Suppose the sample sizes, n_1 and n_2, are each only 15 in this case.

Calculating the CI, you first need to find the t^*-value on the t-distribution with $(15 + 15 - 2) = 28$ degrees of freedom. (Assume the confidence level is still 95 percent.) Using the t-table (in the Appendix), look at the row for 28 degrees of freedom and the column representing a confidence level of 95 percent (see the labels on the last row of the table); intersect them and you see $t^*_{28} = 2.048$. Using the rest of the information you are given, the confidence interval for the difference in mean cob length for the two brands is

$$(8.5-7.5) \pm 2.048\sqrt{\frac{(15-1)(0.4)^2+(15-1)(0.5)^2}{15+15-2}}\sqrt{\frac{1}{15}+\frac{1}{15}}$$
$$= 1.0 \pm 2.048(0.45)(0.365)$$
$$= 1.00 \pm 0.34 \text{ inches}$$

That means that a 95 percent CI for the difference in the mean cob lengths of these two brands of corn in this situation is (0.66, 1.34) inches, with Corn-e-stats coming out on top. (***Note:*** This CI is wider than what was found in Case 1, as expected.)

EXAMPLE

Q. Suppose 100 randomly selected used cars on Lot 1 have an average of 35,328 miles on them, with a standard deviation of 750 miles. A hundred randomly selected used cars on Lot 2 have an average of 30,250 miles on them, with a standard deviation of 500 miles. Find a 95 percent confidence interval for the difference in average miles between all cars on Lots 1 and 2.

A. Using Lot 1 as Group 1 and Lot 2 as Group 2, you know that $\bar{x} = 35{,}328$; $s_1 = 750$; $n_1 = 100$; $\bar{y} = 30{,}250$; $s_2 = 500$; and $n_2 = 100$. Your z^*-value is 1.96 (from Table 14-1). Using the formula for a confidence interval for the difference of two population means, the 95 percent confidence interval is

$$\left(\bar{x}_1 - \bar{x}_2\right) \pm t^*_{n_1+n_2-2}\sqrt{\frac{(n_1-1)s_1^2+(n_2-1)s_2^2}{n_1+n_2-2}}\sqrt{\frac{1}{n_1}+\frac{1}{n_2}}$$

$$= (35{,}328 - 30{,}250) \pm t^*_{198}\sqrt{\frac{99(750)^2+99(500)^2}{198}}\sqrt{\frac{1}{100}+\frac{1}{100}}$$

$$= 5{,}078 \pm 1.96(637.38)(0.14)$$

$$= 5{,}078 \pm 176.67 \text{ miles}$$

Note: After a certain point, the values on the *t*-distribution follow those on the *Z*-distribution, as they do here with 198 degrees of freedom; that's why I used 1.96 for *t*; 198 degrees of freedom is way off the *t*-table.

Interpretation: You're 95 percent confident that the difference in average mileage between all cars on Lots 1 and 2 is $5{,}078 \pm 176.67$ miles (and that cars on Lot 1 have higher mileage than cars on Lot 2).

19 Based on a sample of 100 participants, the average weight loss the first month under a new (competing) weight loss plan is 11.4 pounds with a standard deviation of 5.1 pounds. The average weight loss for the first month for 100 people on the old (standard) weight loss plan is 12.8 pounds, with a standard deviation of 4.8 pounds.

(a) Find a 90 percent confidence interval for the difference in weight loss for the two plans (old minus new).

(b) What's the margin of error for your calculated confidence interval?

20 The average miles per gallon for a sample of 40 cars of model SX last year was 32.1, with a standard deviation of 3.8. A sample of 40 cars from this year's model SX has an average of 35.2 mpg, with a standard deviation of 5.4.

(a) Find a 99 percent confidence interval for the difference in average mpg for this car brand (this year's model minus last year's).

(b) Find a 99 percent confidence interval for the difference in average mpg for last year's model minus this year's. What does the negative difference mean?

21 You want to compare the average number of tines on the antlers of male deer in two nearby metro parks. A sample of 15 deer from the first park shows an average of five tines with a standard deviation of 3. A sample of ten deer from the second park shows an average of six tines with a standard deviation of 3.2.

(a) Find a 95 percent confidence interval for the difference in average number of tines for all male deer in the two metro parks (second park minus first park).

(b) Do the parks' deer populations differ in average size of deer antlers?

22 Suppose a group of 100 people is on one weight-loss plan for one month and then switches to another weight-loss plan for one month. Suppose you want to compare the two plans, and you decide to compare the average weight loss for the first month to the second month. Explain why you can't use this data to make a 95 percent confidence interval for the average difference.

Estimating the Difference of Two Proportions

When a characteristic, such as opinion on an issue (support/don't support), of the two groups being compared is *categorical*, people want to report on the differences between the two population proportions — for example, the difference between the proportion of women who support a four-day work week and the proportion of men who support a four-day work week. How do you do this?

You estimate the difference between two population proportions, $p_1 - p_2$, by taking a sample from each population and using the difference of the two sample proportions, $\hat{p}_1 - \hat{p}_2$, plus or minus a margin of error. The result is called a *confidence interval for the difference of two population proportions*, $p_1 - p_2$.

The formula for a CI for the difference between two population proportions is $(\hat{p}_1 - \hat{p}_2) \pm z^* \sqrt{\frac{\hat{p}_1(1-\hat{p}_1)}{n_1} + \frac{\hat{p}_2(1-\hat{p}_2)}{n_2}}$, where $\hat{p}_1$ and n_1 are the sample proportion and sample size of the first sample, and $\hat{p}_2$ and n_2 are the sample proportion and sample size of the second sample; z^* is the appropriate value from the standard normal distribution for your desired confidence level. (Refer to Table 14-1 for z^*-values.)

To calculate a CI for the difference between two population proportions, do the following:

1. **Determine the confidence level and find the appropriate z^*-value.**

   Refer to Table 14-1.

2. **Find the sample proportion $\hat{p}_1$ for the first sample by taking the total number from the first sample that are in the category of interest and dividing by the sample size, n_1. Similarly, find $\hat{p}_2$ for the second sample.**

3. **Take the difference between the sample proportions, $\hat{p}_1 - \hat{p}_2$.**

4. **Find $\hat{p}_1(1-\hat{p}_1)$ and divide that by n_1. Find $\hat{p}_2(1-\hat{p}_2)$ and divide that by n_2. Add these two results together and take the square root.**

5. **Multiply z^* times the result from Step 4.**

   This step gives you the margin of error.

6. **Take $\hat{p}_1 - \hat{p}_2$ plus or minus the margin of error from Step 5 to obtain the CI.**

   The lower end of the CI is $\hat{p}_1 - \hat{p}_2$ minus the margin of error, and the upper end of the CI is $\hat{p}_1 - \hat{p}_2$ plus the margin of error.

The formula shown here for a CI for $p_1 - p_2$ is used under the condition that both of the sample sizes are large enough for the Central Limit Theorem to kick in and allow you to use a z^*-value (see Chapter 12); this is true when you are estimating proportions using large-scale surveys, for example. For small sample sizes, confidence intervals are beyond the scope of an intro statistics course.

Suppose you work for the Las Vegas Chamber of Commerce, and you want to estimate with 95 percent confidence the difference between the percentage of females who have ever gone to see an Elvis impersonator and the percentage of males who have ever gone to see an Elvis impersonator, in order to help determine how you should market your entertainment offerings.

1. **Because you want a 95 percent confidence interval, your z^*-value is 1.96.**
2. Suppose your random sample of 100 females includes 53 females who have seen an Elvis impersonator, so $\hat{p}_1$ is $53 \div 100 = 0.53$. Suppose also that your random sample of 110 males includes 37 males who have ever seen an Elvis impersonator, so $\hat{p}_2$ is $37 \div 110 = 0.34$.
3. The difference between these sample proportions (females – males) is $0.53 - 0.34 = 0.19$.
4. Take $0.53 \times (1 - 0.53)$ and divide that by 100 to get $0.2491 \div 100 = 0.0025$. Then take $0.34 \times (1 - 0.34)$ and divide that by 110 to get $0.2244 \div 110 = 0.0020$. Add these two results to get $0.0025 + 0.0020 = 0.0045$; the square root is 0.0671.
5. Multiplying $(1.96)(0.0671)$ gives you 0.13, or 13 percent, which is the margin of error.
6. Your 95 percent confidence interval for the difference between the percentage of females who have seen an Elvis impersonator and the percentage of males who have seen an Elvis impersonator is 0.19 or 19 percent (which you got in Step 3), plus or minus 13 percent. The lower end of the interval is $0.19 - 0.13 = 0.06$ or 6 percent; the upper end is $0.19 + 0.13 = 0.32$ or 32 percent.

   To interpret these results within the context of the problem, you can say with 95 percent confidence that a higher percentage of females than males have seen an Elvis impersonator, and the difference in these percentages is somewhere between 6 percent and 32 percent, based on your sample.

Now, I'm thinking there are some guys out there who wouldn't admit they'd ever seen an Elvis impersonator (although they've probably pretended to be one doing karaoke at some point). This may create some bias in the results. (The last time Deborah was in Vegas, she believes she really saw Elvis; he was driving a van taxi to and from the airport.)

TIP

Notice that you could get a negative value for $\hat{p}_1 - \hat{p}_2$. For example, if you had switched the males and females, you would have gotten −0.19 for this difference. That's okay, but you can avoid negative differences in the sample proportions by having the group with the larger sample proportion serve as the first group (here, females).

EXAMPLE

Q. A random sample of 500 male U.S. college students finds that 35 percent watch the Super Bowl every year, and a random sample of 500 female U.S. college students finds that 21 percent watch every year. Find a 95 percent confidence interval for the difference in the proportion of Super Bowl watchers for male versus female U.S. college students.

A. For the males, the sample proportion is 0.35, and $n = 500$. For the females, the sample proportion is 0.21, and $n = 500$. You know $z^* = 1.96$ (from Table 14-1). The 95 percent confidence interval is

$$(\hat{p}_1 - \hat{p}_2) \pm z^* \sqrt{\frac{\hat{p}_1(1-\hat{p}_1)}{n_1} + \frac{\hat{p}_2(1-\hat{p}_2)}{n_2}}$$

$$= (0.35 - 0.21) \pm 1.96\sqrt{\frac{0.35(1-0.35)}{500} + \frac{0.21(1-0.21)}{500}}$$

$$= 0.14 \pm 1.96(0.028)$$

$$= 0.14 \pm 0.055$$

$$= (0.085,\ 0.195)$$

Interpretation: You're 95 percent confident that the difference in the proportion of Super Bowl watchers for male versus female U.S. college students is between 0.085 and 0.195. In other words, between 8.5 percent and 19.5 percent more males watch the Super Bowl than females.

YOUR TURN

23 A random sample of 1,117 domestic students at a U.S. university finds that 915 go home at least once each term, compared to 212 from a random sample of 1,200 international students from the same university. Find a 98 percent confidence interval for the difference in the proportion of students who go home at least once each term.

24 A poll of 1,000 smokers shows that 16 percent approve of a smoking ban in bars and restaurants; 84 percent from a sample of 500 nonsmokers approve of the ban.

(a) What's the margin of error for a 95 percent confidence interval for the difference in proportion for all smokers versus all nonsmokers?

(b) In this case, the sample percents sum to one; does this always happen?

25 Suppose 72 percent of a sample of 1,000 traditional college students drive a used car as opposed to a new car or no car at all. A sample of 1,000 nontraditional students shows 71 percent drive used cars.

(a) Find an 80 percent confidence interval for the difference in the percentages of students who drive used cars for these two populations.

(b) The lower bound of this interval is negative, and the upper bound is positive. How do you interpret these results?

26 Suppose a class of 100 students contains 10 percent Independents and 90 percent party-affiliates. If Bob samples half the students in each group and compares the percentage approving of the president, can he use the formula for finding a confidence interval for the difference of two proportions? Why or why not?

Spotting Misleading Confidence Intervals

When the MOE is small, relatively speaking, you would like to say that these confidence intervals provide accurate and credible estimates of their parameters. This is not always the case, however.

WARNING

Not all estimates are as accurate and reliable as the sources may want you to think. For example, a website survey result based on 20,000 hits may have a small MOE according to the formula, but the MOE means nothing if the survey is only given to people who happened to visit that website, especially if we want to draw conclusions about the general population.

In other words, the sample isn't even close to being a random sample (where every sample of equal size selected from the population has an equal chance of being chosen to participate). Nevertheless, such results do get reported, along with their margins of error that make the study seem truly scientific. Beware of these bogus results! (See Chapter 13 for more on the limits of the MOE.)

REMEMBER

Before making any decisions based on someone's estimate, do the following:

- Investigate how the statistic was created; it should be the result of a scientific process that results in reliable, unbiased, accurate data.
- Look for a margin of error. If one isn't reported, go to the original source and request it.
- Remember that if the statistic isn't reliable or contains bias, the margin of error will be meaningless.

(See Chapter 17 to find out how to evaluate survey data, and see Chapter 18 for criteria for good data in experiments.)

Practice Questions Answers and Explanations

1. **The margin of error doesn't change**, because it doesn't depend on what the average measurements are; it depends on the standard deviation and sample sizes. Also, margin of error doesn't take bias into account, and this scale is a biased scale.

Margin of error measures precision (consistency), but it doesn't measure bias (being systematically over or systematically under the true value).

2. This survey is biased because of the very low number of people who responded. The response rate for this survey (number of respondents divided by total number sent out) is only $1{,}000 \div 10{,}000 = 0.10$ or 10 percent. That means 90 percent of the people who received the survey didn't respond. If you base the results only on those who responded, you would say $80 \div 1{,}000 = 0.08$ or 8 percent of them are in favor of the dog barking ordinance, and the other 92 percent are against it. This may lead you to believe the results are biased, and indeed they are. Those who have the strongest opinions respond to surveys, in general. Suppose that the 9,000 people who didn't respond were actually in favor of the dog ordinance. That would mean a total of $9{,}000 + 80 = 9{,}080$ in favor (and $9{,}080 \div 10{,}000 = 0.908$ or 90.8 percent) and 920 against (that percentage would be $920 \div 10{,}000 = 0.092$ or 9.2 percent). Those percentages change dramatically from the results based only on the respondents.

Bottom line: Respondents and nonrespondents are not alike and should not be assumed to be so. If the response rate of a survey is too low, you should ignore the results; they are likely to be biased toward those with strong opinions.

Don't be misled by seemingly precise studies that offer a small margin of error without looking at the quality of the data. If the data is bad, the results are garbage, even if the formulas don't know that.

3. **No.** Margin of error measures the sampling error only, which is the error due to the random sampling process. The word "error" makes it seem like human error, but it actually means random error (error due to chance alone).

4. **No,** you can never be 100 percent confident, unless you include all possible values in your interval, which renders it useless. The confidence level tells you what percentage of the samples you expect to yield correct intervals. For example, if your confidence level is 95 percent, you can expect that 95 percent of the time your sample produces an interval that contains the true population value, and that 5 percent of the time it doesn't, just by random chance.

5. The margin of error is meaningless because the news channel bases the sample on local area gas stations that don't represent a statewide sample.

6. **Possibly, only because the control may reduce the amount of variability in the results** (as well as eliminate bias, which doesn't affect margin of error). By controlling for certain variables, you can reduce the variability.

7. The margin of error is the part you add and subtract to get the upper and lower boundaries of the confidence interval.

   a. The margin of error here is **plus or minus 35 minutes.**

   b. **The lower boundary is $110 - 35 = 75$ minutes; the upper boundary is $110 + 35 = 145$ minutes.**

8. Increasing confidence increases the width of the confidence interval.

   a. **The 99 percent confidence interval is wider**, because (all else remaining the same) the z^*-value increases from 1.96 to 2.58 (see Table 14-1), which increases the margin of error.

   b. A wider interval means you aren't as precise at estimating your population parameter — **not a good thing**.

9. **No.** You're 100 percent confident that the sample statistic is in the confidence interval, because you find the confidence interval by using the sample statistic plus or minus the margin of error.

10. **No.** It means that 95 percent of all intervals obtained by different samples contain the population parameter. One particular 95 percent confidence interval either contains it or it doesn't.

TIP

As a professor, I'm 95 percent confident that your instructor will ask you what a 95 percent confidence interval means; all statistics professors love to use this doozy on exams, so be ready!

11. Tines can range from 2 to upwards of 50 or more on a male deer.

   a. The 95 percent confidence interval for the average number of tines on deer in the park is **5, plus or minus $t^*_{n-1}\frac{s}{\sqrt{n}}$**, because you're given the sample standard deviation of 3 and the sample size is $n = 30$. Looking at the t-table, you find $t^*_{n-1} = t^*_{29} = 2.04523$ when constructing a 95 percent confidence interval. So the margin of error is

   $$t^*_{n-1}\frac{s}{\sqrt{n}} = 2.04523 \cdot \frac{3}{\sqrt{30}} = 1.12$$

   Interpretation: You're 95 percent confident that the average number of tines on the deer in this park is 5 ± 1.12. That is, between 3.88 and 6.12 tines.

   b. To construct a 98 percent confidence interval from this sample data, you look at the t-table and find $t^*_{n-1} = t^*_{29} = 2.46202$. The 98 percent confidence interval is 5 plus or minus $t^*_{n-1}\frac{s}{\sqrt{n}} = 2.46202 \cdot \frac{3}{\sqrt{30}} = \mathbf{1.35}$. *Interpretation:* You're 98 percent confident that the average number of tines on the deer in this park is 5 ± 1.35. That is, between 3.65 and 6.35 tines.

12. Try to interpret confidence intervals in a way that a layperson can understand.

   a. According to your sample, the average weight loss for everyone in the program in month one is 11.4 pounds, plus or minus 0.51. Your process of sampling will be correct 95 percent of the time.

   b. **The margin of error is plus or minus 0.51 pound.**

13. Margin of error maintains the same units as the original data. Be aware that sample size strongly affects the MOE.

   a. **Miles per gallon.**

   b. **Increase the sample size.**

14. **Yes, by also increasing the sample size.** That will offset the larger z^*-value that goes with the 95 percent confidence level.

REMEMBER

The margin of error increases when the confidence level or standard deviation goes up and decreases when the sample size goes up (refer to Chapter 13).

15) The needed pieces of this problem are $n = 1{,}117$, $\hat{p} = 729 \div 1{,}117 = 0.65$, and $z^* = 2.33$ (from Table 14-1) for a 98 percent confidence level. The confidence interval is

$$\hat{p} \pm z^* \sqrt{\frac{\hat{p}(1-\hat{p})}{n}} = 0.65 \pm 2.33\sqrt{\frac{0.65(1-0.65)}{1{,}117}} = 0.65 \pm 2.33\sqrt{0.0002} = 0.65 \pm 0.03 = (0.62,\ 0.68)$$

Interpretation: You're 98 percent confident that the proportion of all U.S. college students who go home at least once each term is **between 0.62 and 0.68**.

WARNING

Be careful with what you use in the formula for $\hat{p}$. Don't use the total number in the category of interest (in this example, 729). You need to divide the total number by the total sample size to get an actual proportion (in this case, $729 \div 1{,}117 = 0.65$) — a number between zero and one — or you get incorrect calculations. Also, you can't use the percentage in the formula; use the proportion (decimal version of the percent). In this example, you use 0.65, not 65.

16) Here you have $n = 2{,}500$, $\hat{p} = 0.50$, and $z^* = 1.96$ (from Table 14-1), so the margin of error is

$$\pm z^* \sqrt{\frac{\hat{p}(1-\hat{p})}{n}} = \pm 1.96\sqrt{\frac{0.50(1-0.50)}{2{,}500}} = \pm 1.96(0.01) = \pm 0.02$$

17) Confidence levels and sample sizes affect confidence intervals.

a. Here you have $n = 1{,}000$, $\hat{p} = 0.73$, and $z^* = 1.28$ (from Table 14-1), so the 80 percent confidence interval is

$$\begin{aligned} \hat{p} \pm z^* \sqrt{\frac{\hat{p}(1-\hat{p})}{n}} &= 0.73 \pm 1.28\sqrt{\frac{0.73(1-0.73)}{1{,}000}} \\ &= 0.73 \pm 1.28(0.014) \\ &= 0.73 \pm 0.0179 \\ &= (0.71,\ 0.75) \end{aligned}$$

which is **71 percent to 75 percent**.

Interpretation: You're 80 percent confident that the percentage of all U.S. college students who drive a used car is between 71 percent and 75 percent. (***Note:*** I used the decimal versions for all calculations and then changed to percentages at the very end.)

b. **Quadruple the sample size to 4,000.** Increasing the sample size 4 times will reduce that margin of error by $\sqrt{4} = 2$ times. You don't need any fancier calculation than that (refer to Chapter 13).

18) **No.** The group uses a biased sample, so the margin of error is meaningless.

WARNING

Instructors are big on the idea, "Garbage in equals garbage out." It applies to confidence intervals (and basically every other statistical calculation). If you put bad data into a formula, the formula still cranks out an answer, but the answer may be garbage. You have to evaluate results by the stated margin of error and by the manner of data collection.

(19) Keeping track of which group is which, and staying consistent throughout your calculations, will help you a great deal on these "difference of means" problems.

a. Using the notation and letting the old plan be Group 1, you have $\bar{x} = 12.8$; $s_1 = 4.8$; $n_1 = 100$; $\bar{y} = 11.4$; $s_2 = 5.1$; and $n_2 = 100$. Your z^*-value for a 90 percent CI is 1.645 (from Table 14-1). Therefore, the 90 percent confidence interval for the difference in average weight loss (old plan minus competing plan) is

$$(\bar{x}-\bar{y}) \pm t^*_{n_1+n_2-2}\sqrt{\frac{(n_1-1)s_1^2+(n_2-1)s_2^2}{n_1+n_2-2}}\sqrt{\frac{1}{n_1}+\frac{1}{n_2}}$$

$$=(12.8-11.4) \pm 1.66\sqrt{\frac{(99)4.8^2+(99)5.1^2}{198}}\sqrt{\frac{1}{100}+\frac{1}{100}}$$

$$=1.4 \pm 1.66(4.95)(0.14)$$

$$=1.4 \pm 1.15$$

Interpretation: You're 90 percent confident that the difference in average weight loss on the two plans is **1.4 – 1.15 pounds**. That is, those on the standard (old) plan lose an average of between 0.25 and 2.55 more pounds than those on the competing (new) plan.

Note: Depending on which group you choose to make Group 1, your results are completely opposite (in terms of sign) if you switch the groups, which is fine as long as you know which group is which. You may want to always choose the group with the highest mean as Group 1 so the difference between the sample means is positive rather than negative.

TIP

Sometimes, confidence intervals do go into impossible values because of the error involved. You typically ignore the values that don't make sense.

b. The margin of error is the part after the ± sign, which in this case is **1.1 pounds**.

(20) Switching the order of groups switches the sign on the upper and lower bounds of your confidence interval.

a. Using the notation, you have $\bar{x} = 35.2$; $s_1 = 5.4$; $n_1 = 40$; $\bar{y} = 32.1$; $s_2 = 3.8$; and $n_2 = 40$. Your z^*-value for a 99 percent CI is 2.58 (from Table 14-1). So the 99 percent confidence interval for this year minus last year is

$$(\bar{x}-\bar{y}) \pm t^*_{n_1+n_2-2}\sqrt{\frac{(n_1-1)s_1^2+(n_2-1)s_2^2}{n_1+n_2-2}}\sqrt{\frac{1}{n_1}+\frac{1}{n_2}}$$

$$=(35.2-32.1) \pm 2.64\sqrt{\frac{(39)5.4^2+(39)3.8^2}{78}}\sqrt{\frac{1}{40}+\frac{1}{40}}$$

$$=3.10 \pm 2.64(4.67)(0.22)$$

$$=3.10 \pm 2.71 \text{ miles per gallon}$$

Interpretation: You're 99 percent confident that the difference in average mpg for this year's brand versus last year's is **3.10 – 2.71**. That is, this year's model gets on average between 0.45 and 5.75 more miles per gallon than last year's model.

b. In this case, you find last year minus this year, so the 99 percent confidence interval switches the sample means (all else remains the same), which gives you **3.10 − 2.69 miles per gallon**. That is, (−5.75, −0.45). You still get a correct answer; it just may be harder to interpret. The differences are negative in this confidence interval because the first group (last year's model) had fewer miles per gallon than the second group (this year's model).

TIP

If the difference between two numbers is negative, the second number is larger than the first. Use this added piece of information in your interpretation of the results. Not only do you know the groups differ, but you also know which one has the larger value and which one has the smaller value.

21 Just because two sample means are different doesn't mean you should expect their population means to be different. It all depends on what your definition of "different" is.

REMEMBER

Sample results are always going to vary from sample to sample, so their differing is no big deal. The question is, are they different enough that, even taking that variability into account, you can be confident that one will be higher than the other in the population? That's what being statistically significant really means.

a. Use the second park as Group 1 to keep the numbers positive. Using the notation, you have $\bar{x}=6$; $s_1=3.2$; $n_1=10$; $\bar{y}=5$; $s_2=3.0$; and $n_2=15$. The sample sizes are small enough that you need to use the *t*-distribution. Your t^*-value for a 95 percent CI with $n_1+n_2-2=10+15-2=23$ degrees of freedom is $t^*_{n_1+n_2-2}=t^*_{23}=2.06866$ (from Table A-2). So the 95 percent confidence interval for this year minus last year is

$$(\bar{x}-\bar{y})\pm t^*_{n_1+n_2-2}\sqrt{\frac{(n_1-1)s_1^2+(n_2-1)s_2^2}{n_1+n_2-2}}\sqrt{\frac{1}{n_1}+\frac{1}{n_2}}$$

$$=(6-5)\pm 2.06866\sqrt{\frac{(10-1)3.2^2+(15-1)3.0^2}{10+15-2}}\sqrt{\frac{1}{10}+\frac{1}{15}}$$

$$=1\pm 2.06866(3.08)(0.41)$$

$$=1\pm 2.61$$

or (−1.61, 3.61) tines per deer.

Interpretation: You're 95 percent confident that the difference in the average number of tines for all male deer in the two metro parks is **between −1.61 and 3.61**. That is, you can't say the deer in one park have more tines on average than the other because the difference could go either way, depending on the sample that's taken.

b. **No,** because the lower boundary is negative, the upper boundary is positive, and zero is in the interval. Therefore, as samples vary, the estimates for the differences in the two populations are too close to call, so you conclude no significant difference in the two population means.

22 You need two independent samples to use the confidence interval for the difference of two populations, and the two samples aren't independent. The same people participate in both samples.

23) Using the notation, you have $\hat{p}_1 = \frac{915}{1{,}117} = 0.82$; $n_1 = 1{,}117$; $\hat{p}_2 = \frac{212}{1{,}200} = 0.18$; $n_2 = 1{,}200$; and $z^* = 2.33$, so the 98 percent confidence interval for the difference is

$$(\hat{p}_1 - \hat{p}_2) \pm z^* \sqrt{\frac{\hat{p}_1(1-\hat{p}_1)}{n_1} + \frac{\hat{p}_2(1-\hat{p}_2)}{n_2}}$$

$$= (0.82 - 0.18) \pm 2.33 \sqrt{\frac{0.82(1-0.82)}{1{,}117} + \frac{0.18(1-0.18)}{1{,}200}}$$

$$= 0.64 \pm 2.33(0.016)$$

$$= 0.64 \pm 0.04$$

$$= (0.60, 0.68)$$

Interpretation: You're 98 percent confident that the difference in the proportion of all college students who go home at least once each term (domestic versus international students) is **between 0.60 and 0.68**. And because the differences are positive, this result says that the domestic students go home that much more often (60 to 68 percent more) than the international students.

24) This is a standard question that asks you to find a confidence interval for the difference in two population proportions.

a. Assign group 1 to the nonsmokers. Using the notation, you have $\hat{p}_1 = 0.84$; $n_1 = 500$; $\hat{p}_2 = 0.16$; $n_2 = 1{,}000$; and $z^* = 1.96$, so the 95 percent confidence interval for the difference in population proportions is

$$(\hat{p}_1 - \hat{p}_2) \pm z^* \sqrt{\frac{\hat{p}_1(1-\hat{p}_1)}{n_1} + \frac{\hat{p}_2(1-\hat{p}_2)}{n_2}}$$

$$= (0.84 - 0.16) \pm 1.96 \sqrt{\frac{0.84(1-0.84)}{500} + \frac{0.16(1-0.16)}{1{,}000}}$$

$$= 0.68 \pm 1.96(0.02)$$

$$= 0.68 \pm 0.04$$

$$= (0.64,\ 0.72)$$

Interpretation: You're 95 percent confident that the difference in the proportion of smokers versus nonsmokers who approve the smoking ban is **0.68 ± 0.04**. In percentage terms, that means 68 percent more nonsmokers approve of the ban than smokers, plus or minus 0.04 percent.

b. **No.** The proportion from sample 1 that falls under the desired category of interest has nothing to do with the proportion from sample 2 that does. Remember, the samples are independent.

(25) Sometimes confidence intervals can give you seemingly inconclusive results.

a. Assign the traditional students to group 1. Using the notation, you have $\hat{p}_1 = 0.72$; $n_1 = 1{,}000$; $\hat{p}_2 = 0.71$; $n_2 = 1{,}000$; and $z^* = 1.28$, so the 80 percent confidence interval for the difference in population proportions is

$$(\hat{p}_1 - \hat{p}_2) \pm z^* \sqrt{\frac{\hat{p}_1(1-\hat{p}_1)}{n_1} + \frac{\hat{p}_2(1-\hat{p}_2)}{n_2}}$$

$$= (0.72 - 0.71) \pm 1.28 \sqrt{\frac{0.72(1-0.72)}{1{,}000} + \frac{0.71(1-0.71)}{1{,}000}}$$

$$= 0.01 \pm 0.03$$

$$= (-0.02, 0.04)$$

Interpretation: You're 95 percent confident that the difference in the percentages of students who drive used cars among traditional students versus nontraditional students is **between −0.02 and 0.04**. This result means you can't say one group drives used cars more often than the other group.

b. Because zero is included in this interval, the difference is "too close to call," so **you conclude no significant difference between the two populations**.

(26) **No.** Because the total population is 100, and 10 percent are Independent, the class produces only 10 Independents. Because Bob samples half of the 10, his sample size is only 5 Independents. Because $n\hat{p} = (5)(0.10) = 0.5$ isn't at least 5 for the Independent group, the sample size condition for this confidence interval formula isn't met. Bob should take a larger sample of Independents for his study.

If you're ready to test your skills a bit more, take the following chapter quiz that incorporates all the chapter topics.

Whaddya Know? Chapter 14 Quiz

Quiz time! Complete each problem to test your knowledge on the various topics covered in this chapter. You can then find the solutions and explanations in the next section.

1. Anabav wants to estimate the average age of a student at his university using a random sample of 100 students. The average age in his sample is 19.3 years and the standard deviation is 2.1 years. What is the parameter in this situation?

2. You take a random sample of 30 exam scores from a very large introductory statistics course, and you find the sample mean is 82. Assume the population standard deviation for test scores is 10. Estimate the average test score for the whole class with 95% confidence. Assume a normal distribution.

3. You take a random sample of 30 exam scores from a very large introductory statistics course, and you find the sample mean is 82. Assume the population standard deviation for test scores is 10. Estimate the average test score for the whole class with 98% confidence. Assume a normal distribution.

4. You live in a rural county and want to estimate the proportion of people who live on a farm. You take a random sample of 49 people and you find that 65% live on a farm. You want 99% confidence. What is your best estimate?

5. Sydney wants to estimate the price of gas in Ohio with 95% confidence. She wants her margin of error to be $0.05. Assume the standard deviation of gas prices in Ohio is $0.25. How many gas stations should be sampled?

6. An 80% confidence interval for the price of a cart of groceries at the local grocery store is ($50.25, $55.75). What is the sample mean price, what is the width of the interval, and what is the margin of error?

7. Anabav wants to estimate the average age of a student at his university using a random sample of 10 students. The average age in his sample is 19.3 years and the sample standard deviation is 2.1.

8. To compare the average cost of a cart of groceries from this year versus last year, you took a random sample of 50 carts of groceries last year and found the average was $45.75 and you took a random sample of 50 carts of groceries this year and found the average was $53. Assume the population standard deviations for last year and this year are both $10.00. Find a 95% confidence interval for the difference in average grocery cart prices (this year minus last year).

9. To compare the average cost of a cart of groceries from this year versus last year, you took a random sample of 50 carts of groceries last year and found the average was $45.75 and the sample standard deviation was $9.75. You took a random sample of 50 carts of groceries this year and found the average was $53 and the sample standard deviation was $10.25. Find a 95% confidence interval for the difference in average grocery cart prices (this year minus last year).

10. You want to compare the proportion of women who shop at one home improvement store (DIY Central) compared to another home improvement store (DIY Max). You take a random sample of 100 people from each store and find the proportion of women who shop at each store is 55% and 45%, respectively. Find a 99% confidence interval for the difference in the proportion of women at the two stores.

11. Does margin of error take bias into account?

Answers to Chapter 14 Quiz

1. **The parameter is the average age of a student at his university**. It's an unknown value; that's why he's doing a confidence interval to estimate it.

2. Because you know the population standard deviation, σ, you use the formula $\bar{x} \pm z^* \frac{\sigma}{\sqrt{n}} \rightarrow 82 \pm 1.96 \frac{10}{\sqrt{30}} \rightarrow 82 \pm 3.58 = \mathbf{(78.42,\ 85.58)}$. You are 95% confident that the average test score for all the students in this class falls in this range.

3. Again you use the same formula as in Problem 2, except you want 98% confidence instead of 95% confidence, so you use $z^* = 2.33$ instead of 1.96.

   You have $\bar{x} \pm z^* \frac{\sigma}{\sqrt{n}} \rightarrow 82 \pm 2.33 \frac{10}{\sqrt{30}} \rightarrow 82 \pm 4.25 = \mathbf{(77.75, 86.25)}$. You are 98% confident that the average test score for all the students in this class falls in this range.

4. This is a confidence interval about a population proportion, so you use $\hat{p} \pm z^* \sqrt{\frac{\hat{p}(1-\hat{p})}{n}}$, provided the conditions are met to use z: $n\hat{p} = 49 * 0.65 = 31.85 \geq 10$ and $n(1-\hat{p}) = 49 * (1-0.65) = 17.15 \geq 10$. Yes, the conditions are met. That gives you $0.65 \pm 2.58 \sqrt{\frac{0.65(1-0.65)}{49}} \rightarrow 0.65 \pm 0.18 = (\mathbf{0.47},\ \mathbf{0.83})$. You are 99% confident that the total percentage of people in this county that live on a farm is in this range. Note that two things make this interval quite wide: first, the confidence level is high, and second, the sample size isn't that large.

5. Sydney needs to find the sample size required for a margin of error of $0.05 with $z^* = 1.96$ (for 95% confidence) and $\sigma = \$0.25$. The formula is $n = \left(\frac{z^* \sigma}{\text{MOE}}\right)^2 = \left(\frac{1.96(0.25)}{(0.05)}\right)^2 = 9.8^2 = 96.04$. You always round up to meet the margin of error requirement, so **she needs to sample 97 gas stations.**

6. The interval is ($50.25, $55.75). The sample mean must be at the center of the interval, because it's what you build the interval around, so find the midpoint: $(55.75 + 50.25)/2 = \$53$. The width is half of the length of the interval, which is $(55.75 - 50.25)/2 = \$2.75$, and the margin of error is the same as the width, so it's also $2.75.

7. Anabav needs a 90% confidence interval for the population mean, μ, and he doesn't have the population standard deviation, σ, so he uses the sample standard deviation, s, instead. Because of this, he uses a t^*-value instead of a z^*-value to make the interval. The interval is $\bar{x} \pm t^* \frac{s}{\sqrt{n}} \rightarrow 19.3 \pm 1.833 \frac{2.1}{\sqrt{10}} = 19.3 \pm 1.22 = (18.08,\ 20.52)$ years. He is 90% confident that the average age of a student in his university is in this range.

8. $\bar{x}_1 - \bar{x}_2 \pm z^* \sqrt{\frac{\sigma_1^2}{n_1} + \frac{\sigma_2^2}{n_2}} = (53 - 45.75) \pm 1.96 \sqrt{\frac{100}{50} + \frac{100}{50}} = 7.25 \pm 1.96\sqrt{4} = 7.25 \pm 3.92 = \mathbf{(\$3.33,\ \$11.17)}$.

   You are 95% confident that the difference in average grocery cart prices (this year minus last year) is in this range.

9. $$(\bar{x}-\bar{y}) \pm t^*_{n_1+n_2-2}\sqrt{\frac{(n_1-1)s_1^2+(n_2-1)s_2^2}{n_1+n_2-2}}\sqrt{\frac{1}{n_1}+\frac{1}{n_2}}$$

$$=(53-45.75) \pm t^*_{98}\sqrt{\frac{(50-1)10.25^2+(50-1)9.75^2}{50+50-2}}\sqrt{\frac{1}{50}+\frac{1}{50}}$$

$$=7.25 \pm 1.96(10.00)(0.20)$$

$$= \mathbf{\$7.25 - \$3.92}$$

You are 95% confident that the difference in average grocery cart prices (this year minus last year) is in this range. (Note the answers to Questions 8 and 9 are the same; this is because the sample sizes are high.)

10. $$(\hat{p}_1-\hat{p}_2) \pm z^*\sqrt{\frac{\hat{p}_1(1-\hat{p}_1)}{n_1}+\frac{\hat{p}_2(1-\hat{p}_2)}{n_2}}$$

$$=(0.55-0.45) \pm 2.58\sqrt{\frac{0.55(1-0.55)}{100}+\frac{0.45(1-0.45)}{100}}$$

$$=0.10 \pm 2.58(0.07)$$

$$=0.10 \pm 0.18$$

$$= \mathbf{(-0.08, 0.28)}$$

You are 99% confident that the difference in the percentage of women at these stores (DIY Central minus DIY Max) is within this range.

11. **No, it does not.** That means you need to avoid bias if you are doing your own confidence interval and/or check for bias in someone else's data before you believe their confidence interval.

IN THIS CHAPTER

» Testing other people's claims

» Using hypothesis tests to weigh evidence and make decisions

» Recognizing that your conclusions could be wrong

Chapter 15
Claims, Tests, and Conclusions

You hear claims involving statistics all the time; the media has no shortage of them:

- Twenty-five percent of all women in the United States have varicose veins. (Wow, are some claims better left unsaid, or what?)
- Cigarette use in the U.S. continues to drop, with the percentage of all American smokers decreasing by about 2 percent per year over the last ten years.
- A 6-month-old baby sleeps an average of 14 to 15 hours in a 24-hour period. (Yeah, right!)
- A name-brand, ready-mix pie takes only 5 minutes to make.

In today's age of information (and big money), a great deal rides on being able to back up your claims. Companies that say their products are better than the leading brand had better be able to prove it, or they could face lawsuits. Drugs that are approved by the FDA have to show strong evidence that their products actually work without producing life-threatening side effects. Manufacturers have to make sure their products are being produced according to specifications to avoid recalls, customer complaints, and loss of business.

Although many claims are backed up by solid scientific (and statistically sound) research, others are not. In this chapter, you find out how to use statistics to investigate whether a claim is actually valid and get the lowdown on the process that researchers *should* be using to validate claims that they make.

REMEMBER

A *hypothesis test* is a statistical procedure that's designed to test a claim. Before diving into details, I want to give you the big picture of a hypothesis test by showing the main steps involved. These steps are discussed in the following sections:

1. **Set up the null and alternative hypotheses.**
2. **Collect good data using a well-designed study (see Chapters 3 and 18).**
3. **Calculate the test statistic based on your data.**
4. **Find the *p*-value for your test statistic.**
5. **Decide whether or not to reject H_0 based on your *p*-value.**
6. **Understand that your conclusion may be wrong, just by chance.**

Setting Up the Hypotheses

Typically in a hypothesis test, the claim being made is about a population *parameter* (one number that characterizes the entire population). Because parameters tend to be unknown quantities, everyone wants to make claims about what their values may be. For example, the claim that 25 percent (or 0.25) of all women have varicose veins is a claim about the proportion (that's the *parameter*) of all women (that's the *population*) who have varicose veins (that's the *variable* — having or not having varicose veins).

Researchers often challenge claims about population parameters. You may hypothesize, for example, that the actual proportion of women who have varicose veins is lower than 0.25, based on your observations. Or you may hypothesize that due to the popularity of high-heeled shoes, the proportion may be higher than 0.25. Or if you're simply questioning whether the actual proportion is 0.25, your alternative hypothesis is: "No, it isn't 0.25."

Defining the null

Every hypothesis test contains a set of two opposing statements, or hypotheses, about a population parameter. The first hypothesis is called the *null hypothesis*, denoted H_0. The null hypothesis always states that the population parameter is *equal* to the claimed value. For example, if the claim is that the average time to make a name-brand, ready-mix pie is five minutes, the statistical shorthand notation for the null hypothesis in this case would be as follows: H_0: $\mu = 5$. (That is, the population mean is 5 minutes.)

REMEMBER

All null hypotheses include an equal sign in them; there are no $\leq$ or $\geq$ signs in H_0. Not to cop out or anything, but the reason it's always equal is beyond the scope of this book; let's just say you wouldn't pay me to explain it to you.

What's the alternative?

Before actually conducting a hypothesis test, you have to put two possible hypotheses on the table — the null hypothesis is one of them. But, if the null hypothesis is rejected (that is, there is sufficient evidence against it), what's your alternative going to be? Actually, three possibilities exist for the second (or alternative) hypothesis, denoted H_a. Here they are, along with their shorthand notations in the context of the pie example:

- The population parameter is *not equal* to the claimed value (H_a: $\mu \neq 5$).
- The population parameter is *greater than* the claimed value (H_a: $\mu > 5$).
- The population parameter is *less than* the claimed value (H_a: $\mu < 5$).

Which alternative hypothesis you choose in setting up your hypothesis test depends on what you're interested in concluding, should you have enough evidence to refute the null hypothesis (the claim).

For example, if you want to test whether a company is correct in claiming its pie takes five minutes to make and it doesn't matter whether the actual average time is more or less than that, you use the not-equal-to alternative. Your hypotheses for that test would be H_0: $\mu = 5$ versus H_a: $\mu \neq 5$.

If you only want to see whether the time turns out to be greater than what the company claims (that is, whether the company is falsely advertising its quick prep time), you use the greater-than alternative, and your two hypotheses are H_0: $\mu = 5$ versus H_a: $\mu > 5$.

Finally, say you work for the company marketing the pie, and you think the pie can be made in less than five minutes (and could be marketed by the company as such). The less-than alternative is the one you want, and your two hypotheses would be H_0: $\mu = 5$ versus H_a: $\mu < 5$.

TIP

How do you know which hypothesis to put in H_0 and which one to put in H_a? Typically, the null hypothesis says that nothing new is happening; the previous result is the same now as it was before, or the groups have the same average (their difference is equal to zero). In general, you assume that people's claims are true until proven otherwise. So the question becomes: Can you prove otherwise? In other words, can you show sufficient evidence to reject H_0?

Gathering Good Evidence (Data)

After you've set up the hypotheses, the next step is to collect your evidence and determine whether your evidence goes against the claim made in H_0. Remember, the claim is made about the population, but you can't test the whole population; the best you can usually do is take a sample. As with any other situation in which statistics are being collected, the quality of the data is extremely critical. (See Chapter 2 for ways to spot statistics that have gone wrong.)

Collecting good data starts with selecting a good sample. Two important issues to consider when selecting your sample are avoiding bias and being accurate. To avoid bias when selecting a sample, make it a random sample (one that's got the same chance of being selected as every other possible sample of the same size) and choose a large enough sample size so that the results will be accurate. (See Chapter 12 for more information on accuracy.)

Data is collected in many different ways, but the methods used basically boil down to three: surveys, other observational studies, and experiments (controlled studies). Chapter 17 gives all the information you need to design and critique surveys and observational studies, as well as information on selecting samples properly. In Chapter 18, you examine experiments: what they can do beyond an observational study, the criteria for a good experiment, and when you can conclude cause and effect.

Compiling the Evidence: The Test Statistic

After you select your sample, the appropriate number-crunching takes place. Your null hypothesis (H_0) makes a statement about the population parameter — for example, "The proportion of all women who have varicose veins is 0.25" (in other words, $H_0: p = 0.25$); or the average miles per gallon of a U.S.-built light truck is 27 ($H_0: \mu = 27$). The data you collect from the sample measures the variable of interest, and the statistics that you calculate will help you test the claim about the population parameter.

Gathering sample statistics

Say you're testing a claim about the proportion of women with varicose veins. You need to calculate the proportion of women in your sample who have varicose veins, and that number will be your sample statistic. If you're testing a claim about the average miles per gallon of a U.S.-built light truck, your statistic will be the average miles per gallon of the light trucks in your sample. And knowing you want to measure the variability in average miles per gallon for various trucks, you'll want to calculate the sample standard deviation. (See Chapter 5 for all the information you need on calculating sample statistics.)

Measuring variability using standard errors

After you've calculated all the necessary sample statistics, you may think you're done with the analysis part and ready to make your conclusions — but you're not. The problem is you have no way to put your results into any kind of perspective just by looking at them in their regular units. That's because you know that your results are based only on a sample and that sample results are going to vary. That variation needs to be taken into account, or your conclusions could be completely wrong. (How much do sample results vary? Sample variation is measured by the standard error; see Chapter 12 for more on this.)

Suppose the claim is that the percentage of all women with varicose veins is 25 percent, and your sample of 100 women had 20 percent with varicose veins. The standard error for your sample percentage is 4 percent (according to formulas in Chapter 12), which means that your results are expected to vary by about twice that, or about 8 percent, according to the Empirical Rule (see Chapter 5). So a difference of 5 percent, for example, between the claim and your sample result $(25\% - 20\% = 5\%)$ isn't that much, in these terms, because it represents a distance of less than 2 standard errors away from the claim.

However, suppose your sample percentage was based on a sample of 1,000 women, not 100. This decreases the amount by which you expect your results to vary, because you have more information. Again using formulas from Chapter 12, I calculate the standard error to be 0.013 or 1.3 percent. The margin of error (MOE) is about twice that, or 2.6 percent on either side. Now a difference of 5 percent between your sample result (20 percent) and the claim in H_0 (25 percent) is a more meaningful difference; it's way more than 2 standard errors.

Exactly how meaningful are your results? In the next section, you get more specific about measuring exactly how far apart your sample results are from the claim in terms of the number of standard errors. This leads you to a specific conclusion as to how much evidence you have against the claim in H_0.

Understanding standard scores

The number of standard errors that a statistic lies above or below the mean is called a *standard score* (for example, a z-value is a type of standard score; see Chapter 10). In order to interpret your statistic, you need to convert it from original units to a standard score. When finding a standard score, you take your statistic, subtract the mean, and divide the result by the standard error.

In the case of hypothesis tests, you use the value in H_0 as the mean. (That's what you go with unless/until you have enough evidence against it.) The standardized version of your statistic is called a *test statistic*, and it's the main component of a hypothesis test. (Chapter 16 contains the formulas for the most common hypothesis tests.)

Calculating and interpreting the test statistic

The general procedure for converting a statistic to a test statistic (standard score) is as follows:

1. **Take your statistic minus the claimed value (the number stated in H_0).**
2. **Divide by the standard error of the statistic.**

   (Different formulas for standard error exist for different problems; see Chapter 14 for detailed formulas for standard error and Chapter 16 for formulas for various test statistics.)

Your test statistic represents the distance between your actual sample results and the claimed population value, in terms of number of standard errors. In the case of a single population mean or proportion, you know that these standardized distances should at least have an approximate standard normal distribution if your sample size is large enough (see Chapter 12). So, to interpret your test statistic in these cases, you can see where it stands on the standard normal distribution (Z-distribution).

Using the numbers from the varicose veins example in the previous section, the test statistic is found by taking the proportion in the sample with varicose veins, 0.20, subtracting the claimed proportion of all women with varicose veins, 0.25, and then dividing the result by the standard error, 0.04. These calculations give you a test statistic (standard score) of $-0.05 \div 0.04 = -1.25$. This tells you that your sample results and the population claim in H_0 are 1.25 standard errors apart; in particular, your sample results are 1.25 standard errors below the claim. Now is this enough evidence to reject the claim? The next section addresses that issue.

Weighing the Evidence and Making Decisions: *p*-Values

After you find your test statistic, you use it to make a decision about whether to reject H_0. You make this decision by coming up with a number that measures the strength of this evidence (your test statistic) against the claim in H_0. That is, how likely is it that your test statistic could have occurred while the claim was still true? This number you calculate is called the *p-value;* it's the chance that someone could have gotten results as extreme as yours while H_0 was still true. Similarly in a jury trial, the jury discusses how likely it is that all the evidence came out the way it did assuming the defendant was innocent.

This section shows all the ins and outs of *p*-values, including how to calculate them and use them to make decisions regarding H_0.

Connecting test statistics and *p*-values

To test whether a claim in H_0 should be rejected (after all, it's all about H_0), you look at your test statistic taken from your sample and see whether you have enough evidence to reject the claim. If the test statistic is large (in either the positive or negative directions), your data is far from the claim; the larger the test statistic, the more evidence you have against the claim. You determine "how far is far" by looking at where your test statistic ends up on the distribution that it came from. When testing one population mean, under certain conditions the distribution of comparison is the standard normal (Z-) distribution, which has a mean of 0 and a standard deviation of 1; I use it throughout this section as an example. (See Chapter 10 to find out more about the Z-distribution.)

REMEMBER

If your test statistic is close to 0, or at least within that range where most of the results should fall, then you don't have much evidence against the claim (H_0) based on your data. If your test statistic is out in the tails of the standard normal distribution (see Chapter 10 for more on tails), then your evidence against the claim (H_0) is great; this result has a very small chance of happening if the claim is true. In other words, you have sufficient evidence against the claim (H_0), and you reject H_0.

But how far is "too far" from 0? As long as you have a normal distribution or a large enough sample size, you know that your test statistic falls somewhere on a standard normal distribution (see Chapter 12). If the null hypothesis (H_0) is true, most (about 95 percent) of the samples will result in test statistics that lie roughly within 2 standard errors of the claim. If H_a is the not-equal-to alternative, any test statistic outside this range will result in H_0 being rejected. See Figure 15-1 for a picture showing the locations of your test statistic and their corresponding conclusions. In the next section, you see how to quantify the amount of evidence you have against H_0.

TIP

Note that if the alternative hypothesis is the less-than alternative, you reject H_0 only if the test statistic falls in the left tail of the distribution (below −1.64). Similarly, if H_a is the greater-than alternative, you reject H_0 only if the test statistic falls in the right tail (above 1.64).

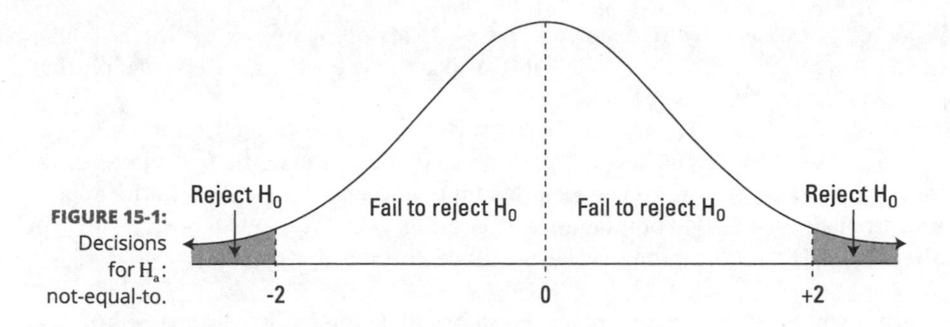

FIGURE 15-1: Decisions for H_a: not-equal-to.

Defining a *p*-value

REMEMBER

A *p*-value is a probability associated with your test statistic. It measures the chance of getting results at least as strong as yours if the claim (H_0) were true. In the case of testing the population mean, the farther out your test statistic is on the tails of the standard normal (*Z*-) distribution, the smaller your *p*-value will be, the less likely your results were to have occurred, and the more evidence you have against the claim (H_0).

Calculating a *p*-value

To find the *p*-value for your test statistic:

1. **Look up your test statistic on the appropriate distribution — in this case, on the standard normal (*Z*-) distribution (see the *Z*-table in the Appendix).**
2. **Find the chance that *Z* is beyond (more extreme than) your test statistic:**
   - If H_a contains a less-than alternative, find the probability that *Z* is less than your test statistic (that is, look up your test statistic on the *Z*-table and find its corresponding probability). This is the *p*-value.
   - If H_a contains a greater-than alternative, find the probability that *Z* is greater than your test statistic (look up your test statistic on the *Z*-table, find its corresponding probability, and subtract it from one). The result is your *p*-value.
   - If H_a contains a non-equal-to alternative, find the probability that *Z* is beyond your test statistic and double it. There are two cases:
     - If your test statistic is negative, first find the probability that *Z* is less than your test statistic (look up your test statistic on the *Z*-table and find its corresponding probability). Then double this probability to get the *p*-value.
     - If your test statistic is positive, first find the probability that *Z* is greater than your test statistic (look up your test statistic on the *Z*-table, find its corresponding probability, and subtract it from one). Then double this result to get the *p*-value.

REMEMBER

Why do you double the probabilities if your H_a contains a not-equal-to alternative? Think of the not-equal-to alternative as the combination of the greater-than alternative and the less-than alternative. If you've got a positive test statistic, its *p*-value only accounts for the greater-than portion of the not-equal-to alternative; double it to account for the less-than portion. (The doubling of one *p*-value is possible because the *Z*-distribution is symmetric.)

Similarly, if you've got a negative test statistic, its p-value only accounts for the less-than portion of the not-equal-to alternative; double it to also account for the greater-than portion.

When testing H_0: $p = 0.25$ versus H_0: $p = 0.25$ in the varicose veins example from the previous section, the p-value turns out to be 0.1056. This is because the test statistic (calculated in the previous section) was −1.25, and when you look this number up on the Z-table (in the Appendix) you find a probability of 0.1056 of it being less than this value. If you had been testing the two-sided alternative, H_a: $p \neq 0.25$, the p-value would be 2×0.1056, or 0.2112.

REMEMBER

If the results are likely to have occurred under the claim, then you fail to reject H_0 (like a jury decides on a verdict of 'not guilty' when there's not enough evidence to convict). If the results are unlikely to have occurred under the claim, then you reject H_0 (like a jury decides on a verdict of 'guilty' when there's lots of evidence). The cutoff point between rejecting H_0 and failing to reject H_0 is another whole can of worms that I dissect in the next section (no pun intended).

EXAMPLE

Q. Suppose you want to test H_0: $\mu = 0$ versus H_a: $\mu > 0$, and your test statistic is $z = 1.96$ (based on a large random sample). What's your p-value?

A. Because the alternative hypothesis is ">", you need to find the area above 1.96. On the Z-table, you find that $p(Z \leq 1.96) = 0.975$, which means the area beyond (in this case, above) 1.96 is $1 - 0.975 = 0.025$. This is the p-value. (When you work with p-values, use them in their decimal form, not as percents.)

YOUR TURN

1 Suppose you want to test H_0: $\mu = 0$ versus H_a: $\mu < 0$, and your test statistic is $z = -2.5$. (Assume a large random sample.) What's your p-value?

Suppose you want to test H_0: $p = \frac{1}{2}$ versus H_a: $p \neq \frac{1}{2}$, and your test statistic is $z = 0.5$. (Assume a large random sample.) What's your p-value?

Suppose you want to test H_0: $p = \frac{1}{2}$ versus H_a: $p \neq \frac{1}{2}$, and your test statistic is $z = -1.2$. (Assume a large random sample.) What's your p-value?

Suppose you want to test H_0: $\mu = 0$ versus H_a: $\mu = 0$, and your test statistic (based on a random sample of 10) is $t = 1.96$. What's your p-value?

Suppose you want to test H_0: $\mu = 0$ versus H_a: $\mu = 0$, and your test statistic (based on a random sample of 10) is $t = -2.5$. What's your p-value?

Suppose you want to test H_0: $\mu = 0$ versus H_a: $\mu = 0$, and your p-value is 0.0446. (Assume a large random sample.) What's the test statistic?

Making Conclusions

To draw conclusions about H_0 (reject or fail to reject) based on a p-value, you need to set a predetermined cutoff point where only those p-values less than or equal to the cutoff will result in rejecting H_0. This cutoff point is called the *alpha level* (α), or *significance level* for the test. While 0.05 is a very popular cutoff value for rejecting H_0, cutoff points and resulting decisions can vary — some people use stricter cutoffs, such as 0.01, requiring more evidence before rejecting H_0, and others may have less strict cutoffs, such as 0.10, requiring less evidence.

If H_0 is rejected (that is, the p-value is less than or equal to the predetermined significance level), the researcher can say they've found a statistically significant result. A result is *statistically significant* if it's chance of occurring is less than a set-aside cutoff value, assuming H_0 is true. If you get a statistically significant result, you have enough evidence to reject the claim, H_0, and conclude that something different or new is in effect (that is, H_a).

REMEMBER

The significance level can be thought of as the highest possible p-value that would reject H_0 and declare the results statistically significant. Following are the general rules for making a decision about H_0 based on a p-value:

- If the p-value is less than or equal to your significance level, then it meets your requirements for having enough evidence against H_0; you reject H_0.
- If the p-value is greater than your significance level, your data failed to show evidence beyond a reasonable doubt; you fail to reject H_0.

However, if you plan to make decisions about H_0 by comparing the p-value to your significance level, you must decide on your significance level ahead of time. It wouldn't be fair to change your cutoff point after you have a sneak peak at what's happening in the data.

WARNING

You may be wondering whether it's okay to say "Accept H_0" instead of "Fail to reject H_0." The answer is a big "no." In a hypothesis test, you are *not* trying to show whether or not H_0 is true (which *accept* implies) — indeed, if you knew whether H_0 was true, you wouldn't be doing the hypothesis test in the first place. You're trying to show whether you have enough evidence to say H_0 is false, based on your data. Either you have enough evidence to say it's false (in which case you reject H_0) or you don't have enough evidence to say it's false (in which case you fail to reject H_0). Also make sure your results are discernible, which means they are large enough to be important, practically speaking.

Setting boundaries for rejecting H_0

These guidelines help you make a decision (reject or fail to reject H_0) based on a *p*-value when your significance level is 0.05:

- If the *p*-value is less than 0.01 (very small), the results are considered highly statistically significant — reject H_0.
- If the *p*-value is between 0.05 and 0.01 (but not super-close to 0.05), the results are considered statistically significant — reject H_0.
- If the *p*-value is really close to 0.05 (like 0.051 or 0.049), the results should be considered marginally significant — the decision could go either way.
- If the *p*-value is greater than (but not super-close to) 0.05, the results are considered non-significant — you fail to reject H_0.

Depending on the field of study, researchers may choose a stricter significance level (like 0.01) or a less strict significance level (like 0.10). Be sure to adjust the preceding ranges based on 0.05 so that your conclusions are reasonable with other significance levels that you may encounter.

WARNING

When you hear a researcher say their results are found to be statistically significant, look for the *p*-value and make your own decision; the researcher's predetermined significance level may be different from yours. If the *p*-value isn't stated, ask for it.

Testing varicose veins

In the varicose veins example in the previous section, the *p*-value was found to be 0.1056. This *p*-value is fairly large and indicates very weak evidence against H_0 by almost anyone's standards because it's greater than 0.05 and even slightly greater than 0.10 (considered to be a very large significance level). In this case, you fail to reject H_0. You don't have enough evidence to say the proportion of women with varicose veins is less than 0.25 (your alternative hypothesis). This isn't declared to be a statistically significant result.

But say your *p*-value is something like 0.026. A reader with a personal cutoff point of 0.05 would reject H_0 in this case because the *p*-value (of 0.026) is less than 0.05. Their conclusion would be that the proportion of women with varicose veins isn't equal to 0.25; according to H_a in this case, you would conclude it's less than 0.25, and the results would be statistically significant. However, a reader whose significance level is 0.01 wouldn't have enough evidence (based on your sample) to reject H_0 because the *p*-value of 0.026 is greater than 0.01. These results wouldn't be statistically significant.

Finally, if the p-value turns out to be 0.049 and your significance level is 0.05, you can go by the book and say because it's less than 0.05 you reject H_0, but you really should say your results are marginal, and let the readers decide. (Maybe they can flip a coin or something — "Heads, we reject H_0; tails, we don't!")

EXAMPLE

Q. Suppose you hear a researcher say they found a "statistically significant result," and in parentheses you see ($p = 0.001$). Explain what "statistically significant result" must mean.

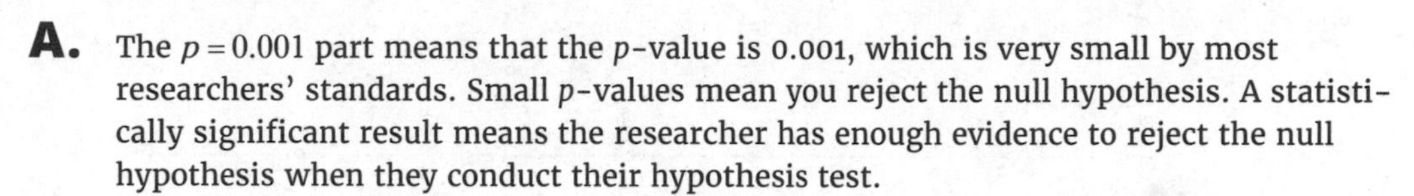

A. The $p = 0.001$ part means that the p-value is 0.001, which is very small by most researchers' standards. Small p-values mean you reject the null hypothesis. A statistically significant result means the researcher has enough evidence to reject the null hypothesis when they conduct their hypothesis test.

Q. Suppose you're doing a statistical study and you find a p-value of 0.026 when you test H_0: $p = 0.25$ versus H_a: $p < 0.25$. You report this result to your fellow researchers.

(a) What would a fellow researcher with $\alpha = 0.05$ conclude?

(b) What would a fellow researcher with $\alpha = 0.01$ conclude?

A. Always compare the p-value to the significance level chosen beforehand, using the rules mentioned earlier.

(a) Because the p-value 0.026 is less than 0.05, this researcher has enough evidence to reject H_0.

(b) Because the p-value 0.026 is greater than 0.01, this researcher doesn't have enough evidence to reject H_0.

YOUR TURN

7 Bob and Teresa each collect their own samples to test the same hypothesis. Bob's p-value turns out to be 0.05, and Teresa's turns out to be 0.01.

(a) Why don't Bob and Teresa get the same p-values?

(b) Who has stronger evidence against the null hypothesis, Bob or Teresa?

8 Does a small p-value mean that the null hypothesis isn't true? Explain.

9. Suppose you want to test H_0: $\mu = 0$ versus H_a: $\mu < 0$, and your *p*-value turns out to be 0.06. What do you conclude? Use $\alpha = 0.05$.

10. Suppose you want to test H_0: $\mu = 0$ versus H_a: $\mu < 0$, and your *p*-value turns out to be 0.60. What do you conclude? Use $\alpha = 0.05$.

11. Suppose you want to test H_0: $\mu = 0$ versus H_a: $\mu < 0$, and your *p*-value turns out to be 0.05. What do you conclude? Use $\alpha = 0.05$.

12. Suppose you want to test H_0: $p = \frac{1}{2}$ versus H_a: $p \neq \frac{1}{2}$, and your *p*-value turns out to be 0.95. What do you conclude? Use $\alpha = 0.05$.

Assessing the Chance of a Wrong Decision

After you make a decision to either reject H_0 or fail to reject H_0, the next step is living with the consequences, in terms of how people respond to your decision.

- If you conclude that a claim isn't true but it actually *is*, will that result in a lawsuit, a fine, unnecessary changes in the product, or consumer boycotts that shouldn't have happened? It's possible.
- If you can't disprove a claim that's wrong, what happens then? Will products continue to be made in the same way as they are now? Will no new law be made, no new action taken, because you showed that nothing was wrong? Missed opportunities to blow the whistle have been known to occur.

REMEMBER

Whatever decision you make with a hypothesis test, you know there is a chance of being wrong; that's life in the statistics world. Knowing the kinds of errors that can happen and finding out how to curb the chance of them occurring are key.

Making a false alarm: Type I errors

Suppose a company claims that its average package delivery time is 2 days, and a consumer group tests this hypothesis, gets a *p*-value of 0.04, and concludes that the claim is false: They believe that the average delivery time is actually more than 2 days. This is a big deal. If the group can stand by its statistics, it has done well to inform the public about the false advertising issue. But what if the group is wrong?

Even if the group bases its study on a good design, collects good data, and makes the right analysis, it can still be wrong. Why? Because its conclusions were based on a sample of packages, not on the entire population. And as Chapter 12 tells you, sample results vary from sample to sample.

Just because the results from a sample are unusual doesn't mean they're impossible. A *p*-value of 0.04 means that the chance of getting your particular test statistic, even if the claim is true, is 4 percent (less than 5 percent). You reject H_0 in this case because that chance is small. But even a small chance is still a chance!

Perhaps your sample, though collected randomly, just happens to be one of those atypical samples whose result ended up far from what was expected. So, H_0 could be true, but your results lead you to a different conclusion. How often does that happen? Five percent of the time (or whatever your given cutoff probability is for rejecting H_0).

Rejecting H_0 when you shouldn't is called a *Type I error* (read as "type one"). I don't really like this name, because it seems so nondescript. I prefer to call a Type I error a *false alarm.* In the case of the packages, if the consumer group made a Type I error when it rejected the company's claim, they created a false alarm. What's the result? A very angry delivery company, I guarantee that!

To reduce the chance of false alarms, set a low cutoff probability (significance level) for rejecting H_0. Setting it to 5 percent or 1 percent will keep the chance of a Type I error in check.

Missing out on a detection: Type II errors

On the other hand, suppose the company really wasn't delivering on its claim. Who's to say that the consumer group's sample will detect it? If the actual delivery time were 2.1 days instead of 2 days, the difference would be pretty hard to detect. If the actual delivery time were 3 days, even a fairly small sample would probably show that something's up. The issue lies with those in-between values, like 2.5 days.

If H_0 is indeed false, you want to find out about it and reject H_0. Not rejecting H_0 when you should have is called a *Type II error* (read as "type two"). I like to call it a *missed detection.*

By trying to make a Type II error smaller, you may make the Type I one bigger, and vice versa. Think of a jury trial situation. If you want to make sure criminals don't get away with crimes, you make it easier to convict (rejecting H_0) and increase the risk of innocent people going to jail. If you want to make sure no innocent people go to jail (failing to reject H_0), you make it harder to convict and increase the risk of more criminals getting away with their crimes. Bigger samples generally reduce Type II errors because more data helps you do a better job of detecting even small deviations from the null hypothesis. (Bigger samples and more data are like having lots of witnesses and evidence at a trial.) And small alpha levels generally reduce Type I errors because they reduce the chance of you rejecting H_0 in the first place.

Sample size is the key to being able to detect situations where H_0 is false and, thus, avoiding Type II errors. The more information you have, the less variable your results will be (see Chapter 12) and the more ability you have to zoom in on detecting problems that exist with a claim made by H_0.

This ability to detect when H_0 is truly false is called the *power* of a test. Power is a pretty complicated issue, but what's important for you to know is that the higher the sample size, the more powerful a test is. A powerful test has a small chance for a Type II error.

TIP

As a preventative measure to minimize the chances of a Type II error, statisticians recommend that you select a large sample size to ensure that any differences or departures that really exist won't be missed.

EXAMPLE

Q. Suppose you want to test H_0: $\mu = 0$ versus H_a: $\mu < 0$, and your *p*-value turns out to be 0.006. You reject H_0. (Your alpha level was 0.05.) Explain what a Type I error would mean in this situation if such an error is possible.

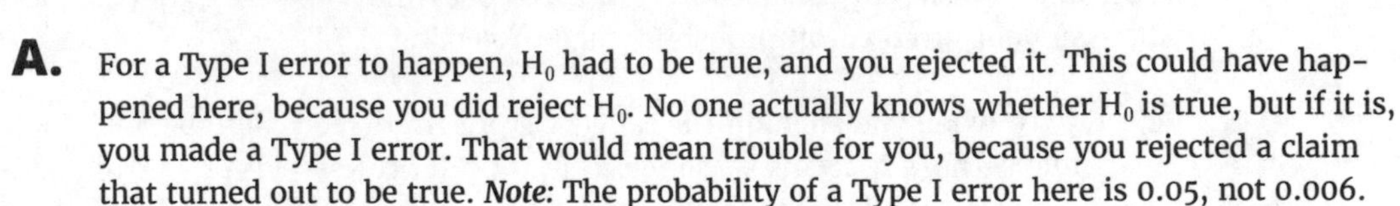

A. For a Type I error to happen, H_0 had to be true, and you rejected it. This could have happened here, because you did reject H_0. No one actually knows whether H_0 is true, but if it is, you made a Type I error. That would mean trouble for you, because you rejected a claim that turned out to be true. ***Note:*** The probability of a Type I error here is 0.05, not 0.006.

Q. In the jury trial example earlier in this section, which error in the decision made by the jury is the Type I error and which is the Type II error?

A. A Type I error means a false alarm; you rejected H_0 when you didn't need to. That's the case where you sent an innocent person to jail. The Type II error is the missed detection. You should have rejected H_0, but you didn't. That means you let a guilty person go free.

YOUR TURN

13 Suppose you want to test H_0: $\mu = 0$ versus H_a: $\mu < 0$, and your *p*-value turns out to be 0.6. You fail to reject H_0. Explain why you couldn't have made a Type I error here.

14 Suppose a gubernatorial candidate claims that they would get 60 percent of the vote if the state held the election today. You test their hypothesis, believing that the actual percentage is less than 60 percent. Suppose you make a Type I error.

(a) Does that mean you rejected H_0 or not?

(b) Is H_0 true or not?

(c) Describe the impact of making a Type I error in this situation.

15 Suppose you test to see if cereal boxes are under-filled (H_0: $\mu = 18$ ounces versus H_a: $\mu < 18$ ounces). Describe what Type I and Type II errors would mean in this situation.

16 Can you make a Type I error and a Type II error at the same time?

17 Suppose your friend claims that a coin is fair, but you don't think so. You decide to test it. Your *p*-value is 0.045 (at alpha level 0.05). You reject H_0: $p = \frac{1}{2}$. Suppose the coin really is fair.

(a) Which type of error did you make?

(b) Describe the impact of your error.

18 Suppose you want to test whether or not a machine makes its widgets according to specifications by taking a sample. H_0 says the machine is fine, and the alpha is 0.05. Your *p*-value is 0.3, so you decide the machine works to specification.

(a) Suppose the machine really doesn't work properly. Which type of error did you make?

(b) What's the impact of your error?

Practice Questions Answers and Explanations

1. Look at the Z-table to find $p(Z \le -2.5) = 0.0062$, which means the area beyond (in this case, below) −2.5 is 0.0062. (You don't subtract from one because the probability already gives you the area below, and when the test statistic is negative, that's what you want.) **The p-value is 0.0062.**

TIP

All p-values are probabilities and must be between 0 and 1. If you find a p-value that's greater than 1 or less than 0, you know that you've made a mistake.

2. Look at the Z-table to find $p(Z \le 0.5) = 0.6915$, which means the area beyond (in this case, above, because 0.5 is positive) 0.5 is $1 - 0.6915 = 0.3085$. Because H_a has a "not equal to" in it, don't forget to double the probability found in the Z-table (see Chapter 16). **The p-value is $2 \times 0.3085 = 0.6170$.**

3. Look at the Z-table to find $p(Z \le -1.2) = 0.1151$. (You take this probability as it is because area "beyond" a negative number means area below it.) Again, because H_a has a "not equal to" in it, double the probability found in the Z-table. **The p-value is $2 \times 0.1151 = 0.2302$.**

4. Here, $t = 1.96$ and the sample size is $n = 10$, so you have to look at a t-distribution with $10 - 1 = 9$ degrees of freedom. Use the t-table (Table A-2 in the Appendix) to look at the row corresponding to 9 degrees of freedom. Your test statistic, 1.96, falls between the columns with the headers 0.05 and 0.025. Your p-value is the area beyond (in this case, above) the test statistic, which means your p-value is between those two values. **The best you can say is that $0.025 < p\text{-value} < 0.05$.**

REMEMBER

Computer software can give you an exact p-value for any test statistic. However, if you calculate p-values by using tables, and you need to use a t-table, chances are you have to settle for a p-value falling between two numbers. But that's still enough information to make a conclusion regarding your hypotheses.

5. Here, $t = -2.5$ and the sample size is $n = 10$, so you have to look at a t-distribution with $10 - 1 = 9$ degrees of freedom. Use the t-table (Table A-2 in the Appendix) to look at the row corresponding to 9 degrees of freedom. Negative test statistics don't appear on the t-table, so you have to use symmetry to get your answer. Look for 2.5, which falls between the columns with the headers 0.025 and 0.01. Your p-value is the area beyond (in this case, below) the test statistic, which means your p-value is between those two values. **In other words, $0.01 < p\text{-value} < 0.025$.**

6. Look in the body of the Z-table to find the value closest to 0.0446. It just so happens that the probability 0.0446 corresponds to a z-value of −1.7. However, your H_a has a ">" sign, indicating that your test statistic is a positive value. **So your test statistic is actually 1.7.** **Note:** The probability of being beyond 1.7, that is, $p(Z > 1.7)$, is equal to $p(Z < -1.7)$ thanks to the symmetry of the normal (Z-) distribution.

7. Remember that because sample results vary, so do their p-values.

   a. **Bob and Teresa take different samples**, which leads to different test statistics, which gives different percentiles from the Z-table (found in Table A-1). The different percentiles lead to different p-values.

   b. **Teresa's p-value is smaller, so her evidence against H_0 is stronger.**

REMEMBER

Always think of a p-value as the strength of the evidence against the null hypothesis. The null hypothesis is on trial, and the p-value is the prosecutor.

(8) **No,** it just means that you have plenty of evidence against the null hypothesis. The null hypothesis can still be true. You never know what the truth is; the best you can do is to make a conclusion based on your sample.

(9) Because the p-value 0.06 is greater than $\alpha = 0.05$, **you don't have quite enough evidence to reject H_0. But you could say that these results are marginally significant.**

(10) Because the p-value 0.60 is greater than $\alpha = 0.05$, **you don't have nearly enough evidence to reject H_0. In other words, the results are nonsignificant**.

(11) This situation resembles a flipped coin that lands on its side; it can go either way. Because the p-value is exactly equal to 0.05, **you can say the results are marginal**. Reporting the p-value is important, so in borderline situations like this, everyone can make a personal decision. (Of course, you can play it like a true statistician and flip a coin to tell you what to do.) But in this situation, you might want to check with your instructor and see what they want you to do.

(12) **Large p-values, such as 0.95, always result in failing to reject H_0.** Make sure you don't get 0.95 (from confidence intervals; see Chapter 14) and 0.05 mixed up and make the wrong conclusion. You need small p-values to reject H_0, and 0.95 isn't a small p-value.

(13) **You can't make a Type I error in this situation, because to make a Type I error, you have to reject H_0. You didn't.** If H_0 really is true in this case, your conclusion is correct.

(14) Every time you make a decision, you can be wrong, and if you are, your mistake makes some kind of impact.

a. **Yes.** In order to make a Type I error, you have to have rejected H_0.

b. **Yes.** If you make a Type I error, that assumes H_0 is true.

c. **The impact is that you conclude they can't get the votes, when in reality they can.** If you're their campaign manager, you're basically giving up on the election too soon and you should probably look for another job.

(15) A Type I error means you make a false alarm. You reject H_0 (the boxes are fine) and conclude H_a (the boxes are under-filled), when in truth, the cereal boxes are just fine. You may get into trouble with the cereal company if you press your point. A Type II error means you miss detection. You fail to reject H_0 when you really should have, because H_a is true: The boxes are under-filled. You let the company get by with cheating its customers.

(16) **No.** In any given situation, you can only make one of the two errors; you can't make both at the same time. Type I errors can happen only when H_0 is true, and Type II errors can generally only happen only when H_0 is false. Both can't happen at the same time. The problem is, you typically can't identify the true situation, so you have to be prepared to identify and discuss both types of errors.

(17) With a problem like this, it helps to first write down H_0 and H_a, even if you put them in words. Here, you have H_0: coin is fair versus H_a: coin isn't fair. (Always put the "status quo" or "innocent until proven guilty" item in the null hypothesis.)

a. Your *p*-value suggests that the coin isn't fair, so you reject H_0. But the coin really is fair, so H_0 turns out to be true. **Rejecting H_0 when H_0 is true is a Type I error.** You're guilty of making a false alarm. Do you plead the fifth?

b. **Your friend may be pretty upset with you**, or you may be the subject of subsequent ridicule, I guess.

TIP

A Type I error is a false alarm, which is much easier to remember than "you reject H_0 when H_0 is true."

(18) First, you write down H_0 and H_a, even if you put them in words. Here, you have H_0: machine is okay versus H_a: machine isn't okay. (Always put the "status quo" or "innocent until proven guilty" item in the null hypothesis.)

a. Your *p*-value suggests that the machine is okay, so you fail to reject H_0. But the machine really isn't okay, so H_0 turns out to be false. **Failing to reject H_0 when H_0 is false is a Type II error.** You missed a chance to detect a problem.

TIP

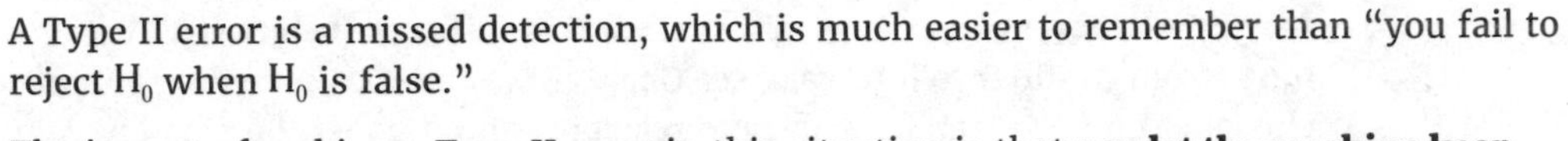

A Type II error is a missed detection, which is much easier to remember than "you fail to reject H_0 when H_0 is false."

b. The impact of making a Type II error in this situation is that **you let the machine keep making widgets that don't meet specifications**. You miss a detection of a problem, which costs people money and time in the long run.

If you're ready to test your skills a bit more, take the following chapter quiz that incorporates all the chapter topics.

Whaddya Know? Chapter 15 Quiz

Quiz time! Complete each problem to test your knowledge on the various topics covered in this chapter. You can then find the solutions and explanations in the next section.

1. The H_0 and H_a hypotheses are always about the parameter in question, not the statistics from the data. True or false?

2. Suppose you hear it reported that the average price of a home in your area is $250,000 and you believe it's more than that; what is your H_a?

3. H_0 always contains an equal sign, true or false?

4. A standard score is always a score off the standard normal distribution. True or false?

5. What are the two steps to forming a test statistic from your sample statistic?

6. If the *p*-value is 0.08 and your significance level is 0.05, what do you conclude about H_0?

7. You could be wrong if you reject H_0. What type of error could you be making?

8. You could be wrong if you fail to reject H_0. What type of error could you be making?

9. What is another name for a Type I error?

10. What is another name for a Type II error?

Answers to Chapter 15 Quiz

1. **True.** Always use parameters in H_0 and H_a, never sample statistics. I like to say no one cares about your sample statistic but you, but everyone cares what you are testing about the population parameter.
2. **$H_a: \mu > \$250,000$.** You believe that the mean is greater than $250,000, so you use ">" in Ha.
3. **True.** We always use an equal sign in H_0, no matter what H_a is.
4. **False.** It could come from a wide range of distributions, one of which is the standard normal distribution.
5. **Step 1: Subtract the value in H_0; Step 2: divide by the standard error of the sample statistic.**
6. **Fail to reject H_0**; not enough evidence against it, because the *p*-value is > significance level.
7. **Type 1 error**. If H_0 is true and you reject H_0, you are wrong, and you have committed a Type I error.
8. **Type II error.** If H_0 is false and you fail to reject H_0, you are wrong, and you have committed a Type II error.
9. **False alarm.** If H_0 is true and you reject H_0, you are wrong, and you have committed a Type I error. You have made a false alarm.
10. **Missed opportunity.** If H_0 is false and you fail to reject H_0, you are wrong, and you have committed a Type II error. You have missed the opportunity to find an issue.

IN THIS CHAPTER

- » Breaking down commonly used hypothesis tests
- » Calculating their test statistics
- » Using the results to make informed decisions

Chapter 16

Commonly Used Hypothesis Tests: Formulas and Examples

From product advertisements to media blitzes on recent medical breakthroughs, you often run across claims made about one or more populations. For example, "We promise to deliver our packages in two days or less" or "Two recent studies show that a high-fiber diet may reduce your risk of colon cancer by 20 percent." Whenever someone makes a claim (also called a null hypothesis) about a population (such as all packages or all adults), you can test the claim by doing what statisticians call a hypothesis test.

A hypothesis test involves setting up your *hypotheses* (a claim and its alternative), selecting a sample (or samples), collecting data, calculating the relevant statistics, and using those statistics to decide whether the claim is true.

In this chapter, I outline the formulas used for some of the most common hypothesis tests, explain the necessary calculations, and walk you through some examples.

TIP

If you need more background information on hypothesis testing (such as setting up hypotheses and understanding test statistics, p-values, significance levels, and Type I and Type II errors), just flip to Chapter 15. All the general concepts of hypothesis testing are developed there. This chapter focuses on their application.

Testing One Population Mean

When the variable is numerical (for example, age, income, time, and so on) and only one population or group (such as all U.S. households or all college students) is being studied, you use the hypothesis test in this section to examine or challenge a claim about the population mean. For example, a child psychologist says that the average time that working mothers spend talking to their children is 11 minutes per day, on average. (For dads, the claim is 8 minutes.) The variable — time — is numerical, and the population is all working mothers. Using statistical notation, μ represents the average number of minutes per day that all working mothers spend talking to their children, on average.

The null hypothesis is that the population mean, μ, is equal to a certain claimed value, μ_0. The notation for the null hypothesis is H_0: $\mu = \mu_0$. So the null hypothesis in this example is H_0: $\mu = 11$ minutes, and μ_0 is 11. The three possibilities for the alternative hypothesis, H_a, are $\mu \neq 11$, $\mu < 11$, or $\mu > 11$, depending on what you are trying to show. (See Chapter 15 for more on alternative hypotheses.) If you suspect that the average time working mothers spend talking with their kids is more than 11 minutes, your alternative hypothesis would be H_a: $\mu > 11$.

To test the claim, you compare the mean you got from your random sample ($\bar{x}$) with the mean shown in $H_0(\mu_0)$. To make a proper comparison, you look at the difference between them and divide by the standard error to take into account the fact that your sample results will vary. (See Chapter 13 for all the info you need on standard error.) This result is your test statistic. In the case of a hypothesis test for the population mean, the test statistic turns out under certain conditions to be a z-value (a value from the Z-distribution; see Chapter 10) and under other conditions to be a t-value (a value from the t-distribution; see Chapter 11).

Then you can look up your test statistic on the appropriate table, and find the chance that this difference between your sample mean and the claimed population mean really could have occurred if the claim were true.

The test statistic for testing one population mean (under certain conditions) is

$$z = \frac{\bar{x} - \mu_0}{\sigma / \sqrt{n}}$$

where $\bar{x}$ is the sample mean, σ is the population standard deviation (assume for this case that this number is known), and z is a value on the Z-distribution. To calculate the z-test statistic, do the following:

1. **Calculate the sample mean, $\bar{x}$.**
2. **Find $\bar{x} - \mu_0$.**
3. **Calculate the standard error: $\sigma / \sqrt{n}$.**
4. **Divide your result from Step 2 by the standard error found in Step 3.**

REMEMBER

The conditions for using this test statistic are that the population standard deviation, σ, is known, and either the population has a normal distribution or the sample size is large enough to use the CLT ($n > 30$); see Chapter 12.

For example, suppose a random sample of 100 working mothers spend an average of 11.5 minutes per day talking with their children. (Assume prior research suggests the population standard deviation is 2.3 minutes.)

1. **You are given that $\bar{x}$ is 11.5, $n = 100$, and σ is 2.3.**
2. **Take $11.5 - 11 = 0.5$.**
3. **Take 2.3 divided by the square root of 100 (which is 10) to get 0.23 for the standard error.**
4. **Divide 0.5 by 0.23 to get 2.17. That's your *z*-test statistic, which means your sample mean is 2.17 standard errors above the claimed population mean.**

REMEMBER

The big idea of a hypothesis test is to challenge the claim that's being made about the population (in this case, the population mean); that claim is shown in the null hypothesis, H_0. If you have enough evidence from your sample against the claim, H_0 is rejected.

To decide whether you have enough evidence to reject H_0, calculate the *p*-value by looking up your test statistic (in this case 2.17) on the standard normal (*Z*-) distribution — see the *Z*-table in the Appendix — and take 1 minus the probability shown. (You subtract from 1 because your H_a is a greater-than hypothesis and the table shows less-than probabilities.)

For this example you look up the test statistic (2.17) on the *Z*-table and find the (less-than) probability is 0.9850, so the *p*-value is $1 - 0.9850 = 0.015$. It's quite a bit less than your (typical) significance level 0.05, which means your sample results would be considered unusual if the claim (of 11 minutes) was true. So reject the claim (H_0: $\mu = 11$ $\mu = 11$ minutes). Your results support the alternative hypothesis H_a: $\mu > 11$. According to your data, the child psychologist's claim of 11 minutes per day is too low; the actual average is greater than that.

For information on how to calculate *p*-values for the less-than or not-equal-to alternatives, also see Chapter 15.

Handling Small Samples and Unknown Standard Deviations: The *t*-Test

In two cases, you can't use the *Z*-distribution for a test statistic for one population mean. The first case is where the sample size is small (and by small, I mean dropping below 30 or so); the second case is when the population standard deviation, σ, is not known, and you have to estimate it using the sample standard deviation, *s*. In both cases, you have less reliable information on which to base your conclusions, so you have to pay a penalty for this by using a distribution with more variability in the tails than a *Z*-distribution has. Enter the *t*-distribution. (See Chapter 11 for all things *t*-distribution, including its relationship with the *Z*.)

A hypothesis test for a population mean that involves the *t*-distribution is called a *t*-test. The formula for the test statistic in this case is

$$t_{n-1} = \frac{\bar{x} - \mu_0}{s / \sqrt{n}}$$

where t_{n-1} is a value from the t-distribution with $n-1$ degrees of freedom.

Note that it is just like the test statistic for the large sample and/or normal distribution case (see the section, "Testing One Population Mean"), except σ is not known, so you substitute the sample standard deviation, s, instead, and use a t-value rather than a z-value.

REMEMBER

Because the t-distribution has fatter tails than the Z-distribution, you get a larger p-value from the t-distribution than one that the standard normal (Z-) distribution would have given you for the same test statistic. A bigger p-value means less chance of rejecting H_0. Having less data and/or not knowing the population standard deviation should create a higher burden of proof.

Putting the *t*-test to work

Suppose a delivery company claims they deliver their packages in two days on average, and you suspect it's longer than that. The hypotheses are H_0: $\mu = 2$ versus H_a: $\mu > 2$. To test this claim, you take a random sample of ten packages and record their delivery times. You find the sample mean is $\bar{x} = 2.3$ days, and the sample standard deviation is 0.35 day. (Because the population standard deviation, is unknown, you estimate it with s, the sample standard deviation.) This is a job for the t-test.

REMEMBER

Because the sample size is small ($n = 10$ is much less than 30) and the population standard deviation is not known, your test statistic has a t-distribution. Its degrees of freedom are $10 - 1 = 9$. The formula for the t-test statistic (often referred to as the *t-value*) is

$$t_{10-1} = \frac{2.3 - 2.0}{0.35/\sqrt{10}} = 2.71$$

To calculate the p-value, you look in the row in the t-table (in the Appendix) for $df = 9$. Your test statistic (2.71) falls between two values in the row for $df = 9$ in the t-table: 2.26 and 2.82 (rounding to two decimal places). To calculate the p-value for your test statistic, find which columns correspond to these two numbers. The number 2.26 appears in the 0.025 column and the number 2.82 appears in the 0.010 column; you now know the p-value for your test statistic lies between 0.025 and 0.010 (that is, $0.010 < p - \text{value} < 0.025$).

Using the t-table you don't know the exact number for the p-value, but because 0.010 and 0.025 are both less than your significance level of 0.05, you reject H_0; you have enough evidence in your sample to say the packages are not being delivered in two days, and in fact the average delivery time is more than two days.

REMEMBER

The t-table (in the Appendix) doesn't include every possible t-value; just find the two values closest to yours on either side, look at the columns they're in, and report your p-value in relation to theirs. (If your test statistic is greater than all the t-values in the corresponding row of the t-table, just use the last one; your p-value will be less than its probability.)

TIP

Of course you can use statistical software, if available, to calculate exact p-values for any test statistic; using software, you get 0.012 for the exact p-value.

Relating t to Z

The next-to-the-last line of the t-table shows the corresponding values from the standard normal (Z-) distribution for the probabilities listed on the top of each column. Now choose a column in the table and move down the column, looking at the t-values. As the degrees of freedom of the t-distribution increase, the t-values get closer and closer to that row of the table where the z-values are.

This confirms a result found in Chapter 11: As the sample size (hence degrees of freedom) increases, the t-distribution becomes more and more like the Z-distribution, so the p-values from their hypothesis tests are virtually equal for large sample sizes. And those sample sizes don't even have to be that large to see this relationship; for $df = 30$ the t-values are already very similar to the z-values shown in the bottom of the table. These results make sense; the more data you have, the less of a penalty you have to pay. (And of course, you can use computer technology to calculate more exact p-values for any t-value you like.)

Handling negative t-values

For a less-than alternative hypothesis (for example, $H_a: \mu < 2$), your t-test statistic would be a negative number (to the left of 0 on the t-distribution). In this case, you want to find the percentage below, or to the left of, your test statistic to get your p-value. Yet negative test statistics don't appear on the t-table (in the Appendix).

Not to worry! The percentage to the left (below) a negative t-value is the same as the percentage to the right (above) the positive t-value, due to symmetry. So to find the p-value for your negative t-test statistic, look up the positive version of your test statistic on the t-table, find the corresponding right tail (greater-than) probability, and use that.

For example, suppose your test statistic is −2.7105 with 9 degrees of freedom and H_a is the less-than alternative. To find your p-value, first look up 2.7105 on the t-table; by the work in the previous section, you know its p-value falls between the column headings 0.025 and 0.010. Because the t-distribution is symmetric, the p-value for −2.7105 also falls somewhere between 0.025 and 0.010. Again you reject H_0 because these values are both less than or equal to 0.05.

Examining the not-equal-to alternative

REMEMBER

To find the p-value when your alternative hypothesis (H_a) is not-equal-to, simply double the probability that you get from the Z-table or t-table when you look up your test statistic. Why double it? Because the distribution tables show only one-sided probabilities (less-than for the Z-table; greater-than for the t-table), which are only half the story. To find the p-value when you have a not-equal-to alternative, you must add the p-values from the less-than and greater-than alternatives. Because both the Z- and t-distributions are symmetric, the less-than and greater-than probabilities are the same, so just double the one you look up on the table and you'll have the p-value for the not-equal-to alternative.

For example, if your t-test statistic is 2.7171 and H_a is a not-equal-to alternative, look up 2.7171 on the t-table ($df = 9$ again), and you find the p-value lies between 0.025 and 0.010, as shown previously. These are the p-values for the greater-than alternative. Now double these values to include the less-than alternative and you find the p-value for your test statistic lies somewhere between $2(0.025) = 0.05$ and $2(0.010) = 0.020$.

Drawing conclusions using the critical value

After you calculate your test statistic, all you have to do is make your conclusion by seeing where the test statistic falls on the Z-distribution (or t-distribution for small samples). Earlier in this chapter and in Chapter 15, I show the p-value method for making your conclusions in a hypothesis test. Another method for drawing conclusions can be done by using critical values.

Under the critical value method, before you collect your data, you set one (or two) cutoff point(s) on the Z- or t-distribution so that if your test statistic falls beyond the cutoff point(s), you reject H_0; otherwise, you fail to reject H_0. The cutoff points are called *critical values.* (A critical value is much like the goal line in football; it's the place you have to reach in order to "score" a significant result and reject H_0.)

The critical value(s) is (are) determined by the significance level (or alpha level) of the test, denoted by α. Alpha levels differ for each situation, but as you've seen, most researchers are happy with an alpha level of 0.05, much in the same way they're happy with a 95 percent confidence level for a confidence interval. (Notice that $1-\alpha$ equals the confidence level of a confidence interval.)

To find a critical value for a given significance level, look up that alpha level in the distribution table just as you would any probability. Then find the corresponding z- or t-value that goes with that probability and the direction of the alternative hypothesis. For example, suppose you are using a significance level of $\alpha = 0.05$. For the Z-distribution, this corresponds to a critical value of $z = -1.64$ if H_a is "less than" and a critical value of $z = 1.64$ if H_a is "greater than." For the t-distribution and, say, 10 degrees of freedom, $\alpha = 0.05$ corresponds to a critical value of either plus or minus 1.812461, depending on the direction of the alternative hypothesis.

If H_a contains a "not-equal-to" statement, then it's two-sided. Remember to split the alpha level in half so that the area is shared between the two extremes. For example, suppose you are using a significance level of $\alpha = 0.05$ and have a "not-equal-to" alternative hypothesis. You split α into two halves, putting 0.025 in each tail. For the Z-distribution, this corresponds to critical values of ± 1.96, and for the t-distribution (and 10 degrees of freedom), it corresponds to critical values of ± 2.22814.

After you set up the critical value(s), if the test statistic falls beyond the critical value(s), your conclusion is "reject H_0 at level α." This means the test statistic falls into the *rejection region.* If the test statistic doesn't go beyond the critical value(s), you conclude that you "fail to reject H_0 at level α." This means the test statistic falls into the *acceptance region.*

REMEMBER

Even if the test statistic falls into the acceptance region, it doesn't mean that you "accept H_0" as the absolute truth. It's just called the acceptance region because that means that the test statistic fell into an acceptable range of values under the assumption that H_0 is correct.

TIP

Statisticians prefer the p-value method because it allows you to make a decision on H_0 and report how strong your evidence actually is. The critical value method only states whether or not you reject H_0 at a certain alpha level.

EXAMPLE

Q. Suppose you hear a claim that the average score on a national exam is 78. You think the average is higher than that, and your sample of 100 students produces an average of 75. You also learn that the standard deviation of last year's scores is 12.

(a) Set up your null and alternative hypotheses.

(b) Does this situation call for using the Z-distribution or the *t*-distribution?

(c) Calculate the test statistic and *p*-value.

(d) What's the critical value if you use $\alpha = 0.01$?

A. Setting up the hypotheses correctly is critical to your success with hypothesis tests.

(a) In this case, you have H_0: $\mu = 78$ versus H_a: $\mu > 78$. ***Note:*** 75 is a sample statistic and doesn't belong in H_0 or H_a. You want to show that the mean is higher than the claim, so the alternative has a ">" sign. You're conducting a right-tailed test.

(b) Because the population standard deviation is known and the sample size is fairly large with $n = 100$, you can use the Z-distribution.

(c) Here, the test statistic is found by $z = \frac{\bar{x} - \mu_0}{\sigma / \sqrt{n}} = \frac{75 - 78}{12 / \sqrt{100}} = \frac{-3}{1.2} = -2.50$. So, the *p*-value will be represented by the area beyond the test statistic, namely $p(Z > -2.50)$. (In this case, beyond means "above" because you conduct a right-tailed test.) From the Z-table in the Appendix, you find that $p(Z \leq -2.50) = 0.0062$, so $p(Z > -2.50) = 1 - p(Z \leq -2.50) = 1 - 0.0062 = 0.9938$. This is way above the traditionally acceptable significance level of 0.05, so the results are about as insignificant as possible. You fail to reject the null hypothesis that $\mu = 78$.

TIP

You actually didn't need to calculate the test statistic or *p*-value to form the correct conclusion. Because the alternative hypothesis is that $\mu > 78$, the only way you could reject the null hypothesis is if you get a sample mean that supports the alternative. But because the observed sample mean was 75, it actually supports the null hypothesis. Anytime your sample statistic doesn't fall in the range of your predetermined alternative hypothesis, you automatically know that you'll fail to reject H_0.

(d) In Answer (b), you figured out that this situation calls for using the Z-distribution. Looking for $\alpha = 0.01$ in the body of the Z-table, you find that the closest probability to this is 0.0099. This corresponds to $z = -2.33$, but because H_a is a "greater than" statement, the critical value is actually 2.33. So that means the test statistic would have to be greater than 2.33 for you to reject H_0. As you figured out in Answer (c), the test statistic is –2.50, so there's no way you're rejecting the null hypothesis.

YOUR TURN

1. Conduct the hypothesis test H_0: $\mu = 7$ versus H_a: $\mu > 7$, where $\bar{x} = 7.5$, $\sigma = 2$, and $n = 30$. Use $\alpha = 0.01$.

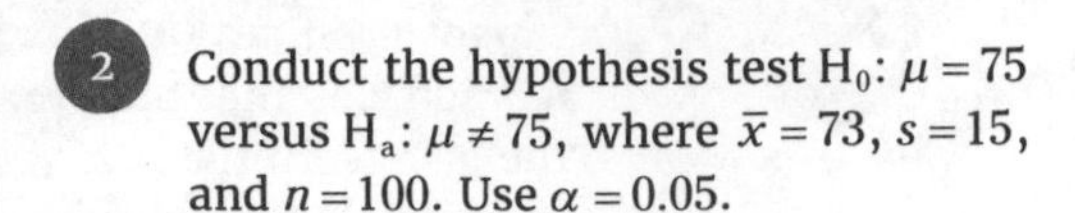

2. Conduct the hypothesis test H_0: $\mu = 75$ versus H_a: $\mu \neq 75$, where $\bar{x} = 73$, $s = 15$, and $n = 100$. Use $\alpha = 0.05$.

3. Conduct the hypothesis test H_0: $\mu = 100$ versus H_a: $\mu > 100$, where $\bar{x} = 105$, $s = 30$, and $n = 10$. Use $\alpha = 0.10$.

4. Suppose your critical value for a left-tailed hypothesis test is –1.96. For what values of the test statistic would you reject H_0?

Testing One Population Proportion

When the variable is categorical (for example, gender or support/oppose) and only one population or group is being studied (for example, all registered voters), you use the hypothesis test in this section to test a claim about the population proportion. The test looks at the proportion (p) of individuals in the population who have a certain characteristic — for example, the proportion of people who carry cellphones. The null hypothesis is H_0: $p = p_0$, where p_0 is a certain claimed value of the population proportion, p. For example, if the claim is that 70 percent of people carry cellphones, p_0 is 0.70. The alternative hypothesis is one of the following: $p > p_0$, $p < p_0$ or $p \neq p_0$. (See Chapter 15 for more on alternative hypotheses.)

The formula for the test statistic for a single proportion (under certain conditions) is

$$z = \frac{\hat{p} - p_0}{\sqrt{\frac{p_0(1 - p_0)}{n}}}$$

where $\hat{p}$ is the proportion of individuals in the sample who have that characteristic and z is a value on the Z-distribution (see Chapter 10). To calculate the test statistic, do the following:

1. **Calculate the sample proportion, $\hat{p}$, by taking the number of people in the sample who have the characteristic of interest (for example, the number of people in the sample carrying cellphones) and dividing that by n, the sample size.**
2. **Find $\hat{p} - p_0$, where p_0 is the value in H_0.**
3. **Calculate the standard error, $\sqrt{\frac{p_0(1-p_0)}{n}}$.**
4. **Divide your result from Step 2 by your result from Step 3.**

To interpret the test statistic, look up your test statistic on the standard normal (Z-) distribution (in the Appendix) and calculate the p-value (see Chapter 15 for more on p-value calculations).

REMEMBER

The conditions for using this test statistic are that $np_0 \geq 10$ and $n(1-p_0) \geq 10$ (see Chapter 10 for details).

For example, suppose Cavifree claims that four out of five dentists recommend Cavifree toothpaste to their patients. In this case, the population is all dentists, and p is the proportion of all dentists who recommended Cavifree. The claim is that p is equal to "four out of five," or p_0 is $4 \div 5 = 0.80$. You suspect that the proportion is actually less than 0.80. Your hypotheses are $H_0: p = 0.80$ versus $H_a: p < 0.80$.

Suppose that 151 out of your sample of 200 dental patients reported receiving a recommendation for Cavifree from their dentist. To find the test statistic for these results, follow these steps:

1. **Start with $\hat{p} = \frac{151}{200} = 0.755$ and $n = 200$.**
2. **Because $p_0 = 0.80$, take $0.755 - 0.80 = -0.045$ (the numerator of the test statistic).**
3. **Next, the standard error equals $\sqrt{\frac{0.80(1-0.80)}{200}} = 0.028$ (the denominator of the test statistic).**
4. **The test statistic is $\frac{-0.045}{0.028} = -1.61$.**

REMEMBER

Because the resulting test statistic is negative, it means your sample results are –1.61 standard errors below (less than) the claimed value for the population. How often would you expect to get results like this if H_0 were true? The chance of being at or beyond (in this case less than) –1.61 is 0.0537. (Keep the negative with the number and look up –1.61 in the Z-table in the Appendix.) This result is your p-value because H_a is a less-than hypothesis. (See Chapter 15 for more on this.)

Because the p-value is greater than 0.05 (albeit not by much), you don't have quite enough evidence for rejecting H_0. According to your data, you conclude that the claim that 80 percent of dentists recommend Cavifree can't be rejected. However, it's important to report the actual p-value too, so others can make their own decisions.

TIP

The letter p is used two different ways in this chapter: p-value and p. The letter p by itself indicates the population proportion, not the p-value. Don't get confused. Whenever you report a p-value, be sure you add "*-value*" so it's not confused with p, the population proportion.

EXAMPLE

Q. Suppose a political candidate claims the percentage of uninsured drivers is 30 percent, but you believe the percentage is more than 30.

(a) Set up your null and alternative hypotheses.

(a) Find the critical value(s), assuming $\alpha = 0.02$ and you have a large sample size.

A. First, make sure you can identify that this is a hypothesis test about a proportion. It has a claim that's being challenged or tested, and the claim is about a percentage (or proportion). That's how you know.

(a) The claim is that $p = 0.30$, so you put that into the null hypothesis. The alternative of interest is the one where the percentage is actually more (>) than 30. So you have $H_0: p = 0.30$ versus $H_a: p > 0.30$.

(b) The critical value is $z = 2.05$. Here's how you get it: Using the Z-table, the area beyond the critical value must be 0.02. (In this case, "beyond" means "above," because you run a right-tailed test.) The closest value to 0.02 you can find in the Z-table is 0.0202. The left-tailed probability of 0.0202 corresponds to $p(Z \leq -2.05)$. But by symmetry of the Z-distribution, $p(Z > 2.05)$ is also 0.0202. So because this is a right-tailed test, you'll reject H_0 if your test statistic is greater than 2.05.

YOUR TURN

5 Carry out the hypothesis test of $H_0: p = 0.5$ versus $H_a: p > 0.5$, where $\hat{p} = 0.60$ and $n = 100$. Use $\alpha = 0.05$.

6 Carry out the hypothesis test of $H_0: p = 0.5$ versus $H_a: p < 0.5$, where $\hat{p} = 0.40$ and $n = 100$. Use $\alpha = 0.05$.

7 Carry out the hypothesis test of $H_0: p = 0.5$ versus $H_a: p \neq 0.5$, with $x = 40$ and $n = 100$, where x is the number of people in the sample who have the characteristic of interest. Use $\alpha = 0.01$.

8 Suppose you want to test the fairness of a single die, so you concentrate on the proportion of 1s that come up. Write down the null and alternative hypotheses for this test.

Comparing Two (Independent) Population Averages

When the variable is numerical (for example, income, cholesterol level, or miles per gallon) and two populations or groups are being compared (for example, men versus women), you use the steps in this section to test a claim about the difference in their averages. (For example, is the difference in the population means equal to zero, indicating their means are equal?) Two independent (totally separate) random samples need to be selected, one from each population, in order to collect the data needed for this test.

The null hypothesis is that the two population means are the same — in other words, that their difference is equal to 0. The notation for the null hypothesis is $H_0: \mu_1 = \mu_2$, where μ_1 represents the mean of the first population and μ_2 represents the mean of the second population.

TIP

You can also write the null hypothesis as $H_0: \mu_1 - \mu_2 = 0$, emphasizing the idea that their difference is equal to zero if the means are the same.

Case 1: Difference of two population means when population standard deviations are known

The formula for the test statistic comparing two means (in the case where the population standard deviations are known) is

$$z = \frac{(\bar{x}_1 - \bar{x}_2) - 0}{\sqrt{\frac{\sigma_1^2}{n_1} + \frac{\sigma_2^2}{n_2}}}$$

To calculate the z statistic, do the following:

1. **Calculate the sample means $\bar{x}_1$ and $\bar{x}_2$.**

   (Assume the population standard deviations, σ_1 and σ_2, are given.) Let n_1 and n_2 represent the two sample sizes (they need not be equal).

   See Chapter 5 for these calculations.

2. **Find the difference between the two sample means: $\bar{x}_1 - \bar{x}_2$.**

   TIP

   Because $\mu_1 - \mu_2$ is equal to 0 if H_0 is true, it doesn't need to be included in the numerator of the test statistic. However, if the difference they are testing is any value other than 0, you subtract that value in the numerator of the test statistic.

3. **Calculate the standard error using the following equation:**

$$\sqrt{\frac{\sigma_1^2}{n_1} + \frac{\sigma_2^2}{n_2}}$$

4. **Divide your result from Step 2 by your result from Step 3.**

REMEMBER

To interpret the test statistic, add the following two steps to the list.

5. **Look up your test statistic on the standard normal (*Z*-) distribution (see the *Z*-table in the Appendix) and calculate the *p*-value.**

   (See Chapter 15 for more on *p*-value calculations.)

6. **Compare the *p*-value to your significance level, such as 0.05. If it's less than or equal to 0.05, reject H_0. Otherwise, fail to reject H_0.**

   (See Chapter 15 for the details on significance levels.)

REMEMBER

The conditions for using this test are that the two population standard deviations are known and either both populations have a normal distribution or both sample sizes are large enough for the Central Limit Theorem (see Chapter 12).

For example, suppose you want to compare the absorbency of two brands of paper towels (call the brands Stats-absorbent and Sponge-o-matic). You can make this comparison by looking at the average number of ounces each brand can absorb before being saturated. H_0 says the difference between the average absorbencies is 0 (nonexistent), and H_a says the difference is not 0. In other words, one brand is more absorbent than the other. Using statistical notation, you have $H_0 = \mu_1 - \mu_2 = 0$ versus $H_0 = \mu_1 - \mu_2 \neq 0$. Here, you have no indication of which paper towel may be more absorbent, so the not-equal-to alternative is the one to use (see Chapter 15).

Suppose you select a random sample of 50 paper towels from each brand and measure the absorbency of each paper towel. (Assume a normal distribution.) Suppose the average absorbency of Stats-absorbent (x_1) for your sample is 3 ounces, and assume the population standard deviation is 0.9 ounce. For Sponge-o-matic (x_2), the average absorbency is 3.5 ounces according to your sample; assume the population standard deviation is 1.2 ounces. Carry out this hypothesis test by following the six steps listed previously, as follows:

1. **Given the information, you know $\bar{x}_1 = 3$, $\sigma_1 = 0.9$, $\bar{x}_2 = 3.5$, $\sigma_2 = 1.2$, $n_1 = 50$, and $n_2 = 50$.**

2. **The difference between the sample means for (Stats-absorbent minus Sponge-o-matic) is $\bar{x}_1 - \bar{x}_2 = (3 - 3.5) = -0.5$ ounce.**

   (A negative difference simply means that the second sample mean was larger than the first.)

3. **The standard error is $\sqrt{\frac{\sigma_1^2}{n_1} + \frac{\sigma_2^2}{n_2}} = \sqrt{\frac{0.9^2}{50} + \frac{1.2^2}{50}} = 0.2121$.**

4. **Divide the difference, −0.5, by the standard error, 0.2121, which gives you −2.36.**

   This is your test statistic.

5. **To find the *p*-value, look up −2.36 on the standard normal (*Z*-) distribution — see the *Z*-table in the Appendix.**

   The chance of being beyond, in this case to the left of, −2.36 is equal to 0.0091. Because H_a is a not-equal-to alternative, you double this percentage to get $2 \times 0.0091 = 0.0182$, your *p*-value. (See Chapter 15 for more on the not-equal-to alternative.)

6. **This *p*-value is quite a bit less than 0.05.**

   That means you have fairly strong evidence to reject H_0.

Your conclusion is that a statistically significant difference exists between the absorbency levels of these two brands of paper towels, based on your samples. And Sponge-o-matic comes out on top because it has a higher average. (Stats-absorbent minus Sponge-o-matic being negative means Sponge-o-matic had the higher value.)

EXAMPLE

Q. A teacher instructs two statistics classes with two different teaching methods (Group 1: computer versus Group 2: pencil/paper). The teacher wants to see whether the computer method works better by comparing average final exam scores for the two groups. They select volunteers to be in each group.

(a) Has the teacher implemented a right-tailed, left-tailed, or two-tailed test?

(b) The teacher doesn't try to control for other factors that can influence their results. Name some of the factors.

(c) How can the teacher change their study to improve the quality of the results?

A. The teacher uses a test for two population means, because they compare the average exam scores, and exam scores are a quantitative variable.

(a) The teacher wants to see whether the computer group does better, and puts these students in Group 1, so they want to show that the mean of Group 1 is greater than (>) Group 2. They implement a right-tailed test.

(b) Some other factors include the intelligence level of the students, how comfortable they are with computers, the quality of the teaching activities, the way the teacher writes the test, and so on.

(c) The teacher can improve their results by randomly assigning the students to groups, which creates a more level playing field, rather than asking for volunteers. They can also match up students according to the variables mentioned in answer (b) and randomly assign one member of each pair to the computer group and the other to the pencil/paper group. This would require that they do a paired *t*-test, which comes up later in this chapter.

YOUR TURN

9 Conduct the hypothesis test $H_0: \mu_1 - \mu_2 = 0$ versus $H_a: \mu_1 - \mu_2 < 0$, where $\bar{x}_1 = 7$, $\bar{x}_2 = 8$, $\sigma_1 = 2$, $\sigma_2 = 2$, $n_1 = 30$, and $n_2 = 30$. Use $\alpha = 0.01$.

10 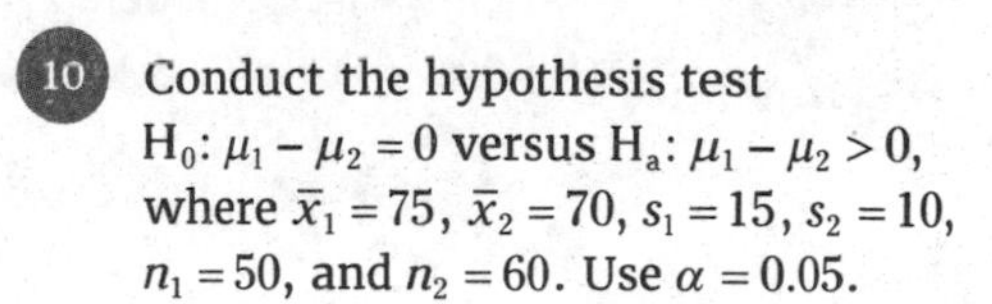
Conduct the hypothesis test $H_0: \mu_1 - \mu_2 = 0$ versus $H_a: \mu_1 - \mu_2 > 0$, where $\bar{x}_1 = 75$, $\bar{x}_2 = 70$, $s_1 = 15$, $s_2 = 10$, $n_1 = 50$, and $n_2 = 60$. Use $\alpha = 0.05$.

11 Conduct the hypothesis test $H_0: \mu_1 = \mu_2$ versus $H_a: \mu_1 \neq \mu_2$, where $\bar{x}_1 = 75$, $\bar{x}_2 = 70$, $s_1 = 15$, $s_2 = 10$, $n_1 = 50$, and $n_2 = 60$. Use $\alpha = 0.05$.

Suppose you conducted a hypothesis test for two means (Group 1 mean minus Group 2 mean) and you reject $H_0: \mu_1 = \mu_2$ versus $H_a: \mu_1 \neq \mu_2$. You conclude that the two population means aren't equal. Can you say a little more? Explain how you can tell from the sign on the test statistic which group probably has the higher mean.

Case 2: Difference of two population means when population standard deviations are unknown

The formula for the test statistic comparing two means (in the case where the population standard deviations are unknown but assumed to be equal) is $t_{n_1+n_2-2} = \dfrac{(\bar{x}_1 - \bar{x}_2) - 0}{\sqrt{\dfrac{(n_1-1)s_1^2 + (n_2-1)s_2^2}{n_1+n_2-2}}\sqrt{\dfrac{1}{n_1} + \dfrac{1}{n_2}}}$, where you have the two means $\bar{x}_1$, $\bar{x}_2$, the two standard deviations s_1, s_2, and the two sample sizes n_1, n_2. This formula is a little more complex that the one in Case 1, but it follows the same idea.

EXAMPLE

For example, suppose you conduct the hypothesis test $H_0: \mu_1 - \mu_2 = 0$ versus $H_a: \mu_1 - \mu_2 < 0$, where $\bar{x}_1 = 7$, $\bar{x}_2 = 8$, $s_1 = 2$, $s_2 = 2$, $n_1 = 30$, and $n_2 = 30$. Use $\alpha = 0.01$. Your test statistic is

$$t_{n_1+n_2-2} = \frac{(\bar{x}_1 - \bar{x}_2) - 0}{\sqrt{\dfrac{(n_1-1)s_1^2 + (n_2-1)s_2^2}{n_1+n_2-2}}\sqrt{\dfrac{1}{n_1} + \dfrac{1}{n_2}}} \rightarrow t_{58} = \frac{(7-8) - 0}{\sqrt{\dfrac{(30-1)2^2 + (30-1)2^2}{30+30-2}}\sqrt{\dfrac{1}{30} + \dfrac{1}{30}}} \rightarrow$$

$$t_{58} = \frac{-1}{\sqrt{4}(0.26)} = -1.92.$$

Compare this value to the bottom row of the t-table in the Appendix, and you see the p-value is between 0.05 and 0.025. The significance level of this test is 0.01, so you fail to reject H_0. There is not enough evidence to say the difference in the averages is less than 0.

Testing for an Average Difference (The Paired *t*-Test)

You can test for an average difference using the test in this section when the variable is numerical (for example, income, cholesterol level, or miles per gallon) and either the individuals in the sample are paired up in some way according to relevant variables (such as age or perhaps weight), or the same people are used twice (for example, using a pre-test and post-test). Paired tests are typically used for studies in which someone is testing to see whether a new treatment, technique, or method works better than an existing method, without having to worry about other factors about the subjects that may influence the results (see Chapter 18 for details).

The average difference (tested in this section) isn't the same as the difference in the averages (tested in the previous section).

- With the difference in averages, you compare the difference in the means of two separate samples to test the difference in the means of two different populations.
- With the average difference, you match up the subjects so they are thought of as coming from a single population, and the set of differences measured for each subject (for example, pre-test versus post-test) are thought of as one sample. The hypothesis test then boils down to a test for one population mean (as I explain earlier in this chapter).

For example, suppose a researcher wants to see whether teaching students to read using a computer game gives better results than teaching with a tried-and-true phonics method. The researcher randomly selects 20 students and puts them into 10 pairs according to their reading readiness level, age, IQ, and so on. They randomly select one student from each pair to learn to read via the computer game method (abbreviated CM), and the other learns to read using the phonics method (abbreviated PM). At the end of the study, each student takes the same reading test. The data are shown in Table 16-1.

TABLE 16-1 Reading Scores for Computer Game Method versus Phonics Method

Student Pair	Computer Game Method	Phonics Method	Difference (CM - PM)
1	85	80	+5
2	80	80	0
3	95	88	+7
4	87	90	-3
5	78	72	+6
6	82	79	+3
7	57	50	+7
8	69	73	-4
9	73	78	-5
10	99	95	+4

The original data are in pairs, but you're really only interested in the difference in reading scores (computer reading score minus phonics reading score) for each pair, not the reading scores themselves. So the *paired differences* (the differences in the pairs of scores) are your new data set. You can see their values in the last column of Table 16-1.

By examining the differences in the pairs of observations, you really only have a single data set, and you only have a hypothesis test for one population mean. In this case, the null hypothesis is that the mean (of the paired differences) is 0, and the alternative hypothesis is that the mean (of the paired differences) is > 0.

If the two reading methods are the same, the average of the paired differences should be 0. If the computer method is better, the average of the paired differences should be positive, meaning that the computer reading score is larger than the phonics score.

The notation for the null hypothesis is $H_0: \mu_d = 0$, where μ_d is the mean of the paired differences for the population. (The *d* in the subscript just reminds you that you're working with the paired differences.)

The formula for the test statistic for paired differences is $t_{n-1} = \frac{\bar{d} - 0}{s_d / \sqrt{n_d}}$, where $\bar{d}$ is the average of all the paired differences found in the sample, and t_{n-1} is a value on the *t*-distribution with $n_d - 1$ degrees of freedom (see Chapter 11).

You use a *t*-distribution here because in most matched-pairs experiments the sample size is small and/or the population standard deviation σ_d is unknown, so it's estimated by s_d. (See Chapter 11 for more on the *t*-distribution.)

To calculate the test statistic for paired differences, do the following:

1. **For each pair of data, take the first value in the pair minus the second value in the pair to find the paired difference.**

   Think of the differences as your new data set.

2. **Calculate the mean, $\bar{d}$, and the standard deviation, s_d, of all the differences.**
3. **Letting n_d represent the number of paired differences that you have, calculate the standard error:**

   $$s_d / \sqrt{n_d}$$

4. **Divide $\bar{d}$ by the standard error from Step 3.**

Because μ_d is equal to 0 if H_0 is true, it doesn't really need to be included in the formula for the test statistic. As a result, you sometimes see the test statistic written like this:

$$\frac{\bar{d} - 0}{s_d / \sqrt{n_d}} = \frac{\bar{d}}{s_d / \sqrt{n_d}}$$

For the reading scores example, you can use the preceding steps to see whether the computer method is better in terms of teaching students to read.

To find the statistic, follow these steps:

1. **Calculate the differences for each pair (they're shown in column 4 of Table 16-1).**

   Notice that the sign on each of the differences is important; it indicates which method performed better for that particular pair.

2. **Calculate the mean and standard deviation of the differences from Step 1.**

   Your calculations find the mean of the differences, $\bar{d} = 2$, and the standard deviation, $s_d = 4.64$. Note that $n_d = 10$ here.

3. **The standard error is $\frac{4.64}{\sqrt{10}} = 1.47$.**

   (Remember that here, n_d is the number of pairs, which is 10.)

4. **Take the mean of the differences (Step 2) divided by the standard error of 1.47 (Step 3) to get 1.36, the test statistic.**

Is the result of Step 4 enough to say that the difference in reading scores found in this experiment applies to the whole population in general? Because the population standard deviation, σ, is unknown and you estimated it with the sample standard deviation (s), you need to use the t-distribution rather than the Z-distribution to find your p-value (see the section, "Handling Small Samples and Unknown Standard Deviations: The t-Test," earlier in this chapter). Using the t-table (in the Appendix), you look up 1.36 on the t-distribution with $10 - 1 = 9$ degrees of freedom to calculate the p-value.

The p-value in this case is greater than 0.05 because 1.36 is smaller than (or to the left of) the value of 1.38 on the table, and therefore its p-value is more than 0.10 (the p-value for the column heading corresponding to 1.38).

Because the p-value is greater than 0.05, you fail to reject H_0; you don't have enough evidence that the mean difference in the scores between the computer method and the phonics method is significantly greater than 0. However, that doesn't necessarily mean a real difference isn't present in the population of all students. But the researcher can't say the computer game is a better reading method based on this sample of ten students. (See Chapter 15 for information on the power of a hypothesis test and its relationship to sample size.)

REMEMBER

In many paired experiments, the data sets are small due to costs and time associated with doing these kinds of studies. That means the t-distribution (see the t-table in the Appendix) is often used instead of the standard normal (Z-) distribution (the Z-table in the Appendix) when figuring out the p-value.

EXAMPLE

Q. Suppose you use a paired t-test using matched pairs to find out whether a certain weight-loss method works. You measure the participants' weights before and after the study, and you take weight-before minus weight-after as your pairs of data.

(a) If you want to show that the program works, what's your H_a?

(b) Explain why measuring the same participants both times rather than measuring two different groups of people (those on the program compared to those not) makes this study much more credible.

A. The signs on the differences are important when making comparisons. If you subtract two numbers and get a positive result, that means the first number is larger than the second. If the result is negative, the second is larger than the first.

(a) If the program works, the weight loss (weight before minus weight after) has to be positive. So, you have H_0: $\mu_d = 0$ versus H_a: $\mu_d > 0$.

(b) If you use two different groups of people, you introduce other variables that can account for the weight differences. Using the same people cuts down on unwanted variability by controlling for possible confounding variables.

Note: If you switch the data around and take weight after minus weight before, you have to switch the sign in the alternative hypothesis to be < (less than). Most people like to use positive values, and greater-than signs (>) produce them. If you have a choice, always order the groups so the one that may have the higher average is Group 1.

YOUR TURN

13 Conduct the hypothesis test H_0: $\mu_d = 0$ versus H_a: $\mu_d > 0$, where $\bar{d} = 2$, $s_d = 5$, and $n_d = 10$. Use $\alpha = 0.05$.

14 Conduct the hypothesis test H_0: $\mu_d = 0$ versus H_a: $\mu_d > 0$, where $\bar{d} = 2$, $s_d = 5$, and $n_d = 30$. Use $\alpha = 0.05$.

Comparing Two Population Proportions

This test is used when the variable is categorical (for example, smoker/nonsmoker, Democrat/Republican, support/oppose an opinion, and so on) and you're interested in the proportion of individuals with a certain characteristic — for example, the proportion of smokers. In this case, two populations or groups are being compared (such as the proportion of female smokers versus male smokers).

In order to conduct this test, two independent (separate) random samples need to be selected, one from each population. The null hypothesis is that the two population proportions are the same; in other words, their difference is equal to 0. The notation for the null hypothesis is H_0: $p_1 = p_2$, where p_1 is the proportion from the first population, and p_2 is the proportion from the second population.

REMEMBER

Stating in H_0 that the two proportions are equal is the same as saying their difference is zero. If you start with the equation $p_1 = p_2$ and subtract p_2 from each side, you get $p_1 - p_2 = 0$. So you can write the null hypothesis either way.

The formula for the test statistic comparing two proportions (under certain conditions) is

$$z = \frac{(\hat{p}_1 - \hat{p}_2) - 0}{\sqrt{\hat{p}(1-\hat{p})\left(\frac{1}{n_1} + \frac{1}{n_2}\right)}}$$

where $\hat{p}_1$ is the proportion in the first sample with the characteristic of interest, $\hat{p}_2$ is the proportion in the second sample with the characteristic of interest, $\hat{p}$ is the proportion in the combined sample (all the individuals in the first and second samples together) with the characteristic of interest, and z is a value on the Z-distribution (see Chapter 10). To calculate the test statistic, do the following:

1. **Calculate the sample proportions $\hat{p}_1$ and $\hat{p}_2$ for each sample. Let n_1 and n_2 represent the two sample sizes (they don't need to be equal).**
2. **Find the difference between the two sample proportions, $\hat{p}_1 - \hat{p}_2$.**
3. **Calculate the overall sample proportion $\hat{p}$, the total number of individuals from both samples who have the characteristic of interest (for example, the total number of smokers, male or female, in the sample) divided by the total number of individuals from both samples $(n_1 + n_2)$.**
4. **Calculate the standard error:**

$$\sqrt{\hat{p}(1-\hat{p})\left(\frac{1}{n_1} + \frac{1}{n_2}\right)}$$

5. **Divide your result from Step 2 by your result from Step 4. This answer is your test statistic.**

To interpret the test statistic, look up your test statistic on the standard normal (Z-) distribution (the Z-table in the Appendix) and calculate the p-value; then make decisions as usual (see Chapter 15 for more on p-values).

Consider those drug ads that pharmaceutical companies put in magazines. The front page of an ad shows a serene picture of the sun shining, flowers blooming, people smiling — their lives changed by the drug. The company claims that its drugs can reduce allergy symptoms, help people sleep better, lower blood pressure, or fix whichever other ailment they're targeted to help. The claims may sound too good to be true, but when you turn the page to the back of the ad, you see all the fine print where the drug company justifies how it's able to make its claims. (This is typically where statistics are buried!) Somewhere in the tiny print, you'll likely find a table that shows adverse effects of the drug when compared to a *control group* (subjects who took a fake drug), for fair comparison to those who actually took the real drug (the *treatment group*; see Chapter 18 for more on this).

For example, the makers of Adderall, a drug for attention deficit hyperactivity disorder (ADHD), reported that 26 of the 374 subjects (7 percent) who took the drug experienced vomiting as a side effect, compared to 8 of the 210 subjects (4 percent) who were on a *placebo* (fake drug). Note that patients didn't know which treatment they were given. In the sample, more people on the drug experienced vomiting, but is this percentage enough to say that the entire population on the drug would experience more vomiting? You can test it to see.

In this example, you have $H_0: p_1 - p_2 = 0$ versus $H_0: p_1 - p_2 > 0$, where p_1 represents the proportion of subjects who vomited using Adderall, and p_2 represents the proportion of subjects who vomited using the placebo.

REMEMBER

Why does H_a contain a ">" sign and not a "<" sign? H_a represents the scenario in which those taking Adderall experience more vomiting than those on the placebo — that's something the FDA (and any candidate for the drug) would want to know about. But the order of the groups is important, too. You want to set it up so the Adderall group is first, so that when you take the Adderall proportion minus the placebo proportion, you get a positive number if H_a is true. If you were to switch the groups, the sign would be negative.

Now calculate the test statistic:

1. **First, determine that $\hat{p}_1 = \frac{26}{374} = 0.070$ and $\hat{p}_2 = \frac{8}{210} = 0.038$.**

   The sample sizes are $n_1 = 374$ and $n_2 = 210$, respectively.

2. **Take the difference between these sample proportions to get $\hat{p}_1 - \hat{p}_2 = 0.070 - 0.038 = 0.032$.**

3. **Calculate the overall sample proportion to get $\hat{p} = \frac{26+8}{374+210} = 0.058$.**

4. **The standard error is $\sqrt{0.058(1-0.058)\left(\frac{1}{374}+\frac{1}{210}\right)} = 0.020$.**

5. **Finally, the test statistic is $0.032 \div 0.020 = 1.60$.**

   Whew!

The *p*-value is the percentage chance of being at or beyond (in this case to the right of) 1.60, which is $1 - 0.9452 = 0.0548$. This *p*-value is just slightly greater than 0.05, so, technically, you don't have quite enough evidence to reject H_0. That means that according to your data, there is not strong evidence that vomiting is experienced any more by those taking this drug when compared to a placebo.

WARNING

A *p*-value that's very close to that magical but somewhat arbitrary significance level of 0.05 is what statisticians call a *marginal result.* In the preceding example, because the *p*-value of 0.0548 is close to the borderline between accepting and rejecting H_0, it's generally viewed as a marginal result and should be reported as such.

The beauty of reporting a *p*-value is that you can look at it and decide for yourself what you should conclude. The smaller the *p*-value, the more evidence you have against H_0, but how much evidence is enough evidence? Each person is different. If you come across a report from a study in which someone found a statistically significant result and that result is important to you, ask for the *p*-value so that you can make your own decision. (See Chapter 15 for more.)

EXAMPLE

Q. Suppose you want to test whether there's a higher percentage of males who are Democrat than females who are Democrat.

(a) Write down your null and alternative hypotheses.

(b) Explain why it doesn't matter what the actual percentage of Democrats is for males or females.

A. Your two populations are males and females, and you compare the percentage of Democrats in each group. This means p_1 equals the percentage of all males who are Democrat, and p_2 equals the percentage of all females who are Democrat.

(a) Your H_0 is that the percentages are the same ($p_1 = p_2$) versus the H_a that $p_1 > p_2$ (because you want to see whether the percentage of males is higher than the percentage of females).

(b) Your only concern should be the difference in the percentage of Democrats for males and females and whether the difference is zero. A zero differential can happen in many ways; for example, both genders have about 40 percent Democrats, or both have 80 percent Democrats. The actual values of the proportions don't matter when you look at their difference ($p_1 - p_2$).

Note: The actual percentages of the two groups do matter when you calculate the overall sample proportion ($\hat{p}$) and then the standard error. The further away from 50 percent the proportions get, the larger the variation you'll see in their estimated difference. For example, if both genders have about 80 percent Democrats, that will result in a larger standard error than if they both had closer to about 40 percent Democrats.

REMEMBER

Saying H_0: $p_1 = p_2$ is the same as saying H_0: $p_1 - p_2 = 0$. Take the first equation and subtract p_2 from each side. The second version gives you a number to put in the null hypothesis (0), which is nice. That's because if the proportions are equal, their difference has to be zero.

YOUR TURN

15 Conduct the hypothesis test H_0: $p_1 - p_2 = 0$ versus H_a: $p_1 - p_2 > 0$, where $\hat{p}_1 = 0.60$, $\hat{p}_2 = 0.50$, $\hat{p} = 0.55$, $n_1 = 100$, and $n_2 = 100$. Use $\alpha = 0.05$.

16 Conduct the hypothesis test H_0: $p_1 - p_2 = 0$ versus H_a: $p_1 - p_2 \neq 0$, where $x_1 = 1{,}000, x_2 = 1{,}100, n_1 = 2{,}500$, and $n_2 = 2{,}500$. Use $\alpha = 0.05$.

Practice Questions Answers and Explanations

1. You have a right-tailed test for one population mean with $\mu_0 = 7$, $\bar{x} = 7.5$, $\sigma = 2$, and $n = 30$. Because you know the population standard deviation (σ) and you have a relatively large enough sample size with $n = 30$, you can use the Z-distribution. The test statistic is $z = \frac{\bar{x} - \mu_0}{\sigma / \sqrt{n}} = \frac{7.5 - 7}{2 / \sqrt{30}} = \frac{0.5}{0.365} = 1.37$. The corresponding p-value is found by looking at the Z-table for $p(Z > 1.37) = 1 - p(Z \le 1.37) = 1 - 0.9147 = 0.0853$.

   Conclusion: You shouldn't reject H_0 because this p-value is greater than the significance level of 0.01. *Interpretation:* According to your data, the mean of this population is 7. You don't have enough evidence to say the mean is more than that.

TIP

Don't stop with the statistically correct conclusion: reject H_0 or fail to reject H_0. Always go back to the question and try to answer it in the context of the problem as best you can. Your instructor will love you for it.

2. You have a two-tailed test for one population mean with $\mu_0 = 75$, $\bar{x} = 73$, $s = 15$, and $n = 100$. You don't know the population standard deviation (σ) and you are just given the sample standard deviation (s). So, the test statistic is $t = \frac{\bar{x} - \mu_0}{s / \sqrt{n}} = \frac{73 - 75}{15 / \sqrt{100}} = \frac{-2}{1.5} = -1.33$. Because the sample size is substantially large and results in a degrees of freedom $(n - 1 = 99)$ that isn't even on the t-table, you can look at the z row in the t-table or just go to the Z-table itself. From the Z-table, you find that the probability of getting a value beyond −1.33 is 0.0918. But because this is a two-tailed test, you need to multiply 0.0918 by 2 to get the final p-value of $2(0.0918) = 0.1836$.

   Conclusion: You shouldn't reject H_0 because this p-value is greater than the significance level of 0.05. *Interpretation:* You don't have enough evidence to say that the mean for this population is anything but 75.

TIP

When doing problems involving hypothesis tests, I recommend you immediately write down what type of test you have and what the given information is, as I do in these solutions. It helps your instructor see where you're going, and it helps you keep it all straight.

3. You have a right-tailed test for one population mean with $\mu_0 = 100$, $\bar{x} = 105$, $s = 30$, and $n = 10$. Notice the sample size is too small to use a Z-distribution, so you have to use the t-distribution. The test statistic is $t = \frac{\bar{x} - \mu_0}{s / \sqrt{n}} = \frac{105 - 100}{30 / \sqrt{10}} = \frac{5}{9.487} = 0.53$. The corresponding p-value is found by looking at the t-table in the row with $n - 1 = 9$ degrees of freedom. In that row, the t-value falls between 0.260955 and 0.702722. Looking up at the column headers tells you that the p-value falls between 0.40 and 0.25.

   Conclusion: You shouldn't reject H_0 because $0.25 < p\text{-value} < 0.40$, which means the p-value has to be greater than the alpha level of 0.10. *Interpretation:* You don't have enough evidence to say that the mean is more than 100.

4. Anything beyond the critical value leads to a rejection of H_0. In this case, the critical value is −1.96, so beyond means "less than"; therefore, any test statistic that comes in less than −1.96 leads you to reject H_0.

5. You have a right-tailed test for one population proportion with $p_0 = 0.50$, $\hat{p} = 0.60$, and $n = 100$. Note the conditions are met to use *z*: $np_o = 100 * (0.50) = 50$, $n(1-p_o) = 100 * (1-0.50) = 50$ are both at least 10.

   The test statistic is $z = \frac{\hat{p}-p_0}{\sqrt{\frac{p_0(1-p_0)}{n}}} = \frac{0.60-0.50}{\sqrt{\frac{0.50(1-0.50)}{100}}} = \frac{0.10}{0.05} = 2.00$. The corresponding *p*-value is found by looking at the Z-table for $p(Z > 2.00) = 1 - p(Z \le 2.00) = 1 - 0.9772 = 0.0228$.

   Conclusion: You should reject H_0 because this *p*-value is less than the significance level of 0.05. *Interpretation:* According to your data, there is strong evidence that the proportion of people in the population who have the characteristic of interest is more than 0.50.

6. You have a left-tailed test for one population proportion with $p_0 = 0.50$, $\hat{p} = 0.40$, and $n = 100$. Note the conditions are met to use *z*: $np_o = 100 * (0.50) = 50$, $n(1-p_o) = 100 * (1-0.50) = 50$ are both at least 10.

   The test statistic is $z = \frac{\hat{p}-p_0}{\sqrt{\frac{p_0(1-p_0)}{n}}} = \frac{0.40-0.50}{\sqrt{\frac{0.50(1-0.50)}{100}}} = \frac{-0.10}{0.05} = -2.00$. The corresponding *p*-value is found by looking at the Z-table for $p(Z \le -2.00) = 0.0228$.

   Conclusion: You should reject H_0 because this *p*-value is less than the significance level of 0.05. *Interpretation:* According to your data, there is strong evidence that the proportion of people in the population who have the characteristic of interest is more than 0.50.

7. You have a two-tailed test for one population proportion with $p_0 = 0.50$, $x = 40$, and $n = 100$. Note the conditions are met to use *z*: $np_o = 100 * (0.50) = 50$, $n(1-p_o) = 100 * (1-0.50) = 50$ are both at least 10. The sample proportion, $\hat{p}$, is the number of people in the sample who have the characteristic of interest divided by *n*, so in this case, $\hat{p} = \frac{x}{n} = \frac{40}{100} = 0.40$. The test statistic is $z = \frac{\hat{p}-p_0}{\sqrt{\frac{p_0(1-p_0)}{n}}} = \frac{0.40-0.50}{\sqrt{\frac{0.50(1-0.50)}{100}}} = \frac{-0.10}{0.05} = -2.00$. The corresponding *p*-value is found by looking at the Z-table for $p(Z \le -2.00) = 0.0228$ and then, because the alternative hypothesis is two-tailed, you double 0.0228 to get a *p*-value of $2(0.0228) = 0.0456$.

   Conclusion: You shouldn't reject H_0 because this *p*-value is greater than the significance level of 0.01. *Interpretation:* You don't have enough evidence to say that the proportion of this population who fall in the group of interest is anything but 0.50.

8. If the die is fair, each face should show up one-sixth of the time. If you let *p* equal the proportion of times this die will show a 1, you have $H_0: p = \frac{1}{6}$. The alternative is that the coin isn't fair, so either $p < \frac{1}{6}$ or $p > \frac{1}{6}$ fits that case, which means you have $H_a: p \ne \frac{1}{6}$.

9. You have a left-tailed test for two population means with $\bar{x}_1 = 7$, $\bar{x}_2 = 8$, $\sigma_1 = 2$, $\sigma_2 = 2$, $n_1 = 30$, and $n_2 = 30$. Because you know the population standard deviations and you have relatively large enough sample sizes with $n_1 = 30$ and $n_2 = 30$, you can use the Z-distribution. The test statistic is $z = \frac{(\bar{x}_1-\bar{x}_2)-0}{\sqrt{\frac{\sigma_1^2}{n_1}+\frac{\sigma_2^2}{n_2}}} = \frac{(7-8)-0}{\sqrt{\frac{2^2}{30}+\frac{2^2}{30}}} = \frac{-1}{0.516} = -1.94$. The corresponding *p*-value is found by looking at the Z-table for $p(Z < -1.94) = 0.0262$.

Conclusion: You shouldn't reject H_0 because this p-value is greater than the significance level of 0.01. *Interpretation:* According to your data, there is not evidence that the difference in the means of these two populations is not 0 (indicating no statistically significant difference between the means of these two populations).

10. You have a right-tailed test for two population means with $\bar{x}_1 = 75$, $\bar{x}_2 = 70$, $\sigma_1 = 15$, $\sigma_2 = 10$, $n_1 = 50$, and $n_2 = 60$. Because you know the population standard deviations and have large enough sample sizes with $n_1 = 50$ and $n_2 = 60$, you can use the Z-distribution. The test statistic is $z = \frac{(\bar{x}_1 - \bar{x}_2) - 0}{\sqrt{\frac{\sigma_1^2}{n_1} + \frac{\sigma_2^2}{n_2}}} = \frac{(75 - 70) - 0}{\sqrt{\frac{15^2}{50} + \frac{10^2}{60}}} = \frac{5}{2.48} = 2.02$. The corresponding p-value is found by looking at the Z-table for $p(Z \geq 2.02) = 1 - p(Z < 2.02) = 1 - 0.9783 = 0.0217$.

    Conclusion: You should reject H_0 because this p-value is less than the significance level of 0.05. *Interpretation:* According to your data, there is evidence that the difference in the means of these two populations is greater than zero (indicating a statistically significant difference between the means of these two populations).

11. This is the same hypothesis test as in the previous problem, except you run a two-tailed test rather than a right-tailed test. Notice that H_0 and H_a appear different than usual in this problem, but notice also that H_0: $\mu_1 = \mu_2$ is the same as H_0: $\mu_1 - \mu_2 = 0$ (just subtract μ_2 from each side of the original equation). This is another way your professor may write these hypotheses, so be ready for it.

    To work the problem, note that, again, you have a two-tailed test for two population means with $\bar{x}_1 = 75$, $\bar{x}_2 = 70$, $\sigma_1 = 15$, $\sigma_2 = 10$, $n_1 = 50$, and $n_2 = 60$. Again, the test statistic is 2.02. From the Z-table, you find that the probability of getting a value beyond 2.02 is 0.0217. However, because this is a two-tailed test, you need to multiply 0.0217 by 2 to get the final p-value of $2(0.0217) = 0.0434$.

    Conclusion: You should still reject H_0 because this p-value is less than the significance level of 0.05. *Interpretation:* According to your data, there is evidence that the difference in the means of these two populations is greater than zero, with the small p-value indicating a statistically significant difference between the means of these two populations. (And because the difference is positive, you can say that the first population has a larger mean than the second population.)

12. After H_0 has been rejected and you conclude the groups don't have the same mean, the numerator of your test statistic $\bar{x}_1 - \bar{x}_2$ tells you which group has the larger mean. If the numerator of your test statistic is positive, $\bar{x}_1 - \bar{x}_2 > 0$, which means $\bar{x}_1 > \bar{x}_2$. In terms of the populations, the mean of Group 1 (which is μ_1) is likely to be larger than the mean of Group 2 (which is μ_2). If the numerator of your test statistic is negative, $\bar{x}_1 - \bar{x}_2 < 0$, which means $\bar{x}_1 < \bar{x}_2$. In terms of the populations, the mean of Group 1 (which is μ_1) is likely to be smaller than the mean of Group 2 (which is μ_2). Of course, these results all depend on the samples being representative of their populations.

13. You have a right-tailed test for the average difference with $\bar{d} = 2$, $s_d = 5$, and $n_d = 10$. You do this test the same way you conduct a test for one population mean. The test statistic is $t = \frac{\bar{d} - 0}{s_d / \sqrt{n_d}} = \frac{2 - 0}{5 / \sqrt{10}} = \frac{2}{1.58} = 1.27$. Because you don't know the population standard deviation

and the sample size is under 30, you compare your test statistic to the t-distribution with $n_d - 1 = 9$ degrees of freedom. The corresponding p-value is found by looking at the t-table in the row with 9 degrees of freedom. In that row, the t-value falls between 0.702722 and 1.383029. Looking up at the column headers tells you that the p-value falls between 0.25 and 0.10.

Conclusion: You shouldn't reject H_0 because $0.10 < p\text{-value} < 0.25$, which means the p-value has to be greater than the alpha level of 0.05. *Interpretation:* According to your data, there is no evidence that the mean difference for this population is not equal to zero (indicating no statistically significant mean difference for this population).

(14) You have the same hypothesis test here as in the previous problem, except the sample size is larger for this test. You have a right-tailed test for the average difference with $\bar{d} = 2$, $s_d = 5$, and $n_d = 30$. You do this test the same way you conduct a test for one population mean. The test statistic is $t = \frac{\bar{d} - 0}{s_d / \sqrt{n_d}} = \frac{2 - 0}{5 / \sqrt{30}} = \frac{2}{0.913} = 2.19$, which is larger than the test statistic in the previous problem. Because you don't know the population standard deviation, you compare your test statistic to the t-distribution with $n_d - 1 = 29$ degrees of freedom. The corresponding p-value is found by looking at the t-table in the row with 29 degrees of freedom. In that row, the t-value falls between 2.04523 and 2.46202. Looking up at the column headers tells you that the p-value falls between 0.025 and 0.01.

Conclusion: You should reject H_0 because $0.01 < p\text{-value} < 0.025$, which means the p-value has to be less than the alpha level of 0.05. *Interpretation:* According to your data, there is evidence that the mean difference for this population is greater than zero (indicating a statistically significant and positive mean difference for this population).

REMEMBER

A larger sample size means that you get a smaller standard error in the denominator of your test statistic. It makes your test statistic more extreme, which increases its chances of crossing over into the rejection region (or similarly, having a smaller p-value).

(15) You have a right-tailed test for two population proportions $\hat{p}_1 = 0.60$, $\hat{p}_2 = 0.50$, $\hat{p} = 0.55$, $n_1 = 100$, and $n_2 = 100$. The test statistic is $z = \frac{(\hat{p}_1 - \hat{p}_2) - 0}{\sqrt{\hat{p}(1 - \hat{p})\left(\frac{1}{n_1} + \frac{1}{n_2}\right)}} = \frac{(0.60 - 0.50) - 0}{\sqrt{0.55(1 - 0.55)\left(\frac{1}{100} + \frac{1}{100}\right)}} = \frac{0.10}{0.07} = 1.43$. The corresponding p-value is found by looking at the Z-table for $p(Z > 1.43) = 1 - p(Z \leq 1.43) = 1 - 0.9236 = 0.0764$.

Conclusion: You shouldn't reject H_0 because this p-value is greater than the significance level of 0.05. *Interpretation:* According to your data, there is no evidence that the difference in the proportions for these two populations is not equal to zero (indicating no statistically significant difference between the proportions for these two populations).

16. You have a two-tailed test for two population proportions with $\hat{p}_1 = \frac{x_1}{n_1} = \frac{1{,}000}{2{,}500} = 0.40$, $\hat{p}_2 = \frac{x_2}{n_2} = \frac{1{,}100}{2{,}500} = 0.44$, and $\hat{p} = \frac{x_1 + x_2}{n_1 + n_2} = \frac{1{,}000 + 1{,}100}{2{,}500 + 2{,}500} = \frac{2{,}100}{5{,}000} = 0.42$. The test statistic is $z = \frac{(\hat{p}_1 - \hat{p}_2) - 0}{\sqrt{\hat{p}(1-\hat{p})\left(\frac{1}{n_1} + \frac{1}{n_2}\right)}} = \frac{(0.40 - 0.44) - 0}{\sqrt{0.42(1-0.42)\left(\frac{1}{2{,}500} + \frac{1}{2{,}500}\right)}} = \frac{-0.04}{0.0140} = -2.86$. The corresponding p-value is found by looking at the Z-table for $p(Z \leq -2.86) = 0.0021$ and then, because the alternative hypothesis is two-tailed, you double 0.0021 to get a p-value of $2(0.0021) = 0.0042$.

    Conclusion: You should reject H_0 because this p-value is less than the significance level of 0.05. *Interpretation:* According to your data, there is evidence that the difference in the proportions for these two populations isn't equal to zero, with the small p-value indicating a statistically significant difference between the proportions for these two populations. (And because the test statistic is negative [taking Group 1 – Group 2], the proportion in the first population who have that characteristic of interest is likely to be lower than the proportion in the second population who have that characteristic of interest.)

If you're ready to test your skills a bit more, take the following chapter quiz that incorporates all the chapter topics.

Whaddya Know? Chapter 16 Quiz

Quiz time! Complete each problem to test your knowledge on the various topics covered in this chapter. You can then find the solutions and explanations in the next section.

1. True or false? Because the t-distribution has fatter tails than the Z-distribution, you get a larger p-value from the t-distribution than one that the standard normal (Z-) distribution would have given you for the same test statistic.

2. Suppose you are testing the difference in the averages, where $H_a = \mu_1 - \mu_2 > 0$, and your p-value is 0.07. What do you conclude? Assume the significance level of the test is 0.05.

3. Suppose you are testing the difference in the averages, where $H_a = \mu_1 - \mu_2 > 0$, and your p-value is 0.01. What do you conclude? If you reject H_0, state which mean is higher. Assume the significance level of the test is 0.05.

4. Suppose you do two hypothesis tests and everything is the same except that in the first test, the sample size is larger than in the second test. Will the test statistics be different, and if so, how?

5. What do you have to do to the probability of being beyond your test statistic when you have a two-tailed test in order to calculate the p-value?

6. Suppose you want to know which of your two classes did better on the final exam. You do not know the population standard deviations so you have to use sample standard deviations. Which technique do you use?

7. True or False? Stating in H_0 that difference in proportions is zero is the same as saying that the proportions are equal.

8. Looking at the average difference is the same as looking at the difference in the averages. True or False?

9. Which hypothesis test does the following test statistic go with: $\frac{\bar{d} - 0}{s_d / \sqrt{n_d}}$?

10. True or False? The beauty of reporting a p-value is that you can look at it and decide for yourself what you should conclude. (Assume no one has set a significance level.)

Answers to Chapter 16 Quiz

1. **True.** The t-distribution has fatter tails than the Z-distribution, so the same test statistic on a t-distribution doesn't go out as far as the Z-distribution. This means you get a larger p-value and it's harder to reject H_0. This is the penalty for not knowing the population standard deviation.

2. Because the p-value is greater than 0.05, **you fail to reject H_0**; you don't have enough evidence that the difference in the averages is significantly greater than 0.

3. Because the p-value is less than 0.05, **you reject H_0**; you have enough evidence that the difference in the averages is significantly greater than 0. This means **the first mean is higher than the second mean,** because their difference is positive.

4. A larger sample size means that you get a smaller standard error in the denominator of your test statistic. **It makes your test statistic more extreme (further from zero), which increases its chances of crossing over into the rejection region** (or similarly, having a smaller p-value).

5. **You double the probability of being beyond your test statistic when you have a two-sided or two-tailed test.** This is because you could have fallen in the lower tail or in the upper tail with your test statistic, and you need to account for both possibilities. So double this "tail probability" before finding the p-value when H_a is $\neq$.

6. You want to know which of your two classes did better on the final exam. This involves looking at the difference in the averages (means) for the two classes. Because you don't know the population standard deviations, **you use the sample standard deviations, and that means using the (two-sample) t-test.**

7. **True.** H_0: $p_1 - p_2 = 0 \rightarrow H_0$: $p_1 = p_2$.

8. **False.** The average difference is the average of the difference, which takes the differences first, then finds their average (as in a paired test), and you have H_0: $\mu_d = 0$. The difference in the averages takes the averages first and looks at their difference, and you have H_0: $\mu_1 - \mu_2 = 0$.

9. **The paired t-test,** which looks at paired data, finds the differences, and tests the average difference.

10. **True.** The smaller the p-value, the more evidence you have against H_0, but how much evidence is enough evidence? Each person is different.

5 Statistical Studies and the Hunt for a Meaningful Relationship

In This Unit . . .

IN THIS CHAPTER

» Realizing the impact of polls and surveys

» Going behind the scenes of polls and surveys

» Detecting biased and inaccurate survey results

Chapter 17 Polls, Polls, and More Polls

Surveys are all the rage amid today's information explosion. Everyone wants to know how the public feels about issues from prescription drug prices and methods of disciplining children to approval ratings of the president and ratings of reality TV shows. Polls and surveys are a big part of American life; they're a vehicle for quickly getting information about how you feel, what you think, and how you live your life, and they're a means of quickly disseminating information about important issues. Surveys highlight controversial topics, raise awareness, make political points, stress the importance of an issue, and educate or persuade the public.

REMEMBER

Survey results can be powerful, because when many people hear that "such and such percentage of the American people do this or that," they accept these results as the truth, and then make decisions and form opinions based on that information. But in fact, many surveys *don't* provide correct, complete, or even fair or balanced information.

In this chapter, I discuss the impact of surveys and how they're used, and I take you behind the scenes of how surveys are designed and conducted so you know what to watch for when examining survey results and how to run your own surveys right. I also talk about how to interpret survey results and how to spot biased and inaccurate information, so that you can determine for yourself which results to believe and which to ignore.

Recognizing the Impact of Polls

A *survey* is an instrument that collects data through questions and answers. It is used to gather information about the opinions, behaviors, demographics, lifestyles, and other reportable characteristics of the population of interest. What's the difference between a poll and a survey? Statisticians don't make a clear distinction between the two, but I've noticed that what people call a poll is typically a short survey containing only a few questions (maybe that's how researchers get more people to respond — they call it a poll rather than a survey!). But for all intents and purposes, surveys and polls are the same thing.

You come into contact with surveys and their results on a daily basis. Compared to other types of studies, such as medical experiments, some surveys can be relatively easy to conduct. They provide quick results that can often make interesting headlines in newspapers or eye-catching stories in magazines. People connect with surveys because they feel that survey results represent the opinions of people just like themselves (even though they may never have been asked to participate in a survey). And many people enjoy seeing how other people feel, what they do, where they go, and what they care about. Looking at survey results makes people feel linked with a bigger group, somehow. That's what *pollsters* (the people who conduct surveys) bank on, and that's why they spend so much time doing surveys and polls and reporting the results of this research.

Getting to the source

Who conducts surveys these days? Pretty much anyone and everyone who has a question to ask. Some of the groups that conduct polls and report the results include the following:

- News organizations
- Political parties and candidates running for office
- Professional polling organizations (such as the Gallup Organization, the Harris Poll, Zogby International, and the National Opinion Research Center [NORC])
- Representatives of magazines, TV shows, and radio programs
- Professional research organizations (like the American Medical Association, Smithsonian Institution, and Pew Research Center for the People and the Press)
- Special-interest groups (such as the National Rifle Association, Greenpeace, and the American Civil Liberties Union)
- Academic researchers
- The United States government
- Joe Six-Pack (who can easily conduct their own survey on the Internet)

REMEMBER

Some surveys are just for fun, and others are more serious. Be sure to check the source of any serious survey in which you're asked to participate and for which you're given results. Groups that have a special interest in the results should either hire an independent organization to conduct (or at least to review) the survey, or offer copies of the survey questions to the public. Groups should also disclose in detail how the survey was designed and conducted, so that the public can make an informed decision about the credibility of the results.

RANKING THE WORST CARS OF THE MILLENNIUM

You may be familiar with a radio show called *Car Talk* (repeat episodes are typically aired Saturday mornings on National Public Radio). The hosts offer wise and wacky advice to callers with strange car problems. The show's website regularly offers "just for fun" surveys on a wide range of car-related topics, such as, "Who has bumper stickers on their cars, and what do they say?" One of their surveys asked the question, "What do you think was the worst car of the millennium?" Thousands upon thousands of folks responded with their votes — but, of course, these folks don't represent all car owners. They represent only those who listen to the radio show, logged on to the website, and answered the survey question.

Just so you won't be left hanging (and I know you're dying to find out!), the results of the survey are shown in the following table. Although you may not be old enough to remember some of these vehicles, it is certainly an easy exercise to search the Internet for a plethora of pictures and stories about them. (Remember, though, that these results represent only the opinions of *Car Talk* fans who took the time to get to the website and take the survey.) Notice that the percentages won't add up to 100 percent because the results in the table represent only the top ten vote-getters.

Rank	Type of Car	Percentage of Votes
1	Yugo	33.7%
2	Chevy Vega	15.8%
3	Ford Pinto	12.6%
4	AMC Gremlin	8.5%
5	Chevy Chevette	7.0%
6	Renault Le Car	4.3%
7	Dodge Aspen / Plymouth Volare	4.1%
8	Cadillac Cimarron	4.0%
9	Renault Dauphine	3.6%
10	Volkswagen (VW) Bus	2.7%

Surveying what's hot

The topics of many surveys are driven by current events, issues, and areas of interest; after all, timeliness and relevance to the public are two of the most attractive qualities of any survey. Here are just a few examples of some of the subjects being brought to the surface by today's surveys, along with some of the results being reported:

- Does celebrity activism influence the political opinions of the American public? (Over 90 percent of the American public says no, according to CBS News.)
- What percentage of Americans have dated a co-worker? (A whopping 40 percent have, according to a career networking website.)
- How many patients surf the web to find health-related information? (An impressive 55 percent do, according to a national medical journal.)

When you read the preceding survey results, do you find yourself thinking about what the results mean to you, rather than first asking yourself whether the results are valid? Some of the preceding survey results are more valid and accurate than others, and you should think about whether to believe the results first, before accepting them without question. Nationally known polling and research organizations such as those mentioned in the previous section are credible sources, as well as journals that are *peer-reviewed* (meaning all papers published in the journal have been reviewed by others in the field and have passed a certain set of standards). And the U.S. government does a good job with their data collection as well. If you are not familiar with a group conducting a survey and the results are important to you, check out the source.

Impacting lives

Whereas some surveys are just fun to look at and think about, other surveys can have a direct impact on your life or your workplace. These life-decision surveys need to be closely scrutinized before action is taken or important decisions are made. Surveys at this level can cause politicians to change or create new laws, motivate researchers to work on the latest problems, encourage manufacturers to invent new products or change business policies and practices, and influence people's behavior and ways of thinking. The following are some examples of survey results that can impact you.

- **Children's healthcare suffers:** A survey of 400 pediatricians by the Children's National Medical Center in Washington, D.C., reported that pediatricians spend, on average, only 8 to 12 minutes with each patient.
- **Teens drink more:** According to the 2009 Partnership Attitude Tracking Study and the CDC, the number of teens in grades 9 through 12 who use alcohol has grown by 21 percent (from 39 percent in 2009 to 60 percent in 2019), reversing the downward trend experienced in the ten years prior to the survey.

REMEMBER

Always look at how researchers define the terms they're using to collect their data. In the preceding example, how did they define "alcohol use"? Does it count if the teenager tried alcohol once? Does it mean they drink alcohol on a consistent basis? Results can be misleading if the range of what or who gets counted is too wide. Find out what questions were actually asked when the data was collected.

- **Crimes go unreported:** The U.S. Bureau of Justice Crime Victimization Survey concludes that only 49.4 percent of violent crimes were reported to police. The reasons victims gave for not reporting crimes to the police are listed in Table 17-1.

The most frequently given reason for not reporting a violent crime to the police was that the victim considered it to be a personal matter (19.2 percent). Note that almost 12 percent of the reasons relate to perception of the reporting process itself (for example, that it would take too much time or that the police would be bothered, biased, or ineffective).

WARNING

By the way, did you notice how large the "Other reasons" category is? This large, unexplained percentage indicates that the survey can be more specific and/or more research can be done regarding why crime victims don't report crimes. Maybe the victims themselves aren't even sure.

TABLE 17-1 Reasons Victims Didn't Report Violent Crimes

Reason for Not Reporting	Percentage of Victims
Considered it to be a personal matter	19.2%
The offender was not successful/didn't complete the crime	15.9%
Reported the crime to another official	14.7%
Didn't consider the crime to be important enough	5.5%
Didn't think police would want to be bothered	5.3%
Lack of proof	5.0%
Fear of reprisal	4.6%
Too inconvenient/time consuming to report it	3.9%
Thought police would be biased/ineffective	2.7%
Property stolen had no ID number	0.5%
Not aware that a crime occurred until later	0.4%
Other reasons	22.3%

Behind the Scenes: The Ins and Outs of Surveys

Surveys and their results are a part of your daily experience, and you use these results to make decisions that affect your life. (Some decisions may even be life changing.) Looking at surveys with a critical eye is important. Before taking action or making decisions based on survey results, you must determine whether those results are credible, reliable, and believable. A good way to begin developing these detective skills is to go behind the scenes and see how surveys are designed, developed, implemented, and analyzed.

The survey process can be broken down into a series of ten steps:

1. **Clarify the purpose of your survey.**
2. **Define the target population.**
3. **Choose the type and timing of the survey.**
4. **Design the introduction with ethics in mind.**
5. **Formulate the questions.**
6. **Select the sample.**
7. **Carry out the survey.**
8. **Follow up, follow up, and follow up.**
9. **Organize and analyze the data.**
10. **Draw conclusions.**

Each step presents its own set of special issues and challenges, but each step is critical in terms of producing survey results that are fair and accurate. This sequence of steps helps you design, plan, and implement a survey, but it can also be used to critique someone else's survey, if those results are important to you.

Planning and designing a survey

The purpose of a survey is to answer questions about a target population. The *target population* is the entire group of individuals that you're interested in drawing conclusions about. In most situations, surveying the entire target population (that is, conducting a full-blown *census*) is impossible because researchers would have to spend too much time or money to do so. Usually, the best you can do is to select a sample of individuals from the target population, survey those individuals, and then draw conclusions about the target population based on the data from that sample.

Sounds easy, right? Wrong. Many potential problems arise after you realize that you can't survey everyone in the entire target population. Then, after a sample is selected, many researchers aren't sure what to do to get the data they need. Unfortunately, many surveys are conducted without taking the time needed to think through these issues, resulting in errors, misleading results, and wrong conclusions. In the following sections, I give specifics for the first five steps in the survey process.

Clarifying the purpose of your survey

This sounds like it should just be common sense, but in reality, many surveys have been designed and carried out that never met their purpose, or that met only some of the objectives, but not all of them. Getting lost in the questions and forgetting what you're really trying to find out is easy to do. In stating the purpose of a survey, be as specific as possible. Think about the types of conclusions you would want to make if you were to write a report, and let that help you determine your goals for the survey.

Lots of researchers can't see the forest for the trees. If Sue, a restaurant manager, wants to determine and compare satisfaction rates for her customers, she needs to think ahead about what kinds of comparisons she wants to make and what information she wants to be able to report on. Questions that pinpoint when the customers came into the restaurant (date and time), or even what table they were at, are relevant. And if she wants to compare satisfaction rates for, say, adults versus families, she needs to ask how many people were in the party and how many were children. But if she simply asks a couple of questions on satisfaction or throws in every question she can think of, without considering in advance why she needs the information, she may end up with more questions than answers.

TIP

The more specific you can be about the purpose of the survey, the more easily you can design questions that meet your objectives, and the better off you'll be when you need to write your report.

Defining the target population

Suppose, for example, that you want to conduct a survey to determine the extent to which people send and receive personal email in the workplace. You may think that the target population is email users in the workplace. However, you want to determine the *extent* to which personal email is used in the workplace, so you can't just ask email users, or your results will be biased against those who don't use email in the workplace. But should you also include those who don't even have access to a computer during their workday? (See how quickly surveys can get tricky?)

The target population that probably makes the most sense here is all the people who use Internet-connected computers in the workplace. Everyone in this group at least has access to email, though only some of those with access to email in the workplace actually use it, and of those who use it, only some use it for personal email. (And that's what you want to find out — how much they use email for that purpose.)

REMEMBER

You need to be clear in your definition of the target population. Your definition is what helps you select the proper sample, and it also guides you in your conclusions, so that you don't overgeneralize your results. If the researcher didn't clearly define the target population, this can be a sign of other problems with the survey.

Choosing the type and timing of the survey

The next step in designing your survey is to choose what type of survey is most appropriate for the situation at hand. Surveys can be done over the phone, through the mail, with door-to-door interviews, or over the Internet. However, not every type of survey is appropriate for every situation. For example, suppose you want to determine some of the factors that relate to illiteracy in the United States. You wouldn't want to send a survey through the mail, because people who can't read won't be able to take the survey. In that case, a telephone interview is more appropriate.

REMEMBER

Choose the type of survey that's most appropriate for the target population, in terms of getting the most truthful and informative data possible. You also have to keep in mind the budget you have to work with; door-to-door interviews are more expensive than phone surveys, for example. When examining the results of a survey, be sure to look at whether the type of survey used is most appropriate for the situation, keeping budget considerations in mind.

Next you need to decide when to conduct the survey. In life, timing is everything, and the same goes for surveys. Current events shape people's opinions all the time, and although some pollsters try to determine how people feel about those events, others take advantage of events, especially negative ones, and use them as political platforms or as fodder for headlines and controversy. For example, surveys about gun control often come up after a shooting takes place. Also take note of other events that are going on at the time of the survey; for example, people may not want to answer their phones during the Super Bowl, on election night, during the Olympics, or around holidays. Improper timing can lead to bias.

In addition to the date, the time of day is also important. If you conduct a telephone survey to get people's opinions on stress in the workplace and you call them at home between the hours of 9 a.m. and 5 p.m., you're going to have bias in your results; those are the hours when the majority of people are at work (busy being stressed out!).

Designing the introduction with ethics in mind

While this rule doesn't apply to little polls that you see on the Internet and in magazines, serious surveys need to provide information pertaining to important ethical issues. First, they should include what pollsters call a *cover letter* — an introduction that explains the purpose of the survey, what will be done with the data, whether the information the respondent supplies will be confidential or anonymous (see the sidebar, "Anonymity versus confidentiality," later in this chapter), and that the person's participation is appreciated but not required. The cover letter should also provide the researcher's contact information for respondents to use if they have questions or concerns.

REMEMBER

If the survey is done by any institution or group that is federally regulated, such as a university, research institute, or hospital, the survey has to be approved in advance by a committee designated to review, regulate, and/or monitor the research to make sure it's ethical, scientific, and follows regulations. Such committees are called institutional review boards (IRBs), independent ethics committees (IECs), or ethical review boards (ERBs). The survey cover letter should explain who has approved the research. If you don't see such information, ask.

Formulating the questions

After the purpose, type, timing, and ethical issues of the survey have been addressed, the next step is to formulate the questions. The way that the questions are asked can make a huge difference in the quality of the data that will be collected. One of the single most common sources of bias in surveys is the wording of the questions. Research shows that the wording of the questions can directly affect the outcome of a survey. *Leading questions,* also called *misleading questions,* are designed to favor a certain response over another. They can greatly affect how people answer the questions, and their responses may not accurately reflect how they truly feel about an issue.

For example, here are two ways that I've seen survey questions worded about a proposed school bond issue (both of which are leading questions):

> *Don't you agree that a tiny percentage increase in sales tax is a worthwhile investment in improving the quality of the education of our children?*
>
> *Don't you think we should stop increasing the burden on the taxpayers and stop asking for yet another sales tax hike to fund the wasteful school system?*

From the wording of each of these leading questions, you can easily see how the pollsters want you to respond. To make matters worse, neither question tells you exactly how much of a tax increase is being proposed, which is also misleading.

REMEMBER

The best way to word a question is in a neutral way, giving the reader the necessary information required to make an informed decision. For example, the tax issue question is better worded this way:

> *The school district is proposing a 0.01 percent increase in sales tax to provide funds for a new high school to be built in the district. What's your opinion on the proposed sales tax? (Possible responses: strongly in favor, in favor, neutral, against, strongly against.)*

If the purpose of a survey is purely to collect information rather than influence or persuade the respondent, the questions should be worded in a neutral and informative way in order to minimize bias. The best way to assess the neutrality of a question is to ask yourself whether you can tell how the person wants you to respond. If the answer is yes, that question is a leading question and can give misleading results.

TIP

If the results of a survey are important to you, ask the researcher for a copy of the questions used on the survey so you can assess the quality of the questions. When conducting your own survey, have others check the questions to verify that the wording is neutral and informative.

EXAMPLE

Q. Why is it important to state the purpose of a survey before you conduct it?

A. I can think of two reasons. First, if you state the purpose of the survey, you give your audience a clear idea of what you want to determine. Second, stating the purpose of the survey helps keep the researchers on task, so they don't go off the track finding information about people or topics that don't adhere to the original survey agenda.

YOUR TURN

1 You want to determine the extent to which employees use personal email in the workplace of a certain company. What's your target population?

Suppose your target population is the homeless population in a certain city, and your purpose is to find out how they live from day to day. Explain what types of surveys are inappropriate and why, and suggest a reasonable format to get the information you need.

Explain what's wrong with the following survey question and how you would fix it: "Don't you agree that we have wasted too many resources trying to find alternative fuels in this country and that we should do something about that?"

What problems can you foresee if you conduct a survey to assess opinions on gun control right after a shooting takes place in a school?

Selecting the sample

After the survey has been designed, the next step is to select people to participate in the survey. Because you typically don't have time or money to conduct a census (a survey of the entire target population), you need to select a subset of the population, called a *sample.* How this sample is selected can make all the difference in terms of the accuracy and the quality of the results.

Three criteria are important in selecting a good sample, as you find out in the following sections.

A good sample represents the target population

To represent the target population, the sample must be selected from the target population, the whole target population, and nothing but the target population. Suppose you want to find out how many hours of TV Americans watch in a day, on average. Asking students in a dorm at a local university to record their TV viewing habits isn't going to cut it. Students represent only a portion of the target population.

Unfortunately, many people who conduct surveys don't take the time or spend the money to select a representative sample of people to participate in the study, and they end up with biased survey results. When presented with survey results, find out how the sample was selected before examining the results of the survey and see how well they match the target population.

A good sample is selected randomly

A *random* sample is one in which every possible sample (of the same size) has an equal chance of being selected from the target population. The easiest example to visualize here is that of a hat (or bucket) containing individual slips of paper, each with the name of a person written on it; if the slips are thoroughly mixed before each slip of paper is drawn out, the result will be a random sample of the target population (in this case, the population of people whose names are in the hat). A random sample eliminates bias in the sampling process.

Reputable polling organizations, such as the Gallup Organization, use a random digit-dialing procedure to telephone the members of their sample. Of course, this excludes people without telephones, but because most American households today have at least one telephone, the bias involved in excluding people without telephones is relatively small.

Beware of surveys that have a large but not randomly selected sample. Internet surveys are the biggest culprit. Someone can say that 50,000 people logged on to a website to answer a survey, and that means the person posting this site has gotten a lot of data. But the information is biased; research shows that people who respond to surveys tend to have stronger opinions than those that don't respond. And if they didn't even select the participants randomly to start with, imagine how strong (and biased) the respondents' opinions would be. If the survey designer sampled fewer people but did so randomly, the survey results would be more accurate.

A good sample is large enough for the results to be accurate

If you have a large sample size, and if the sample is representative of the target population and is selected at random, you can count on that information being pretty accurate. *How* accurate depends on the sample size, but the bigger the sample size, the more accurate the information will be (as long as that information is good information). The accuracy of most survey questions is measured in terms of a percentage. This percentage is called the *margin of error*, and it represents how much the researcher would expect the results to vary if they were to repeat the survey many times using different samples of the same size. Read more about this in Chapter 13.

A quick-and-dirty formula to estimate the minimum amount of accuracy of a survey involving categorical data (such as gender or political affiliation) is to take 1 divided by the square root of the sample size. For example, a survey of 1,000 (randomly selected) people is accurate to within ±0.032, or 3.2 percentage points. (See Chapter 13 for the exact formula for calculating the accuracy of a survey.) In cases where not everyone responded, you should replace the sample size with the number of respondents (see the section, "Following up, following up, and following up," later in this chapter). Remember, these quick-and-dirty estimates of accuracy are conservative; using the precise formulas gives you accuracy rates that are often much better than these estimates (see Chapter 14 for details).

REMEMBER

With large populations (in the thousands, say) it's the size of the sample, not the size of the population, that matters. For example, if you randomly sample 1,000 individuals from a large population, your accuracy level is estimated to be within 3.2 percentage points, no matter whether you sample from a small town of 10,000 people, a state of 1,000,000 people, or all of the United States. It's amazing how accurate you can get with such a comparatively small sample size.

However, with small populations, you have to apply different methods to determine accuracy and sample size. A sample of 10 out of a population of 100 takes a much larger piece out of the pie than a sample of 10 out of 10,000 does, for example. More advanced methods involving a finite population correction handle issues that come up with small populations.

EXAMPLE

Q. Your psychology professor undertakes a study on the impact of negative political ads on potential voters in the general public. They find participants for the study by offering their students extra credit.

(a) What's the professor's target population for the study?

(b) Does this sample represent the target population? (What was the sampled population here?)

(c) What's the impact of using this sample, in terms of the conclusions the professor may make?

A. This kind of situation happens all the time in research, although it should be avoided.

(a) The target population is potential voters in the general public, which means anyone who's eligible to vote.

(b) The sample doesn't represent the target population; it only represents a small part of it. (The sampled population is the students from this particular university who were enrolled in psychology at that time.)

(c) Any results based on this sample of students should be made only about the students, not about the general public, or they will be biased and misrepresent the truth.

YOUR TURN

While you're surfing a news website, a pop-up asks you to participate in a survey to determine how people feel about reality television. What is the target population? What is the sampled population? Do they agree? Explain.

Bob wants to survey post-Christmas shoppers, so he goes to the local mall, walks up to people at random, and asks them to participate in his survey. Is this a random sample?

Sue wants to conduct a telephone survey of people who live in her town. She opens the phonebook to a random page and selects the first 100 names she sees. Is this a random sample?

A study, as described by the media, concludes that a new type of hair dye isn't harmful to your health. It turns out that it was based on only 14 people. Do you believe the results? Why or why not?

Carrying out a survey

The survey has been designed, and the participants have been selected. Now you have to go about the process of carrying out the survey, which is another important step — one where lots of mistakes and biases can occur.

Collecting the data

During the survey itself, the participants can have problems understanding the questions, they may give answers that aren't among the choices (in the case of a multiple-choice question), or they may decide to give answers that are inaccurate or blatantly false; the latter is called *response bias*. (As an example of response bias, think about the difficulties involved in getting people to tell the truth about whether they've cheated on their income-tax forms.)

Some of the potential problems with the data-collection process can be minimized or avoided with careful training of the personnel who carry out the survey. With proper training, any issues that arise during the survey are resolved in a consistent and clear way, and fewer errors are made in recording the data. Problems with confusing questions or incomplete choices for answers can be resolved by conducting a pilot study on a few participants prior to the actual survey and then, based on their feedback, fixing any problems with the questions.

Personnel can also be trained to create an environment in which each respondent feels safe enough to tell the truth; ensuring that privacy will be protected also helps encourage more people to respond. To minimize interviewer bias, the interviewers must follow a script that's the same for each subject.

WARNING

Beware of conflicts of interest that come up with misleading surveys. For example, if you are being asked about the quality of your service by the person who gave you the service, you may not want to respond truthfully. Or, if your physical therapist gives you an "anonymous" feedback survey on your last day and tells you to give it to them when you're done, the survey may have issues of bias.

ANONYMITY VERSUS CONFIDENTIALITY

If you were to conduct a survey to determine the extent of personal email use at work, the response rate would probably be an issue, because many people are reluctant to discuss their use of personal email in the workplace, or at least to do so truthfully. You could try to encourage people to respond by letting them know that their privacy would be protected during and after the survey.

When you report the results of a survey, you generally don't tie the information collected to the names of the respondents, because doing so would violate the privacy of the respondents. You've probably heard the terms *anonymous* and *confidential* before, but what you may not realize is that these two words are completely different in terms of privacy issues. Keeping results *confidential* means that the researchers could tie your information to your name in their report, but they promise that they won't do that. Keeping results *anonymous* means that they have no way of tying your information to your name in their report, even if they wanted to.

If you're asked to participate in a survey, be sure you're clear about what the researchers plan to do with your responses and whether or not your name can be tied to the survey. (Good surveys always make this issue very clear for you.) Then make a decision as to whether you still want to participate.

Following up, following up, and following up

Anyone who has ever thrown away a survey or refused to "answer a few questions" over the phone knows that getting people to participate in a survey isn't easy. If the researcher wants to minimize bias, the best way to handle it is to get as many folks to respond as possible by following up, one, two, or even three times. Offer dollar bills, coupons, self-addressed stamped return envelopes, chances to win prizes, and so on. Every little bit helps.

If only those folks who feel very strongly respond to a survey, that means that only their opinions will count, because the other people who didn't really care about the issue didn't respond, and their "I don't care" vote didn't get counted. Or maybe they did care, but they just didn't take the time to tell anyone. Either way, their vote didn't count.

For example, suppose 1,000 people are given a survey about whether the park rules should be changed to allow dogs without leashes. Most likely, the respondents would be those who strongly agree or disagree with the proposed rules. Suppose only 200 people responded — 100 against and 100 for the issue. That would mean that 800 opinions weren't counted. Suppose none of those 800 people really cared about the issue either way. If you could count their opinions, the results would be $800 \div 1{,}000 = 80\%$ "no opinion," $100 \div 1{,}000 = 10\%$ in favor of the new rules, and $100 \div 1{,}000 = 10\%$ against the new rules. But without the votes of the 800 non-respondents, the researchers would report, "Of the people who responded, 50 percent were in favor of the new rules and 50 percent were against them." This gives the impression of a very different (and a very biased) result from the one you would've gotten if all 1,000 people had responded.

The *response rate* of a survey is a ratio found by taking the number of respondents divided by the number of people who were originally asked to participate. You, of course, want to have the highest response rate you can get with your survey; but how high is high enough to be minimizing bias? The purest of the pure statisticians feel that a good response rate is anything over 70 percent, but I think you need to be a little more realistic. Today's fast-paced society is saturated with surveys; many if not most response rates fall far short of 70 percent. In fact, response rates for today's surveys are more likely to be in the 20 percent to 30 percent range, unless the survey is conducted by a professional polling organization such as Gallup or you are being offered a new car just for filling one out.

REMEMBER

Look for the response rate when examining survey results. If the response rate is too low (much less than 50 percent), then the results are likely to be biased and should be taken with a grain of salt, or even ignored.

REMEMBER

Don't be fooled by a survey that claims to have a large number of respondents but actually has a low response rate; in this case, many people may have responded, but many more were asked and didn't respond.

Note that statistical formulas at this level (including the formulas in this book) assume that your sample size is equal to the number of respondents, so statisticians want you to know how important it is to follow up with people and not end up with biased data due to non-response. However, in reality, statisticians know that you can't always get everyone to respond, no matter how hard you try; indeed, even the U.S. Census doesn't have a 100 percent response rate. One way statisticians combat the non-response problem after the data have been collected is to break down the data to see how well it matches the target population. If it's a fairly good match, they can rest easier on the bias issue.

So which number do you put in for *n* in all those statistical formulas you use so often (such as the sample mean in Chapter 5)? You can't use the intended sample size (the number of people contacted). You have to use the final sample size (the number of people who responded). In the media you most often see only the number of respondents reported, but you also need the response rate (or the total number of respondents) to be able to critically evaluate the results.

REMEMBER

Regarding the quality of results, selecting a smaller initial sample size and following up more aggressively is a much better approach than selecting a larger group of potential respondents and having a low response rate, because of the bias introduced by nonresponse.

EXAMPLE

Q. Discuss two ways that survey participants can give incorrect information on a survey, and explain how you can minimize the problem in each case.

A. They can lie, or they can give the wrong answer by mistake (for example, they may be confused about a question). To minimize lying, make the survey anonymous (no one can link the people to their responses; not even you) or at least confidential (you promise not to divulge the connection between them and their answers). To minimize confusion, train the people asking the questions (for phone surveys or interviews) to be consistent, and make sure that mail surveys have clear questions and directions. Interviewers often have a script to ensure consistency between interviews.

YOUR TURN

9 Response bias occurs when survey participants give biased answers — either systematically higher than the truth or systematically lower than the truth. Give an example of two questions that could lead to these two types of response bias.

Suppose you have two mail surveys; for the first, you mail out 10,000 surveys, and 1,000 people respond. For the second, you mail out 1,500 surveys and send reminder letters, and you ultimately receive 1,000 responses. Which survey do you think will produce more accurate results?

Interpreting results and finding problems

The purpose of a survey is to gain information about your target population; this information can include opinions, demographic information, or lifestyles and behaviors. If the survey has been designed and conducted in a fair and accurate manner with the goals of the survey in mind, the data should provide good information as to what's happening with the target population (within the stated margin of error; see Chapter 13). The next steps are to organize the data to get a clear picture of what's happening; to analyze the data to look for links, differences, or other relationships of interest; and then to draw conclusions based on the results.

Organizing and analyzing

After a survey has been completed, the next step is to organize and analyze the data (in other words, crunch some numbers and make some graphs). Many different types of data displays and summary statistics can be created and calculated from survey data, depending on the type of information that was collected. (Numerical data, such as income, have different characteristics and are usually presented differently than categorical data, such as gender.) For more information on how data can be organized and summarized, see Chapters 4 through 7. Depending on the research question, different types of analyses can be performed on the data, including coming up with population estimates, testing a hypothesis about the population, or looking for relationships, to name a few. See Chapters 14 to 16, and 19 to 20 for more on each of these analyses, respectively.

WARNING

Watch for misleading graphs and statistics. Not all survey data are organized and analyzed fairly and correctly. See Chapter 2 for more about how statistics can go wrong.

Drawing conclusions

The conclusions are the best part of any survey — they're why the researchers do all of the work in the first place. If the survey was designed and carried out properly — the sample was selected carefully and the data were organized and summarized correctly — the results should fairly and accurately represent the reality of the target population. But, of course, not all surveys are done right. And even if a survey is done correctly, researchers can misinterpret or overinterpret results so that they say more than they really should.

You know the saying, "Seeing is believing"? Some researchers are guilty of the converse, which is "Believing is seeing." In other words, they claim to see what they want to believe about the results. This is all the more reason for you to know where the line is drawn between reasonable conclusions and misleading results, and to realize when others have crossed that line.

Here are some common errors made in drawing conclusions from surveys:

- » Making projections to a larger population than the study actually represents
- » Claiming a difference exists between two groups when a difference isn't really there (see Chapter 16)
- » Saying, "These results aren't scientific, but," and then going on to present the results as if they are scientific

To avoid common errors made when drawing conclusions, do the following:

- » Check whether the sample was selected properly and that the conclusions don't go beyond the population presented by that sample.
- » Look for any disclaimers about the survey *before* reading the results. That way, if the results aren't based on a scientific survey (an accurate and unbiased survey), you'll be less likely to be influenced by the results you're reading. You can judge for yourself whether the survey results are credible.
- » Be on the lookout for statistically incorrect conclusions. If someone reports a difference between two groups in terms of survey results, be sure that the difference is larger than the reported margin of error. If the difference is within the margin of error, you should expect the sample results to vary by that much just by chance, and the so-called "difference" can't really be generalized to the entire population (see Chapter 15 for more on this).

Know the limitations of any survey and be wary of any information coming from surveys in which those limitations aren't respected. A bad survey is cheap and easy to do, but you get what you pay for. But don't let big, expensive surveys fool you either — they can be riddled with bias as well! Before looking at the results of any survey, investigate how it was designed and conducted, using the criteria and tips in this chapter, so you can judge the quality of the results and express yourself confidently and correctly about what is wrong.

Q. Suppose a senator wants to introduce a bill to the U.S. Senate, and they use the results of a survey posed to their constituents to help sell the bill to their colleagues. The senator introduces it by saying, "Americans really want this bill." Is this reasonable?

Q. **No.** The senator is generalizing beyond the sample. The survey only represents their constituents, not all American people. If the senator says, "My constituents really want this bill," that statement may be reasonable, assuming they took a large enough random sample.

YOUR TURN

11 Suppose poll results show 48 percent of voters favor Candidate A and 52 percent favor Candidate B. Is it reasonable to say that Candidate B has a clear edge?

12 An evening television news program shows the results of its latest "call-in" poll, which explores how Americans feel about the job the president is doing. The station adds a disclaimer at the end of its piece, admitting that the results aren't scientific. Does this take care of the statistical problems?

13 Explain why a "call-in" poll to an evening news program constitutes an unscientific survey.

14 A survey asks college students a variety of questions about their daily routines and finds out that students who eat breakfast in the morning tend to have higher grades. The survey concludes that eating breakfast causes students to get better grades. Is this a proper conclusion?

Practice Questions Answers and Explanations

1. **The target population is employees who have access to personal email in the workplace** (not just employees who use personal email in the workplace). The target population is the group of individuals that you focus on making conclusions about. Because the extent of the use of personal email is the question, you must consider anyone who has a chance to send personal emails.

REMEMBER

Knowing your target population helps keep the results from being biased. Also, if the actual sample doesn't represent the target population, you know that the survey has a problem.

2. Because you're working with the homeless population, you need to realize that they have no permanent home, so they have no address or phone number for you to contact. A mail or telephone survey is out. Some type of personal interview with them is your best bet, in a non-intimidating setting. Some types of surveys are clearly inappropriate for certain situations, and you need to keep this in mind when you design a survey.

3. This is a leading question; you know exactly how the pollster wants you to answer it. The researcher clearly has an agenda, and the results of their survey are going to be biased. The way to avoid this is to word a poll question in a neutral fashion, such as the following: "What's your opinion on the issue of searching for alternative fuels — strongly agree, agree, no opinion, disagree, or strongly disagree?"

WARNING

Research shows that changing the wording of a survey question even slightly can greatly affect the results. Questions should be worded neutrally so as not to incur bias in the results.

4. Because the shooting just happened, people may have a stronger opinion about gun control than they had before the shooting, and research shows that over time, after the tragedy fades from the mind, many of those folks go back to their original opinions. Although a survey directly after the incident may be informative and newsworthy, you have to put it into a larger timeline, keep the results in perspective, and be careful not to take advantage of certain situations.

REMEMBER

You have to take the timing of a survey under consideration, because it can greatly affect the results. Timing can mean what time of year the survey takes place or what time of day, for that matter. You shouldn't expect accurate results phoning office workers at home during the day, right?

5. The target and sampled populations do not agree. This sample represents only the people who visit the website and would be willing to participate in the survey. The target population is the people the survey implies to represent — all television watchers. The results of this web survey are biased because they don't represent the target population.

6. **No.** The definition of a random sample is a sample selected by giving every sample of the same size within the target population an equal chance to participate. The survey at the mall leaves out anyone who doesn't visit the mall that day. Bob also has no random mechanism for selecting people to participate, so he likely uses his own bias by approaching people who seem more friendly and willing, who walk slower, who don't look busy, and so on.

TIP

A random sample has a strict definition; to determine whether or not a sample is a truly random sample, check to see whether every member of the target population had an equal chance of being selected.

7. **No.** Sue probably wouldn't choose the very first or very last page of the phone book to make her selection, so those pages don't have an equal chance of being selected. Second, choosing all the people from the same page is biased, because certain people may be related or come from a certain ethnic background if they share the same last name.

REMEMBER

A sample can't be "sort of" random or "close to" random. Either a sample is random or it isn't.

8. **No.** Because the study involves only 14 people, the population can't possibly be represented. The results would vary too much from sample to sample (which is what the margin of error measures).

TIP

To get a rough idea of the precision of a survey, take 1 divided by the square root of the sample size. For example, a survey of 400 people is precise to within $\frac{1}{\sqrt{400}} = \frac{1}{20} = 0.05$, or 5 percent. But for only 14 people, the margin of error is roughly 26.7 percent!

9. Many questions are possible. Question 1: "How many fish did you catch on your fishing trip today?" This may entice people to overstate their actual results to avoid embarrassment or to impress people. Question 2: "How much money did you make last year, for income-tax purposes?" This question can tempt people to understate the truth to avoid having to pay higher taxes.

10. **Although both surveys received 1,000 responses, the second survey should probably have much more accurate results,** because $1{,}000 \div 1{,}500$, or 67 percent, of the people surveyed responded, compared to only $1{,}000 \div 10{,}000$, or 10 percent, for the first survey. These percentages are called the *response rates,* and larger response rates typically mean more accurate results, because the survey leaves less chance for bias, and the sample is more representative of the target population.

REMEMBER

People who don't respond to a survey may be different from the people who do, which can bias the results. Always try for a high response rate, because a smaller survey with a high response rate is always better than a larger survey with a low response rate.

11. **Not necessarily.** You should expect the results to vary, and you need to take the margin of error into account. With a margin of error of 2 percent, Candidate A could have between 46 and 50 percent of the vote if you sample again, and Candidate B could have between 50 and 54 percent of the vote. Any larger margin of error causes overlap in the results; you can't say who will win because the percentages from the sample, although numerically different, aren't statistically different. (See Chapter 13 for more on margin of error.)

12. **No.** The results can still affect people; they may take them for the truth, because people regularly ignore disclaimers.

WARNING

Disclaimers about survey results not being scientific don't solve any statistical problems. They merely contaminate the pool of statistical information available. Use such survey results for entertainment purposes only.

13. People who participate in a call-in poll don't represent the target population; they have to have a television, have it turned on, watch that program, and take the initiative to call in with an opinion. They also select themselves to participate in the survey (called a *self-selected sample*). Self-selected samples are biased, non-random samples.

14. **No.** This conclusion cannot be made without further study that controls a host of other factors, such as exercise, sleep habits, dietary habits, study habits, IQ, and so on. This researcher goes too far in terms of trying to make an immediate cause-and-effect connection between certain findings. You can't make these types of conclusions without a study to look at the issue specifically. (See Chapters 18 and 19 for more on cause and effect.)

If you're ready to test your skills a bit more, take the following chapter quiz that incorporates all the chapter topics.

Whaddya Know? Chapter 17 Quiz

Quiz time! Complete each problem to test your knowledge on the various topics covered in this chapter. You can then find the solutions and explanations in the next section.

1. What is the purpose of the American Crime Victimization Survey? (You can look it up online.)

2. Bob selects a random sample of 100 students at his university and wants to study their attitudes on the latest tuition increase. What is Bob's target population?

3. Suppose a gun control group conducts a survey on people's opinions on gun control right after a series of robberies in a large city take place. How might the timing of this survey affect the results?

4. Can the wording of a survey question affect the results? Yes or no?

5. Which is stronger in terms of protecting the subject's information: a confidential survey or an anonymous survey?

6. Which is better: conducting a survey on 100 people and following up to get 100 results, or conducting a survey of 1,000 people and getting 100 results? Or are they the same?

7. Savannah randomly samples 1,000 registered voters in her county and sends them a survey. They all respond. What is the "quick and dirty" measure of the margin of error?

8. Paisley stands on a street corner by the library asking every 10th person to fill out a survey. She gets 100 responses. Does this constitute a random sample?

9. A TV commercial asks viewers to go to a certain sports website and vote for who should go into the Hall of Fame that year. Ten thousand people respond. Is this a good sample?

10. Levi reports on an income survey that he made $50,000 this year when he actually made $150,000. What kind of bias is this?

Answers to Chapter 17 Quiz

1. The purpose of the American Crime Victimization Survey is to do an ongoing study of crime victims, the crimes that happened to them, and whether or not the crimes were reported, among other things.

2. **The target population is all the students at Bob's university.** That's the entire group he's interested in studying.

3. The timing of this survey may tend to bring out more responses in favor of gun control due to all the robberies that just occurred, or more responses in favor of less gun control so that any regular person could have a gun to protect themselves.

4. **Yes.** Even changing one word can affect the outcome of a survey.

5. **An anonymous survey** has no way of connecting the subject to their results and is stronger than a confidential survey, which does.

6. **The first choice is best, choosing a smaller random sample and following up to get the highest response rate possible.** This will minimize biases that can occur when only 100 out of 1,000, or 10% of the individuals, respond.

7. The "quick and dirty" measure of margin of error is **1 divided by the square root of *n*.** In this case, $n = 1{,}000$, so you have a margin of error of about $\frac{1}{\sqrt{1{,}000}} = 0.03$, or 3 percent.

8. **No.** Only the people that happen to be on that street corner next to the library are able to be selected to participate in the survey. That constitutes bias and is not a random sample.

9. **No.** This is what statisticians call a *volunteer survey*, where people choose themselves to participate. It's very biased, and only represents the people who were watching the TV commercial at the time and took it upon themselves to vote.

10. **This is called response bias,** when a person responds to a survey but gives the wrong information. People often do not like to report their actual incomes, and often do not even respond at all (nonresponse bias), or respond with a lower value than they actually make (response bias).

IN THIS CHAPTER

» Distinguishing experiments from observational studies

» Dissecting the criteria for a good experiment

» Understanding how observational studies work

» Watching for misleading results

Chapter 18
Experiments and Observational Studies: Medical Breakthroughs or Misleading Results?

Medical breakthroughs seem to come and go quickly. One day you hear about a promising new treatment for a disease, only to find out later that the drug didn't live up to expectations in the last stage of testing. Pharmaceutical companies bombard TV viewers with commercials for pills, sending millions of people to their doctors clamoring for the latest and greatest cures for their ills, sometimes without even knowing what the drugs are for. Anyone can search the Internet for details about any type of ailment, disease, or symptom and come up with tons of information and advice. But how much can you really believe? And how do you decide which options are best for you if you get sick, need surgery, or have an emergency?

In this chapter, I take you behind the scenes of experiments and observational studies, the driving forces of medical studies and other investigations in which comparisons are made — comparisons that test, for example, which building materials are best, which soft drink teens prefer, and so on. You find out the difference between experiments and observational studies and discover what experiments and observational studies can do for you, how they're supposed to be done, how they can go wrong, and how you can spot misleading results. With so many headlines, sound bytes, and pieces of "expert advice" coming at you from all directions, you need to use all your critical thinking skills to evaluate the sometimes-conflicting information you're presented with on a regular basis.

Boiling Down the Basics of Studies

Although many different types of studies exist, you can basically boil them down to two types: experiments and observational studies. This section examines what exactly makes experiments different from other studies. But before I dive in to the details, I need to lay some jargon on you.

Looking at the lingo of studies

To understand studies, you need to find out what their commonly used terms mean.

- **Participants:** When the study involves people, these are the individuals in the study.
- **Experiment:** A study that deliberately applies treatments to participants/experimental units in a controlled situation and studies their effects on the outcome. Examples include testing types of fertilizer on cornfields, computer chips, types of concrete, or brands of candles and how long they last.
- **Randomized Controlled Trial (RCT):** A type of experiment used often in the health care industry, where participants are randomly placed into groups, given some treatment (or control) and the outcome is compared between the groups. Examples of RCTs are clinical trials that compare the effects of drugs, surgical techniques, medical devices, diagnostic procedures or other medical treatments.
- **Response:** The variable whose outcome is of interest; think of this as the million-dollar question. For example, if researchers want to know what happens to your blood pressure when you take a large amount of ibuprofen each day, the response variable is blood pressure.
- **Factor:** The variable whose effect on the response is being studied. For example, if you want to know whether a particular drug increases blood pressure, your factor is the amount of the drug taken. If you want to know which weight loss program is most effective, your factor is the type of weight loss program used.

  You can have more than one factor in a study; however, in this book I stick with discussing one factor only. For the analysis of two-factor studies, including the use of Analysis of Variance (ANOVA) and multiple comparisons to compare treatment combinations, you can check out *Statistics II For Dummies* by Deborah Rumsey, also published by Wiley.
- **Level:** One possible outcome of a factor. Each factor has a certain number of levels. In the weight-loss example, the factor is the type of weight-loss program and the levels are the specific programs studied (for example, Weight Watchers, South Beach, or the famous Potato Diet). Levels need not be ascending in any way; however, in a study like the drug example, the levels would be the various dosages taken each day, in increasing amounts.

» **Treatment:** A combination of the levels of the factors being studied. If you only have one factor, the levels and the treatments are the same thing. If you have more than one factor, each combination of levels of the factors is called a treatment.

For example, if you want to study the effects of the type of weight-loss program and the amount of water consumed daily, you have two factors: 1) the type of program, with three levels (Weight Watchers, South Beach, Potato Diet); and 2) the amount of water consumed, with, say, three levels (24, 48, and 64 ounces per day). In this case, there are $3 \times 3 = 9$ treatments: Weight Watchers and 24 ounces of water per day; Weight Watchers and 48 ounces of water per day, all the way up to the famous Potato Diet and 64 ounces of water per day. Each subject is assigned to one treatment. (With my luck, I'd get that last treatment.)

» **Cause-and-effect:** A type of relationship in which a change in the factor results in a direct change in the response (for example, increasing calorie intake causes weight gain).

In the following sections, you see the differences between observational studies and experiments, when each is used, and what their strengths and/or weaknesses may be.

Observing observational studies

Just as with tools, you want to find the right type of study for the right job. In certain situations, observational studies are the optimal way to go. The most common observational studies are *polls* and *surveys* (see Chapter 17). When the goal is to find out what people think and to collect some demographic information (such as gender, age, income, and so on), surveys and polls can't be beat, as long as they're designed and conducted correctly.

In other situations, especially those looking for cause-and-effect relationships, simple observational studies may not be optimal. For example, suppose you took a couple of vitamin C pills last week; is that what helped you avoid getting that cold that's going around the office? Maybe the extra sleep you got recently or the extra hand-washing you've been doing helped you ward off the cold. Or maybe you just got lucky this time. With so many variables in the mix, how can you tell which one had an influence on the outcome of your not getting a cold? A simple experiment that takes these other variables into account may be best in this situation. Note that more intricate observational studies can take other variables into account also; it's just the simplest "off the cuff" kind of observational studies like the example I gave we need to be watchful of. Unfortunately, these types of studies are everywhere, and we can be glad that there are strict rules in place for large-scale government studies of this type.

TIP

When looking at the results of any study, first determine what the purpose of the study was and whether the type of study fits the purpose. For example, if an observational study was done instead of an experiment to establish a cause-and-effect relationship, check with a statistician to be sure that the correct methods were used in order to establish that relationship (for example, detailed observational studies done in different situations can bring researchers to a cause-and-effect conclusion, like showing smoking causes lung cancer).

Examining experiments

The object of an experiment is to see if the response changes as a result of the factor you are studying; that is, you are looking for cause and effect. For example, does taking ibuprofen cause blood pressure to increase? If so, by how much? But because results will vary with any experiment, you want to know that your results have a high chance of being repeatable if you found something interesting happening. That is, you want to know that your results were unlikely to be due to chance; statisticians call such results *statistically significant*. That's the objective of any study, observational or experimental. However one must also check that the results are statistically discernible, which means large enough to be important to the problem.

REMEMBER

A good experiment is conducted by creating an environment that mimics a lab (controlled environment) the best you can — so the researcher can pinpoint whether a certain factor or combination of factors causes a change in the response variable, and if so, the extent to which that factor (or combination of factors) influences the response. For example, to gain government approval for a proposed blood pressure drug, pharmaceutical researchers set up randomized control trials to determine whether that drug helps lower blood pressure, what dosage level is most appropriate for each different population of patients, what side effects (if any) occur, and to what extent those side effects occur in each population. It's very difficult to pull this off, but they do a great job, with the help of thousands of researchers across the country and around the world.

EXAMPLE

Q. A study follows two groups for one year. The first group routinely uses antibacterial soap, and the other group routinely uses regular hand soap. The researchers record and compare the number and severity of illnesses for the two groups. Is this an observational study or an experiment?

A. This is an observational study, because the researchers didn't ask the people to use a certain type of soap; the groups used personal preference. The study didn't impose treatments or control other factors.

YOUR TURN

1 Suppose the antibacterial soap group in the example problem experiences fewer and less severe illnesses than the regular hand soap group. Can you conclude that antibacterial soap reduces or even prevents certain illnesses?

Suppose a teacher wants to see whether the use of a computer game helps students to learn to read faster. They divide the class into two groups and teach them the same words, but one group uses the computer game to learn while the other uses a textbook. The results for the two groups are different. Is this an experiment or an observational study?

Suppose you deem an experiment unethical; for example, the experiment forces some people to smoke two packs of cigarettes a day to see whether they develop lung cancer. What should you do in order to show this cause-and-effect relationship in humans?

Suppose a group of people who normally take vitamin C each day experience fewer colds than a group of people who normally don't take vitamin C each day.

(a) What type of study is this?

(b) What other variables may explain the difference in the number of colds for the two groups?

Designing a Good Experiment

How an experiment is designed can mean the difference between good results and garbage. Because most researchers are going to write the most glowing press releases that they can about their experiments, you have to be able to sort through the hype to determine whether to believe the results you're given. To decide whether an experiment is credible, check to see whether it meets *all* the following criteria for a good experiment. A good experiment

- Makes comparisons
- Includes a large enough sample size so that the results are accurate
- Chooses subjects that most accurately represent the target population
- Assigns subjects randomly to the treatment group(s) and the control group
- Appropriately controls for other important variables
- Is ethical, as approved by an Institutional Review Board, or related entity
- Collects good data
- Applies the proper data analysis
- Makes appropriate conclusions

In this section, each of these criteria is explained and illustrated with examples.

Designing the experiment to make comparisons

Every experiment has to make bona fide comparisons where possible to be credible. This seems to go without saying, but researchers often are so gung-ho to prove their results that they forget (or just don't bother) to show that their factor, and not some other factor(s), including random chance, was the actual cause for any differences found in the response.

For example, suppose a researcher is convinced that taking vitamin C prevents colds, and they assign subjects to take one vitamin C pill per day and follow them for six months. Suppose the subjects get very few colds during that time. Can they attribute these results to the vitamin C and nothing else? No; there's no way of knowing whether the subjects would have been just as healthy without the vitamin C, due to some other factor(s), or just by chance. There's nothing to compare the results to.

To tease out the real effect (if any) that your factor has on the response, you need a baseline to compare the results to. This baseline is called the *control.* Different methods exist for creating a control in an experiment; depending on the situation, one method typically rises to the top as being the most appropriate. Three common methods for including control are to administer: 1) a fake treatment; 2) a standard treatment; or 3) no treatment. The following sections describe each method.

When examining the results of an experiment, make sure the researchers established a baseline by creating a control group. Without a control group, you have nothing to compare the results to, and you never know whether the treatment being applied was the real cause of any differences found in the response.

Fake treatments — the placebo effect

A fake treatment (also called a *placebo*) is not distinguishable from a "real" treatment by the subject. For example, when drugs are administered, a subject assigned to the placebo will receive a fake pill that looks and tastes exactly like a real pill; it's just filled with an inert substance like sugar instead of the actual drug. A placebo establishes a baseline measure for what responses would have taken place anyway, in lieu of any treatment (this would have helped the vitamin C study mentioned in the section, "Designing the experiment to make comparisons"). But a fake treatment also takes into account what researchers call the *placebo effect,* a response that people have (or think they're having) because they know they're getting some type of "treatment" (even if that treatment is a fake treatment, such as sugar pills).

Pharmaceutical companies are required to account for the placebo effect when examining both the positive and negative effects of a drug. When you see an ad for a drug in a magazine, you see the positive results of the drug standing out in big, bright, happy, colorful visuals. Then look at the back of the page and you see it's entirely filled in black with words written in 3-point font. Embedded somewhere on that page, you can find one or more tiny tables that show the number and nature of side effects reported by each *treatment group* (subjects who received an actual treatment) as well as the *control group* (subjects who were administered a placebo).

If the control group is on a placebo, you may expect the subjects not to report any side effects, but you would be wrong. If you are taking a pill, you know it could be an actual drug, and you are being asked whether or not you're experiencing side effects, you might be surprised at what your response is.

If you don't take the placebo effect into account, you have to believe that any side effects (or positive results) reported are actually due to the drug. This gives an artificially high number of reported side effects because at least some of those reports are likely due to the placebo effect and not to the drug itself. If you have a control group to compare with, you can subtract the percentage of people in the control group who reported the side effects from the percentage of people in the treatment group who reported the side effects, and examine the magnitude of the numbers that remain. You're, in essence, looking at the net number of reported side effects due to the drug, rather than the gross number of side effects, some of which are due to the placebo effect.

The placebo effect has been shown to be real. If you want to be fair about examining the reported side effects (or positive reactions) of a treatment, you have to also take into account the side effects (or positive reactions) that the control group reports — those reactions that are due to the placebo effect only.

Standard treatments

In some situations, such as when the subjects have very serious diseases for which a treatment is available, offering a fake treatment as an option would be unethical. One would offer the current medical treatment as the control group instead.

When ethical reasons bar the use of fake treatments, the new treatment is compared to at least one existing or standard treatment that is known to be effective. Researchers pick a clinically meaningful effect and use that to determine how many people need to be in the study. Researchers use data to see that one of the treatments is working significantly better (or significantly worse) than the other. They may stop the experiment early and put everyone on the better treatment — again, for ethical reasons if the benefit or harm meets a predetermined threshold before the total sample size is reached.

No treatment

No treatment means the researcher can't help but tell which group the subject is in, due to the nature of the experiment. The subjects in this case aren't receiving any type of intervention in terms of their behavior, but they still serve as a control, establishing a baseline of data to compare their results with those in the treatment group(s). For example, if you want to determine whether speed walking around the block ten times a day lowers a person's resting heart rate after six months, the subjects in your control group know they aren't going to be speed walking — obviously you can't do fake speed walking (although faking exercising and still reaping the benefits would be great, wouldn't it?).

Selecting the sample size

The size of a (good) sample greatly affects the accuracy of the results, so you might think in an experiment the larger the sample size, the better. While it's better to have a larger sample than a smaller sample (where you might not even be able to find results that are really there), we reach a point where we don't want to put more people than necessary in an experiment, so we don't always just use a huge group of participants. There is a balance that has to be reached in experiments and especially randomized control studies between numbers and ethical behavior.

In this section, I hit the highlights; Chapter 15 has the details.

The word *sample* is often attributed to surveys where a random sample is selected from the target population (see Chapter 17). However, in the setting of experiments, a sample refers to the group of participants who have volunteered to do the experiment.

Limiting small samples to small conclusions

You may be surprised at the number of research headlines that have been made regarding large populations that were based on very small samples. Such headlines can be of concern to statisticians, who know that detecting true, statistically significant results in a large population using a small sample is difficult because small data sets have more variability from sample to sample (see Chapter 13). When sample sizes are small and big conclusions have been made by the researchers, either the researchers didn't use the right hypothesis test to analyze their data (for example, using the Z-distribution rather than the *t*-distribution; see Chapter 11) or the difference was so large that it would be very difficult to miss. The latter isn't always the case, however.

Be wary of research conclusions that find significant results based on small sample sizes (especially for experiments involving many treatments but only a few subjects being assigned to each treatment). Statisticians want to see at least five subjects per treatment, but more is generally better, to a point. You do need to be aware of some of the limitations of experiments such as cost and time, as well as ethical issues, and realize that the number of subjects for experiments is often smaller than the number of participants in a survey.

If the results are important to you, ask for a copy of the research report and look to see what type of analysis was done on the data. Also look at the sample of subjects to see whether this sample truly represents the population about which the researchers are drawing conclusions.

Defining sample size

When asking questions about *sample size,* be specific about what you mean by the term. For example, you can ask how many subjects were selected to participate and also ask for the number who actually completed the experiment; these two numbers can be very different. Make sure the researchers can explain any situations in which the research subjects decided to drop out or were unable (for some reason) to finish the experiment.

For example, an article in *The New York Times* titled, "Marijuana Is Called an Effective Relief in Cancer Therapy," says in the opening paragraph that marijuana is "far more effective" than any other drug in relieving the side effects of chemotherapy. When you get into the details, you find out that the results are based on only 29 patients (15 on the treatment, 14 on a placebo). Then you find out that only 12 of the 15 patients in the treatment group actually completed the study. What happened to the other three subjects?

Sometimes researchers draw their conclusions based on only those subjects who completed the study. This can be misleading, because the data don't include information about those who dropped out (and why), which may lead to biased data. In government studies, they have a account for everyone who started the experiment, even if they dropped out or switched treatments for some reason, which often happens.

Choosing the subjects

The first step in carrying out an experiment is selecting the subjects (participants). Although researchers would like their subjects to be selected randomly from their respective populations, in most cases, this just isn't appropriate. For example, suppose researchers wants to test out a new laser surgery on nearsighted people. They need a random sample of subjects, so they randomly select various eye doctors from across the country and randomly select nearsighted patients from these doctors' files. They call up each person selected and say, "We're experimenting with a new laser surgery technique for nearsightedness, and you've been selected at random to participate in our study. When can you come in for the surgery?" Something tells me that this approach wouldn't go over very well with many people receiving the call (although some would probably jump at the chance, especially if they didn't have to pay for the procedure).

REMEMBER

The point is that getting a truly random sample of people to participate in an experiment is generally more difficult than getting a random sample of folks to participate in a survey. However, statisticians can build techniques into the design of an experiment to help minimize the potential bias that can occur.

Making random assignments

One way to minimize bias in an experiment is to introduce some randomness. After the sample has been decided on, the subjects are randomly divided into treatment and control groups. The treatment groups receive the various treatments being studied, and the control group receives the current (or standard) treatment, no treatment, or a placebo. (See the section, "Designing the experiment to make comparisons," earlier in this chapter.)

Making random assignments of subjects to treatments is an extremely critical step toward minimizing bias in an experiment. Suppose a researcher wants to determine the effects of exercise on heart rate. The subjects in their treatment group run 5 miles and have their heart rates measured before and after the run. The subjects in their control group sit on the couch the whole time and watch reruns of old TV shows. Which group would you rather be in? Some health nuts out there would no doubt volunteer for the treatment group. If you're not crazy about the idea of running 5 miles, you may opt for the easy way out and volunteer to be a couch potato. (Or maybe you hate to watch old reruns so much that you'd run 5 miles to avoid that.)

FINDING VOLUNTEERS

To find subjects for their experiments, researchers often advertise for volunteers and offer them incentives such as money, free treatments, or follow-up care for their participation. Medical research on humans is complicated and difficult, but it's necessary in order to really know whether a treatment works, how well it works, what the dosage should be, and what the side effects are. In order to prescribe the right treatments in the right amounts in real-life situations, doctors and patients depend on these studies being representative of the target (intended) population. In order to recruit such representative subjects, researchers have to do a broad advertisement campaign and select enough participants with enough different characteristics to represent a cross section of the populations of folks who will be prescribed these treatments in the future.

What impact would this selective volunteering have on the results of the study? If only the health nuts (who probably already have excellent heart rates) volunteer to be in the treatment group, the researcher will be looking only at the effect of the treatment (running 5 miles) on very healthy and active people. They won't see the effect that running 5 miles has on the heart rates of couch potatoes. This nonrandom assignment of subjects to the treatment and control groups could have a huge impact on the conclusions they draw from this study.

REMEMBER

To avoid major bias in the results of an experiment, subjects must be randomly assigned to treatments by a third party and not be allowed to choose which group they will be in. The goal of random assignment is to make sure that none of the characteristics about the participants determines which group they are in. Random assignment doesn't guarantee the groups will be equal, only in the long run, but not randomizing can cause a lot of bias for certain. Keep this in mind when you evaluate the results of an experiment.

Controlling for confounding variables

Suppose you're participating in a research study that looks at factors influencing whether you catch a cold. If a researcher records only whether you got a cold after a certain period of time and asks questions about your behavior (how many times per day you washed your hands, how many hours of sleep you got each night, and so on), the researcher is conducting an observational study. The problem with this type of observational study is that without controlling for other factors that may have had an influence and without regulating which action you were taking when, the researcher won't be able to single out exactly which of your actions (if any) actually impacted the outcome.

REMEMBER

The biggest challenge of observational studies is that they can't automatically show true cause-and-effect relationships, due to what statisticians call confounding variables. A confounding variable, or confounder, as it is sometimes called, is one what is related to both the factor as well as the outcome, and can influence the results. For example, the number of times someone washes their hands in a vitamin C observational study can be a confounder. People more likely to wash their hands might be also more likely to take vitamin C and may also get fewer colds. It can be hard to tease out what caused what in that case, and makes the study design more challenging if you are trying to adjust for it.

For example, one news headline boasted, "Study links older mothers, long life." The opening paragraph said that women who have a first baby after age 40 have a much better chance of living to be 100, compared to women who have a first baby at an earlier age. When you get into the details of the study (done in 1996), you find out, first of all, that it was based on 78 women in suburban Boston who were born in 1896 and had lived to be at least 100, compared to 54 women who were also born in 1896 but died in 1969 (the earliest year the researchers could get computerized death records). This so-called "control group" lived to be exactly 73, no more and no less. Of the women who lived to be at least 100 years of age, 19 percent had given birth after age 40, whereas only 5.5 percent of the women who died at age 73 had given birth after age 40.

I have a problem with these conclusions. What about the fact that the "control group" was based only on mothers who died in 1969 at age 73? What about all the other mothers who died *before* age 73, or who died between the ages of 73 and 100? What about other variables that may affect both mothers' ages at the births of their children and longer life spans — variables such as financial status, marital stability, or other socioeconomic factors? The women in this study were in their thirties during the Depression; this may have influenced both their life span and if or when they had children.

REMEMBER

How do researchers handle confounding variables in observational studies? They work hard to come up with ways to account for them, and also to quantify how big of an influence a variable needs to have in order to impact the results.

For example, in a study trying to determine the effect of different types and volumes of music on the amount of time grocery shoppers spend in the store (yes, they do think about that), researchers have to anticipate as many confounding variables as possible ahead of time and control for them (in an experiment) or take them into an account (in an observational study). What other factors besides volume and type of music may influence the amount of time you spend in a grocery store? I can think of several factors: gender, age, time of day, whether you have children with you, how much money you have, the day of the week, how clean and inviting the store is, how nice the employees are, and (most importantly) what your motive is — are you shopping for the whole week, or are you just running in to grab a candy bar?

How can researchers begin to control for so many possible confounding factors? Some of them can be controlled for in the design of the study, such as the time of day, day of the week, and reason for shopping. But other factors (such as the perception of the store environment) depend totally on the individual in the study. One possible solution to take care of some of the outside variables is to use pairs of people that are matched according to important variables, or to just use the same person twice: once with the treatment and once without. This type of experiment is called a *matched-pairs design.* (See Chapter 16 for more on this.)

REMEMBER

Before believing any medical headlines (or any headlines with statistics, for that matter), look to see how the study was conducted. In cases where an experiment can't be done (after all, no one can force you to have a baby after or before age 40), make sure the observational study takes other possibly affecting variables into account. It can be done; it just takes some more intricate work. And think about possible confounding variables that may affect the conclusions being drawn.

Respecting ethical issues

The trouble with experiments is that some experimental designs are not ethical. You can't force research subjects to smoke in order to see whether they get lung cancer, for example. That's why more evidence was needed to show that smoking causes lung cancer, and why the tobacco companies only recently had to pay huge penalties to victims.

Although the causes of cancer and other diseases can't be determined ethically by conducting experiments on humans, new treatments for cancer can be (and are) tested using experiments. Medical studies that involve experiments are called *clinical trials.* The U.S. government has a registry of federally and privately supported clinical trials conducted in the United States and around the world; it also has information available on who may participate in various clinical trials. Check out `www.clinicaltrials.gov` for more information.

Serious experiments (such as those funded by and/or regulated by the U.S. government) must pass a huge series of tests that can take years to carry out. The approval of a new drug, for example, goes through a very lengthy, comprehensive, and detailed process regulated and monitored by the Federal Drug Administration (FDA). One reason the cost of prescription drugs is so high is the massive amount of time and money needed to conduct research and development of new drugs, most of which fail to pass the tests and have to be scrapped.

Any experiments involving human subjects are also regulated by the federal government and have to gain approval by a committee created for the purpose of protecting "the rights and welfare of the participants." The committees set up for different organizations have different names (such as Institutional Review Board [IRB], Independent Ethics Committee [IEC], and Ethical Review Board [ERB], to name a few), but they all serve the same purpose. Research conducted on animals is different in terms of regulations and continues to be a topic of debate and controversy around the world.

REMEMBER

Surveys, polls, and other observational studies are fine if you want to know people's opinions, examine their lifestyles without intervention, or examine some demographic variables. If you want to try to determine the cause of a certain outcome or behavior (that is, a reason why something happened), an experiment is a much better way to go. If an experiment isn't possible because of ethical concerns (or because of expense or other reasons), a large body of observational studies examining many different factors and coming up with similar conclusions is the next best thing. (See Chapter 19 for more about cause-and-effect relationships.)

Collecting good data

What constitutes "good" data? Statisticians use three criteria for evaluating data quality; each of the criteria really relates most strongly to the quality of the measurement instrument that's used in the process of collecting the data. To decide whether you're looking at good data from a study, look for these characteristics:

- **The data are reliable — you can get repeatable results with subsequent measurements.** Many bathroom scales give unreliable data. You get on the scale, and it gives you one number. You don't believe the number, so you get off, get back on, and get a different number. (If the second number is lower, you'll most likely quit at this point; if not, you may continue getting on and off until you see a number you like.) Or you can do what some researchers do: Take three measurements, find the average, and use that; at least this will improve the reliability a bit.

  Unreliable data come from unreliable measurement instruments or unreliable data collection methods. Errors can go beyond actual scales to more intangible measurement instruments, like survey questions, which can give unreliable results if they're written in an ambiguous way (see Chapter 17).

  REMEMBER

  Find out how the data were collected when examining the results of a study. If the measurements are unreliable, the data could be inaccurate.

- **The data are valid — they measure what they're supposed to measure.** Checking the validity of data requires you to step back and look at the big picture. You have to ask the question: Do these data measure what they should be measuring? Or should the researchers have been collecting altogether different data? The appropriateness of the measurement instrument used is important. For example, many educators say that a student's transcript is not a valid measure of their ability to perform well in college. Alternatives include a more holistic approach, not only taking grades into account, but also adding weight to elements such as service, creativity, social involvement, extracurricular activities, and the like.

REMEMBER

Before accepting the results of an experiment, find out what data were measured and how they were measured. Be sure the researchers are collecting valid data that are appropriate for the goals of the study.

» **The data are unbiased — they contain no systematic errors that either add to or subtract from the true values.** Biased data are data that systematically overmeasure or undermeasure the true result. Bias can occur almost anywhere during the design or implementation of a study. Bias can be caused by a bad measurement instrument (like that bathroom scale that's "always" 5 pounds over), by survey questions that lead participants in a certain way, or by researchers who know what treatment each subject received and who have preconceived expectations.

REMEMBER

Bias is probably the number-one problem in collecting good data. However, you can minimize bias by using methods similar to those discussed in Chapter 17 for surveys and in the earlier section, "Making random assignments," as well as by making your experiments double-blind whenever possible.

Double-blind means neither the subjects nor the researchers know who got what treatment or who is in the control group. The subjects need to be oblivious to which treatment they're getting so that the researchers can measure the placebo effect. And researchers should be kept in the dark so they don't treat subjects differently by either expecting or not expecting certain responses from certain groups. For example, if a researcher knows you're in the treatment group to study the side effects of a new drug, they may expect you to get sick and therefore may pay more attention to you than if they knew you were in the control group. This can result in biased data and misleading results.

If the researcher knows who got what treatment but the subjects don't know, the study is called a *blind* study (rather than a double-blind study). Blind studies are better than nothing, but double-blind studies are best. In case you're wondering: In a double-blind study, does *anyone* know which treatment was given to which subjects? Relax; typically a third party, such as a lab assistant, does that part.

In some cases, the subjects know which group they're in because it's unconcealable — for example, when comparing the benefits of doing yoga versus jogging. However, bias can be reduced by not telling the subjects the precise purpose of the study. This irregular type of plan would have to be reviewed by an institutional review board to make sure it isn't unethical to conduct; see the earlier section, "Respecting ethical issues."

Analyzing the data properly

After the data have been collected, they're put into that mysterious box called the *statistical analysis* for number crunching. The choice of analysis is just as important (in terms of the quality of the results) as any other aspect of a study. A proper analysis should be planned in advance, during the design phase of the experiment. That way, after the data are collected, you won't run into any major problems during the analysis.

Here's the bottom line when selecting the proper analysis: Ask yourself the question, "After the data are analyzed, will I be able to legitimately and correctly answer the question that I set out to answer?" If the answer is "no," then that analysis isn't appropriate.

Some basic types of statistical analyses include *confidence intervals* (used when you're trying to estimate a population value, or the difference between two population values); *hypothesis tests* (used when you want to test a claim about one or two populations, such as the claim that one drug is more effective than another); and *correlation and regression analyses* (used when you want to show if and/or how one quantitative variable can predict or cause changes in another quantitative variable). See Chapters 14, 16, and 19, respectively, for more on each of these types of analyses.

REMEMBER

When choosing how you're going to analyze your data, you have to make sure that the data and your analysis will be compatible. For example, if you want to compare a treatment group to a control group in terms of the amount of weight lost on a new (versus an existing) diet program, you need to collect data on how much weight each person lost — not just each person's weight at the end of the study.

EXAMPLE

Q. Suppose an experiment shows that a new weight-loss plan works for seven out of ten people who try it. Because it works for 70 percent of the people who try it, are the results meaningful?

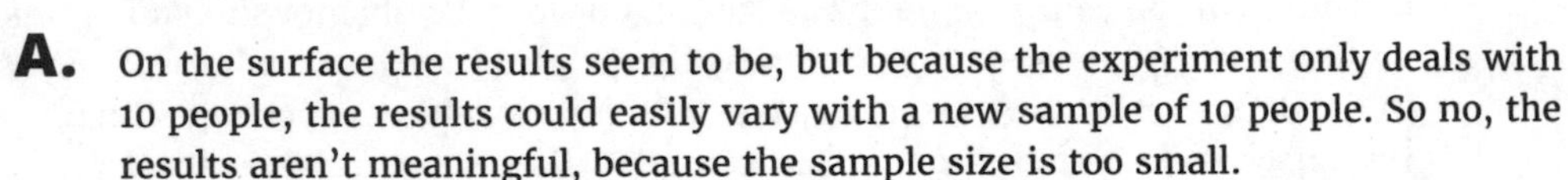

A. On the surface the results seem to be, but because the experiment only deals with 10 people, the results could easily vary with a new sample of 10 people. So no, the results aren't meaningful, because the sample size is too small.

YOUR TURN

5 A study of an experimental drug for people with sleep apnea involves a group of volunteers who have to sleep in the research lab every night for six months. Do you question the representative nature of the sample?

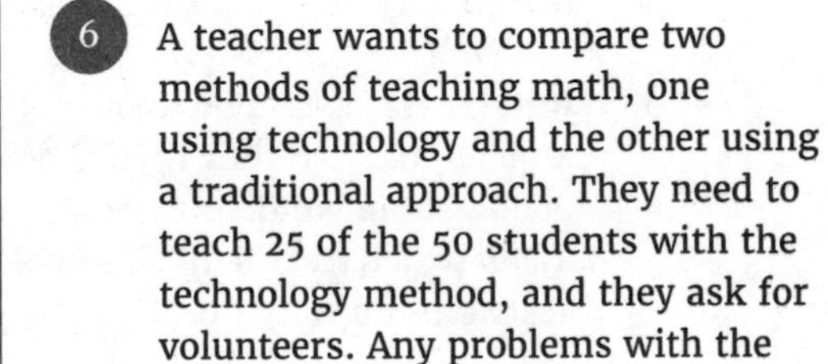

6 A teacher wants to compare two methods of teaching math, one using technology and the other using a traditional approach. They need to teach 25 of the 50 students with the technology method, and they ask for volunteers. Any problems with the design of the experiment?

7. An experiment compares two weight-loss programs (Plans 1 and 2) for overweight people. Of the 100 volunteers, researchers randomly assign half to each program. They record weight changes after six weeks but collect no other data. Plan 1 people lose 25 percent more weight than participants on Plan 2, so researchers deem Plan 1 the better plan. Name a limitation of this study.

8. A researcher is so excited about their experiment that compares blood pressure of dogs on a new drug to a control group, that they watch and record all the information they can on the drug group and ignore the control group. What did they do wrong, and what's the impact of this error?

9. A study examines the effects of a new type of medicine on mild headache sufferers. The control group receives no drug, and the other participants take the new medicine. People on the medicine say they feel better right away. Any problems with this experiment?

10. A scale used in a weight-loss experiment is 5 pounds off in the positive direction. What kind of data problem does this pose, and what impact does the inaccuracy have on the results?

11. And for further proof that the researchers need a new scale, they find that it gives different results when the same person gets off and immediately gets back on. What kind of data problem does this pose, and what impact does this have on the results?

12. A teacher comparing the reading ability of their students using a computer program versus a traditional approach assesses and compares the students' progress by looking at their grades in gym class. What kind of data problem does this create here, and what impact does this have on the results?

Interpreting Experiment Results

Sound experiments have appropriate data analyses for the collected data and answer the posed question. After you analyze the data, you can make conclusions. In any study, you don't want to overstate the results and mistakenly apply them to a larger population than what the sample actually represents. Another common mistake analyzers make is to jump to conclusions about cause and effect prematurely, explaining why they get the results without sufficient data.

Making appropriate conclusions

In my opinion, the biggest mistakes researchers make when drawing conclusions about their studies are the following (discussed in the following sections):

- Overstating their results
- Making connections or giving explanations that aren't backed up by the statistics
- Going beyond the scope of the study in terms of whom the results apply to

Overstating the results

Many times, the headlines in the media overstate actual research results. When you read a headline or otherwise hear about a study, be sure to look further to find out the details of how the study was done and exactly what the conclusions were.

Press releases often overstate results, too. For example, in a recent press release by the National Institute for Drug Abuse, the researchers claimed that use of the street drug Ecstasy was down from the previous year. However, when you look at the actual statistical results in the report, you find that the percentage of teens *from the sample* who said they'd used Ecstasy was lower than those from the previous year, but this difference was not found to be statistically significant when they tried to project it onto the population of *all* teens. This discrepancy means that although fewer teens in the sample used Ecstasy that year, the difference wasn't enough to account for more than chance variability from sample to sample. (See Chapter 15 for more about statistical significance.)

REMEMBER

Headlines and leading paragraphs in press releases and news articles often overstate the actual results of a study. Big results, spectacular findings, and major breakthroughs make the news these days, and reporters and others in the media constantly push the envelope in terms of what is and isn't newsworthy. How can you sort out the truth from exaggeration? The best thing to do is to read the fine print.

Taking the results one step beyond the actual data

A study that links having children later in life to longer life spans illustrates another point about research results. Do the results of this observational study mean that having a baby later in life can make you live longer? "No," said the researchers. Their explanation of the results was that having a baby later in life may be due to women having a "slower" biological clock, which presumably would then result in the aging process being slowed down.

Our question to these researchers is, "Then why didn't you study *that,* instead of just looking at their ages?" The study didn't include any information that would lead us to conclude that women who had children after age 40 aged at a slower rate than other women, so in our view, the researchers shouldn't make that conclusion. Or the researchers should state clearly that this view is only a theory and requires further study. Based on the data in this study, the researchers' theory seems like a leap of faith.

Frequently in a press release or news article, the researcher will give an explanation about *why* they think the results of the study turned out the way they did and what implications these results have for society as a whole when the "why" hasn't been studied yet. These explanations may have been in response to a reporter's questions about the research — questions that were later edited out of the story, leaving only the juicy quotes from the researcher. Many of these after-the-fact explanations are no more than theories that have yet to be tested. In such cases, you should be wary of conclusions, explanations, or links drawn by researchers that aren't backed up by their studies.

WARNING

Be aware that the media wants to make you read the article (they get paid to do that), so they will have strong headlines or will make unconfirmed "cause-effect" statements because it is their job to sell the story. It is *your* job to be wary.

Generalizing results to people beyond the scope of the study

You can make conclusions only about the population that's represented by your sample. If you sample men only, you can't make conclusions about women. If you sample healthy young people, you can't make your conclusions about everyone. But many researchers try to do just that, and it can give misleading results.

Here's how you can determine whether a researcher's conclusions measure up (Chapter 17 has more on samples and populations):

1. **Find out what the target population is (that is, the group that the researcher wants to make conclusions about).**
2. **Find out how the sample was selected and see whether the sample is representative of that target population (and not some more narrowly defined population).**
3. **Check the conclusions made by the researchers and make sure they're not trying to apply their results to a broader population than they actually studied.**

Making informed decisions

Just because someone says they conducted a "scientific study" or a "scientific experiment" doesn't mean it was done right or that the results are credible (not that I'm saying you should discount everything that you see and hear). Unfortunately, I've come across a lot of bad experiments in my days as a statistical consultant. The worst part is that if an experiment was done poorly, you can't do anything about it after the fact except ignore the results — and that's exactly what you need to do.

TIP

Here are some tips that help you make an informed decision about whether to believe the results of an experiment, especially one whose results are very important to you:

- **When you first hear or see the result, grab a pencil and write down as much as you can about what you heard or read, where you heard or read it, who did the research, and what the main results were.**
- **Follow up on your sources until you find the person who did the original research and then ask them for a copy of the report or paper.**
- **Go through the report and evaluate the experiment according to the eight steps for a good experiment described in the earlier section, "Designing a Good Experiment."** (You really don't have to understand everything written in a report in order to do that.)
- **Carefully scrutinize the conclusions that the researcher makes regarding their findings.** Many researchers tend to overstate results, make conclusions beyond the statistical evidence, or try to apply their results to a broader population than the one they studied.
- **Never be afraid to ask questions of the media, the researchers, and even your own experts.** For example, if you have a question about a medical study, ask your doctor. They will be glad that you're an empowered and well-informed patient!
- **And finally, don't get overly skeptical, just because you're now a lot more aware of all the bad practices going on out there.** Not everything is bad. There are many more good researchers, credible results, and well-informed reporters than not. You have to maintain a sense of being cautious and ready to spot problems without discounting everything.

EXAMPLE

Q. A study shows that people who achieve PhDs are less likely to develop memory loss diseases such as Alzheimer's than professionals without a PhD. Does this mean that if you want to avoid getting Alzheimer's, you should get your PhD?

A. **Not necessarily.** Although this study does show that a relationship exists, it doesn't automatically imply cause and effect. This study obviously couldn't have been an experiment, because it isn't ethical to force people to get PhDs or force them not to, so the study should properly control for confounding. You can appropriately state that a link exists, and that researchers need to study it further. Perhaps you can formulate an experiment that mimics this dynamic.

YOUR TURN

13 An experiment concludes that eating an egg a day doesn't raise your cholesterol (as researchers once suspected). The experiment involves healthy, young males on low-fat diets. Explain what's wrong with the conclusions of this study.

14 An experiment shows that rats that receive a big piece of cheese at the end of a maze get to the cheese faster than rats that receive a small piece of cheese. Does this mean you can motivate rats to learn faster, like humans?

Practice Questions Answers and Explanations

1. The words "prevent" and "reduce" imply a cause-and-effect relationship, which can't be determined through a single observational study. Many other variables could account for the difference in results, such as health and hygiene habits, frequency of hand washing, temperature of the water used, length of hand washing, medical history, stress level, and so on.

2. The teacher runs an experiment where they randomly assign students to a treatment (computer game method or textbook method) and compare the results. The teacher controls the content. (Now, whether or not they ran a well-designed experiment is another issue.)

3. In certain cases, conducting an experiment is unethical. Although an observational study isn't always the optimal study to conduct, it may be your only choice. And, the evidence for a cause-and-effect relationship (although not directly proven) does mount as researchers conduct more and more observational studies, trying to explore the relationship between smoking and the occurrence of lung cancer by looking at many different groups of people in many different situations. (This is how researchers eventually showed that smoking causes lung cancer in humans; it wasn't so long ago.)

4. This is typical of many studies that provide information for commercials; take these studies with a grain of salt.

   a. **This is an observational study**, because the people decide for themselves whether or not to take vitamin C. No treatment is imposed.

   b. Other uncontrolled variables that may influence the results include (but aren't limited to): **health and hygiene habits, exercise, level of health consciousness, or diet**.

5. **Yes.** This sample probably doesn't represent the entire population of people with sleep apnea. It includes only the subjects willing to sleep in a research lab for six months. Children, older people, and people who travel a great deal aren't fully represented, which can bias the results.

6. **Yes.** By asking for volunteers, the teacher puts bias in the results. Students who volunteer for the computer method are probably more comfortable with computers than students who don't. The assignments of students to a type of learning model have to be random.

7. One possible limitation is how well subjects adhered to the program. Working with human subjects is hard, as they are only human, after all. It could be that some stopped the diet, never started, switched to the other offering during the study, or didn't follow the rules. Beginning weight or medical issues might affect adherence as well, so that is a possible confounding variable.

8. **The study isn't double blind. Because the researcher knows who gets what treatment, their bias becomes part of the recorded data.** A third party should assign dogs to treatments and not tell the researcher.

REMEMBER

An experiment is blind if the subjects don't know what treatment they receive and double-blind if the researchers also don't know who receives what treatment. Both blinding and double-blinding help reduce bias in the data-collection process.

9. **Yes. Subjects often experience the placebo effect.** That is, when subjects know they've taken a real (non-fake) treatment, the knowledge can affect their expectations and hence the data collected. Their expectations lead to biased results. To fix the problem, the study could use a sugar pill for the control group and not tell them which pill they receive.

Researchers have a control group so they can determine the strength of the placebo effect and take that information into account for the real treatment group. This helps them determine the "real" effect of a real treatment. In cases where it might be unethical to use a fake treatment, a "standard treatment" can be used for comparison. An experiment is stopped in the case where the new treatment is found to be clearly better than the standard or placebo, so everyone suffering can get help, not just the treatment group.

10. **The measurement instrument is biased, which leads to biased data.**

11. **The measurement instrument is unreliable, which leads to imprecise data.**

12. **The measurement is invalid, because grades in gym class don't measure reading ability. The invalidity leads to results that lack credibility.**

13. **The study is based on a sample that represents only young, healthy males on low-fat diets, but the conclusions say eggs don't raise your cholesterol (implying that the diet works for anybody).** This mistake leads to severely biased results and misleading conclusions.

14. **Maybe, but it could also mean that rats have a good sense of smell, and bigger bait smells stronger than smaller bait.** The conclusion about motivation tries to answer the "why" question with insufficient data. Further studies must be done to address the "why" issue.

If you're ready to test your skills a bit more, take the following chapter quiz that incorporates all the chapter topics.

Whaddya Know? Chapter 18 Quiz

Quiz time! Complete each problem to test your knowledge on the various topics covered in this chapter. You can then find the solutions and explanations in the next section.

1. Professor Joe wants to test a new drug on his goats to see if it helps them grow their hair back after significant hair loss. Why is it important to make sure he is making proper comparisons in his study?

2. "Eggs are Good for You!" is an ad put out by an egg-promoting group. They based their headline on a study of 5 men in good health who ate one egg per day and did not experience an increase in cholesterol. What's the problem with the sample size in this situation?

3. "Eggs are Good for You!" is an ad put out by an egg-promoting group. They based their headline on a study of 5 men in good health who ate one egg per day and did not experience an increase in cholesterol. What's the target population, and do the participants represent this target population?

4. Is it ethical to give a fake treatment in a study of a new AIDs drug?

5. What does it mean for a study to be double blind, and why is it important?

6. A teacher wants to see if a computer-based learning system does as good of a job as their own teaching. They have 30 students and ask for 15 volunteers to do the computer-based learning system for 1 month, and assess both groups of students after teaching the remaining students the same topic. What's the main problem here?

7. Granny Eunice believes that her oven is better than her friend Myrtle's oven at baking cookies. They each make a batch of chocolate chip cookies and bake them at 350 degrees for 10 minutes. They have people sample a cookie from each oven and vote for their favorite. Eunice's cookies win. What is one of the many possible confounding variables?

8. What is the main difference between an experiment and an observational study?

9. You want to compare three brands of popcorn to see which one pops the most popcorn. You put 3 ounces of popcorn into a popper and pop for 3 minutes, then count the popped kernels. You do this for each brand. What is the response? What is the factor?

10. A group of researchers test a new blood pressure (BP) drug versus an existing BP drug by randomly assigning 100 people to each drug (new or existing) and to an amount of the drug (100 mg versus 200 mg). They measure blood pressure at the beginning of the experiment and again 7 days later, and write down the amount by which BP changed (up or down). What are the two factors? What is the response? What are the levels, and what are the treatments?

Answers to Chapter 18 Quiz

1. **It's possible that the goats' hair could grow back on its own.**
2. **The sample size is 5, which is very small and won't offer much precision in the results.** If you are testing million-dollar computer chips, maybe you can only afford 5 of them, but you can certainly find more than 5 people to participate in this egg study.
3. **Because the ad says, "Eggs are Good for You," this means the target population is everyone. However, the study was only conducted on men, and only men who were in good health.** This does not represent everyone.
4. **No, it's not ethical to give a fake treatment when there is an existing treatment available.** Researchers in a case like this would use the existing treatment as the control group, and the new treatment as the treatment group, and take it from there.
5. **In a double-blind study, neither the researcher nor the subject knows which treatment they are getting. This prevents bias on either of their parts.**
6. **Students were asked to volunteer for the computer-based learning system; they were not randomly assigned to the system. This constitutes bias.**
7. Some confounding variables include the recipe used to make the cookies; the calibration of the ovens; the rack on which the cookies were placed; and the type of pan they were baked on.
8. **An experiment imposes a treatment; an observational study does not.**
9. **The response is the number of popped kernels; the factor is the brand of popcorn.**
10. The two factors are which drug and the amount of the drug. The response is the change in BP after 7 days. The levels of the drug are existing and new, and the levels of the amount are 100 mg and 200 mg. You have 4 treatment combinations: 1) new drug at 100 mg; 2) new drug at 200 mg; 3) existing drug at 100 mg; and 4) existing drug at 200 mg.

IN THIS CHAPTER

» Exploring statistical relationships between numerical variables

» Looking at correlation and linear regression

» Making predictions based on known relationships

» Considering correlation versus causation

Chapter 19

Looking for Links: Correlation and Regression

Today's media provide a steady stream of information, including reports on all the latest links that have been found by researchers. Just today I heard that increased video game use can negatively affect a child's attention span, the amount of a certain hormone in a woman's body can predict when she will enter menopause, and the more depressed you get, the more chocolate you eat, and the more chocolate you eat, the more depressed you get (how depressing!).

Some studies are truly legitimate and help improve the quality and longevity of our lives. Other studies are not so clear. For example, one study says that exercising 20 minutes three times a week is better than exercising 60 minutes one time a week, another study says the opposite, and yet another study says there is no difference.

If you are a confused consumer when it comes to links and correlations, take heart; this chapter can help. You'll gain the skills to dissect and evaluate research claims and make your own decisions about those headlines and sound bites that you hear each day alerting you to the latest correlation. You'll discover what it truly means for two variables to be correlated, when a cause-and-effect relationship can be concluded, and when and how to predict one variable based on another.

Picturing a Relationship with a Scatterplot

An article in *Garden Gate* magazine caught my eye: "Count Cricket Chirps to Gauge Temperature." According to the article, all you have to do is find a cricket, count the number of times it chirps in 15 seconds, add 40, and voilà! You've just estimated the temperature in Fahrenheit.

The National Weather Service Forecast Office even puts out its own "Cricket Chirp Converter." You enter the number of cricket chirps recorded in 15 seconds, and the converter gives you the estimated temperature in four different units, including Fahrenheit and Celsius.

A fair amount of research does support the claim that frequency of cricket chirps is related to temperature. For the purpose of illustration, I've taken only a subset of some of the data (see Table 19-1).

TABLE 19-1 Cricket Chirps and Temperature Data (Excerpt)

Number of Chirps (in 15 Seconds)	Temperature (Fahrenheit)
18	57
20	60
21	64
23	65
27	68
30	71
34	74
39	77

Notice that each observation is composed of two variables that are tied together: the number of times the cricket chirped in 15 seconds (the X-variable) and the temperature at the time the data was collected (the Y-variable). Statisticians call this type of two-dimensional data *bivariate* data. Each observation contains one pair of data collected simultaneously. For example, Row 1 of Table 19-1 depicts a pair of data (18, 57).

Bivariate data is typically organized in a graph that statisticians call a *scatterplot.* A scatterplot has two dimensions: a horizontal dimension (the x-axis) and a vertical dimension (the y-axis). Both axes are numerical; each one contains a number line. In the following sections, I explain how to make and interpret a scatterplot.

Making a scatterplot

REMEMBER

Placing observations (or points) on a scatterplot is similar to playing the game Battleship. Each observation has two coordinates; the first corresponds to the first piece of data in the pair (that's the x-coordinate; the amount that you go left or right). The second coordinate corresponds to the second piece of data in the pair (that's the y-coordinate; the amount that you go up or down). You place the point representing that observation at the intersection of the two coordinates.

Figure 19-1 shows a scatterplot for the cricket chirps and temperature data listed in Table 19-1. Because I ordered the data according to their x-values, the points on the scatterplot correspond from left to right to the observations given in Table 19-1, in the order listed.

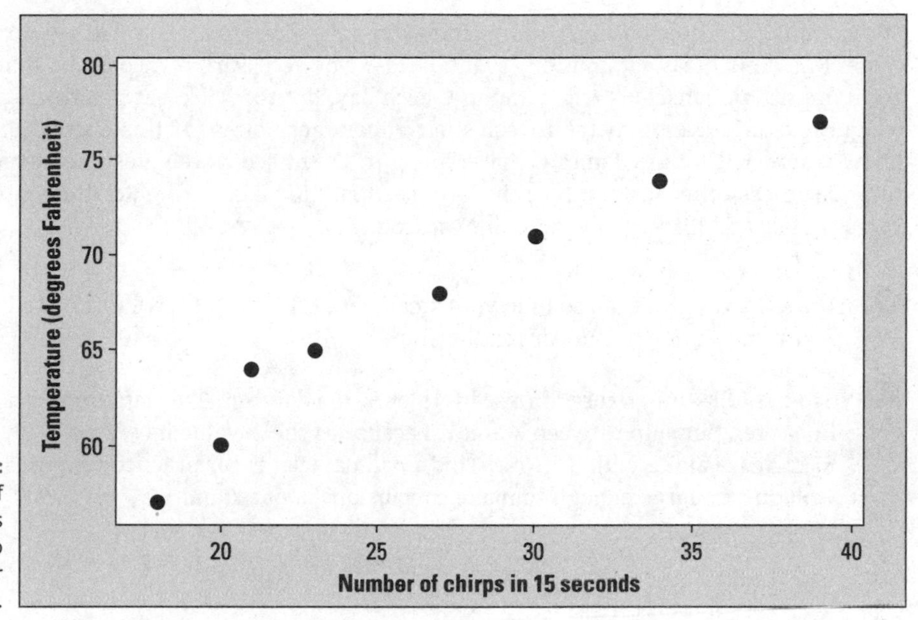

FIGURE 19-1: Scatterplot of cricket chirps in relation to outdoor temperature.

Interpreting a scatterplot

REMEMBER

You interpret a scatterplot by looking for trends in the data as you go from left to right:

- If the data show an uphill pattern as you move from left to right, this indicates a *positive relationship between X and Y.* As the *x*-values increase (move right), the *y*-values increase (move up) a certain amount.
- If the data show a downhill pattern as you move from left to right, this indicates a *negative relationship between X and Y.* As the *x*-values increase (move right) the *y*-values decrease (move down) by a certain amount.
- If the data don't seem to resemble any kind of pattern (even a vague one), then no relationship exists between *X* and *Y.*

One pattern of special interest is a *linear* pattern, where the data has a general look of a line going uphill or downhill. Looking at Figure 19-1, you can see that a positive linear relationship does appear between number of cricket chirps and the temperature. That is, as the cricket chirps increase, the temperature increases as well.

REMEMBER

In this chapter, I explore linear relationships only. A *linear relationship between X and Y* exists when the pattern of *x*- and *y*-values resembles a line, either uphill (with a positive slope) or downhill (with a negative slope). Other types of trends may exist in addition to the uphill/downhill linear trends (for example, curves or exponential functions); however, these trends are beyond the scope of this book. The good news is that many relationships do fall under the uphill/downhill linear scenario.

WARNING

Scatterplots show possible associations or relationships between two variables. However, just because your graph or chart shows something is going on, it doesn't mean that a cause-and-effect relationship exists.

For example, a doctor observes that people who take vitamin C each day seem to have fewer colds. Does this mean vitamin C prevents colds? Not necessarily. It could be that people who are more health conscious take vitamin C each day, but they also eat healthier, are not overweight, exercise every day, and wash their hands more often. If this doctor really wants to know whether it's the vitamin C that's doing it, they need a well-designed experiment that rules out these other factors. (See the later section, "Explaining the Relationship: Correlation versus Cause and Effect," for more information.)

EXAMPLE

Q. Draw a scatterplot of the following small data set: (2, 3), (6, 8), (3, 5). Tell whether *X* and *Y* have a positive linear relationship.

A. See the following figure for a scatterplot of the data set. The data implies a positive linear relationship between *X* and *Y*, because as the *x*-value increases, the *y*-value tends to increase along with it. (***Note***: This small data set is for practice purposes only and wouldn't be large enough to make conclusions about *X* and *Y*.)

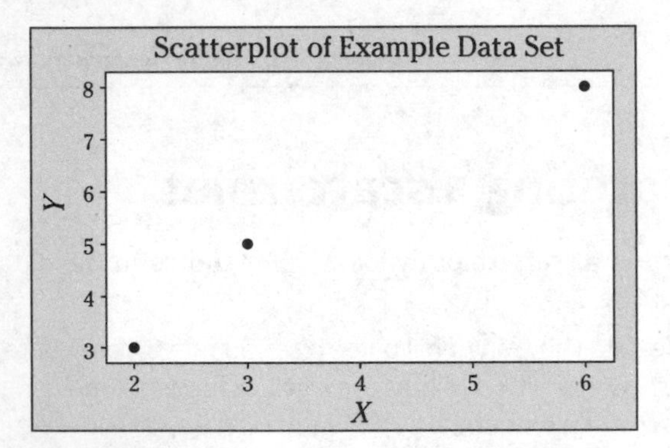

YOUR TURN

Describe the relationship between *X* and *Y* shown by the scatterplot in the following figure.

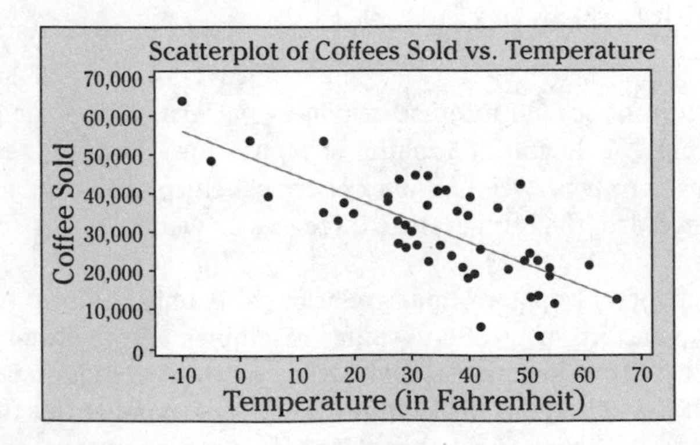

2 Describe the relationship between X and Y shown by the scatterplot in the following figure.

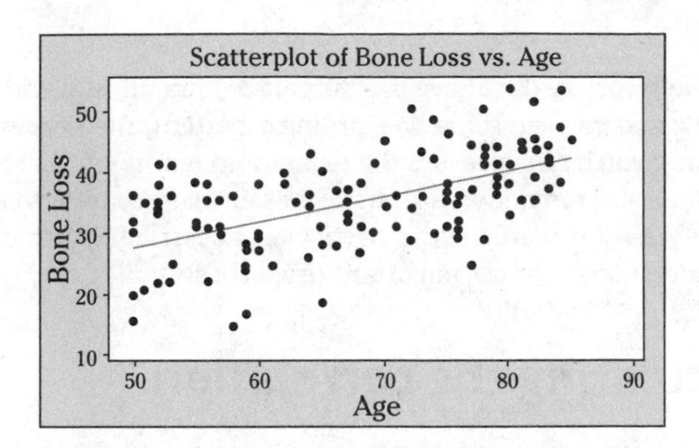

3 Describe the relationship between X and Y shown by the scatterplot in the following figure.

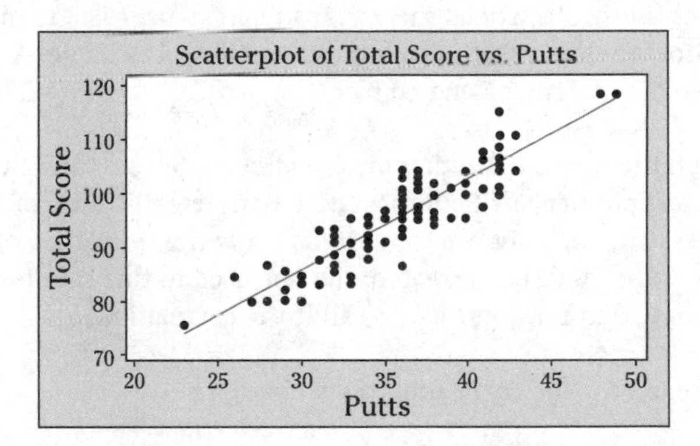

4 What does a scatterplot that shows no linear relationship between X and Y look like?

Quantifying Linear Relationships Using the Correlation

After the bivariate data have been organized graphically with a scatterplot (see the preceding section), and you see some type of linear pattern, the next step is to calculate some statistics that can quantify or measure the extent and nature of the relationship. In the following sections, I discuss *correlation*, a statistic measuring the strength and direction of a linear relationship between two variables; in particular, I describe how to calculate and interpret correlation and understand its most important properties.

Calculating the correlation

In the earlier section, "Interpreting a scatterplot," I say data that resembles an uphill line has a positive linear relationship and data that resembles a downhill line has a negative linear relationship. However, I didn't address the issue of whether or not the linear relationship was strong or weak. The strength of a linear relationship depends on how closely the data resembles a line, and of course varying levels of "closeness to a line" exist.

Can one statistic measure both the strength and direction of a linear relationship between two variables? Sure! Statisticians use the *correlation coefficient* to measure the strength and direction of the linear relationship between two numerical variables *X* and *Y*. The correlation coefficient for a sample of data is denoted by *r*.

REMEMBER

Although the street definition of *correlation* applies to any two items that are related (such as gender and political affiliation), statisticians use this term only in the context of two numerical variables. The formal term for correlation is the *correlation coefficient*. Many different correlation measures have been created; the one used in this case is called the *Pearson correlation coefficient* (but from now on I'll just call it the correlation).

The formula for the correlation (*r*) is

$$r = \frac{1}{n-1}\left(\frac{\sum(x-\bar{x})(y-\bar{y})}{s_x s_y}\right)$$

where n is the number of pairs of data; $\bar{x}$ and $\bar{y}$ are the sample means of all the *x*-values and all the *y*-values, respectively; and s_x and s_y are the sample standard deviations of all the *x*- and *y*-values, respectively.

Use the following steps to calculate the correlation, *r*, from a data set:

1. **Find the mean of all the *x*-values ($\bar{x}$) and the mean of all the *y*-values ($\bar{y}$).**

   See Chapter 5 for more on calculating the mean.

2. **Find the standard deviation of all the *x*-values (call it s_x) and the standard deviation of all the *y*-values (call it s_y).**

   See Chapter 5 to find out how to calculate the standard deviation.

3. **For each (x, y) pair in the data set, take x minus $\bar{x}$ and y minus $\bar{y}$, and multiply them together to get $(x-\bar{x})(y-\bar{y})$.**
4. **Add up all the results from Step 3.**
5. **Divide the sum by $s_x \cdot s_y$.**
6. **Divide the result by $n-1$, where n is the number of (x, y) pairs. (It's the same as multiplying by 1 over $n-1$.)**

   This gives you the correlation, *r*.

For example, suppose you have the data set (3, 2), (3, 3), and (6, 4). You calculate the correlation coefficient *r* via the following steps. (*Note:* for this data, the *x*-values are 3, 3, 6, and the *y*-values are 2, 3, 4.)

1. **$\bar{x}$ is $12 \div 3 = 4$, and $\bar{y}$ is $9 \div 3 = 3$.**
2. **The standard deviations are $s_x = 1.73$ and $s_y = 1.00$.**

   See Chapter 5 for step-by-step calculations.

3. **The difference pairs of each value and its mean ($x-\bar{x}$ and $y-\bar{y}$) multiplied together are: $(3-4)(2-3) = (-1)(-1) = +1$; $(3-4)(3-3) = (-1)(0) = 0$; and $(6-4)(4-3) = (2)(1) = +2$.**
4. **Adding the Step 3 results, you get $1+0+2=3$.**
5. **Dividing by $s_x \cdot s_y$ gives you $3 \div (1.73)(1.00) = 3 \div 1.73 = 1.73$.**
6. **Now divide the Step 5 result by $3-1$ (which is 2), and you get the correlation $r = 0.87$.**

Interpreting the correlation

REMEMBER

The correlation *r* is always between +1 and –1. To interpret various values of *r* (no hard and fast rules here, just a rule of thumb), see which of the following values your correlation is closest to.

- **Exactly –1:** A perfect downhill (negative) linear relationship
- **–0.70:** A strong downhill (negative) linear relationship
- **–0.50:** A moderate downhill (negative) relationship
- **–0.30:** A weak downhill (negative) linear relationship
- **0:** No linear relationship
- **+0.30:** A weak uphill (positive) linear relationship
- **+0.50:** A moderate uphill (positive) relationship
- **+0.70:** A strong uphill (positive) linear relationship
- **Exactly +1:** A perfect uphill (positive) linear relationship

REMEMBER

If the scatterplot doesn't indicate there's at least somewhat of a linear relationship, the correlation doesn't mean much. Why measure the amount of linear relationship if there isn't enough of one to speak of? However, you can take the idea of no linear relationship two ways: 1) If no relationship at all exists, calculating the correlation doesn't make sense because correlation only applies to linear relationships; and 2) If a strong relationship exists but it's not linear, the correlation may be misleading, because in some cases a strong curved relationship exists yet the correlation turns out to be strong. That's why it's critical to examine the scatterplot first.

Figure 19-2 shows examples of what various correlations look like in terms of the strength and direction of the relationship: Figure 19-2a shows a correlation of +1, Figure 19-2b shows a correlation of −0.50, Figure 19-2c shows a correlation of 0.85, and Figure 19-2d shows a correlation of 0.15. Comparing Figures 19-2a and c, you see Figure 19-2a is a perfect uphill straight line, and Figure 19-2c shows a very strong uphill linear pattern. Figure 19-2b is going downhill but the points are somewhat scattered in a wider band, showing a linear relationship is present, but not as strong as in Figures 19-2a and 19-2c. Figure 19-2d doesn't show much of anything happening (and it shouldn't, because its correlation is very close to 0).

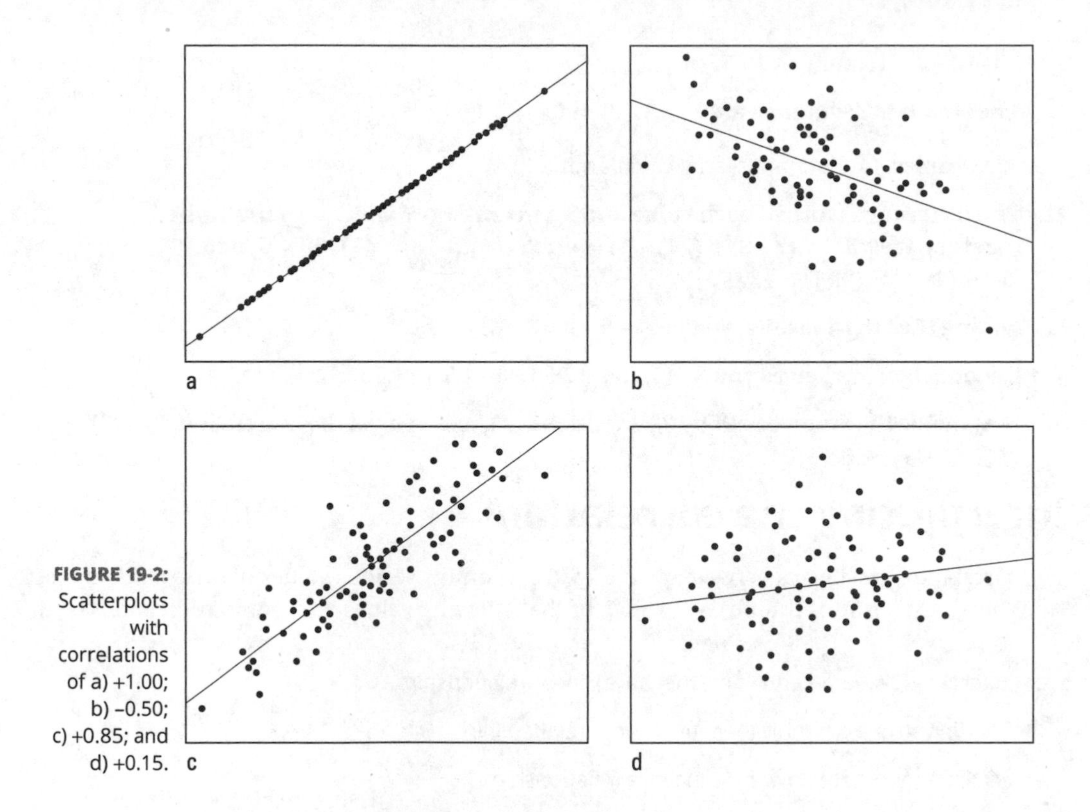

FIGURE 19-2: Scatterplots with correlations of a) +1.00; b) −0.50; c) +0.85; and d) +0.15.

Many folks make the mistake of thinking that a correlation of −1 is a bad thing, indicating no relationship. Just the opposite is true! A correlation of −1 means the data are lined up in a perfect straight line, the strongest linear relationship you can get. The "−" (minus) sign just happens to indicate a negative relationship, a downhill line.

TIP

How close is close enough to −1 or +1 to indicate a strong enough linear relationship? Most statisticians like to see correlations beyond at least +0.5 or −0.5 before getting too excited about them. Don't expect a correlation to always be 0.99, however; remember, this is real data, and real data isn't perfect.

For my subset of the cricket chirps versus temperature data from the earlier section, "Picturing a Relationship with a Scatterplot," I calculated a correlation of 0.98, which is almost unheard of in the real world (these crickets are *good!*).

Examining properties of the correlation

Here are several important properties of the correlation coefficient:

- The correlation is always between –1 and +1, as I explain in the preceding section.
- The correlation is a unitless measure, which means that if you change the units of *X* or *Y*, the correlation won't change. For example, changing the temperature from Fahrenheit to Celsius won't affect the correlation between the frequency of chirps (*X*) and the outside temperature (*Y*).
- The variables *X* and *Y* can be switched in the data set without changing the correlation. For example, if height and weight have a correlation of 0.53, weight and height have the same correlation.

EXAMPLE

Q. Find and interpret the correlation for the following data set: (2, 3), (6, 8), (3, 5).

A. **Step 1:** $\bar{x}$ is 3.67, and $\bar{y}$ is 5.33.

Step 2: The standard deviations are $s_x = 2.08$ and $s_y = 2.52$. (See Chapter 5 if you need help calculating standard deviations.)

Step 3: Taking each x-value minus $\bar{x}$ times each y-value minus $\bar{y}$ gives you $(2-3.67)(3-5.33)=3.891$; $(6-3.67)(8-5.33)=6.221$; and $(3-3.67)(5-5.33)=0.221$.

Step 4: Sum these results to get 10.33.

Step 5: Divide this total by (s_x times s_y) to get $10.33 \div (2.08)(2.52) = 1.97$.

Step 6: Divide 1.97 by 3 – 1 (or 2) to get 0.986. This is the correlation between X and Y for this example.

Interpretation: This correlation indicates a strong positive linear relationship between X and Y. (This corresponds to the scatterplot as well; see the following figure.)

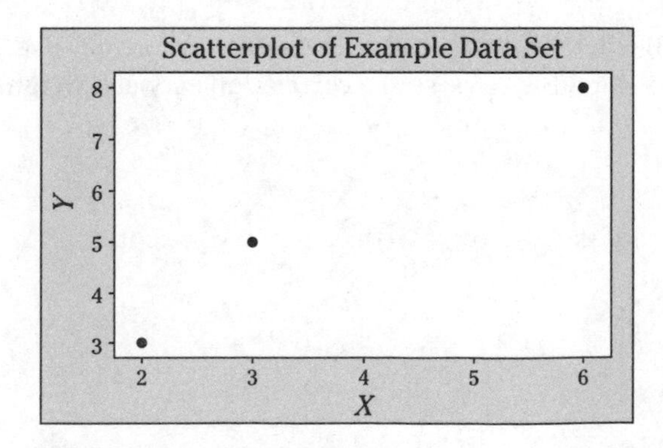

5 Calculate the correlation of the data set (1, 4), (6, 3), (2, 3), (7, 2).

6 Match each of the following correlations to their corresponding scatterplot in the following figure: 0.90, 0.39, −0.74, and 0.57.

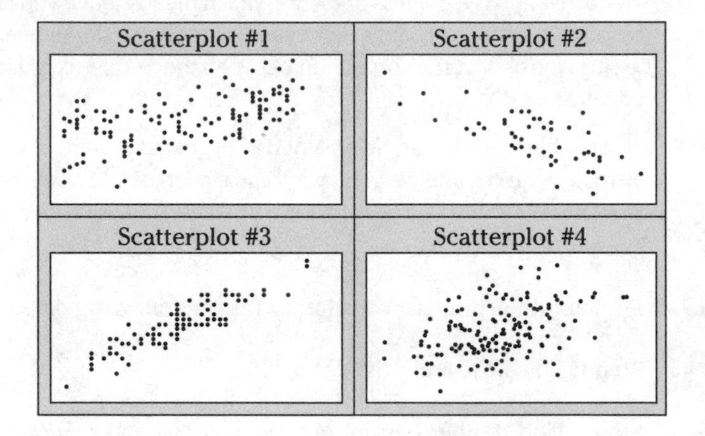

7 Tell whether the following statement is true or false: "The correlation between shoe size and height, in inches, is 0.70. If height is measured in feet, the correlation is $0.70 \div 12 = 0.058$ because 12 inches is equal to 1 foot."

8 Tell whether the following statement is true or false: "The correlation between height and weight is 0.60, so the correlation between weight and height is −0.60."

9 Tell whether the following statement is true or false: "The correlation between gender and political affiliation is 0.65."

10 State whether the following statement is true or false: "The correlation between bushels per acre and annual rainfall is 2.5 inches."

Working with Linear Regression

In the case of two numerical variables *X* and *Y*, when at least a moderate correlation has been established through both the correlation and the scatterplot, you know they have some type of linear relationship. Researchers often use that relationship to predict the (average) value of *Y* for a given value of *X* using a straight line. Statisticians call this line the *regression line*. If you know the slope and the *y*-intercept of that regression line, then you can plug in a value for *X* and predict the average value for *Y*. In other words, you predict (the average) *Y* from *X*. In the following sections, I provide the basics of understanding and using the linear regression equation (I explain how to make predictions with linear regression later in this chapter).

Never do a regression analysis unless you have already found at least a moderately strong correlation between the two variables. (My rule of thumb is it should be at or beyond either positive or negative 0.50, but other statisticians may have different criteria.) I've seen cases where researchers go ahead and make predictions when a correlation is as low as 0.20! By anyone's standards, that doesn't make sense. If the data don't resemble a line to begin with, you shouldn't try to use a line to fit the data and make predictions (but people still try).

Figuring out which variable is *X* and which is *Y*

Before moving forward to find the equation for your regression line, you have to identify which of your two variables is *X* and which is *Y*. When doing correlations (as I explain earlier in this chapter), the choice of which variable is *X* and which is *Y* doesn't matter, as long as you're consistent for all the data. But when fitting lines and making predictions, the choice of *X* and *Y* does make a difference.

REMEMBER

So how do you determine which variable is which? In general, Y is the variable that you want to predict, and X is the variable you are using to make that prediction. In the earlier cricket chirps example, you are using the number of chirps to predict the temperature. So in this case the variable Y is the temperature, and the variable X is the number of chirps. Hence Y can be predicted by X using the equation of a line if a strong enough linear relationship exists.

TIP

Statisticians call the X-variable (cricket chirps in the earlier example) the *explanatory variable,* because if X changes, the slope tells you (or explains) how much Y is expected to change in response. Therefore, the Y variable is called the *response variable.* Other names for X and Y include the *independent* and *dependent* variables, respectively.

Checking the conditions

REMEMBER

In the case of two numerical variables, you can come up with a line that enables you to predict Y from X, if (and only if) the following two conditions from the previous sections are met:

- The scatterplot must form a linear pattern.
- The correlation, *r,* is moderate to strong (typically at or beyond 0.50 or –0.50).

Some researchers actually don't check these conditions before making predictions. Their claims are not valid unless the two conditions are met.

But suppose the correlation is high; do you still need to look at the scatterplot? Yes. In some situations the data have a somewhat curved shape, yet the correlation is still strong; in these cases, making predictions using a straight line is still invalid. Predictions need to be made based on a curve. (This topic is outside the scope of this book; if you are interested, see *Statistics II For Dummies* by Deborah Rumsey [also published by Wiley], which tackles nonlinear relationships.)

Calculating the regression line

For the crickets and temperature data, you can see that the scatterplot in Figure 19-1 shows a linear pattern. The correlation between cricket chirps and temperature was found earlier in this chapter to be very strong ($r = 0.98$). You now can find one line that best fits the data (in terms of having the smallest overall distance to the points). Statisticians call this technique for finding the best-fitting line a *simple linear regression analysis using the least-squares method.*

REMEMBER

The formula for the *best-fitting line* (or *regression line*) is $y = mx + b$, where m is the slope of the line and b is the y-intercept. This equation itself is the same one used to find a line in algebra; but remember, in statistics the points don't lie perfectly on a line — the line is a model around which the data lie if a strong linear pattern exists.

- The *slope* of a line is the change in *Y* over the change in *X*. For example, a slope of $^{10}/_{3}$ means as the *x*-value increases (moves right) by 3 units, the *y*-value moves up by 10 units on average.
- The *y-intercept* is that place on the *y*-axis where the value of *x* is zero. For example, in the equation $2x - 6$, the line crosses the *y*-axis at the point –6. The coordinates of this point are (0, –6); when a line crosses the *y*-axis, the *x*-value is always 0.

REMEMBER

To come up with the best-fitting line, you need to find values for *m* and *b* that fit the pattern of data the best, for your given criteria. Different criteria exist and can lead to other lines, but the criteria I use in this book (and in all introductory level statistics courses in general) is to find the line that minimizes what statisticians call the *sum of squares for error* (SSE). The SSE is the sum of all the squared differences from the points on the proposed line to the actual points in the data set. The line with the lowest possible SSE wins, and its equation is used as the best-fitting line. This process is where the name *the least-squares method* comes from.

You may be thinking that you have to try lots and lots of different lines to see which one fits best. Fortunately, you have a more straightforward option (although eyeballing a line on the scatterplot does help you think about what you'd expect the answer to be). The best-fitting line has a distinct slope and *y*-intercept that can be calculated using formulas (and, I may add, these formulas aren't too hard to calculate).

TIP

To save a great deal of time calculating the best-fitting line, first find the "big five," five summary statistics that you'll need in your calculations:

- The mean of the *x* values (denoted $\bar{x}$)
- The mean of the *y* values (denoted $\bar{y}$)
- The standard deviation of the *x* values (denoted s_x)
- The standard deviation of the *y* values (denoted s_y)
- The correlation between *X* and *Y* (denoted *r*)

Finding the slope

The formula for the slope, *m*, of the best-fitting line is

$$m = r\left(\frac{s_y}{s_x}\right)$$

where *r* is the correlation between X and Y, and s_x and s_y are the standard deviations of the *x*-values and the *y*-values, respectively. You simply divide s_y by s_x and multiply the result by *r*.

Note that the slope of the best-fitting line can be a negative number because the correlation can be a negative number. A negative slope indicates that the line is going downhill. For example, amount of rain and sales per week of certain products move in a linear fashion — the more it rains, the lower sales tend to be; the correlation and hence the slope of the best-fitting line is negative in this case.

TIP

The correlation and the slope of the best-fitting line are not the same. The formula for slope takes the correlation (a unitless measurement) and attaches units to it. Think of $s_y \div s_x$ as the variation (resembling change) in *Y* over the variation in *X*, in units of *X* and *Y*. An example would be variation in temperature (degrees Fahrenheit) over the variation in number of cricket chirps (in 15 seconds).

Finding the y-intercept

The formula for the *y*-intercept, *b*, of the best-fitting line is $b = \bar{y} - m\bar{x}$, where $\bar{x}$ and $\bar{y}$ are the means of the *x*-values and the *y*-values, respectively, and *m* is the slope (the formula for which is given in the preceding section).

REMEMBER

To calculate the y-intercept, b, of the best-fitting line, you start by finding the slope, m, of the best-fitting line using the steps listed in the preceding section. You then multiply m by $\bar{x}$ and subtract your result from $\bar{y}$.

TIP

Always calculate the slope before the y-intercept. The formula for the y-intercept contains the slope!

EXAMPLE

Q. Calculate the equation of the best-fitting line for the following small data set: (2, 3), (6, 8), (3, 5).

A. The "big five" summary statistics for this data set were calculated in the previous example and are shown in the following table.

Variable	Mean	Standard Deviation	Correlation
X	3.67	2.08	0.986
Y	5.33	2.52	

The slope, m, of the best fitting line is $(0.986)(2.52 \div 2.08) = (0.986)(1.21) = 1.19$. Now to find the y-intercept (b), you take the mean of Y and subtract the mean of X times the slope: $5.33 - (1.19)(3.67) = 0.96$. The best-fitting line for this data set is $y = 1.19x + 0.96$.

YOUR TURN

11 A coffee barista records the temperature (degrees Fahrenheit) and number of coffees sold for 50 professional football games; you can see the summary of the statistics in the following table. (See the figure from Practice Question 1 in the earlier section, "Picturing a Relationship with a Scatterplot," for a scatterplot of this data.)

Variable	Mean	Standard Deviation	Correlation
Temperature (*X*)	35.08	16.29	-0.741
Coffees sold (*Y*)	29,913	12,174	

(a) Should you use a line to fit this data? Justify your answer.

(b) Find the equation of the best-fitting regression line.

12 A medical researcher measures bone density and the age of 125 women; you can see the summary of the statistics in the following table. (See the figure from Practice Question 2 in the section, "Picturing a Relationship with a Scatterplot," for a scatterplot of this data.)

Variable	Mean	Standard Deviation	Correlation
Bone loss (*Y*)	35.008	7.684	0.574
Age (*X*)	67.992	10.673	

(a) How well will a line fit this data?

(b) Find the equation of the best-fitting regression line.

 A golf analyst measures the total score and number of putts hit for 100 rounds of golf an amateur plays; you can see the summary of the statistics in the following table. (See the figure from Practice Question 3 in the section, "Picturing a Relationship with a Scatterplot," for a scatterplot of this data.)

Variable	Mean	Standard Deviation	Correlation
Total score (Y)	93.900	7.717	0.896
Putts hit (X)	35.780	4.554	

(a) Is it reasonable to use a line to fit this data? Explain.

(b) Find the equation of the best-fitting regression line.

 Agricultural scientists try to predict corn production by using annual rainfall. They measure crop yields (bushels per acre) and annual rainfall (inches) for 150 one-acre plots; you can see the summary of statistics in the following table.

Variable	Mean	Standard Deviation	Correlation
Rainfall	47.844	9.38	0.608
Corn	150.77	19.76	

(a) Identify the X variable and the Y variable in this problem.

(b) Find and try to interpret the slope in the context of corn and rainfall. (More practice on interpretation in the next section.)

Interpreting the regression line

Even more important than being able to calculate the slope and y-intercept to form the best-fitting regression line is the ability to interpret their values; I explain how to do so in the following sections.

Interpreting the slope

The slope is interpreted in algebra as *rise over run*. If, for example, the slope is 2, you can write this as $^2/_1$ and say that as you move from point to point on the line, as the value of the X variable increases by 1, the value of the Y variable increases by 2. In a regression context, the slope is the heart and soul of the equation because it tells you how much you can expect Y to change as X increases.

In general, the units for slope are the units of the Y variable per units of the X variable. It's a ratio of change in Y per change in X. Suppose in studying the effect of dosage level in milligrams (mg) on systolic blood pressure (mmHg), a researcher finds that the slope of the regression line is −2.5. You can write this as $-2.5/1$ and say that systolic blood pressure is expected to decrease by 2.5 mmHg on average per 1 mg increase in drug dosage.

REMEMBER

Always make sure to use proper units when interpreting slope. If you don't consider units, you won't really see the connection between the two variables at hand. For example if Y is exam score and X = study time, and you find the slope of the equation is 5, what does this mean? Not much without any units to draw from. Including the units, you see you get an increase of 5 points (change in Y) for every 1 hour increase in studying (change in X). Also be sure to watch for variables that have more than one common unit, such as temperature being in either Fahrenheit or Celsius; know which unit is being used.

If using a 1 in the denominator of slope is not super-meaningful to you, you can multiply the top and bottom by any number (as long as it's the same number) and interpret it that way instead. In the systolic blood pressure example, instead of writing slope as $-2.5/1$ and interpreting it as a drop of 2.5 mmHg per 1 mg increase of the drug, you can multiply the top and bottom by 10 to get $-25/10$ and say an increase in dosage of 10 mg results in a decrease in systolic blood pressure of 25 mmHg.

Interpreting the y-intercept

The y-intercept is the place where the regression line $y = mx + b$ crosses the y-axis where $x = 0$, and is denoted by b (see the earlier section, "Finding the y-intercept"). Sometimes the y-intercept can be interpreted in a meaningful way, and sometimes not. This uncertainty differs from slope, which is always interpretable. In fact, between the two elements of slope and y-intercept, the slope is the star of the show, with the y-intercept serving as the less-famous but still noticeable sidekick.

WARNING

At times the y-intercept makes no sense. For example, suppose you use rain to predict bushels per acre of corn. You know if the data set contains a point where rain is 0, the bushels per acre must be 0 as well. As a result, if the regression line crosses the y-axis somewhere else besides 0 (and there is no guarantee it will cross at 0 — it depends on the data), the y-intercept will make no sense. Similarly, in this context a negative value of y (corn production) cannot be interpreted.

Another situation where you can't interpret the y-intercept is when data are not present near the point where $x = 0$. For example, suppose you want to use students' scores on Midterm 1 to predict their scores on Midterm 2. The y-intercept represents a prediction for Midterm 2 when the score on Midterm 1 is 0. You don't expect scores on a midterm to be at or near 0 unless someone didn't take the exam, in which case their score wouldn't be included in the first place.

Many times, however, the y-intercept is of interest to you, it has meaning, and you have data collected in the area where $x = 0$. For example, if you're predicting coffee sales at football games in Green Bay, Wisconsin, using temperature, some games get cold enough to have temperatures at or even below 0 degrees Fahrenheit, so predicting coffee sales at these temperatures makes sense. (As you may guess, they sell more and more coffee as the temperature dips.)

Putting it all together: The regression line for the crickets

In the earlier section, "Picturing a Relationship with a Scatterplot," I introduce the example of cricket chirps related to temperature. The "big five" statistics, which I explain in the section, "Calculating the regression line," are shown in Table 19-2 for the subset of cricket data. (***Note:*** I'm rounding for ease of explanation only.)

TABLE 19-2 **"Big Five" Statistics for the Cricket Data**

Variable	Mean	Standard Deviation	Correlation
Number of chirps (*x*)	$\bar{x} = 26.5$	$s_x = 7.4$	$r = +0.98$
Temp (*y*)	$\bar{y} = 67$	$sy = 6.8$	

The slope, *m*, for the best-fitting line for the subset of cricket chirps versus temperature data is $m = r\left(\frac{s_y}{s_x}\right) = 0.98\left(\frac{6.8}{7.4}\right) = 0.90$. So as the number of chirps increases by 1 chirp per 15 seconds, the temperature is expected to increase by 0.90 degrees Fahrenheit, on average. To get a more meaningful interpretation, you can multiply the top and bottom of the slope by 10 and say as chirps increase by 10 (per 15 seconds) temperature increases by 9 degrees Fahrenheit.

Now, to find the *y*-intercept, *b*, you take $\bar{y} - m\bar{x}$, or $67 - (0.90)(26.5) = 43.15$. So the best-fitting line for predicting temperature from cricket chirps based on the data is $y = 0.90x + 43.15$, or temperature (in degrees Fahrenheit) $= 0.90 \times$ (number of chirps in 15 seconds) $+ 43.2$. Now can you use the *y*-intercept to predict temperature when no chirping is going on at all? Because no data was collected at or near this point, you cannot make predictions for temperature in this area. You can't predict temperature using crickets if the crickets are silent.

EXAMPLE

Q. Consider the toy data set of (2, 3), (6, 8), (3, 5). Using the figure from the first example problem (in the section, "Picturing a Relationship with a Scatterplot") and the table from the third example problem (in the section, "Working with Linear Regression"), for what values of *X* do you feel confident making predictions about *Y*?

A. The correlation is 0.986, so the line fits well. The *X* values go from two to six, so those values of *X* give you the best predictions. However, with only three points, you don't know how well the predictions will hold up when you take another sample.

YOUR TURN

15 Answer the following, using the figures and tables from the temperature versus coffee sales data from Practice Question 1 (from the sections, "Picturing a Relationship with a Scatterplot" and "Working with Linear Regression"):

(a) How many coffees should the manager prepare to make if the temperature is 32°F?

(b) As the temperature increases, how much more/less coffee will consumers purchase? (*Hint:* Use the slope.)

(c) For what temperature values does the regression line make the best predictions?

---

16 Answer the following, using the figures and tables from the age-versus-bone-loss data in Practice Question 2 (from the sections, "Picturing a Relationship with a Scatterplot" and "Working with Linear Regression"):

(a) For what ages is it reasonable to use the regression line to predict bone loss?

(b) Interpret the slope in the context of this problem.

(c) Using the data from the study, can you say that age causes bone loss?

---

17 Referring to the figures and tables from the golf data in Practice Question 3 (from the sections, "Picturing a Relationship with a Scatterplot" and "Working with Linear Regression"), what happens as you keep increasing *X*? Does *Y* increase forever? Explain.

---

18 Using the results from the rainfall-versus-corn-production data in Practice Question 4 from the section, "Working with Linear Regression," answer the following:

(a) Find and interpret the slope in the context of this problem.

(b) Find the *y*-intercept in the context of this problem.

(c) Can the *y*-intercept be interpreted here?

Making Proper Predictions

Before you can make any predictions of Y based on X, you need to check to be sure that the regression line you use to make predictions fits the data well. A good fit is a good indicator that after you take the data away and use the model to predict Y for the next X, the model will do a good job. You also need to take care not to get carried away and make predictions for new data that is out-of-bounds from the original data used to make the model.

Checking the conditions

Here is how you can check the fit of your regression line:

1. **Check the scatterplot to make sure that you see a linear pattern in the data.**
2. **Calculate the correlation and make sure that it's strong enough in either the positive or negative direction.**

   "Strong enough" to most statisticians usually means beyond 0.5 or −0.5, but this is just a general rule.

   You should really do Steps 1 and 2 before you even fit the regression line. If Steps 1 and 2 don't check out, fitting the regression line is not advised!
3. **Create the regression line and draw it on the scatterplot. Make sure it has the right look and fit.**

   That is, make sure you don't find any places where the line is consistently above or consistently below the data, or situations that indicate the data may have some curvature to it and that a line may not be the best model to fit.
4. **Calculate the value of r^2.**

   After you square the value of r, you get a value between zero and one, which you can interpret as a percentage. You interpret r^2 as the percentage of the variability in the Y values that the model between X and Y can explain. In other words, r^2 is the amount of change in Y that can be explained by X when using your model.

   For example, if you use shoe size to predict foot length, your r^2 value should be pretty high (close to one), because after you know a person's shoe size, you know almost everything you need to estimate foot length. (However, if you try to use shoe size to predict grade point average, your value of r^2 is very low, meaning you have a lot more explaining to do.)

Some researchers actually don't check these conditions before making predictions. Their claims are not valid unless the two conditions are met.

But suppose the correlation is high; do you still need to look at the scatterplot? Yes. In some situations the data have a somewhat curved shape, yet the correlation is still strong; in these cases making predictions using a straight line is still invalid. Predictions need to be made based on a curve. (This topic is outside the scope of this book; if you are interested, see *Statistics II For Dummies*, which tackles nonlinear relationships.)

Staying in-bounds

After you have determined a strong linear relationship and you find the equation of the best-fitting line using $y = mx + b$, you use that line to predict (the average) y for a given x-value. To make predictions, you plug the x-value into the equation and solve for y. For example, if your equation is $y = 2x + 1$ and you want to predict y for $x = 1$, then plug 1 into the equation for x to get $y = 2(1) + 1 = 3$.

Keep in mind that you choose the values of X (the explanatory variable) that you plug in; what you predict is Y, the response variable, which totally depends on X. By doing this, you are using one variable that you can easily collect data on to predict a Y variable that is difficult or impossible to measure. This process works well as long as X and Y are correlated. This concept is the big idea of regression.

Using the examples from the previous section, the best-fitting line for the crickets is $y = 0.90x + 43.2$. Say you're camping outside, listening to the crickets, and you remember that you can predict temperature by counting cricket chirps. You count 35 chirps in 15 seconds, put in 35 for x, and find that $y = 0.9(35) + 43.2 = 74.7$. (Yeah, you memorized the formula before you went camping just in case you needed it.) So because the crickets chirped 35 times in 15 seconds, you figure the temperature is probably about 75° Fahrenheit.

WARNING

Just because you have a regression line doesn't mean you can plug in *any* value for X and do a good job of predicting Y. Making predictions using x-values that fall outside the range of your data is a no-no. Statisticians call this *extrapolation;* watch for researchers who try to make claims beyond the range of their data.

For example, in the chirping data, no data is collected for fewer than 18 chirps or more than 39 chirps per 15 seconds (refer to Table 19-1). If you try to make predictions outside this range, you are going into uncharted territory; the farther outside this range you go with your x-values, the more dubious your predictions for y will get. Who's to say the line still works outside of the area where data were collected? Do you really think that crickets will chirp faster and faster without limit? At some point they would either pass out or burn up! And what does a negative number of chirps really mean? (Is this similar to asking what the sound of one hand clapping is?)

REMEMBER

Be aware that not every data point will necessarily fit the regression line well, even if the correlation is high. A point or two may fall outside the overall pattern of the rest of the data; such points are called *outliers*. One or two outliers probably won't affect the overall fit of the regression line much, but in the end you can see that the line didn't do well at those specific points.

The numerical difference between the predicted value of y from the line and the actual y-value you got from your data is called a *residual.* Outliers have large residuals compared to the rest of the points; they are worth investigating to see whether there was an error in the data at those points or whether there is something particularly interesting in the data to follow up on. (There is a much more detailed look at residuals in the book *Statistics II For Dummies* by Deborah Rumsey, also published by Wiley.)

EXAMPLE

Q. Using the information from the toy data set of (2, 3), (6, 8), (3, 5) (see the example problems in all the previous sections), discuss how well the regression line fits the data. Also discuss the limitations that having only three data points presents.

A. A scatterplot with only three points doesn't say much. After all, you can fit a line with any two points, so three points doesn't define any real pattern. You could fit a line to these three points and it wouldn't fit badly. The correlation coefficient of $r = 0.986$ gives you an impressive r^2 value of $(0.986)^2 = 0.972$. This means that the variable X explains 97.2 percent of the change among the values of Y. That's huge! But because you have only three data points, you can't be as confident that your line would be the same if you had another sample. You need more data to develop a more credible model. (***Note:*** Always check the sample size when you look at correlations. After all, you can fit a perfect straight line with only two points, but will you get that same data next time?)

YOUR TURN

19 Comment on the fit of the regression line for the bone loss data, using the information from Practice Question 2 (from the sections, "Picturing a Relationship with a Scatterplot" and "Working with Linear Regression").

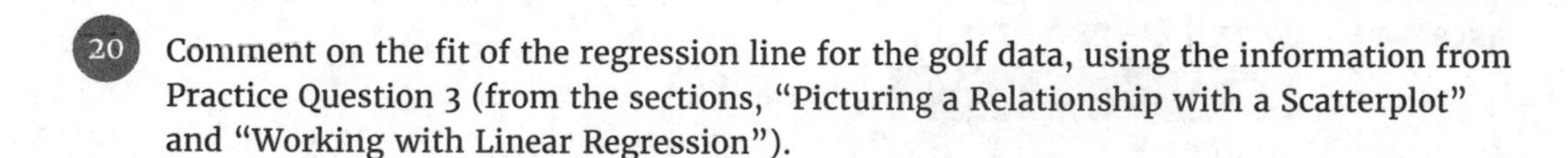

20 Comment on the fit of the regression line for the golf data, using the information from Practice Question 3 (from the sections, "Picturing a Relationship with a Scatterplot" and "Working with Linear Regression").

---

21 Examine the fit of the line used to predict coffee sales from temperature, using the information from Practice Question 1 (from the sections, "Picturing a Relationship with a Scatterplot" and "Working with Linear Regression").

(a) How much can the concessions manager rely on this model to make predictions about coffee sales? (***Hint:*** Use r^2.)

(b) Using only the value of r^2 and the scatterplot, find r.

(c) Explain why knowing r^2 isn't enough to find the correlation.

 Find and interpret the value of r^2 for the rainfall versus corn data, using the table from Practice Question 4 from the section, "Working with Linear Regression."

Regression Analysis: Understanding the Output

Another way to get the results of a regression analysis is to ask the computer software to calculate it directly for you, so you don't have to figure out the formulas yourself for slope and y-intercept (although it's good to know the inner workings of the equations.)

Table 19-3 shows a small data set to use for this purpose. The scatterplot in Figure 19-3 shows what appears to be a strong uphill linear relationship, and the value of the correlation is 0.923, so you can proceed to do the regression analysis.

TABLE 19-3 Small Data Set

x	y
1	2
2	3
3	5
4	4
5	6
6	5
7	7
8	6
9	7
10	8

I asked the computer to do a regression analysis and I got the results in Figure 19-4.

To build the equation of the line using the regression analysis output you go down the first column on numbers. The number on top (see "Constant" row) is the y-intercept, and the number on the bottom (see "x") is the slope. Typically it will say "Constant" in the first row, because the y-intercept is really a constant, it has no x-value associated with it. And the second row will typically be named whatever your x variable is. So if you are using height to predict weight of a toddler for example, your x variable is height, and the output will say "height" in Row 2.

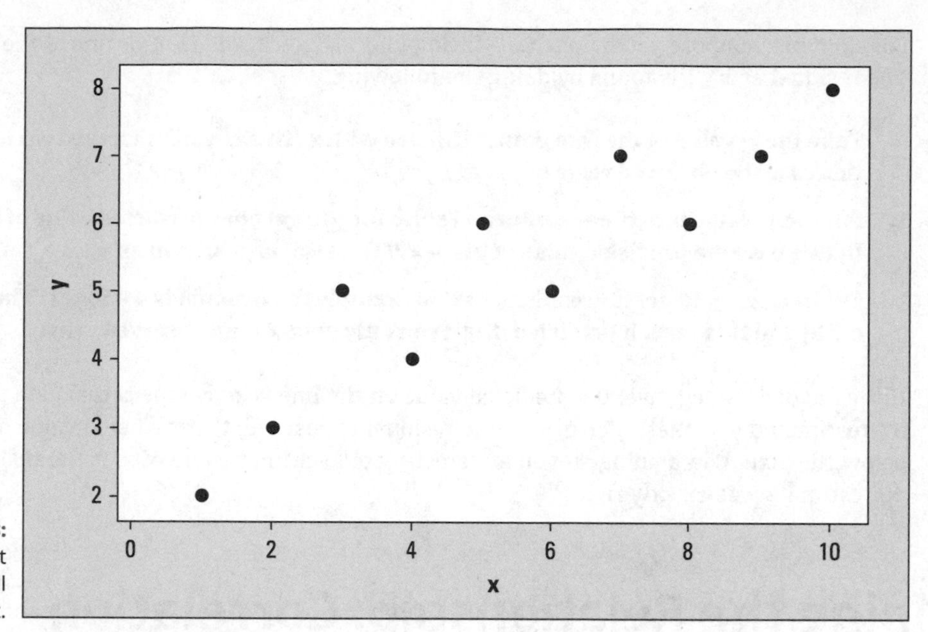

FIGURE 19-3: Scatterplot of the small data set.

FIGURE 19-4: Regression Analysis for the Small Data Set

Regression Analysis

Term	Coef	SE Coef	T-Value	P-Value	VIF
Constant	2.133	0.526	4.05	0.004	
X	0.5758	0.0848	6.79	0.000	1.00

In this case the y-intercept is 2.133 and the slope is 0.5758, moving down the first row of numbers. That tells you that the equation of the best fitting line for the small data set is $y = 2.133 + 0.5759x$.

Residing with Residuals

Residuals are the difference between what you got in your data, and what you expected to get with the regression line when you make a prediction at any value of x where a data point is located. The formula for a residual is $y - \hat{y}$, in other words, the observed (actual) value of y in the data minus the predicted value of y when you plug x into the equation at that point.

Residuals are important for looking at how well a line fits at various points. If you have a point with a large residual, that means the line is quite a bit off at that point and you might want to investigate. If the residual is zero, the data point lies directly on the line.

For example, suppose you have a best-fitting line $y = 2x + 3$, and (1, 4) is one of the data points. The residual at $x = 1$ is found by doing the following:

1. **Take the y-value of the data point affiliated with x. This is your observed value of y; in this case the observed value of y is 4.**
2. **Put the x-value into the equation and solve for y to get your predicted value of y. In this case the predicted value of y is $y = 2(1) + 3 = 5$, also known as $\hat{y}$.**
3. **Subtract $y - \hat{y}$ to get the residual. In this example the residual is $4 - 5 = -1$. The line is off by 1 in this case. It predicted 5, but you only got 4 for an observed value.**

TIP

If the residual is negative, the predicted value on the line is above the actual data point, so you overestimated y at that value of x. If the residual is positive, the predicted value on the line is below the actual data point, so you underestimated y at that value of x. If the residual is zero, our estimate was exactly right.

Explaining the Relationship: Correlation versus Cause and Effect

Scatterplots and correlations identify and quantify relationships between two variables. However, if a scatterplot shows a definite pattern and the data are found to have a strong correlation, that doesn't necessarily mean that a cause-and-effect relationship exists between the two variables. A *cause-and-effect relationship* is one where a change in one variable (in this case X) causes a change in another variable (in this case Y). (In other words, the change in Y is not only associated with a change in X, but also directly caused by X.)

For example, suppose a well-controlled medical experiment is conducted to determine the effects of dosage of a certain drug on blood pressure. (See a total breakdown of experiments in Chapter 18.) The researchers look at their scatterplot and see a definite downhill linear pattern; they calculate the correlation, and it's strong. They conclude that increasing the dosage of this drug causes a decrease in blood pressure. This cause-and-effect conclusion is okay because they controlled for other variables that could affect blood pressure in their experiment, such as other drugs taken, age, general health, and so on.

However, if you made a scatterplot and examined the correlation between ice cream consumption versus murder rates in New York City, you would also see a strong linear relationship (this one is uphill). Yet no one would claim that more ice cream consumption causes more murders to occur.

What's going on here? In the first case, the data were collected through a well-controlled medical experiment, which minimizes the influence of other factors that may affect blood pressure. In the second example, the data were based just on observation, and no other factors were examined. Researchers subsequently found out that this strong relationship exists because increases in murder rates and ice cream sales are both related to increases in temperature. Temperature in this case is called a *confounding variable*; it affects both X and Y but was not included in the study (see Chapter 18).

REMEMBER

Whether two variables are found to be causally associated depends on how the study was conducted. I've seen many instances in which people try to claim cause-and-effect relationships just by looking at scatterplots or correlations. Why would they do this? Because they want to believe it (in other words, for them it's "believing is seeing" rather than the other way around). Beware of this tactic. In order to establish cause and effect, you need to have a well-designed experiment or some very carefully done in-depth observational studies. If someone is trying to establish a cause-and-effect relationship by showing a chart or graph, dig deeper to find out how the study was designed and how the data were collected, and evaluate the study appropriately using the criteria outlined in Chapter 18.

The need for a well-designed experiment or a lot of evidence with in-depth observational studies in order to claim cause and effect is often ignored by some researchers and members of the media, who give us headlines such as "Doctors can lower malpractice lawsuits by spending more time with patients." In reality, it was found that doctors who have fewer lawsuits are the type who spend a lot of time with patients. But that doesn't mean taking a bad doctor and having him spend more time with his patients will reduce his malpractice suits; in fact, spending more time with them may create even more problems.

Practice Questions Answers and Explanations

1. **You can see a fairly strong negative relationship between temperature and the number of coffees sold.** Colder temperatures are associated with more coffees sold, and warmer temperatures are associated with fewer coffees sold.

2. Based on the scatterplot you see here, **you notice a weak to moderate positive relationship between age and bone loss** (the points are farther from the line in general than in Practice Question 1). ***Note:*** Scatterplots can differ in their appearance in terms of scale, so without the correlation itself, you can't be really specific.

3. **You can see a fairly strong positive linear relationship between number of putts and the total score during the golf rounds.** (*Note:* A putt is a stroke that takes place only when you hit on the green and near the hole. Number of putts doesn't count the drives or other shots that take place off the green.)

4. **A scatterplot that shows no linear relationship usually shows a big scattering of points plotted every which way, with no apparent pattern or linear relationship at all.** For example, a woman's shoe size has absolutely no bearing on what her IQ score would be. The following scatterplot demonstrates this by looking like an absolute mess with no pattern to the points. Other "oddball" situations include points that form a perfect box around the point (0, 0) on the *XY*-plane, but such situations are typically figments of an instructor's imagination.

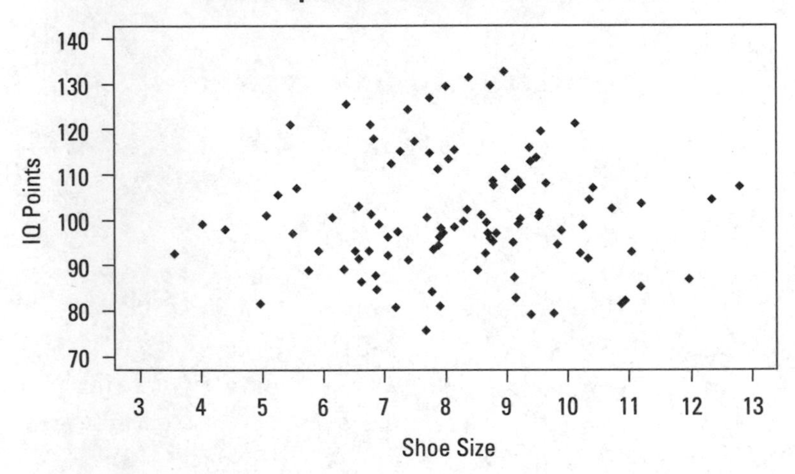

5. $\bar{x}$ is $16 \div 4 = 4$ and $\bar{y}$ is $12 \div 4 = 3$. The standard deviations are $s_x = 2.94$ and $s_y = 0.82$. Steps 3 and 4 give you $(1-4)(4-3)+(6-4)(3-3)+(2-4)(3-3)+(7-4)(2-3) = -3+0+0+-3 = -6$. Dividing the result by (s_x times s_y) gives you $-6\ (2.94)(0.82) = -2.49$. Now divide that result by 4 − 1 (or 3) to get −0.8296, or **−0.83**. *Interpretation:* This (small) data set has a strong negative linear relationship between *X* and *Y*, as shown in the following scatterplot.

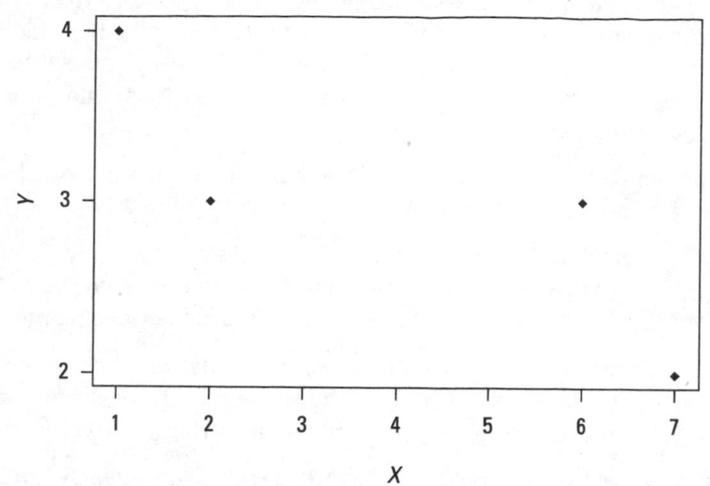

(6) **The correlations for scatterplots 1 to 4, in order, are 0.57, −0.74, 0.90, and 0.39.**

(7) **False.** Correlation doesn't change if you change the units of *X* and/or *Y*.

(8) **False.** Correlation doesn't change if you switch *X* and *Y*.

(9) **False.** Correlation doesn't apply to categorical variables; it applies only to quantitative variables. (Even though you assign a number to a categorical variable, those numbers don't mean anything.)

(10) **False.** Correlation is always between −1 and +1, and it is unitless.

(11) You can follow the example problem to work through the calculations if you want to.

a. **Yes,** the correlation (−0.741) is fairly close to −1, and the scatterplot shows that a line would fit the data well.

TIP

Typically, a correlation at or beyond ±0.60 is a pretty good correlation; the closer to ±1, the better, but data sets with a correlation very close to ±1 are few and far between (and possibly suspect).

b. The slope, *m*, of the best-fitting line is $(-0.741)(12{,}174 \div 16.29) = (-0.741)(747.33)$, which gives you −553.77. To find the *y*-intercept (*b*), you take the mean of *Y* and subtract the slope times the mean of *X*, which becomes $29{,}913 - (-553.77)(35.08) = 29{,}913 + 19{,}426.25 = 49{,}339.25$. The best-fitting line for the data set is $\boldsymbol{y = -553.77x + 49{,}339.25}$. The estimated number of coffees sold equals −553.77 times the temperature (in Fahrenheit) + 49,339.25.

WARNING

Watch for negative slopes in your calculations; they can create problems if you aren't careful.

(12) This problem resembles what you may read about or see in the media, because it's based on a medical study that relates to health issues, something of great interest to the public.

a. **The line doesn't fit the data particularly well; the correlation is only 0.574.** However, in most research circles, 0.574 is an acceptable correlation to work with. In the context of this problem, it means that although age is an important factor, other factors influence bone loss as well.

b. The slope, m, of the best-fitting line is $(0.574)(7.684 \div 10.673) = (0.574)(0.719)$, which gives you 0.413. To find the y-intercept (b), you take the mean of Y and subtract the slope times the mean of X, which becomes $35.008 - (0.413)(67.992) = 35.008 - 28.08 = 6.92$. The best-fitting line for the data set is $\boldsymbol{y = 0.413x + 6.92}$. The estimated amount of bone loss is 0.413 times age plus 6.92.

(13) This problem shows that it's not how hard you hit the ball off the tee; it's the quality of your short game that really matters.

a. The correlation here is 0.896, which shows a moderately strong positive linear relationship. That, added with the fact that the scatterplot shows that the data appear to be linear, means a regression line would provide a fairly reasonable model for this data.

REMEMBER

Don't just look at the correlation to determine whether or not a line would fit the data well. You also need to examine the scatterplot. If the scatterplot isn't linear, the correlation is meaningless. And you can have a "strong" correlation where the data aren't linear.

REMEMBER

When you find correlations only, it doesn't matter which variable you denote X and which you denote Y. However, when you have to find the regression line, it matters a great deal which is which. X is the input variable, the independent variable, and the one that goes into the equation and does the predicting. Y is the output variable, the dependent variable, and the one that X predicts. If you're given a scatterplot in a problem, you have been given a big hint as to what the X and Y variables are. The X variable is on the horizontal (x-) axis and the Y variable is on the vertical (y-) axis.

b. The slope, m, of the best-fitting line is $(0.896)(7.717\ 4.554) = 1.52$. To find the y-intercept (b), you take the mean of Y and subtract the slope times the mean of X, which becomes $93.9 - (1.52)(35.78) = 39.51$. The best-fitting line for the data set is $\boldsymbol{y = 1.52x + 39.51}$. The estimated total score is 1.52 times the number of putts, plus 39.51.

(14) If you want to outdo the *Farmer's Almanac*, you need to collect and analyze a ton of data.

a. The key to identifying X and Y (when a scatterplot is not given) is to look at what the researcher is trying to do. (You can't always assume X is listed first in the statistical information provided.) In the problem, you're told that agricultural scientists try to predict corn production by using annual rainfall. That means they're using annual rainfall to predict corn production. So, **the X variable is annual rainfall** (the variable on which the prediction is based), and **Y is the corn production** (the variable that provides the outcome you are interested in predicting).

b. The slope, m, of the best-fitting line is $(0.608)(19.76 \div 9.38) = \mathbf{1.28}$. The slope of the regression line is *rise over run* — the expected increase in Y (corn production) for every one unit increase in X (rainfall). So, **when the rainfall increases one more inch, the corn production goes up by 1.28 bushels per acre.** (*Note:* You can't tell whether a line would do well here without looking at the scatterplot, as noted in the previous golf question. The formulas allow you to plug numbers in, however. It's up to you to make the right decision on whether or not to use those formulas to make good predictions.)

(15) What good is doing regression if you can't use it to talk about football?

a. This question translates to finding the expected value of Y when $x = 32$. Plug into the equation $y = -553.77x + 49{,}339.25$ (from a previous exercise) to get $y = 32{,}618.61$ coffees. (Better make it **32,619** just to be sure!) This value of Y makes sense if you look at the scatterplot (see the figure in the original Practice Question 1; after you make a prediction for Y, always look at the scatterplot to see whether the value makes sense).

b. This question basically asks how X and Y are related, which is through the slope. The slope of this line is –553.77, which means that as the temperature goes up, the company sells less coffee. How much less? For every 1° increase in temperature, the company should expect to sell 553.77 fewer cups of coffee, on average.

c. The recorded temperatures for the 50 football games ranged **from –10°F to 70°F,** so you can feel comfortable making predictions for coffee sales if the temperature falls within this range. Outside of it, who knows?

TIP

To interpret any slope, put the value of the slope over 1. As you increase x by one unit, the y-value increases or decreases by whatever the slope is. For example, a slope of 2 means $\frac{2}{1}=\frac{\text{rise}}{\text{run}}$ so increasing x by one is associated with an increase of two in y. A slope of –2 says that increasing x by one is associated with a decrease of two in y.

(16) Properly interpreting the results of a medical study is very important. (The media often make mistakes; go figure.)

a. The researcher collects data on women **between the ages of 50 and about 85,** as you see on the scatterplot, so you can feel comfortable with this range for X.

b. The best-fitting line is $y = 0.413x + 6.92$ (from a previous exercise), so the slope is 0.413, or $\frac{0.413}{1}$. **For each year women age, their average bone loss increases by 0.413.** (Y is the amount of bone loss, not the amount of bone density, which is why the relationship is positive and not negative.)

c. **Not necessarily, because the study isn't a controlled experiment.** (How could it be?) Other factors may influence bone loss, such as diet (which age also affects). The longer you go without calcium, the more bone loss you may experience, for example. (See Chapter 18 for more information on experiments and observational studies.)

(17) Here X = number of putts and Y = total score. You can only putt so much, so X will only get so large, depending on the rules of the golf course. That means the score will only get so large. Therefore, the line doesn't go on forever.

(18) You don't have a scatterplot here (oftentimes you don't), but you can still tell plenty from the statistics given.

a. **The slope is 1.28**, which you calculate in a previous exercise. **See my answer for Practice Question 4 from the section, "Working with Linear Regression," for an interpretation.**

b. To find the y-intercept (b), you take the mean of Y and subtract the slope times the mean of X, which becomes $150.77-(1.28)(47.84)=$ **89.53 bushels.**

c. To see whether the y-intercept is interpretable, you need to look at the scatterplot and see whether the scientists collected any data during years without rainfall (in other words, where $X = 0$, do they have any data for Y?). You can safely assume, however, that **with zero rainfall, the corn simply wouldn't grow, making the y-intercept non-interpretable.**

REMEMBER

Scatterplots plot the data at their location, and they often don't show everything starting at $x = 0$. You notice this on the corn data scatterplot in the following figure. (Yeah, I decided to break down and show it to you.) Not having the data for $x = 0$ makes it harder to visualize where the line crosses the y-axis because that part of the graph where $x = 0$ isn't shown. Be aware of this issue when drawing a line on your scatterplot, and don't try to interpret the y-intercept if it's too far away from where the data were actually collected, because you don't know in that case if a line even still fits in that area.

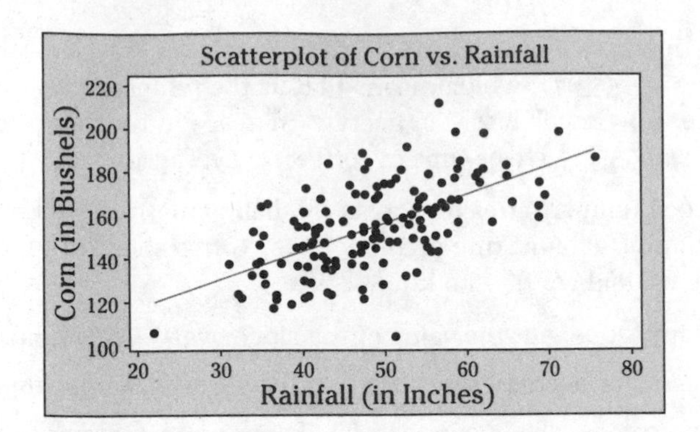

19. The scatterplot shows a "weak to moderate" uphill relationship, and the correlation is indeed weak to moderate, at 0.574. The value of $r^2 = (0.574)^2$ is only 0.329, meaning that a person's age explains 32.9 percent of the change in bone loss. That r^2 is only at 32.9 percent indicates that a lot of other variables besides age need to be taken into account to assess bone loss.

20. The golf data has a strong correlation, and the scatterplot shows no problems. The data seem to fall very close to the line. The value of r^2 is $(0.896)^2$, which equals 0.803, or 80.3 percent. So the number of putts explains an outstanding 80.3 percent of the variability in total score. (No wonder Tiger Woods spends so much time on the greens.)

21. Situations where the value of r is negative can present special problems of their own.

    a. The value of r is −0.741, so squaring it gives you 0.549, or 54.9 percent. So temperature explains 54.9 percent of the variability in coffees sold. **The manager can rely on *temperature* to explain about half the variability in coffee sales,** but the rest has to come from other information (time of day, age of the people going to the game, and so on).

REMEMBER

Because r is between −1 and +1, squaring it makes the number closer to 0. An r of 0.7, which is pretty good, gives you only 0.49 after you square it. Just something to be aware of when interpreting r^2.

    b. In this case, because you know r^2 is 0.549, you can take the square root and get $r = 0.741$. But this isn't the correlation! Why? Because it isn't negative, and you know that it should be negative because you saw it in a previous exercise. You have to look at the scatterplot (or the sign of the slope of the regression line) to see that the relationship is downhill. Include the minus sign to get $\mathbf{r = -0.741}$.

    c. If you have the value of r^2 and you need r, you have to check the scatterplot or the slope of the line to see whether you need to make r positive or negative after you take the square root.

TIP

Question (c) is a very popular test question. Be on the lookout and study up!

22. The correlation for rainfall and corn yield is 0.608 (see the earlier exercise). Squaring 0.608 gives you $\mathbf{r^2 = 0.37}$, or 37 percent. Therefore, **rainfall can explain 37 percent of the changes in corn yield.** (Other factors explain the rest of the variability.)

TIP

The value of r^2 is most easily interpreted as a percent because the value is always between 0 percent and 100 percent.

If you're ready to test your skills a bit more, take the following chapter quiz that incorporates all the chapter topics.

Whaddya Know? Chapter 19 Quiz

Quiz time! Complete each problem to test your knowledge on the various topics covered in this chapter. You can then find the solutions and explanations in the next section.

1. Suppose Mr. Boon collects data on the length of time taking an exam and the score on the exam for his high school math class. He wants to use length of time to try to predict exam score. Which variable is *X* and which is *Y*? Does it matter?

2. Bob does a survey of 100 randomly selected adults and asks them to measure their left hand and their right hand and submit the results. He finds the correlation between left and right hand length to be 0.73. Interpret the correlation.

3. What does SSE stand for and what part does it play in finding the best-fitting regression line?

4. A teacher named Elaine calculated the mean and standard deviation for her students' first exam and found the mean was 70 and the standard deviation was 10. The mean of her students' second exam was 75, with a standard deviation of 7. She wants to see if she can use the first exam score to predict the second exam score. What other information does Elaine need to calculate the best-fitting regression line using the formulas from this chapter?

5. Suppose Elaine (from Problem 4) found the correlation of her students' two exams to be 0.6. The scatterplot of exam 1 scores and exam 2 scores shows a moderately strong positive linear relationship. What is the equation of the best-fitting regression line? (Use the information in Problem 4 to help you.)

6. True or false: All you have to do to determine whether a regression analysis is okay to do on a data set is to find the correlation.

7. Suppose you are looking for a house. You collect data to see whether there is a correlation between square footage of a house and house price. You find a strong positive correlation and the scatterplot shows a strong positive linear relationship. Can you interpret the *y*-intercept here?

8. Can you find the value of the correlation (r) if you are given the value of r^2?

9. For the small data set in Table 19-3, find the residual at the point where $x = 6$. (***Note:*** Figure 19-4 gives you what you need to find the best-fitting regression line.)

10. If *X* and *Y* have a strong correlation, that means a change in *X* causes a change in *Y*. True or false?

Answers to Chapter 19 Quiz

1. If Mr. Boon is just doing correlation, it doesn't matter because switching X and Y gives you the same correlation. But, if you switch X and Y and find the best-fitting line, everything changes, so you need to get X and Y right. In this situation, Mr. Boon is using exam time to predict exam score, so exam time is X (the input variable) and exam score is Y (the output variable).

2. A correlation of $r = 0.73$ means there is a strong positive linear relationship between left- and right-hand measurements.

3. SSE stands for the **sum of squares for error.** Your best-fitting regression line is the one that minimizes SSE.

4. Elaine needs the two means (which she has), the two standard deviations (which she has), and the correlation (which she needs).

5. The slope is $m = r\frac{s_y}{s_x} = 0.6\left(\frac{7}{10}\right) = 0.42$, and the y-intercept is $\bar{y} - m\bar{x} = 75 - 0.42(70) = 45.6$. Putting these all together, the best-fitting line is $y = 45.6 + 0.42x$, where $x =$ exam 1 score and $y =$ exam 2 score.

6. **False.** You need to also look at the scatterplot. Sometimes the scatterplot will show a curved relationship while the correlation looks strong.

7. **Not in this case.** The y-intercept lies where the x value (square footage) is zero, and there are no houses with zero square footage. Additionally, there won't be houses near the area where x (square footage) is zero, so you don't interpret the y-intercept in this situation. It's there, to tell you where the best-fitting line crosses the y-axis; you just can't discuss what it means in terms of house price and square footage.

8. **No;** you take the square root first, of course, but then you need to know whether the sign on r is positive or negative, and you need more information for that, such as the sign on the slope or the direction of the scatterplot. If the data have a negative slope, then the sign on r is negative. If the data have a positive slope, then the sign on r is positive.

9. The value of y at $x = 6$ is 5, so that is the observed value of y. To get the predicted value of y, you put $x = 6$ into the best-fitting line equation and solve for y. The best-fitting line equation, looking at Figure 19-4 and the discussion surrounding it, is $y = 2.133 + 0.5759x$. Putting $x = 6$ into this equation, you get $y = 2.133 + 0.5759(6) = 5.5884$, so the predicted value of y at $x = 6$ is 5.5884. Subtract observed − predicted to get $5 - 5.5884 = \mathbf{-0.884}$.

REMEMBER

The units of a residual are the same as the units of y because you are taking an observed value of y (in units of y) minus the predicted value of y (in units of y). So if you are using height in inches to predict weight in pounds of a toddler, the residual will be in pounds.

10. **False.** It depends on how the study was done. X and Y being correlated means: 1) lower values of X are associated with lower values of Y and higher values of X are associated with higher values of Y if the correlation is positive; or 2) lower values of X are associated with higher values of Y and higher values of X are associated with lower values of Y if the correlation is negative. It does not automatically mean that if you change a value of X, then that will change its value of Y. It depends on how the study was conducted. If they used a well-designed experiment or an in-depth observational study, then yes; otherwise, no.

IN THIS CHAPTER

» Setting up two-way tables with categorical variables

» Delving into marginal, joint, and conditional distributions

» Checking for independence and dependence

» Having perspective on the results of two-way tables

Chapter 20 Two-Way Tables and Independence

Categorical variables place individuals into groups based on certain characteristics, behaviors, or outcomes, such as whether you ate breakfast this morning (yes, no) or political affiliation (Democrat, Republican, Independent, Other). Oftentimes people look for relationships between two categorical variables; hardly a day goes by that you don't hear about another relationship that's reported to have been found.

Here are just a few examples I found on the Internet recently:

- Dog owners are more likely to take their animal to the vet than cat owners.
- Heavy use of social-networking websites in teens is linked to depression.
- Children who play more video games do better in science classes.

With all this information being given to you about variables that are related, how do you decide what to believe? For example, does heavy use of social-networking websites cause depression, or is it the other way around? Or perhaps a third variable out there is related to both of them, such as problems in the home.

In this chapter, you see how to organize and analyze data from two categorical variables. You find out how to use proportions to make comparisons and look at overall patterns and how to check for independence of two categorical variables. You see how to describe dependent relationships appropriately and to evaluate results claiming to indicate cause-and-effect relationships, making predictions, and/or projecting their results to a population.

Organizing a Two-Way Table

To explore links between two categorical variables, you first need to organize the data that's been collected, and a table is a great way to do that. A *two-way table* classifies individuals into groups based on the outcomes of two categorical variables (for example, gender and opinion).

Suppose your local community developers are building a campground, and they've decided pets will be allowed as long as they're on a leash. They are now trying to decide whether the campground should have a separate section for pets. You have a hunch that non-pet campers in the area may be more in favor of a separate pet area than pet campers, so you decide to find out what the members of the camping community think. You randomly select 100 campers from the local area and conduct a pet camping survey, recording each person's opinion on having a pet section (yes, no) and whether they camp with pets (yes, no). You now have a spreadsheet with 100 rows of data, one for each person you surveyed. Each row has two pieces of data: one column for whether the person is a pet camper (yes, no) and one column for that person's opinion on having a pet section (support, oppose). Suppose the first ten rows of your data set look like what's shown in Table 20-1.

TABLE 20-1 **First Ten Rows of Data from the Pet Camping Survey**

Person	Pet Camper?	Opinion on a Separate Pet Section
1	Yes	Oppose
2	Yes	Oppose
3	Yes	Support
4	No	Support
5	No	Support
6	Yes	Support
7	No	Oppose
8	No	Support
9	Yes	Support
10	No	Oppose

From this small portion of your data set, you can start to break it down yourself. For example, looking at Column 2 results, you see that half the respondents ($5 \div 10 = 0.50$) camp with pets and the other half do not. Of those who camp with pets (that is, of those five people who have a yes in Column 2), three of them (60 percent) support having a separate section; and the same results are true for non-pet campers. These results from these 10 campers likely don't apply to all 100 campers surveyed; however, if you tried to examine the raw data from all 100 rows of this data set by hand, you wouldn't make much progress in seeing patterns without a lot of hard work.

In order to get a handle on what's happening in a large data set when you are examining two categorical variables, you organize your data into a two-way table. The following sections take you through it.

Setting up the cells

A two-way table organizes categorical data from two variables by using rows to represent one variable (such as pet camping — yes or no) and columns to represent the other variable (such as opinion on a pet section — support or oppose). Each person appears exactly once in the table.

Continuing with the camping example I start earlier in this chapter, in Table 20-2, I summarize the results from all 100 campers surveyed.

TABLE 20-2 **Two-Way Table of Pet Camping Survey Data (All 100 Rows)**

	Support Separate Pet Section	Oppose Separate Pet Section
Pet Camper	20	10
Non-Pet Camper	55	15

Table 20-2 has $2 \times 2 = 4$ numbers in it. These numbers represent the *cells* of the two-way table; each one represents an intersection of a row and column. The cell in the upper-left corner of the table represents the 20 people who are pet campers supporting a pet section. In the upper-right cell, 10 people are pet campers opposing a pet section. In the lower-left cell are the 55 non-pet campers who want a pet section; the 15 people in the lower-right cell are non-pet campers opposing a pet section.

Figuring the totals

Before getting to the nitty-gritty analysis of a two-way table in the later section, "Interpreting Two-Way Tables," you calculate some totals and add them to the table for later reference. You summarize each variable separately by calculating the *marginal totals,* which represent the total number in each row (for the first variable) and the total number in each column (for the second variable). The *marginal row totals* form an additional column on the right side of the table, and the *marginal column totals* form an additional row on the bottom of the table.

For example, in Table 20-2 in the preceding section, the marginal row total for Row 1, the number of pet campers, is $20 + 10 = 30$; the marginal row total for non-pet campers (Row 2) is $55 + 15 = 70$. The marginal column total for those wanting a pet section (Column 1) is $20 + 55 = 75$; and the marginal column total for those not wanting a separate section (Column 2) is $10 + 15 = 25$.

REMEMBER

The *grand total* is the total of all the cells in the table and is equal to the sample size. (***Note:*** the marginal totals are not included in the grand total, only the cells.) The grand total sits in the lower right-hand corner of the two-way table. In this example, the grand total is $20 + 10 + 55 + 15 = 100$. Table 20-3 shows the marginal row and column totals and the grand total for the pet camping survey data.

The marginal row totals always sum to the grand total, because everyone in the survey either camps with a pet or they don't. In the last column of Table 20-3, you see that $30+70=100$. Similarly, the marginal column totals always sum to the grand total; everyone in the survey either wants a pet section or they don't; in the last row of Table 20-3, you see $75+25=100$.

TABLE 20-3 **Two-Way Table of Pet Camping Survey Data, Including Marginal Totals**

	Support Separate Pet Section	Oppose Separate Pet Section	Marginal Row Totals
Pet Camper	20	10	$20+10=30$
Non-Pet Camper	55	15	$55+15=70$
Marginal Column Totals	$20+55=75$	$10+15=25$	**Grand total** $=100$ $(20+10+55+15)$

REMEMBER

When organizing a two-way table, always include the marginal totals and the grand total. It gets you off on the right foot when analyzing the data.

EXAMPLE

Q. Suppose a researcher divides a sample of 100 cars into groups according to their number of bumper stickers (three or less versus more than three) and the ages of the cars (5 years old or less versus more than 5 years old) and summarizes the results in the following table.

Number of Bumper Stickers	Age of Car ≤ 5 Years	Age of Car >5 Years
0–3 bumper stickers	30	15
>3 bumper stickers	20	35

(a) Describe each cell in the two-way table and the number it contains.

(b) Write in the marginal totals and the grand total and interpret them.

A. Here's how all the bumper-sticker info breaks down:

(a) Thirty cars in the sample have few bumper stickers and are newer cars. Fifteen cars have few bumper stickers and are older cars. Twenty cars have a lot of bumper stickers and are newer cars. Thirty-five cars have a lot of bumper stickers and are older cars.

(b) The marginal totals are included in the following table. The total number of cars in the sample with fewer bumper stickers (fewer than 3) is 45 $(30+15)$; the total number of cars with a lot of bumper stickers (3 or more) is $50(20+35)$. The total number of newer cars (less than 5 years old) is $50(15+35)$; the total number of older cars (5 years old or more) is 50 $(15+35)$. The grand total of all cars in the sample is 100.

Number of Bumper Stickers	Age of Car ≤ 5 Years	Age of Car > 5 Years	Totals (Stickers)
0–3 bumper stickers	30	15	45
>3 bumper stickers	20	35	55
Totals (Age)	50	50	100 (grand total)

1 A medical researcher measures the dominant hand against gender for a group of 42 toddlers and shows the results in the following table.

Gender	Left-Handed	Right-Handed
Male	4	24
Female	2	12

(a) How many toddlers are male?

(b) How many toddlers are right-handed?

(c) How many toddlers are right-handed and male?

2 For the gender and political affiliation data summarized in the following table, find the marginal totals for both variables and interpret them.

The Gender and Political Affiliation of 200 Survey Participants

Gender	Democrat	Republican	Independent	Totals (Gender)
Male	35	55	10	100
Female	55	35	10	100
Totals (Political Affiliation)	90	90	20	200 (grand total)

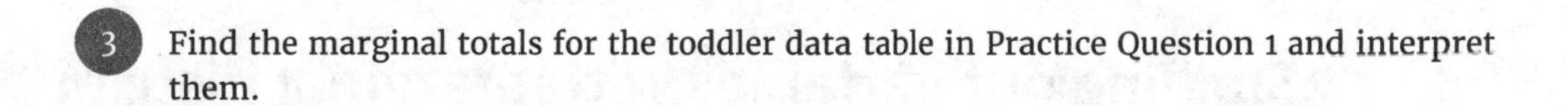

3 Find the marginal totals for the toddler data table in Practice Question 1 and interpret them.

4 It's important to keep the big picture of a two-way table in mind and not get lost in the individual cell-by-cell breakdowns.

(a) If you sum all the marginal totals for the rows of a two-way table, what should you get?

(b) If you sum all the marginal totals for the columns of a two-way table, what should you get?

Interpreting Two-Way Tables

After the two-way table is set up (with the help of the information in the previous section), you calculate percents to explore the data to answer your research questions. Here are some questions of interest from the camping data earlier in this chapter (each question is handled in the following sections, respectively):

- What percentage of the campers are in favor of a pet section?
- What percentage of the campers are pet campers who support a pet section?
- Do more non-pet campers support a pet section, compared to pet campers?

The answers to these (and any other) questions about the data come from finding and working with the proportions, or percentages, of individuals within certain parts of the table. This process involves calculating and examining what statisticians call *distributions*. A distribution in the case of a two-way table is a list of all the possible outcomes for one variable or a combination of variables, along with their corresponding proportions (or percentages).

For example, the distribution for the pet camping variable lists the percentages of people who do and do not camp with pets. The distribution for the combination of the pet camping variable (yes, no) and the opinion variable (support, oppose) lists the percentages of: 1) pet campers who support a pet section; 2) pet campers who oppose a pet section; 3) non-pet campers who support a pet section; and 4) the non-pet campers who oppose a pet section.

REMEMBER

For any distribution, all the percentages must sum to 100 percent. If you're using proportions (decimals), they must sum to 1.00. Each individual has to be somewhere, and they can't be in more than one place at one time.

In the following sections, you see how to find three types of distributions, each one helping you to answer its corresponding question in the preceding list.

Singling out variables with marginal distributions

If you want to examine one variable at a time in a two-way table, you don't look in the cells of the table, but rather in the margins. As seen in the earlier section, "Figuring the totals," the marginal totals represent the total number in each row (or column) separately. In the two-way table for the pet camping survey (refer to Table 20-3), you see the marginal totals for the pet camping variable (yes/no) in the right-hand column, and you find the marginal totals for the opinion variable (support/oppose) in the bottom row.

If you want to make comparisons between two groups (for example, pet campers versus non-pet campers), however, the results are easier to interpret if you use proportions instead of totals. If 350 people were surveyed, visualizing a comparison is easier if you're told that 60 percent are in Group A and 40 percent are in Group B, rather than saying 210 people are in Group A and 140 are in Group B.

To examine the results of a two-way table based on a single variable, you find what statisticians call the *marginal distribution* for that variable. In the following sections, I show you how to calculate and graph marginal distributions.

Calculating marginal distributions

To find a marginal distribution for one variable in a two-way table, you take the marginal total for each row (or column) divided by the grand total.

- If your variable is represented by the rows (for example, the pet camping variable in Table 20-3), use the marginal row totals in the numerators and the grand total in the denominators. Table 20-4 shows the marginal distribution for the pet camping variable (yes, no).
- If your variable is represented by the columns (for example, opinion on the pet section policy, shown in Table 20-3), use the marginal column totals for the numerators and the grand total for the denominators. Table 20-5 shows the marginal distribution for the opinion variable (support, oppose).

TABLE 20-4 Marginal Distribution for Pet Camping Variable

Pet Camping	Proportion
Yes	$30 \div 100 = 0.30$
No	$70 \div 100 = 0.70$
Total	1.00

TABLE 20-5 Marginal Distribution for the Opinion Variable

Opinion	Proportion
Support pet section	$75 \div 100 = 0.75$
Oppose pet section	$25 \div 100 = 0.25$
Total	1.00

TIP

In either case, the sum of the proportions for any marginal distribution must be 1 (subject to rounding). All results in a two-way table are subject to rounding error; to reduce rounding error, keep at least 2 digits after the decimal point throughout.

Graphing marginal distributions

You graph a marginal distribution using either a pie chart or a bar graph. Each graph shows the proportion of individuals within each group for a single variable. Figure 20-1a is a pie chart summarizing the pet camping variable, and Figure 20-1b is a pie chart showing the breakdown of the opinion variable. You see that the results of these two pie charts correspond with the marginal distributions in Tables 20-4 and 20-5, respectively.

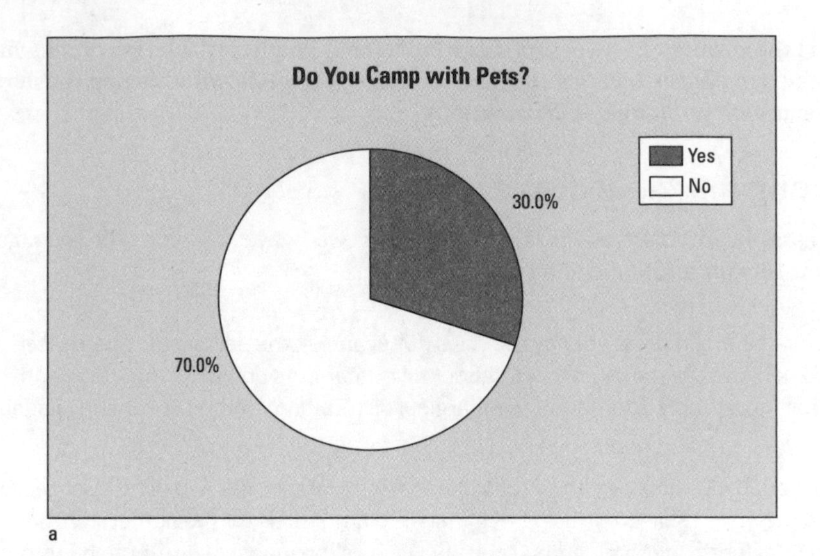

a

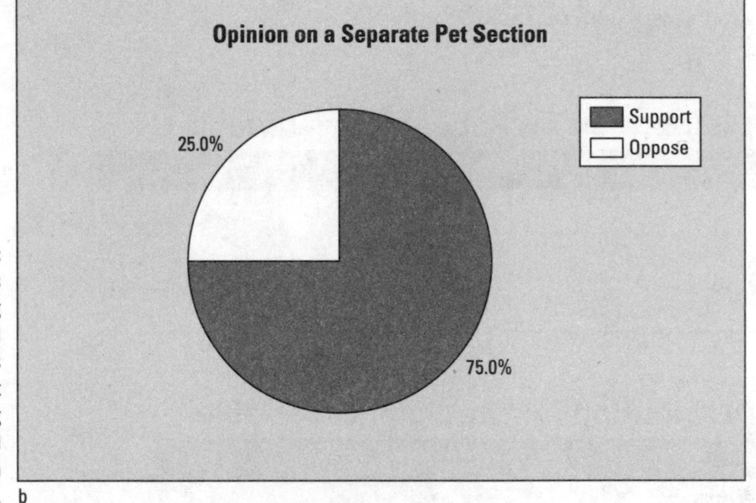

b

FIGURE 20-1: Pie charts showing marginal distributions for a) pet camping variable; and b) opinion variable.

From the results of the two separate marginal distributions for the pet camping and opinion variables, you say that the majority of all the campers in this sample are non–pet campers (70 percent) and the majority of all the campers in this sample (75 percent) support the idea of having a pet section.

REMEMBER

While marginal distributions show how each variable breaks down on its own, they don't tell you about the connection between two variables. For the camping example, you know what percentage of all campers support a new pet section, but you can't distinguish the opinions of the pet campers from the non–pet campers. Distributions for making such comparisons are found in the later section, "Comparing groups with conditional distributions."

EXAMPLE

Q. Using the data from the table, "The Gender and Political Affiliation of 200 Survey Participants," found in Practice Question 2 (in the section, "Figuring the totals") and the appropriate probability notation (for example, unions, intersections, and conditional probabilities), identify and calculate the following:

(a) What's the probability that a participant is Republican?

(b) What proportion of the sample are female?

(c) What percentage of the sample are Democrats?

A. Marginal probabilities are found by taking the number in the group of interest divided by the grand total (the total sample size).

(a) The probability that a participant is Republican, $p(\text{R})$, is equal to the total number of Republicans (a marginal total, 90) divided by the grand total (200): $p(\text{R}) = 90 \div 200 = 0.45$, or 45 percent.

(b) The proportion of females is the same as the probability that a participant is a female, $p(\text{F})$. You take the total number of females (a marginal total, 100) divided by the grand total (200): $p(\text{F}) = 100 \div 200 = 0.50$, or 50 percent.

(c) You're looking for $p(\text{D})$, which is the number of Democrats (90) divided by the grand total (200), which gives you $90 \div 200 = 0.45$, or 45 percent.

YOUR TURN

5 Using the car data from the bumper sticker/age of car table (in the earlier section, "Figuring the totals") and the appropriate probability notation, identify and calculate the following:

(a) What percentage of the cars are newer cars?

(b) What's the proportion of older cars?

(c) What percentage of the cars have a lot of bumper stickers?

(d) What's the probability that a car doesn't have a lot of bumper stickers?

---

6 Using the toddler data from the table in Practice Question 1 (in the section, "Figuring the totals") and the appropriate probability notation, identify and calculate the following:

(a) What percentage of the toddlers are right-handed?

(b) What percentage of the toddlers are female?

(c) What proportion of the toddlers are left-handed?

(d) What's the chance of finding a male toddler from the sample?

Using the same toddler data, describe the toddlers with marginal probabilities only.

Which marginal probabilities that you find in a two-way table should sum to one?

Examining all groups — a joint distribution

Story time: A certain auto manufacturer conducted a survey to see what characteristics customers prefer in their small pickup trucks. They found that the most popular color for these trucks was red and the most popular option was four-wheel drive. In response to these results, the company started making more of their small pickup trucks red with four-wheel drive.

Guess what? They struck out; people weren't buying those trucks. Turns out that the customers who bought the red trucks were more likely to be women, and women didn't use four-wheel drive as often as men did. Customers who bought the four-wheel drive trucks were more likely to be men, and they tended to prefer black ones over red ones. So the most popular outcome of the first variable (color) paired with the most popular outcome of the second variable (options on the vehicle) doesn't necessarily add up to the most popular combination of the two variables.

REMEMBER

To figure out which combination of two categorical variables contains the highest proportion, you need to compare the cell proportions (for example, the color and vehicle options together) rather than the marginal proportions (the color and vehicle options separately). The *joint distribution* of both variables in a two-way table is a listing of all possible row and column combinations and the proportion of individuals within each group. You use it to answer questions involving two characteristics, such as "What proportion of the voters are Democrat and female?" or "What percentage of the campers are pet campers who support a pet section?" In the following sections, I show you how to calculate and graph joint distributions.

Calculating joint distributions

A joint distribution shows the proportion of the data that lies in each cell of the two-way table. For the pet camping example, the four row-column combinations are:

- All campers who camp with pets and support a pet section
- All campers who camp with pets and oppose a pet section

» All campers who don't camp with pets and support a pet section

» All campers who don't camp with pets and oppose a pet section

TIP

The key phrase in all of the proportions mentioned in the preceding list is *all campers*. You are taking the entire group of all campers in the survey and breaking them into four separate groups. When you see the word *all*, think joint distribution. Table 20-6 shows the joint distribution for all campers in the pet camping survey.

TABLE 20-6 Joint Distribution for the Pet Camping Survey Data

	Support Separate Pet Section	Oppose Separate Pet Section
Camp with Pets	$20 \div 100 = 0.20$	$10 \div 100 = 0.10$
Don't Camp with Pets	$55 \div 100 = 0.55$	$15 \div 100 = 0.15$

To find a joint distribution for a two-way table, you take the cell count (the number of individuals in a cell) divided by the grand total, for each cell in the table. The total of all these proportions should be 1 (subject to rounding error).

To get the numbers in the cells of Table 20-6, take the cells of Table 20-3 and divide by their corresponding grand total (100, in this case). Using the results listed in Table 20-6, you report the following:

» 20 percent of all campers surveyed camp with pets and support a pet section. (See the upper left-hand cell of the table.)

» 10 percent of all campers surveyed camp with pets and oppose a pet section. (See the upper right-hand cell of the table.)

» 55 percent of all campers surveyed don't camp with pets and do support the pet section policy. (See the lower left-hand cell of the table.)

» 15 percent of all campers surveyed don't camp with pets and oppose the pet section policy. (See the lower right-hand cell of the table.)

Adding all the proportions shown in Table 20-6, you get $0.20 + 0.10 + 0.55 + 0.15 = 1.00$. Every camper shows up in one and only one of the cells of the table.

Graphing joint distributions

To graph a joint distribution from a two-way table, you make a single pie chart with four slices, representing each proportion of the data that falls within a row-column combination. Groups containing more individuals get a bigger piece of the overall pie, and hence get more weight when all the votes are counted up. Figure 20-2 is a pie chart showing the joint distribution for the pet camping survey data.

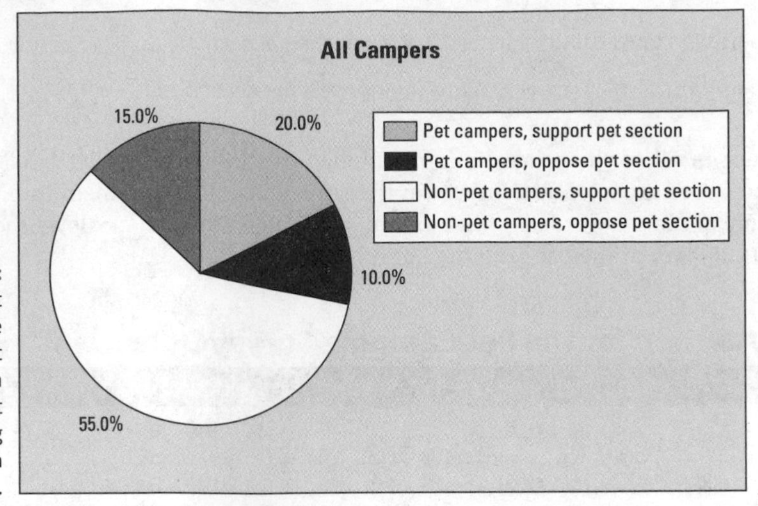

FIGURE 20-2: Pie chart showing the joint distribution of the pet camping and opinion variables.

From the pie chart shown in Figure 20-2, you see some results that stand out. The majority of campers in this sample (0.55 or 55 percent) don't camp with pets and support a separate section for pets. The smallest slice of the pie represents those campers who camp with pets and are opposed to a separate section for pets (0.10 or 10 percent).

A joint distribution gives you a breakdown of the entire group by both variables at once and allows you to compare the cells to each other and to the whole group. The results in Figure 20-2 show that if they were asked to vote today as to whether or not to have a pet section, when all the votes were added up, most of the weight would be placed on the opinions of non-pet campers, because they make up the majority of campers in the survey (70 percent, according to Table 20-4), and the pet campers would have less of a voice, because they are a smaller group (30 percent).

WARNING

A limitation of a joint distribution is that you can't fairly compare two groups to each other (for example, pet campers versus non-pet campers) because the joint distribution puts more weight on larger groups. The next section shows how to fairly compare the groups in a two-way table.

Comparing groups with conditional distributions

You need a different type of distribution other than a joint distribution to compare the results from two groups (for example, comparing opinions of pet campers versus non-pet campers). *Conditional distributions* are used when looking for relationships between two categorical variables; the individuals are first split into the groups you want to compare (for example, pet campers and non-pet campers); then the groups are compared based on their opinion on a pet section (yes, no). In the following sections, I explain how to calculate and graph conditional distributions.

Calculating conditional distributions

To find conditional distributions for the purpose of comparison, first split the individuals into groups according to the variable you want to compare. Then for each group, take the cell count (the number of individuals in a particular cell) divided by the marginal total for that group. Do this for all the cells in that group. Now repeat for the other group, using its marginal total as the denominator and the cells within its group as the numerators. (See the earlier section, "Figuring the totals," for more about marginal totals.) You now have two conditional distributions, one for each group, and you fairly compare the results for the two groups.

For the pet camping survey data example (earlier in this chapter), you compare the opinions of two groups: pet campers and non-pet campers; in statistical terms you want to find the conditional distributions of opinion based on the pet camping variable. That means you split the individuals into the pet camper and non-pet camper groups, and then for each group, you find the percentages of who supports and opposes the new pet section. Table 20-7 shows these two conditional distributions in table form (working off Table 20-3).

TABLE 20-7 Conditional Distributions of Opinion for Pet Campers and Non-Pet Campers

	Support Pet Section Policy	Oppose Pet Section Policy	Total
Pet Campers	$20 \div 30 = 0.67$	$10 \div 30 = 0.33$	1.00
Non-Pet Campers	$55 \div 70 = 0.79$	$15 \div 70 = 0.21$	1.00

REMEMBER

Notice that Table 20-7 differs from Table 20-6 in the earlier section, "Calculating joint distributions," in terms of how the values in the table add up. This represents the key difference between a joint distribution and a conditional distribution that allows you to make fair comparisons using the conditional distribution:

- In Table 20-6, the proportions in the cells of the entire table sum to 1 because the entire group is broken down by both variables at once in a joint distribution.
- In Table 20-7, the proportions in each row of the table sum to 1 because each group is treated separately in a conditional distribution.

Graphing conditional distributions

One effective way to graph conditional distributions is to make a pie chart for each group (for example, one for pet campers and one for non-pet campers) where each pie chart shows the results of the variable being studied (opinion: yes or no).

Another method is to use a stacked bar graph. A *stacked bar graph* is a special bar graph where each bar has a height of 1 and represents an entire group (one bar for pet campers and one bar for non-pet campers). Each bar shows how that group breaks down regarding the other variable being studied (opinion: yes or no).

Figure 20-3 is a stacked bar graph showing two conditional distributions. The first bar is the conditional distribution of opinion for the pet camping group (Row 1 of Table 20-7) and the second bar represents the conditional distribution of opinion for the non-pet camping group (Row 2 of Table 20-7).

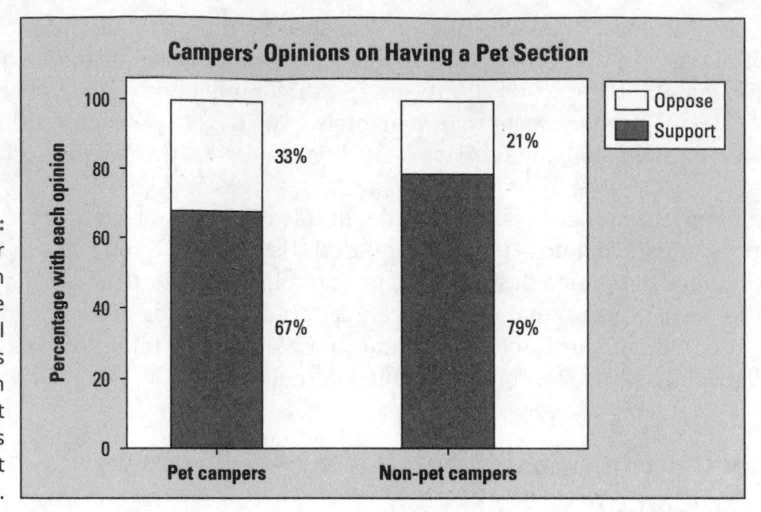

FIGURE20-3: Stacked bar graph showing the conditional distributions of opinion for pet campers and non-pet campers.

Using Table 20-7 and Figure 20-3, first look at the opinions of each group. More than 50 percent of the pet campers support the pet section (the exact number rounds to 67 percent), so you say the majority of pet campers support a pet section. Similarly, the majority of non-pet campers (about 79 percent, way more than half) support a pet section.

Now you compare the opinions of the two groups by comparing the percentage of supporters in the pet camping group (67 percent) to the percentage of supporters in the non-pet camping group (79 percent). While both groups have a majority of supporters of the pet section, you see more of the non-pet campers support the policy than pet campers (because 79 percent > 67 percent). By comparing the conditional distributions, you've found that a relationship appears to exist between opinion and pet camping, and your original hunch that non-pet campers in the area may be more in favor of a separate pet area than pet campers is correct, based on this data.

REMEMBER

The difference in the results found in Figure 20-3 isn't as large as you may have thought by looking at the joint distribution in Figure 20-2. The conditional distribution takes into account and adjusts for the number in each group being compared, while the joint distribution puts everyone in the same boat. That's why you need conditional distributions to make fair comparisons.

When making your conclusions regarding the pet-camping data, the operative words you use are "a relationship *appears* to exist." The results of the pet camping survey are based on only your sample of 100 campers. To be able to generalize these results to the whole population of pet campers and non-pet campers in this community (which is really what you want to do), you need to take into account that these sample results will vary. And when they do vary, will they still show the same kind of difference? That's what a hypothesis test will tell you (all the details are in Chapter 15).

To conduct a hypothesis test for a relationship between two categorical variables (when each variable has only two categories, like yes/no or male/female), you do either a test for two proportions (see Chapter 16) or a Chi-square test (which is covered in the book *Statistics II For Dummies* by Deborah Rumsey, also published by Wiley). If one or more of your variables have more than two categories, such as Democrats/Republicans/Other, you must use the Chi-square test to test for independence in the population.

WARNING

Be mindful that you may run across a report in which someone is trying to give the appearance of a stronger relationship than really exists, or trying to make a relationship less obvious by how the graphs are made. With pie charts, the sample size often is not reported, leading you to believe the results are based on a large sample when they may not be. With bar graphs, they stretch or shrink the scale to make differences appear larger or smaller, respectively. (See Chapter 6 for more information on misleading graphs of categorical data.)

EXAMPLE

Q. Using the data from the table, "The Gender and Political Affiliation of 200 Survey Participants," in Practice Question 2 (in the section, "Figuring totals") and the appropriate probability notation, identify and calculate the following:

(a) What's the probability that a participant is Republican given that she is female?

(b) What proportion of the females are Independents?

(c) Let D = Democrat and F = female. Find and interpret $p(F|D)$ and $p(D|F)$.

A. I word these problems slightly differently, but they all require a conditional probability.

(a) Here you choose an individual from the group of females, so you condition on the participant being female. You want to know the chance the female you select is a Republican, so the probability you want to find is $p(R|F)$, or the probability of being a Republican given that the individual is female. The denominator of this conditional probability is the total number of females (marginal row total, 100), and the numerator is the number of Republicans in that group (row, which is 35). Thus, $p(R|F) = 35 \div 100 = 0.35$, or 35 percent.

Note: You can also use the definition of conditional probability and say $p(R|F) = \frac{p(R \cap F)}{p(F)} = \frac{35/200}{100/200} = \frac{35}{100}$, noticing that the 200s cancel out here (not a coincidence). It is easier in this case to do it just by using the numbers in the cells of the table straightaway. Each problem is different; remember to look at what you're given and what you're asked to find, and choose an approach that relates those two items somehow.

(b) This problem states "of the females," which means you know you select only from the females. You want to know the chance that the female you select is an Independent, so you need to look for $p(I|F)$ The denominator is the number of females, 100 (marginal row total), and the numerator is the number of individuals in that row who are Independents, which is 10. Thus, $p(I|F) = 10 \div 100 = 0.10$, or 10 percent.

(c) The expression $p(F|D)$ means you select from among the Democrats in an effort to know the chance that the Democrat you select is female. The denominator is the total number of Democrats (marginal column total), 90, and the numerator is the number of individuals in this row who are females, 55. Thus, $p(F|D) = 55 \div 90 = 0.61$, or 61 percent. Conversely, $p(D|F)$ means the opposite; you select from among the females (of which you have 100), and you want the probability that the female you select is a Democrat. Of the Democrats, 55 are female. Thus, $p(D|F) = 55100 = 0.55$, or 55 percent.

Note: The denominator of a conditional probability is always smaller than the grand total because you look at a particular subset of the entire group, and the subset is smaller than the grand total. However, be careful which total you work with. For example, $p(A|B)$ uses the total in Group B as the denominator, and $P(B|A)$ uses the total in Group A as the denominator. In general, $p(A|B)$ isn't equal to $p(B|A)$.

YOUR TURN

9 Using the bumper sticker/age of cars data from Example Question 1 and appropriate probability notation, identify and calculate the following:

(a) Let O = older cars and B = a lot of bumper stickers. Find and interpret $p(B|O)$.

(b) What percentage of the older cars have a lot of bumper stickers?

(c) Of the older cars, what percentage have a lot of bumper stickers?

(d) What's the probability that a car has a lot of bumper stickers, given its old age?

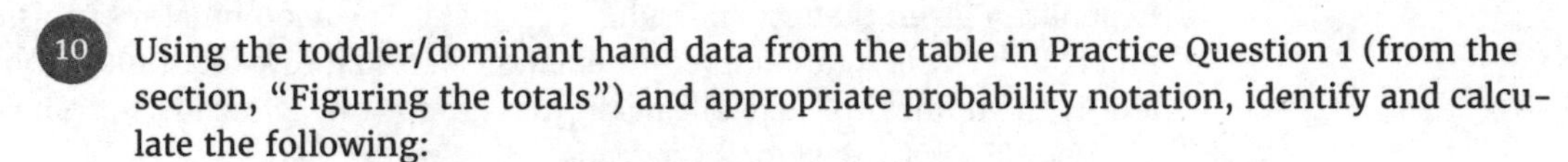

10 Using the toddler/dominant hand data from the table in Practice Question 1 (from the section, "Figuring the totals") and appropriate probability notation, identify and calculate the following:

(a) What percentage of the male toddlers are right-handed?

(b) What percentage of the female toddlers are right-handed?

(c) What percentage of the right-handed toddlers are male?

(d) What percentage of the right-handed toddlers are female?

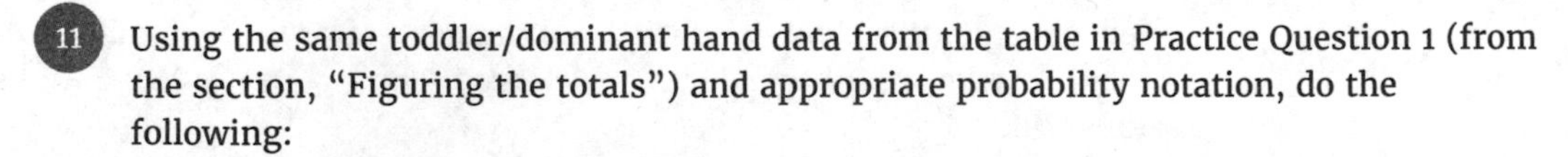

11 Using the same toddler/dominant hand data from the table in Practice Question 1 (from the section, "Figuring the totals") and appropriate probability notation, do the following:

(a) Compare the female and male right-handed toddlers in the sample, using conditional probabilities (only).

(b) Compare the right-handed toddlers in the sample, using conditional probabilities (only).

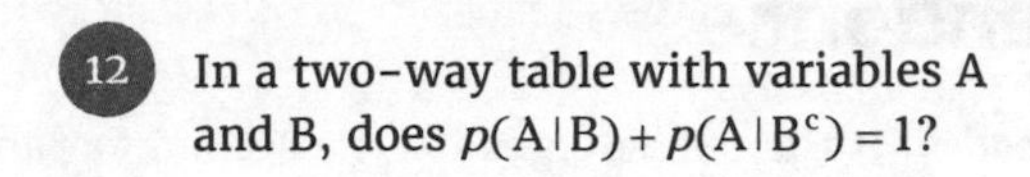

12 In a two-way table with variables A and B, does $p(A \mid B) + p(A \mid B^c) = 1$?

13 In a two-way table with variables A and B, does $p(A \mid B) + p(A^c \mid B) = 1$?

14 Explain why conditional probabilities allow you to compare two groups regarding a second variable where joint and marginal probabilities don't.

Checking Independence and Describing Dependence

The main reason why researchers collect data on two categorical variables is to explore possible relationships or connections between the variables. For example, if a survey finds that more females than males voted for the incumbent president in the last election, then you conclude that gender and voting outcome are related. If a relationship between two categorical variables has been found (that is, the results from the two groups are different), then statisticians say they're *dependent.*

However, if you find that the percentage of females who voted for the incumbent is the same as the percentage of males who voted for the incumbent, then the two variables (gender and voting for the incumbent) have no relationship and statisticians say those two variables are *independent.* In this section, you find out how to check for independence and describe relationships found to be dependent.

Checking for independence

Two categorical variables are *independent* if the percentages for the second variable (typically representing the results you want to compare, such as support or oppose) do not differ based on the first variable (typically representing the groups you want to compare, such as men versus women). You can check for independence with the methods that I cover in this section.

Comparing the results of two conditional distributions

Two categorical variables are *independent* if the conditional distributions are the same for all groups being compared. The variables are independent because breaking them down and comparing them by group doesn't change the results. In the election example I introduce at the beginning of the earlier section, "Checking Independence and Describing Dependence," independence means the conditional distribution for opinion is the same for the males and the females.

Suppose you do a survey of 200 voters to see whether gender is related to whether they voted for the incumbent president, and you summarize your results in Table 20-8.

TABLE 20-8 Results of Election Survey

	Voted for Incumbent President	Didn't Vote for Incumbent President	Marginal Row Totals
Males	44	66	110
Females	36	54	90
Marginal Column Totals	80	120	Grand total = 200

To see whether gender and voting are independent, you find the conditional distribution of voting pattern for the males and the conditional distribution of voting pattern for the females. If they're the same, you've got independence; if not, you've got dependence. These two conditional distributions have been calculated and appear in Rows 1 and 2, respectively, of Table 20-9. (See the earlier section, "Comparing groups with conditional distributions," for details.)

TABLE 20-9 Results of Election Survey with Conditional Distributions

	Voted for Incumbent President	Didn't Vote for Incumbent President	Total
Males	$44 \div 110 = 0.40$	$66 \div 110 = 0.60$	1.00
Females	$36 \div 90 = 0.40$	$54 \div 90 = 0.60$	1.00

To get the numbers in Table 20-9, I started with Table 20-8 and divided the number in each cell by its marginal row total to get a proportion. Each row in Table 20-9 sums to 1 because each row represents its own conditional distribution. (If you're male, you either voted for the incumbent or you didn't — same for females.)

Row 1 of Table 20-9 shows the conditional distribution of voting pattern for males. You see 40 percent voted for the incumbent and 60 percent did not. Similarly, Row 2 of the table shows the conditional distribution of voting pattern for females; again, 40 percent voted for the incumbent and 60 percent did not. Because these distributions are the same, men and women voted the same way; gender and voting pattern are independent.

Figure 20-4 shows the conditional distributions of voting pattern for males and females using a graph called a stacked bar chart. Because the bars look exactly alike, you conclude that gender and voting pattern are independent.

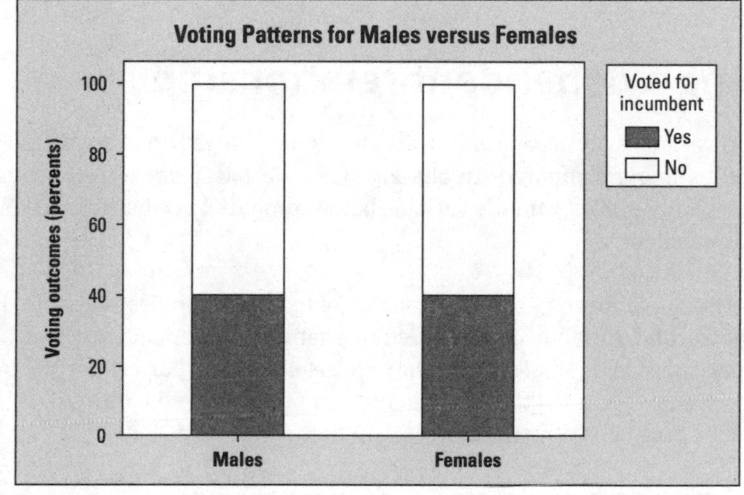

FIGURE 20-4: Bar graph showing the conditional distributions of voting patterns for males versus females.

REMEMBER

To have independence, you don't need the percentages within each bar to be 50-50 (for example, 50 percent males in favor and 50 percent males opposed). It's not the percentages within each bar (group) that have to be the same; it's the percentages across the bars (groups) that need to match (for example, 60 percent of males in favor and 60 percent of females in favor).

REMEMBER

Instead of comparing rows of a two-way table to determine independence, you can compare the columns. In the voting example, you'd be comparing the gender breakdowns for the group who voted for the incumbent to the gender breakdowns for the group who didn't vote for the incumbent. The conclusion of independence would be the same as what you found previously, although the percentages you'd calculate would be different.

Comparing marginal and conditional to check for independence

Another way to check for independence is to see whether the marginal distribution of voting pattern (overall) equals the conditional distribution of voting pattern for each of the gender groups (males and females). If these distributions are equal, then gender doesn't matter. Again, gender and voting pattern are independent.

Looking at the voting pattern example, you find the conditional distribution of voting pattern for the males (first bar in Figure 20-4) is 40 percent yes and 60 percent no. To find the marginal (overall) distribution of voting pattern (males and females together), take the marginal column totals in the last row of Table 20-8 (80 yes and 120 no) and divide through by 200 (the grand total). You get $80 \div 200 = 0.40$ (or 40 percent) yes, and $120 \div 200 = 0.60$ (or 60 percent) no. (See the earlier section, "Calculating marginal distributions," in this chapter for more explanation.) The marginal distribution of overall voting pattern matches the conditional distribution of voting pattern for males, so voting pattern is independent of gender.

Here's where a small table with only two rows and two columns cuts you a break. You have to compare only one of the conditionals to the marginal because you have only two groups to compare. If the voting pattern for the males is the same as the overall voting pattern, then the same will be true for the females. To check for independence when you have more than two groups, you use a Chi-square test (discussed in the book *Statistics II For Dummies*).

Describing a dependent relationship

Two categorical variables are *dependent* if the conditional distributions are different for at least two of the groups being compared. In the election example from the previous section, the groups are males and females, and the variable being compared is whether the person voted for the incumbent president.

Dependence in this case means knowing that the outcome of the first variable does affect the outcome of the second variable. In the election example, if dependence had been found, it would mean that males and females didn't have the same voting pattern for the incumbent (for example, more males voting for the incumbent than females). (Pollsters use this kind of data to help steer their campaign strategies.)

Other ways of saying two variables are dependent are to say they are related, or associated. However, statisticians don't use the term *correlation* to indicate relationships between categorical variables. The word *correlation* in this context applies to the linear relationship between two numerical variables (such as height and weight), as seen in Chapter 19. (This mistake occurs in the media all the time, and it drives us statisticians crazy!)

Here's an example to help you better understand dependence: A recent press release put out by The Ohio State University Medical Center caught my attention. The headline said that aspirin can prevent polyps in colon-cancer patients. Having had a close relative who succumbed to this disease, I was heartened at the prospect that researchers are making progress in this area and decided to look into it.

The researchers studied 635 colon-cancer patients; they randomly assigned approximately half of them to an aspirin regimen (317 people) and the other half to a placebo (fake pill) regimen (318 people). They followed the patients to see which ones developed subsequent polyps and which did not. The data from the study are summarized in Table 20-10.

TABLE 20-10 Summary of Aspirin and Polyps Study Results

	Developed Subsequent Polyps	Didn't Develop Subsequent Polyps	Total
Aspirin	54 (17%)	263 (83%)	317 (100%)
Placebo	86 (27%)	232 (73%)	318 (100%)
Total	140	495	635

Comparing the results in the rows of Table 20-10 to check for independence means finding the conditional distribution of outcomes (polyps or not) for the aspirin group and comparing it to the conditional distribution of outcomes for the placebo group. Making these calculations, you find that $54 \div 317 = 17$ percent of patients in the aspirin group developed polyps (the rest, 83 percent, did not), compared to $86 \div 318 = 27$ percent of the placebo group who developed subsequent polyps (the rest, 73 percent, did not).

Because the percentage of patients developing polyps is much smaller for the aspirin group compared to the placebo group (17 percent versus 27 percent), a dependent relationship appears to exist between aspirin-taking and the development of subsequent polyps among the colon-cancer patients in this study. (But does it carry over to the population? You find out in the section, "Projecting from sample to population," later in this chapter.)

EXAMPLE

Q. Suppose 50 percent of the employees of a company say they would work at home if they could, and you want to see whether these results are independent of gender. You break it down and discover that 75 percent of the females and only 25 percent of the males say they would work at home if they could. Are gender and work preference independent? Describe your results by using statistical notation and the appropriate probabilities.

A. **No,** gender and work preference are dependent. Let H refer to the employees who want to work at home, and let F denote females. According to the information given, $p(H) = 50$ percent for the whole group, but when you look only at the females, the proportion who want to work at home rises to 75 percent, so $p(H|F) = 75$ percent. When you look only at males, the proportion who want to work at home lowers to 25 percent, so $p(H|M) = 25$ percent. Because $p(H)$ isn't equal to $p(H|F)$, gender and work preference are dependent (you don't even have to look at the males).

YOUR TURN

15 Using the data from the table, "The Gender and Political Affiliation of 200 Survey Participants," from Practice Question 2 (in the section, "Figuring the totals"), are gender and political party independent for this group?

16 Using the bumper sticker/age of cars data in the first example problem in this chapter, are car age and number of bumper stickers related? (In other words, are these two events dependent?)

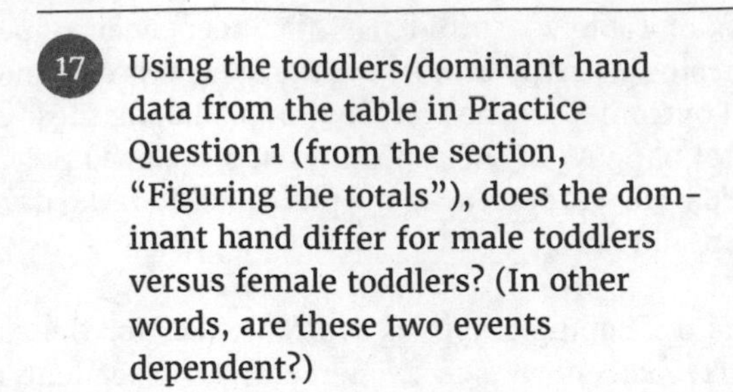

17 Using the toddlers/dominant hand data from the table in Practice Question 1 (from the section, "Figuring the totals"), does the dominant hand differ for male toddlers versus female toddlers? (In other words, are these two events dependent?)

18 Suppose A and B are independent, and $p(A) = 0.6$, and $p(B) = 0.2$. What is $p(A \text{ and } B)$?

19 Suppose you flip a fair coin two times, and the flips are independent.

(a) What is the probability that you will get two heads in a row?

(b) What is the probability that you will get exactly one head?

(c) How does your answer to Problem 19 (b) change if the chance of a head is 0.75 (the coin is not fair)?

20 Suppose you roll a single die two times, and the trials are independent. What is the chance of rolling two 1s?

21 Suppose A and B are independent, and $p(A) = 0.3$, and $p(B) = 0.2$. Find $p(A \cap B)$, also known as $p(A \text{ and } B)$.

22 Suppose medical researchers collect data from an experiment comparing a new drug to an existing drug (call this the treatment variable), regarding whether it made patients' symptoms improve (call this the outcome variable). A check for independence shows that the outcome is related to the treatment the patients receive.

(a) Are treatment and outcome independent or dependent in this case?

(b) Do the results mean that the new medicine causes the symptoms to improve? Explain your answer.

Cautiously Interpreting Results

It's easy to get carried away when a relationship between two variables has been found; you see this happen all the time in the media. For example, a study reports that eating eggs doesn't affect your cholesterol as once thought; in the details of the report you see the study was conducted on a total of 20 men who were all in excellent health, were all on low-fat diets, and who exercised several times a week. Ten men in good health ate two eggs a day and their cholesterol didn't change much, compared to ten men who didn't eat two eggs per day. Do these results carry over to the entire population? Can't tell — the subjects in the study don't represent the rest of us. (See Chapter 18 for the scoop on evaluating experiments.)

In this section, you see how to put the results from a two-way table into proper perspective in terms of what you can and can't say and why. This basic understanding gives you the ability to critically evaluate and make decisions about results presented to you (not all of which are correct).

Checking for legitimate cause and effect

Researchers studying two variables often look for links that indicate a cause-and-effect relationship. A *cause-and-effect relationship* between two categorical variables means as you change the value of one variable and all else remains the same, it causes a change in the second variable — for example, if being on an aspirin regimen decreases the chance of developing subsequent polyps in colon-cancer patients.

However, just because two variables are found to be related (dependent) doesn't mean they have a cause-and-effect relationship. For example, observing that people who live near power lines are more likely to visit the hospital in a year's time due to illness doesn't necessarily mean the power lines caused the illnesses.

One of the most effective ways to conclude a cause-and-effect relationship is by conducting a well-designed experiment (where possible). All the details are laid out in Chapter 18, but I touch on the main points here. A well-designed experiment meets the following three criteria:

- It minimizes *bias* (systematic favoritism of subjects or outcomes).
- It repeats the experiment on enough subjects so the results are reliable and repeatable by another researcher.
- It controls for other variables that may affect the outcome that weren't included in the study.

In the earlier section, "Describing a dependent relationship," I discuss a study involving the use of aspirin to prevent polyps in cancer patients. Because of the way the data was collected for this study, you can be confident about the conclusions drawn by the researchers; this study was a well-designed experiment, according to the criteria established in Chapter 18. To avoid problems, the researchers in this study did the following:

- Randomly chose who took the aspirin and who received a fake pill
- Had large enough sample sizes to obtain accurate information
- Controlled for other variables by conducting the experiment on patients in similar situations with similar backgrounds

Because their experiment was well-designed, the researchers concluded that a cause-and-effect relationship was found for the patients in this study. The next test is to see whether they can project these results to the population of all colon-cancer patients. If so, they are truly entitled to the headline, "Aspirin Prevents Polyps in Colon-Cancer Patients." The next section walks you through the test.

Whether two related variables are found to be causally associated depends on how the study was conducted. A well-designed experiment is the most convincing way to establish cause and effect. In cases where an experiment would be unethical (for example, proving that smoking causes lung cancer by forcing people to smoke), a large number of convincing observational studies (where you collect data on people who smoke and people who don't) with more complex methods is needed to show that an association between two variables crosses over into a cause-and-effect relationship. But it can be done.

Projecting from sample to population

In the aspirin/polyps experiment discussed in the earlier section, "Describing a dependent relationship," I compare the percentage of patients developing subsequent polyps for the aspirin group versus the non-aspirin group and got the results 17 percent and 27 percent, respectively. For this sample, the difference is quite large, so I'm cautiously optimistic that these results will carry over to the population of all cancer patients. But what if the numbers were closer, such as 17 percent and 20 percent? Or 17 percent compared to 19 percent? How different do the proportions have to be in order to signal a meaningful association between the two variables?

Percentages compared using data from your sample reflect relationships within your sample. However, you know that results change from sample to sample. To project these conclusions to the population of all colon-cancer patients (or any population being studied), the sample needs to be representative of the target population, the difference should be clinically meaningful, and the difference in the percentage found in the sample should be statistically significant. Statistical significance means that the result we found is very unlikely if there were truly no difference. This takes into account the variation expected between samples.

I analyzed the data from the aspirin/polyps study using a hypothesis test for the difference of two proportions (found in Chapter 16). The proportions being compared were the proportion of patients taking aspirin who developed subsequent polyps and the proportion of patients not taking aspirin who developed subsequent polyps. Looking at these results, my p-value is less than 0.0024. (A *p-value* measures how likely you were to have gotten the results from your sample if the populations really had no difference; see Chapter 15 to get the scoop on p-values.)

Because this p-value is so small, the difference in proportions between the aspirin and non-aspirin groups is declared to be statistically significant, and I conclude that a relationship exists between taking aspirin and developing fewer subsequent polyps.

You can't make conclusions about relationships between variables in a population based only on the sample results in a two-way table. You must take into account the fact that results change from sample to sample. A hypothesis test gives limits for how different the sample results can be to still say the variables are independent. Beware of conclusions based only on sample data from a two-way table.

Making prudent predictions

A common goal of research (especially medical studies) is to make predictions, recommendations, and decisions after a relationship between two categorical variables is found. However, as a consumer of information, you have to be very careful when interpreting results; some studies are better designed than others.

The colon-cancer study from the previous section shows that patients who took aspirin daily had a lower chance of developing subsequent polyps (17 percent compared to 27 percent for the non-aspirin group). Because this was a well-designed experiment and the hypothesis test for generalizing to the population was significant, making predictions and recommendations for the population of colon-cancer patients based on these sample results is appropriate. They've indeed earned the headline of their press release: "Aspirin Prevents Polyps in Colon-Cancer Patients."

Resisting the urge to jump to conclusions

Try not to jump to conclusions when you hear or see a relationship being reported regarding two categorical variables. Take a minute to figure out what's really going on, even when the media wants to sweep you away with a dramatic result.

For example, as I write this, a major news network reports that men are 40 percent more likely to die from cancer than women. If you're a man, you may think you should panic. But when you examine the details, you find something different. Researchers found that men are much less likely to go to the doctor than women, so by the time cancer is found, it's more advanced and difficult to treat. As a result, men were more likely to die of cancer after its diagnosis. (They aren't necessarily more likely to *get* cancer; that's for a different study.) This study was meant to promote early detection as the best protection and encourage men to keep their annual checkups. The message would have been clearer had the media reported it correctly (but that's not as exciting or dramatic). So men, don't panic — go to the doctor!

Practice Questions Answers and Explanations

1. To get the right numbers from a two-way table, identify which row and/or column you're in and work from there. But before you start, it's always a good idea to figure and write down the marginal totals and grand totals (see the following table).

Gender	Left-Handed	Right-Handed	Totals (Gender)
Male	4	24	28
Female	2	12	14
Totals (Hand)	6	36	42 (grand total)

a. To get the total number of male toddlers, you sum the values across the "male" row to get $4 + 24 = \mathbf{28}$.

b. To get the total number of right-handed toddlers, you sum the values down the "right-handed" column to get $24 + 12 = \mathbf{36}$.

c. The number of right-handed male toddlers is the number of toddlers in the intersection of the "male" row and the "right-handed" column: **24**.

2. Summing the values across Row 1 gives you the total number of males: $35 + 55 = 100$. Summing the values across Row 2 gives you the total number of females: $55 + 35 + 10 = 100$. Summing the values down Column 1 gives you the total number of Democrats: $35 + 55 = 90$. Summing the values down Column 2 gives you the total number of Republicans: $55 + 35 = 90$. Summing the values down Column 3 gives you the total number of Independents: $10 + 10 = 20$.

Interpretation: This survey contains an equal number of men and women. Democrats and Republicans are equal in number and each outweighs Independents (90 Democrats and 90 Republicans compared to 20 Independents).

3. Summing the values across Row 1 gives you the total number of male toddlers: $4 + 24 = 28$. Summing the values across Row 2 gives you the total number of female toddlers: $2 + 12 = 14$. Summing the values down Column 1 gives you the total number of left-handed toddlers: $4 + 2 = 6$. Summing the values down Column 2 gives you the total number of right-handed toddlers: $24 + 12 = 36$.

Interpretation: This data set contains twice as many male toddlers as female toddlers. There are many more right-handed toddlers than left-handed toddlers (36 versus 6).

4. Marginal totals add up to the grand total, regardless of the direction in which you sum them (rows or columns).

a. If you sum all the marginal totals for the rows of a two-way table, **you get the total sample size (otherwise known as the grand total)**. For example, in Practice Question 3, the row totals are 28 (total male toddlers) and 14 (total female toddlers). These values sum to 42, or the total number of toddlers. The result makes sense because every toddler has to fall into one of the two categories in terms of his or her gender.

b. If you sum all the marginal totals for the columns of a two-way table, **you also get the total sample size (the grand total)**. For example, in Practice Question 3, the column totals are 6 (left-handed toddlers) and 36 (right-handed toddlers); the total size of 42 again makes sense because every toddler in the sample falls into one of these two categories in terms of his or her hand dominance. The table from Answer 1 shows what the marginal totals look like if you place them into the two-way table. (Where else would marginal totals be but out in the margins?)

TIP

Your first task when you look at a two-way table is to write down all the marginal totals in their proper places and make sure they add up to the grand total in the lower right-hand corner. The grand total is the total sample size.

5. Let B = cars with a lot of bumper stickers (so B^C represents cars without a lot of bumper stickers as the complement), and let N = newer cars (so N^C represents older cars as the complement).

   a. The question asks for $p(N)$. Take the marginal column total for the newer cars $(30+20=50)$ and divide by the grand total (100) to get 0.50, or **50 percent**.

   b. The percentage of older cars is 100 minus the percentage of newer cars, which is $100-50=50$ percent, because the groups are complements. (Or you can find the marginal column total for the older cars $[15+35=50]$ divided by the grand total [100], which is 0.50, or 50 percent.) Note, however, that the question asks for a proportion, so you should divide your percents by 100. In other words, the proportion of older cars is **0.50**. Refer to Chapter 8 for more information on probability rules, complements, and so on.

REMEMBER

Some instructors are really picky about the proportion versus percentage thing, and others are not. Technically, they are different units to represent the same quantity. Suppose you have 20 people out of 40. This can be written as 20 divided by 40 equals 0.50, which is a proportion (because proportions are decimals between 0 and 1). Or, it can be written as 20 divided by 40×100 percent, which is 50 percent (because percentages are always between 0 and 100). To be safe, ask your instructor how picky they are about this issue, and which formats are acceptable for answers.

   c. The question asks for $p(B)$. Take the marginal row total for the cars with a lot of bumper stickers $(20+35=55)$ and divide by the grand total (100) to get 0.55, or **55 percent**.

   d. You take $1-p(B)$, or $1-0.55=\mathbf{0.45}$, or **45 percent**, again by complements. Note that I word this problem differently in the sense that it asks for a probability, but its meaning is the same as the other problems in this section.

TIP

Be on the lookout for parts of problems that are really just complements of previous parts. Taking one minus a previous answer can save a lot of time. (This practice is a common one among teachers.)

6. Let M = male, F = female, R = right-handed, and L = left-handed. (You can also use complement notation, but why confuse things?)

   a. You want to find $p(R)$. Take the marginal column total for the right-handers $(24+12=36)$ and divide by the grand total (42) to get 0.857, or **85.7 percent**.

   b. You want to find $p(F)$. Take the marginal row total for the females $(2+12=14)$ and divide by the grand total (42) to get 0.333, or **33.3 percent**.

   c. You want to find $p(L)$, which is the same as $p(R^C)$. You can take one minus the answer to Part (a) to get $1-0.857=\mathbf{0.143}$.

   d. You want to find $p(M)$, which is the same as $p(F^C)$. You can take one minus the answer to Part (b) to get $1-0.333=0.667$, or **66.7 percent**.

7. From the table, you can say that the percentage of female toddlers is 33.3 percent, the percentage of males is 66.7 percent, the percentage of right-handers is 85.7 percent, and the percentage of left-handers is 14.3 percent.

REMEMBER

Marginal probabilities discuss only individual variables separately without examining the connection, so you have limited interpretation ability. Beware of people in the media reporting statistics from individual variables without examining the connection between them. If you don't examine all the cells of the two-way table, you miss a lot of information.

(8) The complements in a two-way table have probabilities that sum to one. Let A and A^C be the row values in the table (for example, males and females): $p(A)+p(A^C)=1$. Let B and B^C be the column values in the table (for example, right-handers and left-handers): $p(B)+p(B^C)=1$.

(9) Did you realize that all four parts of this problem are asking for the same thing, just with different wording? (Statistics is such an exact science, isn't it?) This problem is designed to help you get ready for possible wordings (and rewordings) of problems on your exams.

a. Here, $p(B|O)$ means that given an older car, what's the chance of it having a lot of bumper stickers? The denominator of this probability is the total number of older cars (marginal column total, 50), and the numerator is the number of cars in the older-car column that have a lot of bumper stickers (35). So you have $p(B|O)=35\div50=0.70$, or 70 percent.

b. The word "of" tells you you've got a conditional probability, and that you know the car is old, so that's what you're conditioning on (or putting in the back part of the formula). So again, you have $p(B|O)=35\div50=0.70$, or **70 percent**. I can understand why someone would want to know this; don't you? Those old cars have that wild and windblown look about them.

c. Now the wording of this question sounds as if you're setting aside the older cars and examining their bumper stickers. In the end, you get the same answer as for the other parts of this problem, because that's exactly what conditional probability does — it sets aside the group. (I typically draw a circle around the row or column in a conditional probability to remind myself of that.) So again, the answer is $p(B|O)=35\div50=0.70$, or **70 percent**. Now, doesn't it make sense to set older cars aside? Their bumper stickers are probably much more interesting, saying things like, "Honk if you love statistics!"

d. This wording is that warm and fuzzy old standby, using the word "given." Older stats books still use this notation, but most professors have come to realize that people don't talk like that in the real world and have moved on to wording like that found in Parts (a) to (c) in this problem. However, did I say that your statistics class was the real world? So again, you have $p(B|O)=35\div50=0.70$, or **70 percent**.

TIP

Notice that I sometimes change notation throughout this chapter to describe the same event. That's because as the focus changes in a problem, you may want to change the notation. It's your notation after all, and it might as well be notation that you want. If the problem focuses on the fact that the cars are older, and you want other probabilities relating to that, you can use O to indicate older cars. If the problem is focusing on the newer cars for the most part, except for a quick switch to old ones, you can use N^C to indicate older cars. You should try to use whatever notation works and feels most comfortable for you — just be clear in defining it so your instructor knows what you're talking about.

(10) Let M = male, F = female, R = right-handed, and L = left-handed. (Like I said earlier, you can also use complement notation, but just make sure you can understand your notation.)

a. You want to find $p(R|M)$ because of the phrase "of the male toddlers," which means you select from among the males only (so M appears after the "|" sign in the probability). The denominator of this probability is the number of males (marginal row total, 28), and the numerator is the number of right-handed individuals in the row $(24): p(R \mid M) = 24 \div 28 = 0.857$, or **85.7 percent.**

b. You want to find $p(R|F)$ because of the phrase "of the female toddlers," which means you select from the females only (so F appears after the "|" sign in the probability). The denominator of this probability is the number of females (marginal row total, 14), and the numerator is the number of right-handed individuals in the row $(12): p(R \mid F) = 12 \div 14 = 0.857$, or **85.7 percent**. (The percentage of right-handers is the same for females and males — gender and dominant hand aren't related.)

Keep in mind that you can't use $p(R|M)$ to help you find $p(R|F)$ because the two groups (males and females) are independent, and conditional probabilities with different "denominators" don't add up.

c. Now you go the other way with your conditional probabilities — compared to Part (a) — by dividing the toddlers into the left- and right-handed groups and selecting from there. In this part, you want to find $p(M|R)$ because of the phrase "of the right-handed toddlers," which means you select from the right-handers only (so R appears after the "|" sign in the probability). The denominator of this probability is the number of right-handers (marginal column total, 36), and the numerator is the number of male individuals in the column $(24): p(M \mid R) = 24 \div 36 = 0.67$, or **67 percent**.

d. You want to find $p(F|R)$ because of the phrase "of the right-handed toddlers," which means you select from the right-handers only (so R appears after the "|" sign in the probability). The denominator of this probability is the number of right-handers (marginal column total, 36), and the numerator is the number of female individuals in the column $(12): p(F \mid R) = 12 \div 36 = 0.33$, or **33 percent**.

Note: Your answers to Parts (c) and (d) sum to one because (c) and (d) are complements. Knowing that you're in the right-hander group, you have to be either a male or a female, so you could take 1 – 0.67 to get your answer for (d).

REMEMBER

In general, $p(A|B)$ isn't equal to $p(B|A)$, as you can see by comparing your answers to Parts (a) and (c) of this problem. Knowing which group is which and what the notation means is very important to solving these problems correctly.

11 This problem is asking you to do exactly what you did in the previous problem. The trick is knowing that's what the question is asking you to do.

a. See the answers to Problems 10(a) and 10(b) from this section.

b. See the answers to Problems 10(c) and 10(d) from this section.

12 **No.** The probabilities $p(A|B)$ and $p(A \mid B^C)$ aren't related because they condition on two different groups, like males (B) and females (B^C). The two groups are independent, and you can't assume they're related. You know that $p(A|B)$ and $p(A \mid B^C)$ aren't complements because they don't condition on being in the same group. See the answers to Practice Questions 10(a) and 10(b) in this section for examples.

13 **Yes,** because both probabilities condition on the same group, and A and A^C are complements. The answers to Practice Questions 10(c) and 10(d) in this section show examples of events that are complements.

14 Marginal probabilities discuss only individual variables separately without examining the connection, so you have limited interpretation. For example, if you know that 50 percent of all people approve of a smoking ban and 20 percent of people are smokers, can you conclude

that 50 percent of smokers approve of the smoking ban and 50 percent of smokers don't? No. All the smokers may oppose the ban, and all the nonsmokers may approve of it. Conditional probabilities break down the groups and compare them. You need conditional probabilities if you want to examine relationships among two categorical variables.

15. **No.** You can discover this in a couple of different ways. Suppose you condition on gender and look at the percentage of Democrats, Republicans, and Independents. If gender and political party are independent, the percentages are the same for males and females, and the percentage equals the overall percentage of Democrats, Republicans, and Independents for the entire group. You know $p(\text{D}) = 90 \div 200 = 0.45$ and $p(\text{D} \mid \text{F}) = 55 \div 100 = 0.55$, so you know the two aren't independent. The percentages don't match up.

16. **Yes,** age and bumper stickers are related (or dependent). You can show their dependence in two different ways, depending on which variable you condition. First, the percentage of older cars, $p(\text{O})$, is $50 \div 100 = 50$ percent. Of cars with a lot of bumper stickers, the percentage of old cars is $p(\text{O} \mid \text{B}) = 35 \div 55 = 64$ percent. More of the cars loaded with bumper stickers are older, so bumper stickers and car age are dependent. Another way you can show dependence is to look at the percentage of all cars with a lot of bumper stickers, $p(\text{B})$, which is 55 percent, and compare it to the percentage of older cars with a lot of bumper stickers: $p(\text{B} \mid \text{O}) = 35 \div 50 = 70$ percent. More of the older cars are loaded with bumper stickers compared to the entire group.

TIP

To check to see if A and B are independent, you can examine whether or not $p(\text{A} \mid \text{B}) = p(\text{A})$, or you can examine whether or not $p(\text{B} \mid \text{A}) = p(\text{B})$. Choose the situation that you feel comfortable working with. Most people prefer working by conditioning on the row variables rather than the column variables. It just seems easier.

17. Always watch for subtle changes in wording that mean exactly the same thing. Here I use a different kind of wording for the same type of independence problem. If dominant hand differs for the males versus the females, then dominant hand is related to, or dependent upon, gender. So, you need to determine whether gender and dominant hand are independent. Check to see if $p(\text{R}|\text{M})$ is equal to $p(\text{R})$, and if $p(\text{R}|\text{F})$ is equal to $p(\text{R})$. In this case, $p(\text{R} \mid \text{M}) = 24 \div 28 = 0.857$, and $p(\text{R}) = 36 \div 42 = 0.857$. Now you can see that $p(\text{R} \mid \text{F}) = 12 \div 14 = 0.857$ and $p(\text{R}) = 36 \div 42 = 0.857$. The percentage of right-handers for the male and female groups is the same as the percentage of right-handers in the entire group. Therefore, gender and dominant hand are independent for this group. **That means there's no difference between the groups in terms of dominant hand.** Gender and dominant hand are not related.

18. By the definition of independence of events A and B, $p(\text{A and B}) = p(\text{A})p(\text{B}) = (0.6)(0.2) = \mathbf{0.12}$.

19. Let H = heads and H = heads. Note that the probability of getting a head on a fair coin is 50 percent or 0.50, and the same with tails.

TIP

Knowing that the probability of heads is 0.5 is important information that you had to figure out from the problem (it was not explicitly given to you). Instructors love to put these kinds of problems on exams, so watch for them and remember that if you aren't given the probability in the problem, it means you should be able to figure it out from the information given.

a. Here you want to get two heads. So, what you are looking for is $p(\text{Heads on 1st toss} \cap \text{eads on 2nd toss}) = p(\text{HH})$. Because the flips are independent, you know that $p(\text{HH}) = p(\text{H}) \times p(\text{H}) = (0.5)(0.5) = \mathbf{0.25}$. This means that, in the long run, one-fourth of the time you'll see two heads when you flip a coin twice.

b. The probability of getting exactly one head in two tosses is p(HT or TH). By the addition rule, this equals $p(\text{HT}) + p(\text{TH})$ because these events have no intersection. Now by independence of the two flips, you can say this equals $p(\text{H}) \times p(\text{T}) + p(\text{T}) \times p(\text{H})$ $= (0.5)(0.5) + (0.5)(0.5) = 0.25 + 0.25 = \mathbf{0.50}$. So, 50 percent of the time you should expect exactly one head when you flip a coin twice.

WARNING

In Answer (b), resist the urge to say the answer is simply 0.50 because the probability of getting one head is 0.50. You haven't taken into account the fact that you had two tosses, and what might have happened on that second toss. The sample space of this experiment is the four outcomes — HH, HT, TH, and TT — and you have to deal with both items in the pair, not just the one you're interested in.

c. The work to solve this part of problem is the same as for Answer (b) until you get to the last step, where you plug in the probabilities of heads and tails. There, you use $p(\text{H}) = 0.75$ and $p(\text{T}) = 1 - p(\text{H}) = 1 - 0.75 = 0.25$. You have p (exactly one head in two tosses of the unfair coin) is $(0.75)(0.25) + (0.25)(0.75) = 0.1875 + 0.1875 = \mathbf{0.375}$.

(20) This problem is similar to Practice Question 5; you aren't explicitly given any probabilities in the problem, but don't let that get you stuck. Remember that the probability of getting a 1 on a fair die is 1 out of 6, so use $\frac{1}{6}$ as your probability. The question asks you to find the probability of rolling two 1s when you roll a fair die twice, so you need $p(1 \text{ and } 1) = p(1) \times p(1)$ because the rolls are independent. Each $p(1)$ is $\frac{1}{6}$, so multiply them to get $\frac{1}{6} \cdot \frac{1}{6} = \frac{1}{36} = \mathbf{0.028}$. Or you could have just realized that there are 6 times $6 = 36$ outcomes when you roll a die two times, and one of those outcomes is the "roll 1, then 1" outcome.

(21) You're looking for p(A or B), which means you want $p(\text{A} \cup \text{B}) = p(\text{A}) + p(\text{B}) - p(\text{A} \cap \text{B})$. This equals $0.3 + 0.2 - p(\text{A and B})$. You might think you're stuck here, but you aren't. A and B are independent, so p(A and B) is equal to $p(\text{A}) \times p(\text{B}) = (0.3)(0.2) = 0.06$. Substituting this in the problem, you get $0.3 + 0.2 - 0.06 = \mathbf{0.44}$, the probability of A or B.

TIP

You're usually given the information for a reason. In this problem, you knew A and B were independent. Without realizing that means you can multiply p(A) times p(B) to get p(A and B), you would have been sunk. Don't let this happen to you on an exam. Have all those tools (the formulas, properties, definitions, and so on) ready to pull out when you need them. And be able to identify when you need them.

(22) The manner of data collection directly affects what conclusions you can make.

a. Because the outcome is related to treatment, the two variables aren't independent; **they are, therefore, dependent**.

b. As long as the experiment is well designed (see Chapter 18), you can imply a cause-and-effect relationship. If this study is observational, or badly designed, the answer is no.

REMEMBER

The check for independence of two categorical variables is somewhat limited, in that you can really make conclusions only about the sample. You can't make broader conclusions to the general population without doing a hypothesis test. However, a hypothesis test for two proportions is equivalent to a hypothesis test for independence. So, if you want to show whether the variables have a statistically significant relationship (or not), do a hypothesis test for two proportions. See Chapter 16 for more information.

If you're ready to test your skills a bit more, take the following chapter quiz that incorporates all the chapter topics.

Whaddya Know? Chapter 20 Quiz

Quiz time! Complete each problem to test your knowledge on the various topics covered in this chapter. You can then find the solutions and explanations in the next section.

For questions 1 to 10, use the following information. Jose goes on quite a few trips throughout the year for his job. Some trips are long and some are short. Sometimes he checks in luggage, sometimes he doesn't. For 50 random trips that Jose made, he recorded whether it was a long or short trip, and whether he checked luggage or not.

1. The first 10 lines of data are shown here. Organize the first 10 lines of data into a two-way table with rows = length and columns = checked bags (yes/no).

Trip	Length	Checked Bags?
1	Short	yes
2	Long	yes
3	Short	yes
4	Short	no
5	Long	no
6	Short	no
7	Short	no
8	Long	yes
9	Long	yes
10	Short	no

2. Following is the full data set for 50 randomly selected trips, organized in a two-way table. Find the marginal distributions of length and checked bags (separately):

	Checked bag: YES	Checked bag: NO
Short trip	10	20
Long trip	14	6

3. Graph and interpret the marginal distributions found in Question 2.

4. What percentage of the time does Jose check a bag when he's on a short trip? What percentage of the time does Jose check a bag when he's on a long trip?

5. Find the conditional distribution of checked bag (yes/no) for the short trips. Find the conditional distribution of checked bag (yes/no) for the long trips.

6. Suppose Jose checks his bag; do you think he's more or less likely to go on a long trip (or is it the same for long and short trips?)

7. Graph the two conditional distributions from Question 5.

8. Looking at your answer to Question 5, are checked bag and trip length independent or dependent?

9. How might the results be affected if you selected 50 trips in a row for your data set rather than 50 randomly chosen trips?

10. What other variable(s) might be coming into play here that can affect whether Jose checks a bag, besides length of trip?

Answers to Chapter 20 Quiz

1

	Checked bag: YES	Checked bag: NO	TOTAL
Short Trip	2	4	6
Long Trip	3	1	4
TOTAL	5	5	10

2

Marginal Distribution for Checked Bag	Marginal Distribution for Length of Trip
Yes: 24 / 50 = 48%	Short: 30 / 50 = 60%
No: 26 / 50 = 52%	Long 20 / 50 = 40%

3 Interpretation: Jose took a checked bag 48% of the time, and he didn't take a checked bag 52% of the time. Sixty percent of his trips were short and 40% of his trips were long.

Graphs:

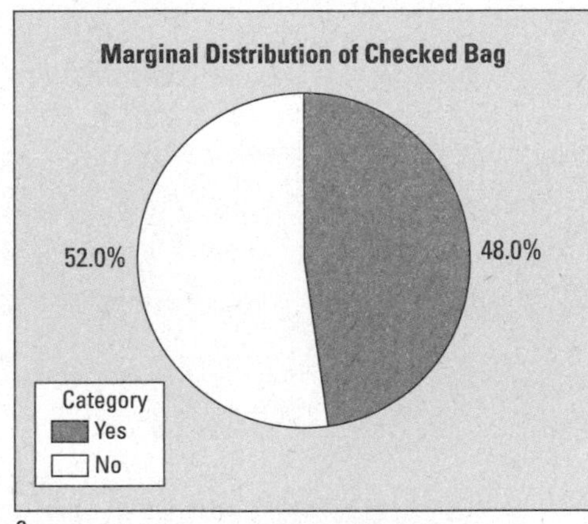

a

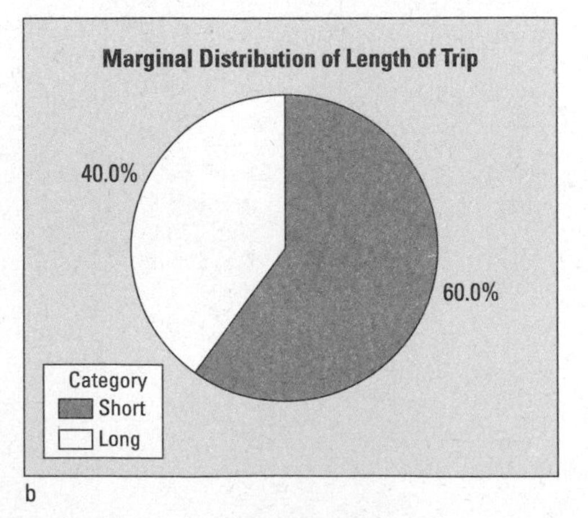

b

4. Percentage of the time Jose checks a bag when he's on a **short trip** = p(checked bag | short trip) = 10 / 30 = **33.33**%

   Percentage of the time Jose checks a bag when he's on a **long trip** = p(checked bag | long trip) = 14 / 20 = **70**%

5. Conditional distribution of checked bag (yes/no) for short trips:

   p(checked bag | short trip) = 10 / 30 = **33.33**%

   p(no checked bag | short trip) = 20 / 30 = **66.67**%

   Conditional distribution of checked bag (yes/no) for long trips:

   p(checked bag | long trip) = 14 / 20 = **70**%

   p(no checked bag | long trip) = 6 / 20 = **30**%

6. You find that p(long | checked) = 14 / 24 = 58.33%; because this probability is higher than 50%, **it's more likely that he is going on a long trip (than a short trip) when he checks his bag.** Note that you could also find that p(short | checked) = 41.67% is lower than p(long|checked).

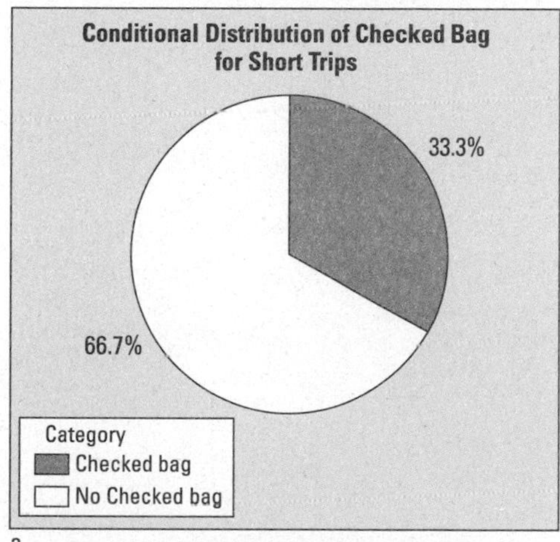

a

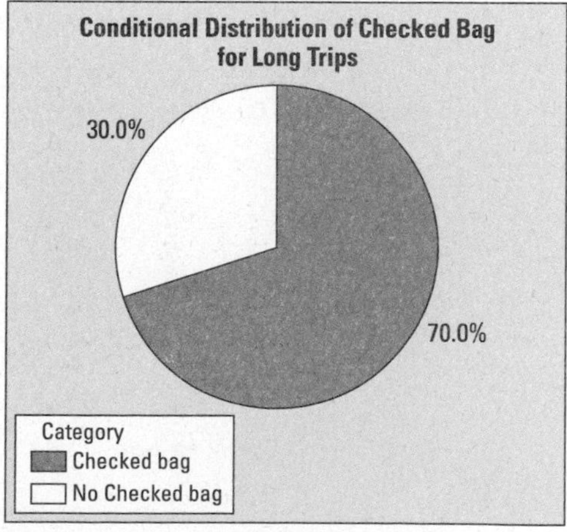

b

8. Checking a bag and length of trip are not independent; **they are dependent**. When he's on a short trip, he's less likely to check his bag (33.33%), but when he's on a long trip, he's more likely to check his bag (70%).

9. If you selected 50 trips in a row, and the sample is not random, you might not get repeatable results. It could be that he gets into a habit of checking his bag for several trips in a row, regardless of the length of trip, for example. Or, he could have had a bad experience and did not check his bag for a long time after that, regardless of the length of the trip.

10. **One big variable is whether Jose has a connecting flight.** A lot of people don't check bags unless they have a direct flight.

Appendix

Tables for Reference

This Appendix includes tables for finding probabilities and/or critical values for the three distributions used in this book: the Z-distribution (standard normal), the t-distribution, and the binomial distribution.

The Z-Table

Table A-1 shows less-than-or-equal-to probabilities for the Z-distribution; that is, $p(Z \le z)$ for a given z-value. (See Chapter 10 for calculating z-values for a normal distribution; see Chapter 12 for calculating z-values for a sampling distribution.) To use Table A-1, do the following:

1. **Determine the z-value for your particular problem.**

   The z-value should have one leading digit before the decimal point (positive, negative, or zero) and two digits after the decimal point; for example, $z = 1.28$, -2.69, or 0.13.

2. **Find the row of the table corresponding to the leading digit and first digit after the decimal point.**

   For example, if your z-value is 1.28, look in the "1.2" row; if $z = -1.28$, look in the "−1.2" row.

3. **Find the column corresponding to the second digit after the decimal point.**

   For example, if your z-value is 1.28 or −1.28, look in the "0.08" column.

4. **Intersect the row and column from Steps 2 and 3.** This number is the probability that Z is less than or equal to your z-value. In other words, you've found $p(Z \le z)$. For example, if $z = 1.28$, you see $p(Z \le 1.28) = 0.8997$. For $z = -1.28$, you see $p(Z \le -1.28) = 0.1003$.

TABLE A-1 **The *Z*-Table**

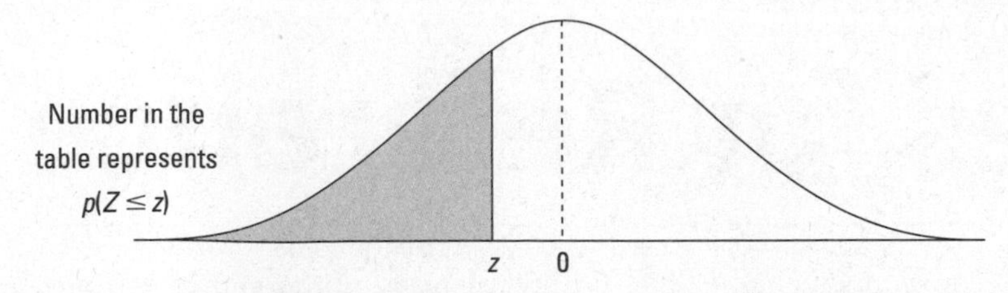

z	0.00	0.01	0.02	0.03	0.04	0.05	0.06	0.07	0.08	0.09
–3.6	.0002	.0002	.0001	.0001	.0001	.0001	.0001	.0001	.0001	.0001
–3.5	.0002	.0002	.0002	.0002	.0002	.0002	.0002	.0002	.0002	.0002
–3.4	.0003	.0003	.0003	.0003	.0003	.0003	.0003	.0003	.0003	.0002
–3.3	.0005	.0005	.0005	.0004	.0004	.0004	.0004	.0004	.0004	.0003
–3.2	.0007	.0007	.0006	.0006	.0006	.0006	.0006	.0005	.0005	.0005
–3.1	.0010	.0009	.0009	.0009	.0008	.0008	.0008	.0008	.0007	.0007
–3.0	.0013	.0013	.0013	.0012	.0012	.0011	.0011	.0011	.0010	.0010
–2.9	.0019	.0018	.0018	.0017	.0016	.0016	.0015	.0015	.0014	.0014
–2.8	.0026	.0025	.0024	.0023	.0023	.0022	.0021	.0021	.0020	.0019
–2.7	.0035	.0034	.0033	.0032	.0031	.0030	.0029	.0028	.0027	.0026
–2.6	.0047	.0045	.0044	.0043	.0041	.0040	.0039	.0038	.0037	.0036
–2.5	.0062	.0060	.0059	.0057	.0055	.0054	.0052	.0051	.0049	.0048
–2.4	.0082	.0080	.0078	.0075	.0073	.0071	.0069	.0068	.0066	.0064
–2.3	.0107	.0104	.0102	.0099	.0096	.0094	.0091	.0089	.0087	.0084
–2.2	.0139	.0136	.0132	.0129	.0125	.0122	.0119	.0116	.0113	.0110
–2.1	.0179	.0174	.0170	.0166	.0162	.0158	.0154	.0150	.0146	.0143
–2.0	.0228	.0222	.0217	.0212	.0207	.0202	.0197	.0192	.0188	.0183
–1.9	.0287	.0281	.0274	.0268	.0262	.0256	.0250	.0244	.0239	.0233
–1.8	.0359	.0351	.0344	.0336	.0329	.0322	.0314	.0307	.0301	.0294
–1.7	.0446	.0436	.0427	.0418	.0409	.0401	.0392	.0384	.0375	.0367
–1.6	.0548	.0537	.0526	.0516	.0505	.0495	.0485	.0475	.0465	.0455
–1.5	.0668	.0655	.0643	.0630	.0618	.0606	.0594	.0582	.0571	.0559
–1.4	.0808	.0793	.0778	.0764	.0749	.0735	.0721	.0708	.0694	.0681
–1.3	.0968	.0951	.0934	.0918	.0901	.0885	.0869	.0853	.0838	.0823
–1.2	.1151	.1131	.1112	.1093	.1075	.1056	.1038	.1020	.1003	.0985
–1.1	.1357	.1335	.1314	.1292	.1271	.1251	.1230	.1210	.1190	.1170
–1.0	.1587	.1562	.1539	.1515	.1492	.1469	.1446	.1423	.1401	.1379
–0.9	.1841	.1814	.1788	.1762	.1736	.1711	.1685	.1660	.1635	.1611
–0.8	.2119	.2090	.2061	.2033	.2005	.1977	.1949	.1922	.1894	.1867
–0.7	.2420	.2389	.2358	.2327	.2296	.2266	.2236	.2206	.2177	.2148
–0.6	.2743	.2709	.2676	.2643	.2611	.2578	.2546	.2514	.2483	.2451
–0.5	.3085	.3050	.3015	.2981	.2946	.2912	.2877	.2843	.2810	.2776
–0.4	.3446	.3409	.3372	.3336	.3300	.3264	.3228	.3192	.3156	.3121
–0.3	.3821	.3783	.3745	.3707	.3669	.3632	.3594	.3557	.3520	.3483
–0.2	.4207	.4168	.4129	.4090	.4052	.4013	.3974	.3936	.3897	.3859
–0.1	.4602	.4562	.4522	.4483	.4443	.4404	.4364	.4325	.4286	.4247
–0.0	.5000	.4960	.4920	.4880	.4840	.4801	.4761	.4721	.4681	.4641

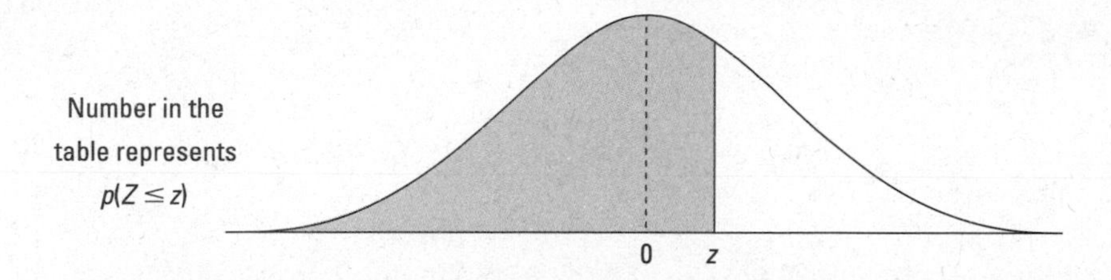

z	0.00	0.01	0.02	0.03	0.04	0.05	0.06	0.07	0.08	0.09
0.0	.5000	.5040	.5080	.5120	.5160	.5199	.5239	.5279	.5319	.5359
0.1	.5398	.5438	.5478	.5517	.5557	.5596	.5636	.5675	.5714	.5753
0.2	.5793	.5832	.5871	.5910	.5948	.5987	.6026	.6064	.6103	.6141
0.3	.6179	.6217	.6255	.6293	.6331	.6368	.6406	.6443	.6480	.6517
0.4	.6554	.6591	.6628	.6664	.6700	.6736	.6772	.6808	.6844	.6879
0.5	.6915	.6950	.6985	.7019	.7054	.7088	.7123	.7157	.7190	.7224
0.6	.7257	.7291	.7324	.7357	.7389	.7422	.7454	.7486	.7517	.7549
0.7	.7580	.7611	.7642	.7673	.7704	.7734	.7764	.7794	.7823	.7852
0.8	.7881	.7910	.7939	.7967	.7995	.8023	.8051	.8078	.8106	.8133
0.9	.8159	.8186	.8212	.8238	.8264	.8289	.8315	.8340	.8365	.8389
1.0	.8413	.8438	.8461	.8485	.8508	.8531	.8554	.8577	.8599	.8621
1.1	.8643	.8665	.8686	.8708	.8729	.8749	.8770	.8790	.8810	.8830
1.2	.8849	.8869	.8888	.8907	.8925	.8944	.8962	.8980	.8997	.9015
1.3	.9032	.9049	.9066	.9082	.9099	.9115	.9131	.9147	.9162	.9177
1.4	.9192	.9207	.9222	.9236	.9251	.9265	.9279	.9292	.9306	.9319
1.5	.9332	.9345	.9357	.9370	.9382	.9394	.9406	.9418	.9429	.9441
1.6	.9452	.9463	.9474	.9484	.9495	.9505	.9515	.9525	.9535	.9545
1.7	.9554	.9564	.9573	.9582	.9591	.9599	.9608	.9616	.9625	.9633
1.8	.9641	.9649	.9656	.9664	.9671	.9678	.9686	.9693	.9699	.9706
1.9	.9713	.9719	.9726	.9732	.9738	.9744	.9750	.9756	.9761	.9767
2.0	.9772	.9778	.9783	.9788	.9793	.9798	.9803	.9808	.9812	.9817
2.1	.9821	.9826	.9830	.9834	.9838	.9842	.9846	.9850	.9854	.9857
2.2	.9861	.9864	.9868	.9871	.9875	.9878	.9881	.9884	.9887	.9890
2.3	.9893	.9896	.9898	.9901	.9904	.9906	.9909	.9911	.9913	.9916
2.4	.9918	.9920	.9922	.9925	.9927	.9929	.9931	.9932	.9934	.9936
2.5	.9938	.9940	.9941	.9943	.9945	.9946	.9948	.9949	.9951	.9952
2.6	.9953	.9955	.9956	.9957	.9959	.9960	.9961	.9962	.9963	.9964
2.7	.9965	.9966	.9967	.9968	.9969	.9970	.9971	.9972	.9973	.9974
2.8	.9974	.9975	.9976	.9977	.9977	.9978	.9979	.9979	.9980	.9981
2.9	.9981	.9982	.9982	.9983	.9984	.9984	.9985	.9985	.9986	.9986
3.0	.9987	.9987	.9987	.9988	.9988	.9989	.9989	.9989	.9990	.9990
3.1	.9990	.9991	.9991	.9991	.9992	.9992	.9992	.9992	.9993	.9993
3.2	.9993	.9993	.9994	.9994	.9994	.9994	.9994	.9995	.9995	.9995
3.3	.9995	.9995	.9995	.9996	.9996	.9996	.9996	.9996	.9996	.9997
3.4	.9997	.9997	.9997	.9997	.9997	.9997	.9997	.9997	.9997	.9998
3.5	.9998	.9998	.9998	.9998	.9998	.9998	.9998	.9998	.9998	.9998
3.6	.9998	.9998	.9999	.9999	.9999	.9999	.9999	.9999	.9999	.9999

The t-Table

Table A-2 shows right-tail probabilities for selected t-distributions (see Chapter 11 for more on the t-distribution).

Follow these steps to use Table A-2 to find right-tail probabilities and p-values for hypothesis tests involving t (see Chapter 16):

1. **Find the t-value (call it t) for which you want the right-tail probability ($p[T \geq t]$), and find the sample size (for example, n).**
2. **Find the row corresponding to the degrees of freedom (df) for your problem (for example, $n - 1$). Go across that row to find the two t-values between which your t falls.**

   For example, if your t is 1.60 and your n is 7, you look in the row for $df = 7 - 1 = 6$. Across that row, you find that your t lies between t-values 1.44 and 1.94.

3. **Go to the top of the columns containing the two t-values from Step 2.**

   The right-tail (greater-than) probability for your t-value is somewhere between the two values at the top of these columns. For example, your $t = 1.60$ is between t-values 1.44 and 1.94 ($df = 6$); so the right-tail probability for your t is between 0.10 (column heading for $t = 1.44$) and 0.05 (column heading for $t = 1.94$).

Tip: The row near the bottom with Z in the *df* column gives right-tail (greater-than) probabilities from the Z-distribution (Chapter 11 shows Z's relationship with t).

Use Table A-2 to find t^*-values (critical values) for a confidence interval involving t (see Chapter 14):

1. **Determine the confidence level you need (as a percentage).**
2. **Determine the sample size (for example, n).**
3. **Look at the bottom row of the table where the percentages are shown. Find your percent confidence level there.**
4. **Intersect this column with the row representing your degrees of freedom (df).** This is the t^*-value you need for your confidence interval. For example, a 95 percent confidence interval with $df = 6$ has $t^* = 2.45$. (Find 95 percent on the last line and go up to row 6.)

TABLE A-2 The *t*-Table

Numbers in each row of the table are values on a *t*-distribution with (*df*) degrees of freedom for selected right-tail (greater-than) probabilities (*p*).

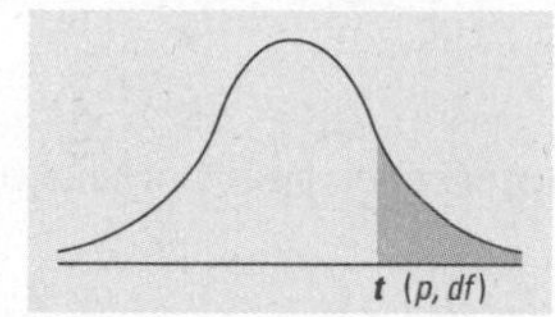

df/p	0.40	0.25	0.10	0.05	0.025	0.01	0.005	0.0005
1	0.324920	1.000000	3.077684	6.313752	12.70620	31.82052	63.65674	636.6192
2	0.288675	0.816497	1.885618	2.919986	4.30265	6.96456	9.92484	31.5991
3	0.276671	0.764892	1.637744	2.353363	3.18245	4.54070	5.84091	12.9240
4	0.270722	0.740697	1.533206	2.131847	2.77645	3.74695	4.60409	8.6103
5	0.267181	0.726687	1.475884	2.015048	2.57058	3.36493	4.03214	6.8688
6	0.264835	0.717558	1.439756	1.943180	2.44691	3.14267	3.70743	5.9588
7	0.263167	0.711142	1.414924	1.894579	2.36462	2.99795	3.49948	5.4079
8	0.261921	0.706387	1.396815	1.859548	2.30600	2.89646	3.35539	5.0413
9	0.260955	0.702722	1.383029	1.833113	2.26216	2.82144	3.24984	4.7809
10	0.260185	0.699812	1.372184	1.812461	2.22814	2.76377	3.16927	4.5869
11	0.259556	0.697445	1.363430	1.795885	2.20099	2.71808	3.10581	4.4370
12	0.259033	0.695483	1.356217	1.782288	2.17881	2.68100	3.05454	43178
13	0.258591	0.693829	1.350171	1.770933	2.16037	2.65031	3.01228	4.2208
14	0.258213	0.692417	1.345030	1.761310	2.14479	2.62449	2.97684	4.1405
15	0.257885	0.691197	1.340606	1.753050	2.13145	2.60248	2.94671	4.0728
16	0.257599	0.690132	1.336757	1.745884	2.11991	2.58349	2.92078	4.0150
17	0.257347	0.689195	1.333379	1.739607	2.10982	2.56693	2.89823	3.9651
18	0.257123	0.688364	1.330391	1.734064	2.10092	2.55238	2.87844	3.9216
19	0.256923	0.687621	1.327728	1.729133	2.09302	2.53948	2.86093	3.8834
20	0.256743	0.686954	1.325341	1.724718	2.08596	2.52798	2.84534	3.8495
21	0.256580	0.686352	1.323188	1.720743	2.07961	2.51765	2.83136	3.8193
22	0.256432	0.685805	1.321237	1.717144	2.07387	2.50832	2.81876	3.7921
23	0.256297	0.685306	1.319460	1.713872	2.06866	2.49987	2.80734	3.7676
24	0.256173	0.684850	1.317836	1.710882	2.06390	2.49216	2.79694	3.7454
25	0.256060	0.684430	1.316345	1.708141	2.05954	2.48511	2.78744	3.7251
26	0.255955	0.684043	1.314972	1.705618	2.05553	2.47863	2.77871	3.7066
27	0.255858	0.683685	1.313703	1.703288	2.05183	2.47266	2.77068	3.6896
28	0.255768	0.683353	1.312527	1.701131	2.04841	2.46714	2.76326	3.6739
29	0.255684	0.683044	1.311434	1.699127	2.04523	2.46202	2.75639	3.6594
30	0.255605	0.682756	1.310415	1.697261	2.04227	2.45726	2.75000	3.6460
z	0.253347	0.674490	1.281552	1.644854	1.95996	2.32635	2.57583	3.2905
CI	——	——	80%	90%	95%	98%	99%	99.9%

The Binomial Table

Table A-3 shows probabilities for the binomial distribution (see Chapter 9).

To use Table A-3, do the following.

1. **Find these three numbers for your particular problem:**
   - The sample size, n
   - The probability of success, p
   - The x-value for which you want $p(X = x)$
2. **Find the section of Table A-3 that's devoted to your n.**
3. **Look at the row for your x-value and the column for your p.**
4. **Intersect that row and column to find $p(X = x)$.**
5. **To get the probability of being less than, greater than, greater than or equal to, less than or equal to, or between two values of X, add the appropriate values of Table A-3 using the steps found in Chapter 9.**

   For example, if $n = 10$, $p = 0.6$, and you want $p(X = 9)$, go to the $n = 10$ section, the $x = 9$ row, and the $p = 0.6$ column to find 0.04.

TABLE A-3 The Binomial Table

Numbers in the table represent $p(X=x)$ for a binomial distribution with n trials and probability of success p.

Binomial probabilities:

$\binom{n}{x} p^x (1-p)^{n-x}$

		p										
n	*x*	0.1	0.2	0.25	0.3	0.4	0.5	0.6	0.7	0.75	0.8	0.9
1	0	0.900	0.800	0.750	0.700	0.600	0.500	0.400	0.300	0.250	0.200	0.100
	1	0.100	0.200	0.250	0.300	0.400	0.500	0.600	0.700	0.750	0.800	0.900
2	0	0.810	0.640	0.563	0.490	0.360	0.250	0.160	0.090	0.063	0.040	0.010
	1	0.180	0.320	0.375	0.420	0.480	0.500	0.480	0.420	0.375	0.320	0.180
	2	0.010	0.040	0.063	0.090	0.160	0.250	0.360	0.490	0.563	0.640	0.810
3	0	0.729	0.512	0.422	0.343	0.216	0.125	0.064	0.027	0.016	0.008	0.001
	1	0.243	0.384	0.422	0.441	0.432	0.375	0.288	0.189	0.141	0.096	0.027
	2	0.027	0.096	0.141	0.189	0.288	0.375	0.432	0.441	0.422	0.384	0.243
	3	0.001	0.008	0.016	0.027	0.064	0.125	0.216	0.343	0.422	0.512	0.729
4	0	0.656	0.410	0.316	0.240	0.130	0.063	0.026	0.008	0.004	0.002	0.000
	1	0.292	0.410	0.422	0.412	0.346	0.250	0.154	0.076	0.047	0.026	0.004
	2	0.049	0.154	0.211	0.265	0.346	0.375	0.346	0.265	0.211	0.154	0.049
	3	0.004	0.026	0.047	0.076	0.154	0.250	0.346	0.412	0.422	0.410	0.292
	4	0.000	0.002	0.004	0.008	0.026	0.063	0.130	0.240	0.316	0.410	0.656
5	0	0.590	0.328	0.237	0.168	0.078	0.031	0.010	0.002	0.001	0.000	0.000
	1	0.328	0.410	0.396	0.360	0.259	0.156	0.077	0.028	0.015	0.006	0.000
	2	0.073	0.205	0.264	0.309	0.346	0.313	0.230	0.132	0.088	0.051	0.008
	3	0.008	0.051	0.088	0.132	0.230	0.313	0.346	0.309	0.264	0.205	0.073
	4	0.000	0.006	0.015	0.028	0.077	0.156	0.259	0.360	0.396	0.410	0.328
	5	0.000	0.000	0.001	0.002	0.010	0.031	0.078	0.168	0.237	0.328	0.590
6	0	0.531	0.262	0.178	0.118	0.047	0.016	0.004	0.001	0.000	0.000	0.000
	1	0.354	0.393	0.356	0.303	0.187	0.094	0.037	0.010	0.004	0.002	0.000
	2	0.098	0.246	0.297	0.324	0.311	0.234	0.138	0.060	0.033	0.015	0.001
	3	0.015	0.082	0.132	0.185	0.276	0.313	0.276	0.185	0.132	0.082	0.015
	4	0.001	0.015	0.033	0.060	0.138	0.234	0.311	0.324	0.297	0.246	0.098
	5	0.000	0.002	0.004	0.010	0.037	0.094	0.187	0.303	0.356	0.393	0.354
	6	0.000	0.000	0.000	0.001	0.004	0.016	0.047	0.118	0.178	0.262	0.531
7	0	0.478	0.210	0.133	0.082	0.028	0.008	0.002	0.000	0.000	0.000	0.000
	1	0.372	0.367	0.311	0.247	0.131	0.055	0.017	0.004	0.001	0.000	0.000
	2	0.124	0.275	0.311	0.318	0.261	0.164	0.077	0.025	0.012	0.004	0.000
	3	0.023	0.115	0.173	0.227	0.290	0.273	0.194	0.097	0.058	0.029	0.003
	4	0.003	0.029	0.058	0.097	0.194	0.273	0.290	0.227	0.173	0.115	0.023
	5	0.000	0.004	0.012	0.025	0.077	0.164	0.261	0.318	0.311	0.275	0.124
	6	0.000	0.000	0.001	0.004	0.017	0.055	0.131	0.247	0.311	0.367	0.372
	7	0.000	0.000	0.000	0.000	0.002	0.008	0.028	0.082	0.133	0.210	0.478

(continued)

Numbers in the table represent $p(X=x)$ for a binomial distribution with n trials and probability of success p.

Binomial probabilities:

$\binom{n}{x} p^x(1-p)^{n-x}$

		p										
n	*x*	0.1	0.2	0.25	0.3	0.4	0.5	0.6	0.7	0.75	0.8	0.9
8	0	0.430	0.168	0.100	0.058	0.017	0.004	0.001	0.000	0.000	0.000	0.000
	1	0.383	0.336	0.267	0.198	0.090	0.031	0.008	0.001	0.000	0.000	0.000
	2	0.149	0.294	0.311	0.296	0.209	0.109	0.041	0.010	0.004	0.001	0.000
	3	0.033	0.147	0.208	0.254	0.279	0.219	0.124	0.047	0.023	0.009	0.000
	4	0.005	0.046	0.087	0.136	0.232	0.273	0.232	0.136	0.087	0.046	0.005
	5	0.000	0.009	0.023	0.047	0.124	0.219	0.279	0.254	0.208	0.147	0.033
	6	0.000	0.001	0.004	0.010	0.041	0.109	0.209	0.296	0.311	0.294	0.149
	7	0.000	0.000	0.000	0.001	0.008	0.031	0.090	0.198	0.267	0.336	0.383
	8	0.000	0.000	0.000	0.000	0.001	0.004	0.017	0.058	0.100	0.168	0.430
9	0	0.387	0.134	0.075	0.040	0.010	0.002	0.000	0.000	0.000	0.000	0.000
	1	0.387	0.302	0.225	0.156	0.060	0.018	0.004	0.000	0.000	0.000	0.000
	2	0.172	0.302	0.300	0.267	0.161	0.070	0.021	0.004	0.001	0.000	0.000
	3	0.045	0.176	0.234	0.267	0.251	0.164	0.074	0.021	0.009	0.003	0.000
	4	0.007	0.066	0.117	0.172	0.251	0.246	0.167	0.074	0.039	0.017	0.001
	5	0.001	0.017	0.039	0.074	0.167	0.246	0.251	0.172	0.117	0.066	0.007
	6	0.000	0.003	0.009	0.021	0.074	0.164	0.251	0.267	0.234	0.176	0.045
	7	0.000	0.000	0.001	0.004	0.021	0.070	0.161	0.267	0.300	0.302	0.172
	8	0.000	0.000	0.000	0.000	0.004	0.018	0.060	0.156	0.225	0.302	0.387
	9	0.000	0.000	0.000	0.000	0.000	0.002	0.010	0,040	0.075	0.134	0.387
10	0	0.349	0.107	0.056	0.028	0.006	0.001	0.000	0.000	0.000	0.000	0.000
	1	0.387	0.268	0.188	0.121	0.040	0.010	0.002	0.000	0.000	0.000	0.000
	2	0.194	0.302	0.282	0.233	0.121	0.044	0.011	0.001	0.000	0.000	0.000
	3	0.057	0.201	0.250	0.267	0.215	0.117	0.042	0.009	0.003	0.001	0.000
	4	0.011	0.088	0.146	0.200	0.251	0.205	0.111	0.037	0.016	0.006	0.000
	5	0.001	0.026	0.058	0.103	0.201	0.246	0.201	0.103	0.058	0.026	0.001
	6	0.000	0.006	0.016	0.037	0.111	0.205	0.251	0.200	0.146	0.088	0.011
	7	0.000	0.001	0.003	0.009	0.042	0.117	0.215	0.267	0.250	0.201	0.057
	8	0.000	0.000	0.000	0.001	0.011	0.044	0.121	0.233	0.282	0.302	0.194
	9	0.000	0.000	0.000	0.000	0.002	0.010	0.040	0.121	0.188	0.268	0.387
	10	0.000	0.000	0.000	0.000	0.000	0.001	0.006	0.028	0.056	0.107	0.349
11	0	0.314	0.086	0.042	0.020	0.004	0.000	0.000	0.000	0.000	0.000	0.000
	1	0.384	0.236	0.155	0.093	0.027	0.005	0.001	0.000	0.000	0.000	0.000
	2	0.213	0.295	0.258	0.200	0.089	0.027	0.005	0.001	0.000	0.000	0.000
	3	0.071	0.221	0.258	0.257	0.177	0.081	0.023	0.004	0.001	0.000	0.000
	4	0.016	0.111	0.172	0.220	0.236	0.161	0.070	0.017	0.006	0.002	0.000
	5	0.002	0.039	0.080	0.132	0.221	0.226	0.147	0.057	0.027	0.010	0.000
	6	0.000	0.010	0.027	0.057	0.147	0.226	0.221	0.132	0.080	0.039	0.002
	7	0.000	0.002	0.006	0.017	0.070	0.161	0.236	0.220	0.172	0.111	0.016
	8	0.000	0.000	0.001	0.004	0.023	0.081	0.177	0.257	0.258	0.221	0.071
	9	0.000	0.000	0.000	0.001	0.005	0.027	0.089	0.200	0.258	0.295	0.213
	10	0.000	0.000	0.000	0.000	0.001	0.005	0.027	0.093	0.155	0.236	0.384
	11	0.000	0.000	0.000	0.000	0.000	0.000	0.004	0.020	0.042	0.086	0.314

Numbers in the table represent $p(X=x)$ for a binomial distribution with n trials and probability of success p.

Binomial probabilities:

$$\binom{n}{x} p^x(1-p)^{n-x}$$

		p										
n	*x*	0.1	0.2	0.25	0.3	0.4	0.5	0.6	0.7	0.75	0.8	0.9
12	0	0.282	0.069	0.032	0.014	0.002	0.000	0.000	0.000	0.000	0.000	0.000
	1	0.377	0.206	0.127	0.071	0.017	0.003	0.000	0.000	0.000	0.000	0.000
	2	0.230	0.283	0.232	0.168	0.064	0.016	0.002	0.000	0.000	0.000	0.000
	3	0.085	0.236	0.258	0.240	0.142	0.054	0.012	0.001	0.000	0.000	0.000
	4	0.021	0.133	0.194	0.231	0.213	0.121	0.042	0.008	0.002	0.001	0.000
	5	0.004	0.053	0.103	0.158	0.227	0.193	0.101	0.029	0.011	0.003	0.000
	6	0.000	0.016	0.040	0.079	0.177	0.226	0.177	0.079	0.040	0.016	0.000
	7	0.000	0.003	0.011	0.029	0.101	0.193	0.227	0.158	0.103	0.053	0.004
	8	0.000	0.001	0.002	0.008	0.042	0.121	0.213	0.231	0.194	0.133	0.021
	9	0.000	0.000	0.000	0.001	0.012	0.054	0.142	0.240	0.258	0.236	0.085
	10	0.000	0.000	0.000	0.000	0.002	0.016	0.064	0.168	0.232	0.283	0.230
	11	0.000	0.000	0.000	0.000	0.000	0.003	0.017	0.071	0.127	0.206	0.377
	12	0.000	0.000	0.000	0.000	0.000	0.000	0.002	0.014	0.032	0.069	0.282
13	0	0.254	0.055	0.024	0.010	0.001	0.000	0.000	0.000	0.000	0.000	0.000
	1	0.367	0.179	0.103	0.054	0.011	0.002	0.000	0.000	0.000	0.000	0.000
	2	0.245	0.268	0.206	0.139	0.045	0.010	0.001	0.000	0.000	0.000	0.000
	3	0.100	0.246	0.252	0.218	0.111	0.035	0.006	0.001	0.000	0.000	0.000
	4	0.028	0.154	0.210	0.234	0.184	0.087	0.024	0.003	0.001	0.000	0.000
	5	0.006	0.069	0.126	0.180	0.221	0.157	0.066	0.014	0.005	0.001	0.000
	6	0.001	0.023	0.056	0.103	0.197	0.209	0.131	0.044	0.019	0.006	0.000
	7	0.000	0.006	0.019	0.044	0.131	0.209	0.197	0.103	0.056	0.023	0.001
	8	0.000	0.001	0.005	0.014	0.066	0.157	0.221	0.180	0.126	0.069	0.006
	9	0.000	0.000	0.001	0.003	0.024	0.087	0.184	0.234	0.210	0.154	0.028
	10	0.000	0.000	0.000	0.001	0.006	0.035	0.111	0.218	0.252	0.246	0.100
	11	0.000	0.000	0.000	0.000	0.001	0.010	0.045	0.139	0.206	0.268	0.245
	12	0.000	0.000	0.000	0.000	0.000	0.002	0.011	0.054	0.103	0.179	0.367
	13	0.000	0.000	0.000	0.000	0.000	0.000	0.001	0.010	0.024	0.055	0.254
14	0	0.229	0.044	0.018	0.007	0.001	0.000	0.000	0.000	0.000	0.000	0.000
	1	0.356	0.154	0.083	0.041	0.007	0.001	0.000	0.000	0.000	0.000	0.000
	2	0.257	0.250	0.180	0.113	0.032	0.006	0.001	0.000	0.000	0.000	0.000
	3	0.114	0.250	0.240	0.194	0.085	0.022	0.003	0.000	0.000	0.000	0.000
	4	0.035	0.172	0.220	0.229	0.155	0.061	0.014	0.001	0.000	0.000	0.000
	5	0.008	0.086	0.147	0.196	0.207	0.122	0.041	0.007	0.002	0.000	0.000
	6	0.001	0.032	0.073	0.126	0.207	0.183	0.092	0.023	0.008	0.002	0.000
	7	0.000	0.009	0.028	0.062	0.157	0.209	0.157	0.062	0.028	0.009	0.000
	8	0.000	0.002	0.008	0.023	0.092	0.183	0.207	0.126	0.073	0.032	0.001
	9	0.000	0.000	0.002	0.007	0.041	0.122	0.207	0.196	0.147	0.086	0.008
	10	0.000	0.000	0.000	0.001	0.014	0.061	0.155	0.229	0.220	0.172	0.035
	11	0.000	0.000	0.000	0.000	0.003	0.022	0.085	0.194	0.240	0.250	0.114
	12	0.000	0.000	0.000	0.000	0.001	0.006	0.032	0.113	0.180	0.250	0.257
	13	0.000	0.000	0.000	0.000	0.000	0.001	0.007	0.041	0.083	0.154	0.356
	14	0.000	0.000	0.000	0.000	0.000	0.000	0.001	0.007	0.018	0.044	0.229

(continued)

Binomial probabilities:

$\binom{n}{x} p^x(1-p)^{n-x}$

Numbers in the table represent $p(X=x)$ for a binomial distribution with n trials and probability of success p.

n	x	p = 0.1	0.2	0.25	0.3	0.4	0.5	0.6	0.7	0.75	0.8	0.9
15	0	0.206	0.035	0.013	0.005	0.000	0.000	0.000	0.000	0.000	0.000	0.000
	1	0.343	0.132	0.067	0.031	0.005	0.000	0.000	0.000	0.000	0.000	0.000
	2	0.267	0.231	0.156	0.092	0.022	0.003	0.000	0.000	0.000	0.000	0.000
	3	0.129	0.250	0.225	0.170	0.063	0.014	0.002	0.000	0.000	0.000	0.000
	4	0.043	0.188	0.225	0.219	0.127	0.042	0.007	0.001	0.000	0.000	0.000
	5	0.010	0.103	0.165	0.206	0.186	0.092	0.024	0.003	0.001	0.000	0.000
	6	0.002	0.043	0.092	0.147	0.207	0.153	0.061	0.012	0.003	0.001	0.000
	7	0.000	0.014	0.039	0.081	0.177	0.196	0.118	0.035	0.013	0.003	0.000
	8	0.000	0.003	0.013	0.035	0.118	0.196	0.177	0.081	0.039	0.014	0.000
	9	0.000	0.001	0.003	0.012	0.061	0.153	0.207	0.147	0.092	0.043	0.002
	10	0.000	0.000	0.001	0.003	0.024	0.092	0.186	0.206	0.165	0.103	0.010
	11	0.000	0.000	0.000	0.001	0.007	0.042	0.127	0.219	0.225	0.188	0.043
	12	0.000	0.000	0.000	0.000	0.002	0.014	0.063	0.170	0.225	0.250	0.129
	13	0.000	0.000	0.000	0.000	0.000	0.003	0.022	0.092	0.156	0.231	0.267
	14	0.000	0.000	0.000	0.000	0.000	0.000	0.005	0.031	0.067	0.132	0.343
	15	0.000	0.000	0.000	0.000	0.000	0.000	0.000	0.005	0.013	0.035	0.206
20	0	0.122	0.012	0.003	0.001	0.000	0.000	0.000	0.000	0.000	0.000	0.000
	1	0.270	0.058	0.021	0.007	0.000	0.000	0.000	0.000	0.000	0.000	0.000
	2	0.285	0.137	0.067	0.028	0.003	0.000	0.000	0.000	0.000	0.000	0.000
	3	0.190	0.205	0.134	0.072	0.012	0.001	0.000	0.000	0.000	0.000	0.000
	4	0.090	0.218	0.190	0.130	0.035	0.005	0.000	0.000	0.000	0.000	0.000
	5	0.032	0.175	0.202	0.179	0.075	0.015	0.001	0.000	0.000	0.000	0.000
	6	0.009	0.109	0.169	0.192	0.124	0.037	0.005	0.000	0.000	0.000	0.000
	7	0.002	0.055	0.112	0.164	0.166	0.074	0.015	0.001	0.000	0.000	0.000
	8	0.000	0.022	0.061	0.114	0.180	0.120	0.035	0.004	0.001	0.000	0.000
	9	0.000	0.007	0.027	0.065	0.160	0.160	0.071	0.012	0.003	0.000	0.000
	10	0.000	0.002	0.010	0.031	0.117	0.176	0.117	0.031	0.010	0.002	0.000
	11	0.000	0.000	0.003	0.012	0.071	0.160	0.160	0.065	0.027	0.007	0.000
	12	0.000	0.000	0.001	0.004	0.035	0.120	0.180	0.114	0.061	0.022	0.000
	13	0.000	0.000	0.000	0.001	0.015	0.074	0.166	0.164	0.112	0.055	0.002
	14	0.000	0.000	0.000	0.000	0.005	0.037	0.124	0.192	0.169	0.109	0.009
	15	0.000	0.000	0.000	0.000	0.001	0.015	0.075	0.179	0.202	0.175	0.032
	16	0.000	0.000	0.000	0.000	0.000	0.005	0.035	0.130	0.190	0.218	0.090
	17	0.000	0.000	0.000	0.000	0.000	0.001	0.012	0.072	0.134	0.205	0.190
	18	0.000	0.000	0.000	0.000	0.000	0.000	0.003	0.028	0.067	0.137	0.285
	19	0.000	0.000	0.000	0.000	0.000	0.000	0.000	0.007	0.021	0.058	0.270
	20	0.000	0.000	0.000	0.000	0.000	0.000	0.000	0.001	0.003	0.012	0.122

Index

Symbols and Numerics

A

B

C

D

E

F

J

L

M

N

O

P

Q

R

S

T

U

V

W

X

Y

Z

About the Author

Deborah J. Rumsey, PhD, is a Statistics Education Specialist and member of the Teaching Faculty in the Department of Statistics at The Ohio State University. Dr. Rumsey is a Fellow of the American Statistical Association. She has won the Presidential Teaching Award from Kansas State University and has been inducted into the Wall of Inspiration at her high school alma mater, Burlington High School, in Burlington, Wisconsin. She is also the author of *Statistics For Dummies*, *Statistics II For Dummies*, *Statistics Workbook For Dummies*, *Probability For Dummies*, and *Statistics Essentials For Dummies.* She has published numerous papers and given many professional presentations and workshops on the subject of statistics education. She is the original conference designer of the biennial United States Conference on Teaching Statistics (USCOTS). Her passions include being with her family, fishing and bird watching, getting seat time on her Kubota tractor, and cheering the Ohio State Buckeyes on to their next national championship.

Dedication

This book is dedicated to my husband Eric Johnson and my son Clint Johnson. Without them, nothing else I do really matters. And to all the students out there taking a statistics class, may this book find you and help you achieve your goals.

Author's Acknowledgments

Thanks to Lindsay Lefevere for giving me this opportunity and for supporting me throughout, and to Chrissy Guthrie, whose editing helped me to realize my dream of writing. Thanks to Sara Conroy for her technical reading and support, and to Marylouise Wiack for her great copy editing.

Publisher's Acknowledgments

Executive Editor: Lindsay Sandman Lefevere

Project Manager and Development Editor: Christina N. Guthrie

Managing Editors: Kelsey Baird, Kristie Pyles

Copy Editor: Marylouise Wiack

Technical Editor: Sara Conroy, PhD

Production Editor: Saikarthick Kumarasamy

Cover Photos: © Vasya Kobelev/Shutterstock

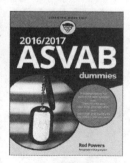

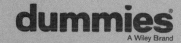